# AN INTRODUCTION TO MATHEMATICAL BIOLOGY

Linda J. S. Allen

*Department of Mathematics and Statistics*
*Texas Tech University*

PEARSON
Prentice Hall

Upper Saddle River, NJ 07458

Original edition, entitled INTRODUCTION TO MATHEMATICAL BIOLOGY, AN, 1st Edition, by ALLEN, LINDA J.S., published by Pearson Education, Inc, Copyright © 2007.







ISBN 978-89-450-4922-3

This book is dedicated to my husband, Edward,
and to my daughter, Anna.

# CONTENTS

## 4 LINEAR DIFFERENTIAL EQUATIONS: THEORY AND EXAMPLES 141



## 5 NONLINEAR ORDINARY DIFFERENTIAL EQUATIONS: THEORY AND EXAMPLES 176

# PREFACE

My goal in writing this book is to provide an introduction to a variety of mathematical models for biological systems, and to present the mathematical theory and techniques useful in the analysis of these models. Classical mathematical models from population biology are discussed, including the Leslie matrix model, the Nicholson-Bailey model, and the Lotka-Volterra predator-prey model. In addition, more recent models are discussed, such as a model for the Human Immunodeficiency Virus (HIV) and a model for flour beetles. Many of the biological applications come from population biology and epidemiology due to personal preference and expertise. However, there are examples from population genetics, cell biology, and physiology as well. The focus of this book is on deterministic mathematical models, models formulated as difference equations or ordinary differential equations. They form the basis for Chapters 1 through 6. The emphasis is on predicting the qualitative solution behavior over time.

The topics in this book are covered in a one-semester graduate course offered by the Department of Mathematics and Statistics at Texas Tech University. The course is open to beginning graduate students and advanced undergraduate students in mathematics, engineering, and biology. The motivation for this course is to provide students with a solid background in the mathematics behind modeling in biology and to expose them to a wide variety of mathematical models in biology. Mathematical prerequisites include undergraduate courses in calculus, linear algebra, and differential equations.

This book is organized according to the mathematical theory rather than the biological application. A review of the basic theory of linear difference equations and linear differential equations is contained in Chapters 1 and 4, respectively. This review material can be covered very briefly or in more detail, depending on the students' background. Difference equation models are presented in Chapters 1, 2, and 3. Ordinary differential equation models are covered in Chapters 4, 5, and 6. The last chapter, Chapter 7, is an introduction to partial differential equation models in biology. Applications of the mathematical theory to biological examples are presented in each chapter. Similar biological applications may appear in more than one chapter. For example, epidemic models and predator–prey models are formulated in terms of difference equations in Chapters 2 and 3 and as differential equations in Chapter 6. In this way, the advantages and disadvantages of the various model formulations can be compared. Chapters 3 and 6 are devoted primarily to biological applications. The instructor may be selective about the applications covered in these two chapters. Exercises at the end of each chapter reinforce concepts discussed in each chapter. To visualize the dynamics of various models, students are encouraged to use the MATLAB® or the Maple programs provided in the appendices. These programs can be modified for other types of models or adapted to other programming languages. In addition, research topics assigned on current biological models that have appeared in the literature can be part of an individual or a group research project. Because mathematical biology is a rapidly growing field, there are many excellent references that can be consulted for additional biological applications. Some of these references are listed at the end of each chapter.

I took my first course in mathematical ecology around 1978 and became very excited about mathematical applications to the field of biology. That first course was taught by my Ph.D. advisor, Thomas G. Hallam, University of Tennessee. Sources of reference were E. C. Pielou's book *Mathematical Ecology,* Robert M. May's book *Theoretical Ecology Principles and Applications,* and excellent notes prepared by Tom Hallam. I taught my first course in mathematical biology in 1989 and used Leah Edelstein-Keshet's wonderful textbook, *Mathematical Models in Biology,* for that first year and for many years thereafter. Another excellent book that I used as a reference was James D. Murray's book *Mathematical Biology.* Over the years, I wrote my own notes for a course in mathematical biology, relying on these excellent sources of reference. This book is a result of that effort.

I would like to acknowledge and thank many individuals who have contributed to the completion of this book. My husband Edward Allen provided support throughout the long process of writing and rewriting, helped with graphing the bifurcation diagrams in Chapter 2, and critiqued early drafts of this book. My daughter Anna Allen offered encouragement when I needed it. Thomas G. Hallam, University of Tennessee, introduced me to the field of mathematical ecology and supported me throughout my academic career. I would like to thank the Prentice Hall reviewers for their insightful comments, helpful suggestions, and corrections on a preliminary draft of this book: Michael C. Reed, Duke University; Gail S. K. Wolkowicz, McMaster University; and Xiuli Chao, North Carolina State University. My friends and colleagues checked many of the exercises for accuracy, identified typographical errors, and made suggestions for improvements in preliminary drafts of this book: Azmy Ackleh, University of Louisiana at Lafayette; Jesse Fagan, Stephen F. Austin State University (Chapters 1–4); Sophia R.-J. Jang, University of Louisiana at Lafayette (Chapters 1–6); and Lih-Ing Roeger, Texas Tech University (Chapters 1–6). In addition, David Gilliam, Texas Tech University, assisted me with some of the technical details associated with LaTeX and MATLAB. The MATLAB program pplane6, written by John C. Polking, Rice University, was used to graph direction fields in the phase plane for the figures in Chapters 5, 6, and 7. Many graduate students in my biomathematics courses helped eliminate some of the errors in the exercises. I am pleased to acknowledge the following individuals who helped with many of the exercises: Armando Arciniega, David Atkinson, Amy Burgin, Garry Block, Amy Drew, Channa Navaratna, Menaka Navaratna, Keith Emmert, Matthew Gray, Kiyomi Kaskela, Jake Kesinger, Nadarajah Kirupaharan, Rachel Koskodan, Karen Lawrence, Robert McCormack, Shelley McGee, Wayne McGee, Penelope Misquitta, Rathnamali Palamakumbura, Niranjala Perera, Sarah Stinnett, Edward Swim, David Thrasher, Ashley Trent, Curtis Wesley, Nilmini Wijeratne, and Yaji Xu.

A special note of thanks to Patrick de Leenheer of the University of Florida for his assistance with the accuracy check of the book. I thank my colleagues at Texas Tech University for providing a friendly and supportive environment in which to teach and to do research. Finally, and most importantly, I give thanks and praise to my Lord and Savior for his ever-present help and guidance. After countless suggestions and revisions from my friends and colleagues, I assume full responsibility for any omissions and errors in the final draft of this book.

Chapter 1

# Linear Difference Equations, Theory, and Examples

## 1.1 Introduction

There are three basic steps in mathematical modeling of biological systems. These steps include (1) formulation of a mathematical model to represent accurately the underlying biological process or system being studied, (2) application of mathematical techniques to understand the model behavior, and (3) interpretation of the model results to determine whether meaningful biological results are obtained. All three of these steps, formulation, analysis, and interpretation, are important to the study of biological systems. In order to apply these three steps, the underlying mathematical theory, tools, and techniques must be carefully applied and thoroughly understood.

Mathematical models of biological processes and systems are often expressed in terms of difference or differential equations. The reason for these types of models is that biological processes are dynamical, changing with respect to time, space, or stage of development. Therefore, the three steps in mathematical modeling require a good understanding of the mathematical theory for both difference and differential equations.

Difference equations, studied in Chapters 1–3, are relationships between quantities as they change over discrete intervals of time, space, and so on. In many of the biological applications studied in this textbook, the discrete intervals represent time intervals (e.g., $t = 0, 1, 2, \dots$), where the time interval is unity. On the other hand, differential equations, studied in Chapters 4–7, describe changes in quantities over continuous intervals (e.g., $t \in [0, \infty)$). For example, if the quantity is changing with respect to time, then the rate of change or derivative with respect to time is specified. The quantities modeled by the difference or differential equations are referred to as the *states* of the system.

The model equations can become quite complex if there are several interacting states whose dynamics depend on time, age, and spatial location. If the temporal dynamics are of interest and not the age or spatial location, then the modeling format is *ordinary* difference or differential equations. But if, in addition, age or spatial location are important to the dynamics, then *partial* difference or differential equations should be used. Although difference and differential equations are the primary modeling formats considered here, other types of models are discussed, including delay difference equations (Chapter 2) and delay differential equations (Chapters 4 and 5). In addition, integrodifference and integrodifferential equations are mentioned briefly in Chapter 7 and in Chapters 4 and 5, respectively.

Difference equations are applied frequently to populations whose generations do not overlap. For example, when adults die and are replaced by their progeny, the population size from one generation to the next, $x_t$ to $x_{t+1}$, can be modeled by difference equations. Discrete time intervals often coincide naturally with periodic data collection used in the laboratory or in the field. Difference equations have been used in modeling age, stage, and size-structured populations (Caswell, 2001; Cushing, 1998; Kot, 2001). For example, the discrete time interval may represent the time required for a transition to occur in the population such as from one age group to another age group or from one stage of development to another stage.

Differential equations are applied when changes in the states occur continuously such as when there is continuous reproduction and deaths. Differential equations have been applied to many types of biological systems ranging from populations to epidemics to physiological systems (Brauer and Castillo-Chávez, 2001; Britton, 2003; Edelstein-Keshet, 1988; Keener and Sneyd, 1998; Murray, 1993, 2002, 2003; Thieme, 2003).

There are many good textbooks on modeling in mathematical biology. The preceding references represent only a fraction of them. Please consult the list of references after each chapter for pertinent research articles and additional textbooks.

To begin our study of mathematical models of biological systems, we consider the simplest type of modeling construct: linear difference equations. Knowledge of the solution behavior for linear difference equations and the solution techniques will be useful in the study of nonlinear difference equations. First, we introduce some terminology and classify different types of difference equations. Then, we give a brief review of some methods for solving linear difference equations and systems. Lots of examples are provided to illustrate the various methods. In Section 1.6, we present a well-known example of a system of linear difference equations known as the Leslie matrix model. This model keeps track of the age structure of a population over time. Important properties associated with the Leslie matrix model are discussed in Section 1.7. These properties will be useful in the study of more complex structured models in Chapter 3.

## 1.2 Basic Definitions and Notation

In difference equations, the changes in states of a system are modeled over discrete intervals. In most cases, the discrete intervals represent time intervals and therefore the letter $t$ is used to denote the time variable. In general, the length of the discrete time interval is some fixed length (one day, one week, etc.), which can be denoted as $\Delta t$. Then the states of a system are modeled at the discrete times $t = 0, \Delta t, 2\Delta t, \ldots$. For ease of notation, the time interval is often simplified so that $\Delta t = 1$.

Denote the state of the system at time $t$ as $x_t$ or $x(t)$, where the variable $x$ is a function of $t$. Generally, lowercase letters $x$ will denote real variables and capital letters $X$ vectors of real variables. The notation $x_t$ is preferred over $x(t)$ when only a few states are modeled (e.g., $x_t$ and $y_t$). The notation $X_t$ or $X(t)$ is preferred when $X$ is a vector [e.g., $X(t) = (x_1(t), x_2(t), \ldots, x_m(t))^T$ denotes an

$m$-column vector with components $x_i(t)$, $i = 1, 2, \ldots, m$, at time $t$]. The superscript $T$ on a vector or a matrix means the transpose of that vector or matrix. The meaning of the notation should be clear from the context.

**Example 1.1** The following difference equation,

$$x_{t+1} = atx_t + bt^2x_{t-1} + \sin(t), \tag{1.1}$$

shows that the state at time $t + 1$ is a linear combination of the state at two previous times $t$ and $t - 1$ plus the sine function at time $t$. ■

**Definition 1.1.** A *difference equation of order k* has the form

$$f(x_{t+k}, x_{t+k-1}, \ldots, x_{t+1}, x_t, t) = 0, \quad t = 0, 1, \ldots, \tag{1.2}$$

where $f$ is a real-valued function of the real variables $x_t$ through $x_{t+k}$ and $t$. In particular, $f$ must depend on $x_t$ and $x_{t+k}$; otherwise the order of the difference equation may be different from $k$. The difference equation (1.2) is called *autonomous* if $f$ does not depend explicitly on $t$ and it is called *nonautonomous* otherwise.

A form of the difference equation that will be encountered most often is

$$x_{t+k} + a_1x_{t+k-1} + \cdots + a_{k-1}x_{t+1} + a_kx_t = b_t, \quad t = 0, 1, \ldots. \tag{1.3}$$

The order of the difference equation (1.3) is $k$, provided $a_k \neq 0$. We will always assume that the coefficients are real and the functions are real valued. The coefficients $a_j$, $j = 1, \ldots, k$, can be functions of $t$ and $x_i$ for $i = t, \ldots, t + k - 1$. The function $b_t$ on the right side of equation (1.3) may depend on $t$ but not on the state variables.

**Definition 1.2.** If the coefficients $a_j$, $j = 1, \ldots, k$ in equation (1.3) are constant or depend on $t$ but do not depend on the state variables, then the difference equation (1.3) is said to be *linear*; otherwise, it is said to be *nonlinear*. In addition, if the difference equation (1.3) is linear and $b_t = 0$ for all $t$, then it is said to be *homogeneous*; otherwise, it is said to be *nonhomogeneous*.

The difference equation (1.1) in Example 1.1 is a second-order, linear difference equation (assuming $a$ and $b$ are constants). In addition, the difference equation (1.1) is nonautonomous and nonhomogeneous.

The preceding terminology applies to systems of difference equations as well. A system of $k$ *first-order difference equations* can be expressed in the form

$$x_i(t + 1) = f_i(x_1(t), x_2(t), \ldots, x_k(t), t), \quad i = 1, 2, \ldots, k. \tag{1.4}$$

**Definition 1.3.** If the functions $f_i$ in (1.4) do not depend explicitly on $t$ for $i = 1, 2, \ldots, k$, then the first-order system is said to be *autonomous*. Otherwise, it is said to be *nonautonomous*.

Suppose the system (1.4) can be expressed as follows:

$$x_i(t + 1) = \sum_{j=1}^{k} a_{ij}(t)x_j(t) + b_i(t), \quad i = 1, 2, \ldots, k. \tag{1.5}$$

If the coefficients $a_{ij}$ and $b_i$ do not depend on the state variables $x_i$, $i, j = 1, 2, \ldots, k$, then system (1.5) is said to be *linear*; otherwise it is said to be *nonlinear*. If the system (1.5) is linear and $b_i(t) \equiv 0$ for $i = 1, 2, \ldots, k$, then the system is said to be *homogeneous*; otherwise, it is said to be *nonhomogeneous*. System (1.5) can be expressed in matrix notation,

$$X(t+1) = A(t)X(t) + B(t),$$

where vector $X = (x_1, x_2, \ldots, x_k)^T$, matrix $A = (a_{ij})_{i,j=1}^k$, and vector $B = (b_1, \ldots, b_k)^T$.

Some examples illustrating these definitions are discussed.

**Example 1.2** Suppose the difference equation is given by $x_{t+1} = ax_t^2 + bx_{t-1}$, where $a$ and $b$ are nonzero constants. Then the difference equation is nonlinear, autonomous, and second order. ■

**Example 1.3** Suppose the system of difference equations $X(t+1) = AX(t)$ satisfies $X(t) = (x_1(t), x_2(t))^T$ and

$$A = \begin{pmatrix} g_1(x_1, x_2) & g_2(x_1, x_2) \\ g_3(x_1, x_2) & g_4(x_1, x_2) \end{pmatrix},$$

where the $g_i$, $i = 1, 2, 3, 4$, are nonconstant functions of $x_1$ and $x_2$ but do not depend explicitly on $t$. Then the first-order system is autonomous and nonlinear. ■

**Example 1.4** Let the size of a population in generation $t$ be denoted as $x_t$. Assume that each individual in the population reproduces $a$ individuals per generation, then dies. This assumption could apply to a population of cells or an annual plant population. The population can be modeled by

$$x_{t+1} = ax_t. \tag{1.6}$$

The difference equation (1.6) is first-order, linear, and homogeneous. The solution to (1.6) can be found by repeated substitution,

$$x_1 = ax_0, \quad x_2 = ax_1 = a^2x_0.$$

Equation (1.6) is a recursion formula. In general, a solution to the difference equation (1.6) has the form $x_t = a^t x_0$. ■

**Definition 1.4.** A *solution* to the difference equation (1.2) is a function $x_t$, $t = 0, 1, 2, \ldots$, such that when substituted into the equation makes it a true statement. A *solution* to the system of difference equations (1.4) is a set of functions $\{x_i(t)\}$, $i = 1, 2 \ldots, k$, often represented in vector form as $X(t) = (x_1(t), \ldots, x_k(t))^T$ such that when substituted into the equations makes each of them a true statement.

In the case of the difference equation (1.6), $x_t = a^t x_0$ is a solution. If the value of $x_0$ is known, the solution is *unique*. The solution $a^t x_0$ uniquely determines the behavior of the population in generation $t$ given the population initially is $x_0$. If $0 < a < 1$, then $\lim_{t\to\infty} x_t = 0$. If $a = 1$, then $x_t = x_0$ for all $t$. If $a > 1$, then $\lim_{t\to\infty} x_t = \infty$, $x_0 > 0$. What happens in the special cases $a = -1$ and $a = 0$? In general, if $|a| > 1$, then solutions diverge (either they approach infinity

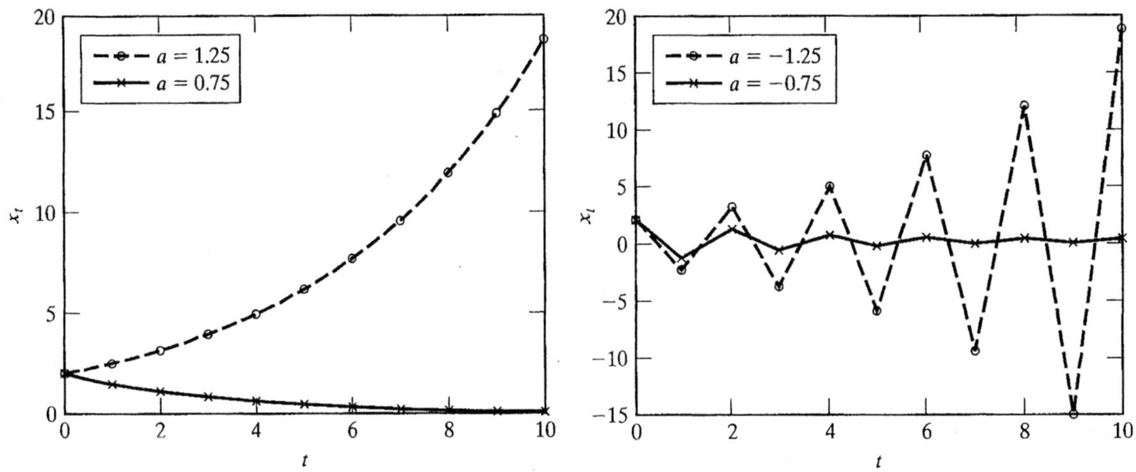


**Figure 1.1** Graphs of the solutions to $x_{t+1} = ax_t$ for various values of $a$.

or they are oscillatory) and if $|a| < 1$, then solutions converge to zero. See Figure 1.1. The solution points $(t, x_t)$, $t = 0, 1, \ldots, 10$ are connected by straight lines in Figure 1.1. However, note that the graph of a solution to a difference equation is a set of discrete points, $\{(t, x_t)\}$, $t = 0, 1, \ldots$.

The following example describes a model for growth and survival of a stage-structured population.

**Example 1.5** Suppose adult females of a species produce offspring at a fixed period of time each year. A proportion of the offspring (juveniles) survives to adulthood, reproduces, and then dies (nonoverlapping generations). Let

$j_t$ = number of juveniles in year $t$.

$a_t$ = number of adult females in year $t$.

$p_j$ = fraction of juveniles that survive in year $t$.

$f$ = number of offspring produced per female.

$r$ = ratio of females to adults.

Each individual can belong to one of two stages: juvenile or adult. Assume census-taking occurs immediately after the offspring are born. Then

$$j_{t+1} = fa_{t+1}, \quad a_{t+1} = rp_j j_t$$

and therefore,

$$j_{t+1} = [rp_j f] j_t.$$

Let $a = rp_j f$. If $r$, $p_j$, and $f$ are constants, then it is clear that the solutions to these difference equations have the same form as the solution in Example 1.4, $j_t = [rp_j f]^t j_0$.

Now suppose offspring or juveniles do not become reproductive adults until their second year. In addition, assume that all adults do not die. Let $p_a$ denote the fraction of female adults that survive each year. Then the model has the form

$$j_{t+1} = fp_a a_t \quad \text{and} \quad a_{t+1} = rp_j j_t + p_a a_t.$$

These equations can be expressed as a first-order system of difference equations, $X_{t+1} = AX_t$, where $X_t = (j_t, a_t)^T$ and matrix

$$A = \begin{pmatrix} 0 & fp_a \\ rp_j & p_a \end{pmatrix}.$$

This latter system is autonomous. In addition, it is linear and homogeneous. There is a unique solution for $X_t$ provided that the initial value $X_0 = (j_0, a_0)^T$ is known. This first-order system can be expressed as a second-order equation in the variable $a$ or $j$,

$$a_{t+2} = p_a a_{t+1} + rp_j f p_a a_t \quad \text{or} \quad j_{t+2} = p_a j_{t+1} + fp_a rp_j j_t. \tag{1.7}$$

The second-order equations (1.7) have unique solutions provided the initial values $a_0$ and $a_1$ (or $j_0$ and $j_1$) are known. ■

In general, for a linear difference equation of the form (1.3), it can be shown that a solution $x_t$ exists and is unique. It can be seen that $x_k$ can be expressed in terms of $x_0, x_1, \ldots, x_{k-1}$, and $b_0$ at $t = 0$,

$$x_k = -(a_1 x_{k-1} + \cdots + a_k x_0) + b_0.$$

Then, it follows at $t = 1$,

$$x_{k+1} = -(a_1 x_k + \cdots + a_k x_1) + b_1.$$

Since $x_k$ is a function of $x_0, x_1, \ldots, x_{k-1}$, so is $x_{k+1}$. By induction, it follows that $x_t$ for $t = k, k+1, k+2, \ldots$, can be expressed as a function of $x_{t-1}, x_{t-2}, \ldots, x_{t-k}$, and $b_{t-k}$. Given the initial conditions, $x_0, x_1, \ldots, x_{k-1}$ and the nonhomogeneous terms $b_t, t = 0, 1, \ldots$, the solution is *uniquely* determined. For example, the unique solution to the first-order, linear, and homogeneous difference equation $x_{t+1} = c_t x_t$ is

$$x_t = c_{t-1} \ldots c_1 c_0 x_0 = x_0 \left( \prod_{j=0}^{t-1} c_j \right).$$

## 1.3 First-Order Equations

A first-order, linear difference equation of the form (1.3) can be written as

$$x_{t+1} = -a_t x_t + b_t = c_t x_t + b_t,$$

where $c_t = -a_t$. When the initial condition $x_0$ is known and the coefficients are known, this difference equation has a unique solution. The solution to this first-order equation can be found by repeated substitution of the values into the equation. First, $x_1 = c_0 x_0 + b_0$, then

$$x_2 = c_1(c_0 x_0 + b_0) + b_1 = c_1 c_0 x_0 + c_1 b_0 + b_1.$$

Substituting these values into $x_3$ yields

$$x_3 = c_2(c_1 c_0 x_0 + c_1 b_0 + b_1) + b_2 = c_2 c_1 c_0 x_0 + c_2 c_1 b_0 + c_2 b_1 + b_2.$$

The solution $x_{t+1}$ for $t \geq 1$ can be expressed as follows:

$$x_{t+1} = \left( \prod_{i=0}^{t} c_i \right) x_0 + b_t + \sum_{i=0}^{t-1} \left[ b_i \prod_{j=i+1}^{t} c_j \right]. \tag{1.8}$$

For constant coefficients, $c_t \equiv c$ and $b_t \equiv b$, $x_{t+1} = cx_t + b$. The solution to (1.8), in this case, can be expressed as

$$x_{t+1} = c^{t+1} x_0 + b \sum_{i=0}^{t} c^i. \tag{1.9}$$

The summation is a geometric series and for the cases $c = 1$ and $c \neq 1$, the solution can be expressed as follows:

$$x_{t+1} = \begin{cases} x_0 c^{t+1} + b\dfrac{1 - c^{t+1}}{1 - c} & c \neq 1, \\ x_0 + (t + 1)b, & c = 1. \end{cases} \tag{1.10}$$

Finding the solution through repeated substitution is cumbersome for higher-order difference equations and often does not give a general formula for the solution. There are methods that can be applied to find the general solution. We demonstrate one method that can be applied when the coefficients are constant. This method is illustrated in the following example and it is generalized to higher-order equations in the next section.

**Example 1.6** Let

$$x_{t+1} = ax_t + b, \tag{1.11}$$

where $a$ and $b$ are constants. The method for finding a general solution to this equation is based on the fact that the general solution is a *superposition* of two solutions, a general solution to the homogeneous equation, and a particular solution to the nonhomogeneous equation. The sum of these two solutions forms a general solution to the nonhomogeneous equation. This can be seen in the solution (1.9), where the first term is the general solution to the homogeneous equation and the second term is a particular solution to the nonhomogeneous equation. For equation (1.11), it follows from Example 1.4 that the general solution to the homogeneous linear difference equation, $x_{t+1} = ax_t$, is $ca^t$, where $c$ is an arbitrary constant. Sometimes, one can "guess" the form of the particular solution. For example, since $b$ is constant, it is reasonable to assume that the particular solution is constant $x_p = k$. To find the constant $k$, substitute $k$ into the difference equation,

$$k = ak + b,$$

which implies $k = b/(1 - a)$ if $a \neq 1$. If $a = 1$, then this "guess" does not work. The reason it does not work is that $x_t = k$ is a solution to the homogeneous equation. Another guess for the particular solution is $x_p = tk$, where $k$ is a constant. Substitution of this guess into the difference equation yields $(t + 1)k = atk + b = tk + b$, which implies $k = b$. Hence, a particular solution to the nonhomogeneous difference equation is $x_p = b/(1 - a)$, if $a \neq 1$ or $x_p = tb$, if $a = 1$. Note that neither of these particular solutions is unique; another particular solution in the case $a = 1$ is $x_t = tb + 3$. Adding a solution of the homogeneous equation to the particular solution gives another particular solution. This method of finding the particular solution is known as the *method of undetermined coefficients*. The *general solution* to the nonhomogeneous equation involves one arbitrary constant $c$,

$$x_t = \begin{cases} ca^t + \dfrac{b}{1 - a}, & a \neq 1, \\ c + tb, & a = 1. \end{cases}$$

If the initial condition $x_0$ is known, then the solution to the linear difference equation is uniquely determined. The constant $c$ is found by substituting $t = 0$ and $x_0$ into the general solution to obtain

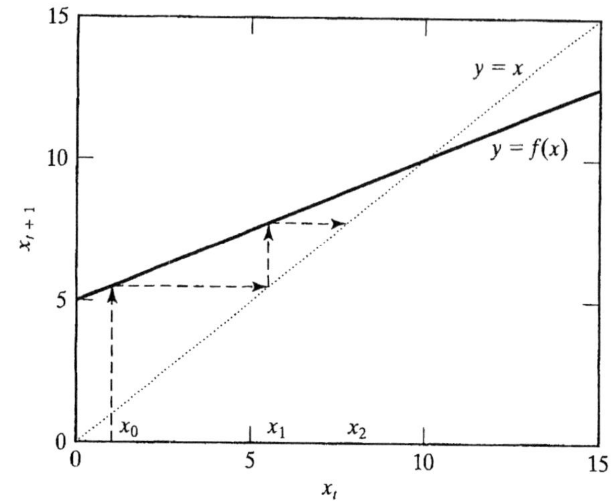


**Figure 1.2** Cobwebbing method applied to the first-order difference equation, $x_{t+1} = f(x_t)$.

$$x_t = \begin{cases} x_0 a^t + b\dfrac{1 - a^t}{1 - a}, & a \neq 1, \\ x_0 + tb, & a = 1. \end{cases} \tag{1.12}$$

The solution (1.12) agrees with the solution (1.10) obtained via repeated substitution where the constant $c$ in (1.10) equals $a$. ■

An important method for studying the behavior of solutions to first-order difference equations is an iterative method known as the *cobwebbing method*. This method is particularly useful when studying nonlinear first-order difference equations and will be discussed in more detail in Chapter 2.

Consider the first-order difference equation, $x_{t+1} = f(x_t)$. The equation $y = f(x)$ is called the *reproduction curve*. The reproduction curve and the line $y = x$ are graphed in the $x$-$y$ plane. The $x$-axis and $y$-axis represent $x_t$ and $x_{t+1}$, respectively. The cobwebbing method uses the curves $y = x$ and $y = f(x)$ as follows: Given $x_0$, locate the point $(x_0, 0)$ on the $x$-axis, then go vertically upward to the point $(x_0, x_1)$ located on the reproduction curve. Go horizontally (left or right) to the point $(x_1, x_1)$, on the line $y = x$; the value $x_1$ on the $x$-axis is the next iteration. Next, go vertically (up or down) from the point $(x_1, x_1)$ to the point $(x_1, x_2)$ on the reproduction curve, then go horizontally (left or right) to the point $(x_2, x_2)$ on the line $y = x$. The value $x_2$ on the $x$-axis is the next iteration. This procedure is continued. Figure 1.2 illustrates the cobwebbing method.

The term "cobwebbing" will become apparent when more complicated reproduction curves are studied in Chapter 2. For example, the vertical and horizontal segments in the diagram in Figure 1.2 can form a spiral of nested rectangles. The rectangles look like "cobwebs." In Figure 1.2, the vertical and horizontal lines converge to the intersection point of $y = x$ and $y = f(x)$. For the reproduction curve shown in Figure 1.2, it follows that $\lim_{t\to\infty} x_t = \bar{x}$, where $\bar{x} = f(\bar{x})$.

The method applied in Example 1.6 to solve a first-order difference equation will be extended to second- and higher-order difference equations in the next section.

## 1.4 Second-Order and Higher-Order Equations

We discuss a method for finding the general solution to the $k$th-order, nonhomogeneous, linear difference equation (1.3). The general solution is a function of $k$

arbitrary constants. These $k$ constants can be uniquely determined if the $k$ initial conditions, $x_0, x_1, \ldots, x_{k-1}$, are specified. The method involves the following steps.

(a) Find the *general solution* to the homogeneous difference equation ($b_t \equiv 0$), a solution depending on $k$ arbitrary constants and $k$ linearly independent solutions of (1.3). (Recall that the $k$ solutions $x^1, \ldots, x^k$ are *linearly independent* if $\sum_{i=1}^{k} \alpha_i x^i = 0$ implies $\alpha_i = 0$ for $i = 1, \ldots, k$.) Denote the general solution as $x_h$. Then

$$x_h = c_1 x^1 + \cdots + c_k x^k = \sum_{i=1}^{k} c_i x^i,$$

where $c_1, \ldots, c_k$ are constants and $x^1, \ldots, x^k$ are $k$ linearly independent solutions of the homogeneous equation. (*Note:* The notation $x^i$ means the $i$th solution and *not* $x$ to the power $i$.) Any linear combination of solutions to the linear homogeneous difference equation is also a solution. This property is known as the *superposition principle*. The key to finding the general solution is to identify the $k$ linearly independent solutions.

(b) Find a *particular solution* to the nonhomogeneous difference equation, a solution with no arbitrary constants. Denote this particular solution as $x_p$.

(c) The sum of the general solution $x_h$ and the particular solution $x_p$ gives the *general solution* to the nonhomogeneous linear difference equation,

$$x_t = x_h + x_p = \sum_{i=1}^{k} c_i x^i + x_p.$$

If initial conditions, $x_0, x_1, \ldots, x_{k-1}$, are prescribed, then the constants $c_1, c_2, \ldots, c_k$ can be uniquely determined.

We apply this method to a homogeneous, second-order, linear difference equation with constant coefficients. Consider the difference equation,

$$x_{t+2} + a x_{t+1} + b x_t = 0, \tag{1.13}$$

where $a$ and $b$ are constants. To find two linearly independent solutions, $x^1$ and $x^2$, assume the solution has the form $x_t = \lambda^t$, where $\lambda \neq 0$. Then

$$\lambda^{t+2} + a\lambda^{t+1} + b\lambda^t = 0,$$
$$\lambda^t(\lambda^2 + a\lambda + b) = 0.$$

**Definition 1.5.** The equation $\lambda^2 + a\lambda + b = 0$ is called the *characteristic equation* and $\lambda^2 + a\lambda + b$ is called the *characteristic polynomial* of the difference equation (1.13). The two solutions or roots, $\lambda_1$ and $\lambda_2$, of the characteristic equation are called the *eigenvalues*.

The form of the solutions depends on the eigenvalues and can be separated into three different cases. Note for equation (1.13) to be second order, $b \neq 0$. Therefore, none of the eigenvalues of the characteristic equation can be zero.

**Case 1:** Suppose the eigenvalues are real and distinct, $\lambda_1 \neq \lambda_2$. Then the two linearly independent solutions are $x^1 = \lambda_1^t$ and $x^2 = \lambda_2^t$. The general solution is

$$x_t = c_1\lambda_1^t + c_2\lambda_2^t.$$

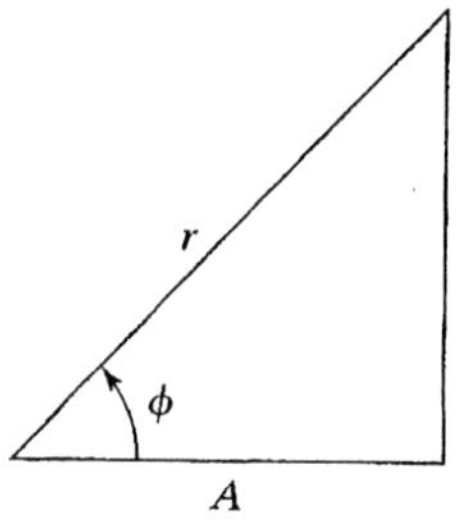


**Figure 1.3** Relation between $r, A = r\cos(\phi)$, and $B = r\sin(\phi)$.

**Case 2:** Suppose the eigenvalues are real and equal, $\lambda_1 = \lambda_2$. Then the two linearly independent solutions are $x^1 = \lambda_1^t$ and $x^2 = t\lambda_1^t$ (the second solution is found by multiplying by $t$). It is easy to verify $x^2$ is a solution and that $x^1$ and $x^2$ are linearly independent. (See Exercise 6.) The general solution is

$$x_t = c_1\lambda_1^t + c_2 t\lambda_1^t.$$

**Case 3:** Suppose the eigenvalues $\lambda_i$ are complex conjugates, $\lambda_{1,2} = A \pm iB = r(\cos\phi \pm i\sin\phi) = re^{\pm i\phi}$, where $r = \sqrt{A^2 + B^2}$, $\phi = \arctan(B/A)$. See Figure 1.3. Then the solution $x_t = k_1(A + Bi)^t + k_2(A - Bi)^t = k_1 r^t e^{it\phi} + k_2 r^t e^{-it\phi}$ can be expressed as

$$x_t = c_1 r^t\cos(t\phi) + c_2 r^t\sin(t\phi),$$

where $c_1 = k_1 + k_2$ and $c_2 = i(k_1 - k_2)$. The two linearly independent solutions can be expressed as the two real functions $x^1 = r^t\cos(t\phi)$ and $x^2 = r^t\sin(t\phi)$. ●

In each case, it can be shown that the solutions $x^1$ and $x^2$ are linearly independent (e.g., $0 = \alpha_1 x^1 + \alpha_2 x^2$ implies $\alpha_1 = 0 = \alpha_2$). Another check for independence involves a quantity known as the Casoratian. The Casoratian plays a role similar to the Wronskian in differential equations (see Elaydi, 1999).

**Definition 1.6.** The *Casoratian of $x^1(t)$ and $x^2(t)$* is

$$C(t) = C(x^1(t), x^2(t)) = \det\begin{pmatrix} x^1(t) & x^2(t) \\ x^1(t+1) & x^2(t+1) \end{pmatrix}$$
$$= x^1(t)x^2(t+1) - x^2(t)x^1(t+1),$$

where the expression "det" means the determinant of a matrix.

The definition of Casoratian can be extended to more than two solutions. It can be shown that if $x^1(t)$ and $x^2(t)$ are two solutions to a second-order linear difference equation and $C(t) \neq 0$ for some $t = 0, 1, 2, \ldots$, then the solutions $x^1$ and $x^2$ are linearly independent (Elaydi, 1999). For example, in Case 1, the Casoratian satisfies

$$C(\lambda_1^t, \lambda_2^t) = \det\begin{pmatrix} \lambda_1^t & \lambda_2^t \\ \lambda_1^{t+1} & \lambda_2^{t+1} \end{pmatrix} = (\lambda_1\lambda_2)^t(\lambda_2 - \lambda_1).$$

$C(\lambda_1^t, \lambda_2^t) \neq 0$ for all $t$ because $\lambda_1 \neq \lambda_2$ and $\lambda_1$ and $\lambda_2$ are nonzero. The independence of the solutions in Case 2 is verified in Exercise 6.

Consider a $k$th-order, homogeneous, linear difference equation with constant coefficients,

$$x_{t+k} + a_1 x_{t+k-1} + \cdots + a_k x_t = 0.$$

Let $x_t = \lambda^t$ and $\lambda \neq 0$. The *characteristic equation* for this difference equation is

$$\lambda^k + a_1\lambda^{k-1} + \cdots + a_k = 0.$$

The characteristic equation has $k$ roots or eigenvalues: $\lambda_i$, $i = 1, \ldots, k$. If the roots are real and distinct, then the general solution has the following form:

$$x_t = c_1\lambda_1^t + c_2\lambda_2^t + \cdots + c_k\lambda_k^t,$$

where the constants $c_i$, $i = 1, 2, \ldots, k$, are arbitrary. The $k$ solutions are linearly independent. (The Casoratian in this case is the determinant of a $k \times k$ matrix.) General solutions for the other cases, repeated eigenvalues and complex conjugate eigenvalues, can be determined once the eigenvalues and their multiplicity are determined. For example, if there is a real eigenvalue $\lambda_1$ of multiplicity $m$, then $m$ linearly independent solutions can be formed by multiplying by increasing powers of $t$:

$$\lambda_1^t, t\lambda_1^t, t^2\lambda_1^t, \ldots, t^{m-1}\lambda_1^t.$$

If there are complex eigenvalues $\lambda_{1,2} = r[\cos(\phi) \pm i\sin(\phi)]$ of multiplicity $m$, then there are $2m$ linearly independent solutions:

$$r^t\cos(t\phi), r^t\sin(t\phi), tr^t\cos(t\phi), tr^t\sin(t\phi), \ldots, t^{m-1}r^t\cos(t\phi), t^{m-1}r^t\sin(t\phi).$$

**Example 1.7** Consider the third-order, homogeneous equation

$$x_{t+3} + x_{t+2} + x_{t+1} + x_t = 0.$$

The characteristic equation satisfies $\lambda^3 + \lambda^2 + \lambda + 1 = (\lambda + 1)(\lambda^2 + 1) = 0$. The eigenvalues are $\lambda_{1,2,3} = -1, \pm i$. Thus, the general solution to the homogeneous equation is given by

$$x_t = c_1(-1)^t + c_2\cos(t\pi/2) + c_3\sin(t\pi/2).$$

If the difference equation is nonhomogeneous,

$$x_{t+3} + x_{t+2} + x_{t+1} + x_t = 4t,$$

then to find the general solution, a particular solution of the nonhomogeneous equation is required. We assume the particular solution has the same form as the right-hand side, that is, it is a linear polynomial in $t$:

$$x_p = k_1 t + k_2.$$

Substituting $x_p$ into the nonhomogeneous difference equation yields

$$k_1(t+3) + k_2 + k_1(t+2) + k_2 + k_1(t+1) + k_2 + k_1 t + k_2 = 4t.$$

Simplifying the left side,

$$4k_1 t + 6k_1 + 4k_2 = 4t.$$

Equating coefficients, $4k_1 = 4$ and $6k_1 + 4k_2 = 0$, yields the coefficients $k_1 = 1$ and $k_2 = -3/2$. Hence, the general solution to the nonhomogeneous difference equation is

$$x_t = x_h + x_p = c_1(-1)^t + c_2\cos(t\pi/2) + c_3\sin(t\pi/2) + t - 3/2.$$

The preceding method to find $x_p$ is known as the *method of undetermined coefficients*. ■

To find a particular solution to a nonhomogeneous difference equation, the method of undetermined coefficients can be applied only if $b_t$ has a particular form (functions of the form $a^t$, polynomials in $t$, cosines or sines, products of a polynomial in $t$ with cosines or sines or with $a^t$, or linear combinations of these forms). For example, suppose the linear difference equation is

$$x_{t+3} + a_1x_{t+2} + a_2x_{t+1} + a_3x_t = 8t\cos(t).$$

Then, as an initial guess for the particular solution, it is assumed that the particular solution has the form

$$x_p = (k_1 + k_2t)\cos(t) + (k_3 + k_4t)\sin(t).$$

If this solution appears in the homogeneous solution, then it is multiplied by a power of $t$ until none of the terms in the particular solution are solutions of the homogeneous equation. If the homogeneous solution is $x_h = c_1 + c_2 t + c_3\cos(t) + c_4\sin(t)$, then the particular solution must be multiplied by $t$ because $\cos(t)$ and $\sin(t)$ are solutions to the homogeneous equation,

$$x_p = t(k_1 + k_2 t)\cos(t) + t(k_3 + k_4 t)\sin(t).$$

The coefficients $k_1$, $k_2$, $k_3$, and $k_4$ are determined by substituting $x_p$ into the difference equation.

The method of undetermined coefficients, as noted earlier, cannot always be applied. When the method of undetermined coefficients cannot be applied, there are other methods that may work (e.g., method of variation of parameters). However, we do not consider these other methods. Please consult the references for more information about solutions to difference equations.

An important question that we will address with respect to biological models is, "What is the asymptotic or long-term behavior of the model?" For models formulated in terms of linear difference equations, the asymptotic behavior depends on the eigenvalues, whether the eigenvalues are real or complex and the magnitude of the eigenvalues. To address this question, it is generally not necessary to find explicit solutions. In cases where there exists an eigenvalue whose magnitude exceeds all others, referred to as a strictly dominant eigenvalue, then this eigenvalue is an important determinant of the dynamics. Recall that the *magnitude* of a real eigenvalue, $\lambda = a$, is its absolute value, $|\lambda| = |a|$. The *magnitude* of a complex eigenvalue, $\lambda = a + bi$, is $|\lambda| = |a + bi| = \sqrt{a^2 + b^2}$.

**Definition 1.7.** Suppose the $k$ eigenvalues of a characteristic equation are $\lambda_1, \lambda_2, \ldots, \lambda_k$. An eigenvalue $\lambda_i$ such that $|\lambda_i| \geq |\lambda_j|$ for all $j \neq i$ is called a *dominant eigenvalue*. If the inequality is strict, $|\lambda_i| > |\lambda_j|$ for all $j \neq i$, then $\lambda_i$ is called a *strictly dominant eigenvalue*.

The magnitude of the eigenvalues determine whether solutions are unbounded or bounded. The types of eigenvalues, either real or complex, determine whether solutions oscillate or whether solutions converge or diverge monotonically. Of particular interest is whether the magnitude of all of the eigenvalues is less than one. For example, if there exists a dominant eigenvalue $\lambda_1$ and $|\lambda_1| < 1$, then solutions to the difference equation converge to zero.

**Example 1.8** Suppose $x_{t+2} + x_t = 0$. The characteristic equation is $\lambda^2 + 1 = 0$. The eigenvalues are complex conjugates, $\lambda_{1,2} = \pm i$, $r = 1$, and $\phi = \pi/2$ so that the general solution is

$$x_t = c_1\cos(t\pi/2) + c_2\sin(t\pi/2).$$

Since the magnitude $|\lambda_{1,2}| = 1$, there is a dominant eigenvalue but no strictly dominant eigenvalue. Because the eigenvalues are complex, solutions oscillate. However, the magnitude of the eigenvalues is one, so that solutions are bounded but they do not converge to any particular solution. This behavior can be seen in the following solution.

Suppose the initial conditions satisfy $x_0 = 0$ and $x_1 = 1$; then the solution is $x_t = \sin(t\pi/2)$, which generates the following oscillating sequence:

$$0, 1, 0, -1, 0, 1, 0, -1, \ldots.$$

The solution is periodic of period 4. The sequence $\{x_t\}$ oscillates and does not converge. ∎

**Example 1.9** Consider the fourth-order linear difference equation $x_{t+4} - 6x_{t+3} + 13x_{t+2} - 12x_{t+1} + 4x_t = 0$. The characteristic equation is given by $(\lambda - 1)^2(\lambda - 2)^2 = 0$. Since $\lambda_{1,2} = 1$ and $\lambda_{3,4} = 2$, the eigenvalues are real and the general solution is

$$x_t = c_1 + c_2 t + 2^t(c_3 + c_4 t).$$

Because the dominant eigenvalue is greater than one, solutions may increase indefinitely. For example, suppose $x_0 = 0 = x_1$, $x_2 = 2$, and $x_3 = 8$; then the constants, $c_1, \ldots, c_4$, are found by solving the following linear system:

$$\begin{aligned} c_1 + c_3 &= 0 \\ c_1 + c_2 + 2c_3 + 2c_4 &= 0 \\ c_1 + 2c_2 + 4c_3 + 8c_4 &= 2 \\ c_1 + 3c_2 + 8c_3 + 24c_4 &= 8. \end{aligned}$$

The unique solution to this linear system is $c_1 = -2 = c_2$, $c_3 = 2$, and $c_4 = 0$, so that

$$x_t = -2 - 2t + 2^{t+1}.$$

Therefore, in this case, $\lim_{t\to\infty} x_t = \infty$. ■

Computer software may be used to help solve systems of linear difference equations. For example, the command `rsolve` in the computer algebra system Maple solves linear recurrence relations. However, one must be careful; sometimes the form of the solution given by Maple is not simplified. In Maple, the general solution to a second-order difference equation is expressed in terms of $x_0$ and $x_1$ instead of arbitrary constants, $c_1$ and $c_2$, and when the eigenvalues are complex, solutions are written in terms of the complex eigenvalues instead of sines and cosines. The commands in Maple to find the general solution and to find the unique solution satisfying the set of initial conditions in Example 1.9 are given by the following statements:

```
>rsolve(x(t+4)-6*x(t+3)+13*x(t+2)-12*x(t+1)+4*x(t)=0,x(t));

>rsolve({x(t+4)-6*x(t+3)+13*x(t+2)-12*x(t+1)+4*x(t)=0,
  x(0)=0,x(1)=0,x(2)=2,x(3)=8},x(t));
```

**Example 1.10** Consider the second-order difference equation in Example 1.5 which models the size of the adult population,

$$a_{t+2} = p_a a_{t+1} + f p_a r p_j a_t.$$

The characteristic equation is $\lambda^2 - p_a\lambda - f p_a r p_j = 0$. The eigenvalues are

$$\lambda_{1,2} = \frac{p_a \pm \sqrt{p_a^2 + 4rfp_ap_j}}{2}.$$

The eigenvalues are real. There is a strictly dominant eigenvalue $\lambda_1$ and $|\lambda_1| < 1$ if and only if (iff)

$$p_a(1 + rfp_j) < 1 \tag{1.14}$$

(see Exercise 9). If condition (1.14) holds, then the adult population size will decrease to zero (extinction). If fecundity, $f$, or juvenile and adult survival, $p_j$ and $p_a$, are sufficiently large, then extinction will not occur, since $\lambda_1 > 1$. ■

## 1.5 First-Order Linear Systems

A higher-order linear difference equation can be converted to a first-order linear system. Thus, a higher-order linear difference equation can be considered a special case of a first-order linear difference system. Consider the $k$th-order linear difference equation (1.3),

$$x(t+k) + a_1 x(t+k-1) + \cdots + a_{k-1}x(t+1) + a_k x(t) = b(t), \quad (1.15)$$

where for convenience $x_t$ is denoted as $x(t)$, $b_t$ as $b(t)$, and so on. To transform this $k$th-order equation to a first-order system, we define the following vector of states. Let $Y(t)$ be a $k$-dimensional vector, $Y(t) = (y_1(t), y_2(t), \dots, y_{k-1}(t), y_k(t))^T$, satisfying the following:

$$\begin{aligned} y_1(t) &= x(t) \\ y_2(t) &= x(t+1) \\ &\vdots \\ y_{k-1}(t) &= x(t+k-2) \\ y_k(t) &= x(t+k-1). \end{aligned}$$

The first element $y_1(t)$ is the solution $x(t)$. Next, a first-order difference equation in $y$ is formed,

$$\begin{aligned} y_1(t+1) &= y_2(t), \\ y_2(t+1) &= y_3(t), \\ &\vdots \\ y_{k-1}(t+1) &= y_k(t), \\ y_k(t+1) &= -a_1 y_k(t) - \cdots - a_{k-1}y_2(t) - a_k y_1(t) + b(t). \end{aligned}$$

The last equation is obtained from the difference equation (1.15). The first-order system can be written in matrix form,

$$Y(t+1) = AY(t) + B,$$

where

$$A = \begin{pmatrix} 0 & 1 & 0 & \cdots & 0 \\ 0 & 0 & 1 & \cdots & 0 \\ \vdots & \vdots & \vdots & \ddots & \vdots \\ 0 & 0 & 0 & \cdots & 1 \\ -a_k & -a_{k-1} & -a_{k-2} & \cdots & -a_1 \end{pmatrix} \quad \text{and} \quad B = \begin{pmatrix} 0 \\ 0 \\ \vdots \\ 0 \\ b(t) \end{pmatrix}. \quad (1.16)$$

Note that matrix $A$ has 1's along the superdiagonal and has the coefficients of the higher-order difference equation along the last row but the signs are reversed. Matrix $A$ in (1.16) is referred to as the *companion matrix* of the $k$th-order scalar difference equation in (1.15).

**Example 1.11** Suppose $x(t+2) - 2x(t+1) + x(t) = \cos(t)$. Let $y_1(t) = x(t), y_2(t) = x(t+1)$. Then the second-order difference equation can be expressed as the first-order system,

$$y_1(t+1) = y_2(t),$$
$$y_2(t+1) = -y_1(t) + 2y_2(t) + \cos(t),$$

or $Y(t+1) = AY(t) + B$, where

$$A = \begin{pmatrix} 0 & 1 \\ -1 & 2 \end{pmatrix} \quad \text{and} \quad B = \begin{pmatrix} 0 \\ \cos(t) \end{pmatrix}.$$

■

There are various methods to find solutions to first-order linear systems. Some solution methods are similar to the methods for higher-order linear equations. The underlying theory for first-order linear systems is similar to higher-order linear equations. In particular, a solution to a first-order linear difference system $X(t+1) = AX(t) + B$ is the superposition of two solutions, the general solution $X_h$ to the homogeneous system, $X_h(t+1) = AX_h(t)$, and a particular solution $X_p$ to the nonhomogeneous system, $X_p(t+1) = AX_p(t) + B$. Hence, the general solution to the nonhomogeneous system is

$$X(t) = X_h(t) + X_p(t).$$

We will concentrate only on the homogeneous solution in the case that the system is autonomous, that is, the system $X(t+1) = AX(t)$.

Let $X(t+1) = AX(t)$, where $X(t) = (x_1(t), x_2(t), \ldots, x_k(t))^T$ and $A = (a_{ij})$ is a $k \times k$ constant matrix. This homogeneous system has $k$ linearly independent solutions, $\{X_i(t)\}_{i=1}^k$, just as in the case of a $k$th-order homogeneous scalar equation. The general solution is

$$X(t) = \sum_{i=1}^{k} c_i X_i(t).$$

There are some direct and indirect methods to find these linearly independent solutions. We shall demonstrate how to find these solutions directly. Indirect methods of finding these solutions use the fact that the solution can be written as $X(t) = A^t X(0)$. Then if a general expression can be found for $A^t$, the solution is known. Elaydi and Harris (1998) present several methods to compute $A^t$. One method is based on the Putzer algorithm from differential equations, which is used to compute $e^{At}$. Please consult the references for these particular methods and algorithms.

Let $X(t+1) = AX(t)$, where $A = (a_{ij})$ is an $k \times k$ constant matrix. Assume that $X(t) = \lambda^t V$, where $V$ is a nonzero $k$-column vector and $\lambda$ is a constant. Substituting $\lambda^t V$ into the linear system yields $\lambda^{t+1} V = A(\lambda^t V)$. Simplifying, the following equation is obtained:

$$(A - \lambda I)V = \mathbf{0}, \tag{1.17}$$

where $I$ is the identity matrix and $\mathbf{0}$ is the zero vector. The zero solution, $V = \mathbf{0}$, is a trivial solution of equation (1.17). We seek nonzero solutions to the original system. Recall from linear algebra that if $\det(A - \lambda I) \neq 0$, then (1.17) has a unique solution. This solution is the zero solution, $V = \mathbf{0}$. Hence, nonzero solutions $V$ are obtained iff the matrix $A - \lambda I$ is singular iff

$$\det(A - \lambda I) = 0. \tag{1.18}$$

**Definition 1.8.** Equation (1.18) is referred to as the *characteristic equation* of matrix $A$. The nonzero solutions $V$ to equation (1.17) are called the *eigenvectors* of matrix $A$ and the values of $\lambda$ corresponding to the nonzero solutions $V$ are called the *eigenvalues* of matrix $A$.

Before continuing the solution method for first-order systems, we note the relationship between the $k$th-order homogeneous scalar equation (1.15) and the linear system $X(t+1) = AX(t)$, where $A$ is given by (1.16). The characteristic equation for (1.15) is the same as the characteristic equation for matrix $A$ (Exercise 12). This should not be a surprise because the solution $y_1(t) = x(t)$. The eigenvalues of matrix $A$ are the same as the eigenvalues associated with the corresponding $k$th-order, homogeneous scalar equation (1.15). This relationship is demonstrated for a third-order difference equation in the next example.

**Example 1.12** The characteristic equation for $x(t+3) + a_1x(t+2) + a_2x(t+1) + a_3x(t) = 0$ is

$$\lambda^3 + a_1\lambda^2 + a_2\lambda + a_3 = 0.$$

The companion matrix corresponding to the third-order difference equation is

$$A = \begin{pmatrix} 0 & 1 & 0 \\ 0 & 0 & 1 \\ -a_3 & -a_2 & -a_1 \end{pmatrix}.$$

We compute the determinant of $(A - \lambda I)$ by expanding the last row:

$$\det \begin{pmatrix} -\lambda & 1 & 0 \\ 0 & -\lambda & 1 \\ -a_3 & -a_2 & -a_1-\lambda \end{pmatrix} = -a_3(1) + a_2(-\lambda) - (a_1 + \lambda)(-\lambda)^2$$

$$= -(\lambda^3 + a_1\lambda^2 + a_2\lambda + a_3).$$

Setting this last expression to zero and multiplying by negative one, the two characteristic equations agree. ■

Now, we return to the method for finding the solution to $X(t+1) = AX(t)$. Summarizing, a $k \times k$ matrix $A$ has $k$ eigenvalues, $\lambda_1, \lambda_2, \ldots, \lambda_k$, that are solutions of the characteristic equation $\det(A - \lambda I) = 0$. The eigenvectors $V_i$ corresponding to the eigenvalue $\lambda_i$ are found by solving $(A - \lambda_i I)V_i = \mathbf{0}$. The general solution to $X(t+1) = AX(t)$ is a linear combination of $k$ linearly independent solutions $\{X_i(t)\}_{i=1}^k$:

$$X(t) = \sum_{i=1}^{k} c_i X_i(t).$$

As noted earlier, the solutions $X_i(t) = \lambda_i^t V_i$, $i = 1, 2, \ldots, k$, may not yield $k$ linearly independent solutions. In the case where the number of linearly independent eigenvectors corresponding to an eigenvalue is less than the multiplicity of that eigenvalue, then the solutions $\lambda_i^t V_i$ do not generate $k$ linearly independent solutions. In addition, when the eigenvalues are complex, the solutions are generally written in another form containing cosines and sines. We do not discuss all of these cases but mention only two special cases.

**Case 1:** Suppose the eigenvalues are real and the solutions, $\lambda_i^t V_i$, $i = 1, 2, \ldots, k$, are linearly independent. Then the general solution to the linear system $X(t+1) = AX(t)$ can be expressed as

$$X(t) = \sum_{i=1}^{k} c_i \lambda_i^t V_i. \tag{1.19}$$

(Note that the eigenvectors are also real valued because $A$ has real entries.)

**Case 2:** Suppose matrix $A$ has $k$ distinct eigenvalues. Then it follows that there are $k$ linearly independent eigenvectors (Ortega, 1987) and the solution can still be written as in (1.19). The entries $\lambda_i^t V_i$ may not have real values.

**Case 3:** In the other cases, when the eigenvalues are complex or the eigenvectors are not independent, the general solution includes such terms as $t^n \lambda^t V$ or $t^n r^t \cos(\phi t)V$ or $t^n r^t \sin(\phi t)V$. ●

It is important to note that the asymptotic behavior of the solution (1.19) does not require knowledge of the eigenvectors. The asymptotic behavior is determined by the eigenvalues and, in particular, their magnitude. For example, if $|\lambda_i| < 1, i = 1, \ldots, k$, then $\lim_{t\to\infty} X(t) = \mathbf{0}$, the zero vector. The largest magnitude of all of the eigenvalues of a matrix $A$ is referred to as the spectral radius of $A$.

**Definition 1.9.** Suppose the $k \times k$ matrix $A$ has $k$ eigenvalues $\lambda_1, \lambda_2, \ldots, \lambda_k$. The *spectral radius* of matrix $A$ is denoted as $\rho(A)$ and is defined as

$$\rho(A) = \max_{i \in \{1, 2, \ldots, k\}} \{|\lambda_i|\}.$$

We state the result concerning the asymptotic behavior $A^t$ as a theorem. For a proof of this result, please consult Elaydi (1999).

**Theorem 1.1** *Let $A$ be a constant $k \times k$ matrix. Then the spectral radius of $A$ satisfies $\rho(A) < 1$ iff $\lim_{t\to\infty} A^t = \mathbf{0}$, zero matrix.* □

As a consequence of this theorem, the solution $X(t) = A^t X(0)$ of $X(t+1) = AX(t)$ approaches the zero solution when $\rho(A) < 1$ as $t \to \infty$. In the special case of (1.19) it is easy to see why this result holds.

**Example 1.13** Suppose $X(t+1) = AX(t)$, where

$$A = \begin{pmatrix} 2 & 3 \\ 2 & 1 \end{pmatrix}. \tag{1.20}$$

The eigenvalues of $A$ are the solutions of the characteristic equation,

$$\det(A - \lambda I) = \det\begin{pmatrix} 2-\lambda & 3 \\ 2 & 1-\lambda \end{pmatrix} = \lambda^2 - 3\lambda - 4 = 0.$$

The characteristic equation can be factored into $(\lambda + 1)(\lambda - 4) = 0$. Hence, the eigenvalues of $A$ are $\lambda_{1,2} = -1, 4$ and the spectral radius of $A$ is $\rho(A) = 4$. The eigenvectors $V$ are solutions to $(A - \lambda I)V = \mathbf{0}$. For $\lambda_1 = -1$ and $V = (v_1, v_2)^T$, eigenvector $V$ is a solution of

$$3v_1 + 3v_2 = 0$$
$$2v_1 + 2v_2 = 0.$$

Hence, $v_1 = -v_2$ and an eigenvector $V$ corresponding to $\lambda_1 = -1$ has the form $V_1 = (1, -1)^T$. (Recall any constant multiple of $V$ is also an eigenvector.) For $\lambda_2 = 4$, to find $V = (v_1, v_2)^T$ we solve the system

$$-2v_1 + 3v_2 = 0$$
$$2v_1 - 3v_2 = 0.$$

In this case, $v_1 = (3/2)v_2$ so that if $v_2 = 2$ and $v_1 = 3$, then $V = (3, 2)^T$. The general solution to the system $X(t + 1) = AX(t)$ is

$$X(t) = c_1(-1)^t\begin{pmatrix}1\\-1\end{pmatrix} + c_2(4)^t\begin{pmatrix}3\\2\end{pmatrix} = \begin{pmatrix}c_1(-1)^t + 3c_2(4)^t\\c_1(-1)^{t+1} + 2c_2(4)^t\end{pmatrix}.$$ ■

**Example 1.14** For the matrix $A$ defined in (1.20), let $B = A/c$, where $c > 0$. Then the eigenvalues of $B$ satisfy $BV = AV/c = (\lambda/c)V$, where $\lambda$ is an eigenvalue of $A$ and $V$ its corresponding eigenvector. Hence, the eigenvalues of $B$ are $4/c$ and $-1/c$ and the corresponding eigenvectors of $B$ are the same as $A$. If $c > 4$ , then $\rho(B) < 1$. ■

As a final example on linear systems of difference equations, an important biological model, known as the Leslie matrix model, is discussed.

## 1.6 An Example: Leslie's Age-Structured Model

The Leslie matrix model is a linear, first-order system of difference equations that models the dynamics of age-structured populations. The name "Leslie" refers to Patrick Holt Leslie (1900–1974), one of the first scientists to study age-structured population dynamics using matrix theory.

An age-structured model is one of many types of structured models. The term "structure" in population models refers to an organization or division of the population into various parts such as age, size, or stage. Example 1.5 describes an example of a stage-structured model, where the population is organized into two developmental stages, juveniles and adults. Stage-structured models may include several developmental stages. For insects, the developmental stages could be egg, larva, pupa, and adult. In an age-structured model, the population is subdivided into age groups. In human demography, the age groups may be 5 years in length, 0–5, 5–10, and so on. In a size-structured model, individuals in the population are grouped according to size, which may be measured by length or weight. In fish populations, size is often the structuring variable. The dynamic interactions among the stages, ages, or sizes determine how the population structure changes over time. Early contributors to the theory of structured models include Bernadelli (1941), Lewis (1942), Leslie (1945), and Lefkovitch (1965). Caswell (2001) gives a brief history of the development of age-structured models.

Other ways of structuring populations, in addition to age, size, and stage, include sex and spatial structure. Models with spatial structure are studied in Chapter 7. Here, we concentrate on a simple model with age structure. The simplest model with age structure is the linear matrix model developed by Leslie (1945) and referred to as the *Leslie matrix model*.

Assume the population is closed to migration and only the females are modeled. Males are present, but are not specifically modeled. Often their numbers can be computed from the female population size. When the sex ratio of males to females is $a : b$ and the survival rate per age group is the same for males and females, then the number of males equals the number of females times $a/b$. For a two-sex model, consult Caswell (2001). Also, see Section 3.8.4 and Exercises 17 and 18 in Chapter 3.

Let the total number of age groups equal $m$ ($m$ is the last reproductive age). During the interval of time $t$ to $t + 1$, individuals "age" from $i$ to $i + 1$ (i.e., the time interval coincides with the age interval). More specifically, let

$x_i(t)$ = the number of females in the $i$th age group at time $t$.

$b_i$ = the average number of newborn females produced by one female in the $i$th age group that survive through the time interval in which they were born, $b_i \geq 0$.

$s_i$ = the fraction of the $i$th age group that live to the $(i + 1)$st age, $0 < s_i \leq 1$.

The first age group $x_1$ consists of offspring from the other age groups, that is,

$$x_1(t + 1) = b_1x_1(t) + b_2x_2(t) + \cdots + b_mx_m(t) = \sum_{i=1}^{m} b_ix_i(t).$$

The number of individuals in the $i$th age group that survive to age $i + 1$ is

$$x_{i+1}(t + 1) = s_ix_i(t).$$

Using matrix notation, the model can be expressed as

$$X(t+1) = \begin{pmatrix} x_1(t+1) \\ x_2(t+1) \\ x_3(t+1) \\ \vdots \\ x_m(t+1) \end{pmatrix} = \begin{pmatrix} b_1 & b_2 & \cdots & b_{m-1} & b_m \\ s_1 & 0 & \cdots & 0 & 0 \\ 0 & s_2 & \cdots & 0 & 0 \\ \vdots & \vdots & \ddots & \vdots & \vdots \\ 0 & 0 & \cdots & s_{m-1} & 0 \end{pmatrix} \begin{pmatrix} x_1(t) \\ x_2(t) \\ x_3(t) \\ \vdots \\ x_m(t) \end{pmatrix} = LX(t), \quad (1.21)$$

where $L$ is the *Leslie matrix*, also referred to as the *projection matrix*. In general, a Leslie matrix has the fertilities or fecundities on the first row and survival probabilities on the subdiagonal. All other entries in the Leslie matrix are zero. One can project forward in time by repeated multiplication by the Leslie matrix, $X(1) = LX(0)$, $X(2) = LX(1) = L^2X(0)$, and in general,

$$X(t) = L^tX(0).$$

The relationships among the age groups in the Leslie matrix can be represented by a *life cycle graph* or *directed graph* (*digraph*) with nodes representing each age group and directed edges representing a relation between the two groups (see Figure 1.4). A node is one of the state variables $x_i$, represented in

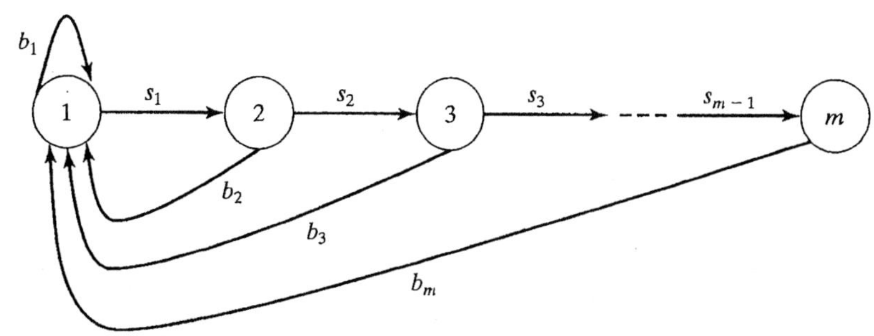


**Figure 1.4** Life cycle graph of the Leslie matrix. The numbers represent the $m$ age classes.

the graph as $i$ enclosed within a circle. An arrow connects the node $j$ to $i$ if the $ij$th element in the Leslie matrix $L$ is nonzero.

## 1.7 Properties of the Leslie Matrix

A matrix $A$ whose entries are nonnegative (positive) is referred to as a *nonnegative (positive) matrix*, denoted $A \geq 0$ $(A > 0)$. Nonnegative and positive matrices have some very nice properties. The Leslie matrix is a nonnegative matrix. An important theorem known as the Frobenius Theorem can be applied to the Leslie matrix, provided the matrix has some additional properties. The Frobenius Theorem gives sufficient conditions that guarantee the Leslie matrix has one positive strictly dominant eigenvalue. Before stating the theorem some definitions are needed.

**Definition 1.10.** A square $m \times m$ matrix $A = (a_{ij})$ is said to be *reducible* if the index set $1, 2, \ldots, m$ can be split into two nonempty complementary sets $S_1$ and $S_2$ (without common indices) $S_1 = \{i_1, \ldots, i_\mu\}$; $S_2 = \{k_1, \ldots, k_\nu\}$, where $\mu + \nu = m$, such that

$$a_{i_\alpha k_\beta} = 0 \; (\alpha = 1, 2, \ldots, \mu; \beta = 1, 2, \ldots, \nu). \tag{1.22}$$

Otherwise, matrix $A$ is said to be *irreducible.*

See Gantmacher (1964) for many equivalent definitions of a reducible matrix.

**Example 1.15** Suppose matrix $A$ satisfies

$$A = \begin{pmatrix} a_{11} & a_{12} & 0 \\ a_{21} & 0 & 0 \\ 0 & a_{32} & a_{33} \end{pmatrix},$$

where $a_{ij} > 0$. Note that $A$ does not have the form of a Leslie matrix due to the $a_{33}$ term. Matrix $A$ is reducible because the complementary sets $S_1 = \{1, 2\}$ and $S_3 = \{3\}$ satisfy condition (1.22) in Definition 1.10. ■

Another way to verify whether a matrix is irreducible is through its directed graph (digraph). Recall that a directed edge from $i$ to $j$ is represented by an arrow from $i$ to $j$. A *directed path* from node $i$ to node $j$ is a set of directed edges *connecting* $i$ to $j$, that is, there exists a set of indices $k_1, k_2, \ldots, k_l$ with corresponding nonzero entries $a_{k_1 i}, a_{k_2 k_1}, \ldots, a_{jk_l}$ in the matrix $A$. For example, in Figure 1.4, there exists a directed path from $m$ to $j$ for every node $j = 1, 2, \ldots, m$.

**Definition 1.11.** If, in the digraph associated with matrix $A$, there exists a directed path from node $i$ to node $j$ for every node $i$ and $j$ in the digraph, then the digraph is said to be *strongly connected.*

A strongly connected digraph associated with a matrix $A$ is equivalent to the irreducibility of $A$ (Ortega, 1987).

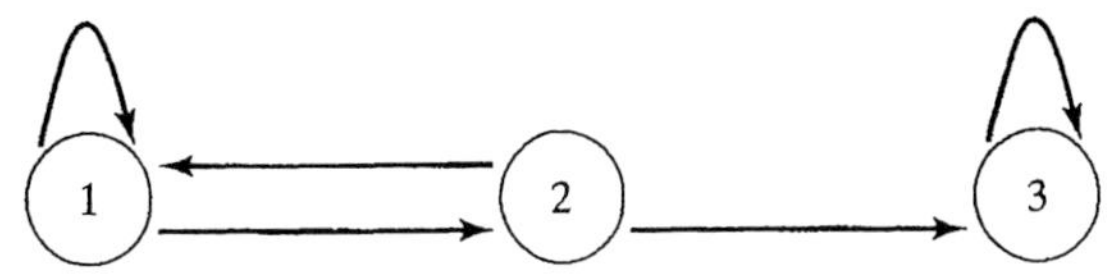


**Figure 1.5** Digraph of Examples 1.15 and 1.16.

**Theorem 1.2** *The digraph of matrix A is strongly connected iff matrix A is irreducible.* □

The life cycle graph of the Leslie matrix $L$ in Figure 1.4 (when $b_m > 0$) shows that the digraph of $L$ is strongly connected, so that $L$ is irreducible. However, if the directed path from $m$ to 1 is removed, $a_{1m} = b_m = 0$, then the Leslie matrix is reducible. Thus, a necessary condition for a Leslie matrix to be irreducible is $b_m \neq 0$. In fact, because it is assumed that the subdiagonal elements of matrix $L$, $s_i = a_{i+1,i} > 0$ for $i = 1, 2, \ldots, m - 1$, matrix $L$ is strongly connected if $b_m = a_{1m} > 0$. Therefore, $b_m > 0$ is a necessary and sufficient condition for the Leslie matrix $L$ to be irreducible.

**Example 1.16** The digraph associated with matrix $A$ in Example 1.15 is graphed in Figure 1.5. It can be easily verified that the digraph corresponding to matrix $A$ is *not* strongly connected. Hence, matrix $A$ is reducible. ■

The following theorem states that a nonnegative, irreducible matrix always has a unique positive eigenvalue that is greater than or equal to the magnitude of all of the other eigenvalues (Frobenius Theorem). If the matrix is positive, then the unique positive eigenvalue has a magnitude that exceeds all of the other eigenvalues (Perron Theorem). See Gantmacher (1964) for a proof of these two theorems.

**Theorem 1.3** **(Frobenius Theorem).** *An irreducible, nonnegative matrix A always has a positive eigenvalue $\lambda$ that is a simple root (multiplicity one) of the characteristic equation. The value of $\lambda$ is greater than or equal to the magnitude of all of the other eigenvalues. To the eigenvalue $\lambda$ there corresponds an eigenvector with positive coordinates.* □

The Perron Theorem applies to positive matrices.

**Theorem 1.4** **(Perron Theorem).** *A positive matrix A always has a real, positive eigenvalue $\lambda$ that is a simple root of the characteristic equation and exceeds the magnitude of all of the other eigenvalues. To the eigenvalue $\lambda$ there corresponds an eigenvector with positive coordinates.* □

A similar theory can be applied to negative matrices (see Exercise 13). However, for our study in this chapter and in the next two chapters, the theory is applied to nonnegative matrices.

The Leslie matrix is not a positive matrix so that the Perron Theorem does not apply. For a nonnegative, irreducible matrix, there may be eigenvalues of the same magnitude. To show that the eigenvalue $\lambda$ in the Frobenius Theorem is strictly dominant, that is, it exceeds all of the other eigenvalues in magnitude, a nonnegative, irreducible matrix needs one more property. This property is known as primitivity.

**Definition 1.12.** If an irreducible, nonnegative matrix $A$ has $h$ eigenvalues $\lambda_1, \ldots, \lambda_h$ of maximum modulus ($|\lambda_1| = |\lambda_i|, i = 1, \ldots, h$), then $A$ is called *primitive* if $h = 1$ and *imprimitive* if $h > 1$. The value of $h$ is called the *index of imprimitivity*.

The *index of imprimitivity* is the number of eigenvalues of matrix $A$ with maximum modulus (with magnitude equal to $\rho(A)$). According to Definition 1.12 a primitive matrix must be irreducible. The converse is not true, as will be seen in the examples. The following theorem gives a simple criterion that can be used to check whether a matrix is primitive.

**Theorem 1.5** *A nonnegative matrix $A$ is primitive iff some power of $A$ is positive (i.e., $A^p > 0$ for some integer $p \geq 1$).* □

If the Leslie matrix satisfies $L^p > 0$ for some positive integer $p$, then $L$ is primitive. Then, the Frobenius Theorem states that, in this case, $L$ has a unique strictly dominant eigenvalue, $\lambda_1$, satisfying $\lambda_1 > |\lambda_j|$ for $j \neq 1$. Associated with the strictly dominant eigenvalue $\lambda_1$ is a positive eigenvector $V_1$. The associated eigenvector $V_1$ is referred to as a *stable age distribution*.

**Example 1.17** Let the Leslie matrix satisfy

$$L = \begin{pmatrix} 0 & b_2 \\ s_1 & 0 \end{pmatrix},$$

where $b_2 > 0$ and $s_1 > 0$. In this case, there are only two ages or stages. The first stage is nonreproductive. Matrix $L$ is nonnegative and irreducible, but $L^p$ is not positive for any positive integer $p$. For even positive integers $p$, $L^p$ has positive diagonal entries and zeros elsewhere and for odd positive integers $p$, $L^p$ has zero diagonal entries and positive entries elsewhere. Hence, according to Theorem 1.5, $L$ is imprimitive. The index of imprimitivity is two since the eigenvalues of $L$ are $\lambda_{1,2} = \pm\sqrt{b_2 s_1}$. ■

**Example 1.18** Let the Leslie matrix satisfy

$$L = \begin{pmatrix} b_1 & b_2 \\ s_1 & 0 \end{pmatrix},$$

where $b_1$, $b_2$, and $s_1$ are positive constants. In this case, there are only two stages but both stages are reproductive. It is easy to check that $L$ is primitive by applying Theorem 1.5; $L^2$ is a positive matrix. In addition, the eigenvalues of $L$ satisfy

$$\lambda_{1,2} = \frac{b_1 \pm \sqrt{b_1^2 + 4b_2 s_1}}{2}.$$

The positive eigenvalue $\lambda_1$ is strictly dominant. ■

The significance of a stable age distribution, vector $V_1$, is illustrated in the following computation. Suppose $\lambda_i$ are the $m$ eigenvalues and $V_i$ are associated eigenvectors, $i = 1, 2, \ldots, m$. Assume matrix $L$ is primitive (and irreducible) and the $m$ eigenvectors, $\{V_i\}_{i=1}^m$, form a linearly independent set. Then the solution to $X(t + 1) = LX(t)$ can be written as

$$X(t) = L^t X(0) = \sum_{i=1}^{m} c_i \lambda_i^t V_i = c_1 \lambda_1^t V_1 + c_2 \lambda_2^t V_2 + \cdots + c_m \lambda_m^t V_m,$$

where $\lambda_1$ is the strictly dominant eigenvalue. Dividing the solution by $\lambda_1^t$ yields

$$\frac{X(t)}{\lambda_1^t} = \frac{L^t X(0)}{\lambda_1^t} = c_1 V_1 + c_2 \left(\frac{\lambda_2}{\lambda_1}\right)^t V_2 + \cdots + c_m \left(\frac{\lambda_m}{\lambda_1}\right)^t V_m.$$

Because $|\lambda_i/\lambda_1| < 1, (\lambda_i/\lambda_1)^t \to 0$ as $t \to \infty$. In the limit,

$$\lim_{t\to\infty} \frac{X(t)}{\lambda_1^t} = \lim_{t\to\infty} \frac{L^t X(0)}{\lambda_1^t} = c_1 V_1.$$

Hence, $X(t) = L^t X(0) \approx c_1 \lambda_1^t V_1$ after many generations. The population size either increases ($\lambda_1 > 1$) or decreases ($\lambda_1 < 1$) geometrically as time increases. The population distribution $X(t)/\lambda_1^t$ approaches a constant multiple of the eigenvector $V_1$. It is for this reason that $V_1$ is referred to as a stable age distribution. Also, note that $V_1 > 0$ is guaranteed by the Frobenius Theorem.

The restriction that the eigenvectors $V_i$ be linearly independent is not necessary to verify the above relationship. Let $P(t)$ denote the total population size, the sum of all of the age groups at time $t$, $P(t) = \sum_{i=1}^m x_i(t)$, and let $v_1 = \sum_{i=1}^m v_{i1}$ be the sum of the entries of $V_1 = (v_{11}, v_{21}, \ldots, v_{m1})^T$. It can be shown that if $L$ is a nonnegative, irreducible, and primitive matrix, and $X(t+1)$ satisfies $X(t+1) = LX(t)$, where $X(0)$ is a nonnegative and nonzero vector, then

$$\lim_{t\to\infty} \frac{X(t)}{P(t)} = \frac{V_1}{v_1}. \tag{1.23}$$

In addition, if $\lambda_1 < 1$, then $\lim_{t\to\infty} P(t) = 0$ and if $\lambda_1 > 1$, then $\lim_{t\to\infty} P(t) = \infty$ (see Cushing, 1998). The limit (1.23) holds for many types of structured models. We emphasize that the limit (1.23) is valid provided that matrix $L$ is primitive.

An explicit expression for $V_1$ in the case of a Leslie matrix is given by Pielou (1977). The form of the positive eigenvector is

$$V_1 = \begin{pmatrix} 1 \\ \dfrac{s_1}{\lambda_1} \\ \vdots \\ \dfrac{s_1 s_2 \cdots s_{m-2}}{\lambda_1^{m-2}} \\ \dfrac{s_1 s_2 \cdots s_{m-1}}{\lambda_1^{m-1}} \end{pmatrix}. \tag{1.24}$$

The first entry in this formula for $V_1$ is normalized to one. For $L = \begin{pmatrix} 1 & 2 \\ 3/8 & 0 \end{pmatrix}$ the positive eigenvector has the form $V = (1, 1/4)^T$.

The characteristic equation associated with the Leslie matrix $L$ has a particularly nice form. The characteristic equation for the Leslie matrix satisfies $\det(L - \lambda I) = 0$ or

$$\det \begin{pmatrix} b_1 - \lambda & b_2 & \cdots & b_{m-1} & b_m \\ s_1 & -\lambda & \cdots & 0 & 0 \\ 0 & s_2 & \cdots & 0 & 0 \\ \vdots & \vdots & \ddots & \vdots & \vdots \\ 0 & 0 & \cdots & s_{m-1} & -\lambda \end{pmatrix} = 0.$$

After expansion, the following characteristic equation is obtained:

$$p_m(\lambda) = \lambda^m - b_1\lambda^{m-1} - b_2 s_1 \lambda^{m-2} - \cdots - b_m s_1 \cdots s_{m-1} = 0. \tag{1.25}$$

Since there is only one change in sign in the polynomial $p_m(\lambda)$, Descartes's Rule of Signs implies that $p_m(\lambda)$ has one positive real root. [Descartes's Rule of Signs states that if there are $k$ sign changes in the coefficients of the characteristic polynomial (with real coefficients), then the number of positive real roots (counting multiplicities) equals $k$ or is less than this number by a positive even integer. If $\lambda$ is replaced by $-\lambda$, a similar rule applies for negative real roots.] A simple proof of Descartes's Rule of Signs is given by Wang (2004). Because there is only one sign change, there is only one positive real root. This positive root is the dominant eigenvalue, $\lambda_1$.

A simple check on whether the dominant eigenvalue is less than or greater than one involves the inherent net reproductive number. The characteristic polynomial, $p_m$, satisfies $p_m(\lambda) \to \infty$ as $\lambda \to \infty$, $p_m(0) < 0$, and $p_m(\lambda)$ crosses the positive $\lambda$ axis only once at $\lambda_1$. Therefore, by examination of $p_m(1)$ it can be seen that the dominant eigenvalue $\lambda_1 > 1$ iff $p_m(1) < 0$ and $\lambda_1 < 1$ iff $p_m(1) > 0$. See Figure 1.6. However, $p_m(1) = 1 - b_1 - b_2 s_1 - \cdots - b_m s_1 \cdots s_{m-1}$. Let

$$R_0 = b_1 + b_2 s_1 + b_3 s_1 s_2 + \cdots + b_m s_1 \cdots s_{m-1} \tag{1.26}$$

so that $p_m(1) = 1 - R_0$. The quantity $R_0$ is referred to as the *inherent net reproductive number*, the expected number of offspring per individual per lifetime (Cushing, 1998). Then $p_m(1) > 0$ iff $R_0 < 1$ and $p_m(1) < 0$ iff $R_0 > 1$. Thus, we have verified the following result:

$$\lambda_1 > 1 \quad \text{iff} \quad R_0 > 1 \quad \text{and} \quad \lambda_1 < 1 \quad \text{iff} \quad R_0 < 1.$$

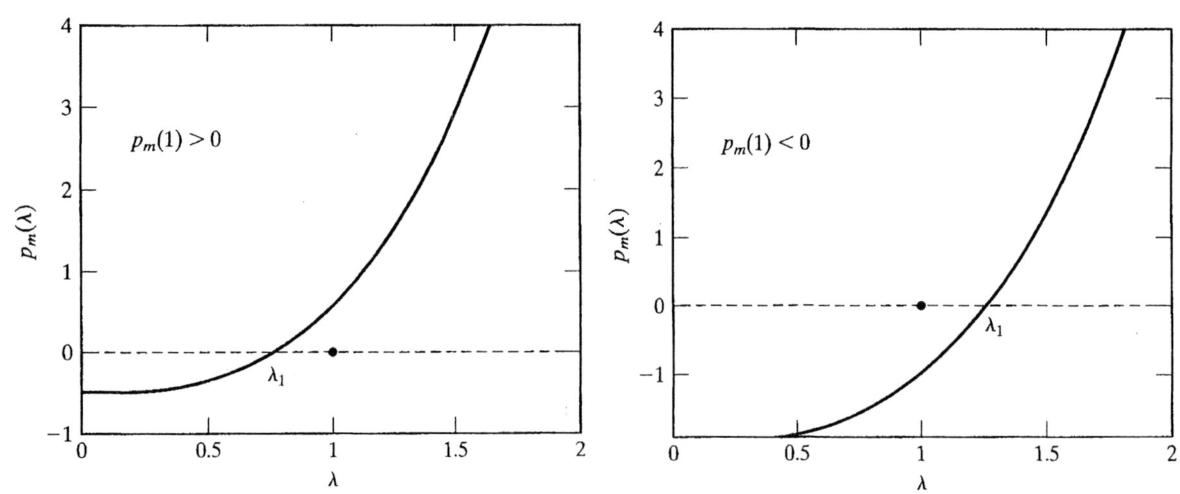


**Figure 1.6** The characteristic polynomial $p_m(\lambda)$ when $p_m(1) > 0$ and $\lambda_1 < 1$ and when $p_m(1) < 0$ and $\lambda_1 > 1$.

The results concerning the Leslie matrix model are summarized in the following theorem.

**Theorem 1.6** *Assume the Leslie matrix $L$ given in (1.21) is irreducible and primitive. The characteristic polynomial of $L$ is given by (1.25). Matrix $L$ has a strictly dominant eigenvalue $\lambda_1 > 0$ satisfying the following relationships:*

$$\lambda_1 = 1 \quad \textit{iff} \quad R_0 = 1,$$

$$\lambda_1 < 1 \quad \textit{iff} \quad R_0 < 1,$$

$$\lambda_1 > 1 \quad \textit{iff} \quad R_0 > 1,$$

*where $R_0$ is the inherent net reproductive number defined by (1.26). In addition, the stable age distribution $V_1$ satisfies (1.24).* □

We mention one additional check for primitivity which applies *only* to Leslie matrices. Sykes (1969) verified the following result.

**Theorem 1.7** *An irreducible Leslie matrix $L$ is primitive iff the birth rates satisfy the following relationship:*

$$\text{g.c.d.}\{i|b_i > 0\} = 1,$$

*where g.c.d. denotes the greatest common divisor of the index of the birth rates $b_i$ that are positive.* □

Remember, this result only applies to the Leslie matrix (see Exercise 19).

**Example 1.19** For the Leslie matrix with $b_1 > 0$ and $b_m > 0$, it follows from Sykes's Theorem 1.7 that g.c.d.$\{1, m\} = 1$ and the Leslie matix is primitive. If two adjacent birth rates are nonzero, $b_j > 0$ and $b_{j+1} > 0$, and $b_m > 0$, then it is also the case that g.c.d.$\{j, j+1, m\} = 1$ so that the corresponding Leslie matrix is primitive. ■

**Example 1.20** Suppose the Leslie matrix satisfies

$$L = \begin{pmatrix} 0 & b_2 \\ s_1 & 0 \end{pmatrix}.$$

Matrix $L$ satisfies the Frobenius Theorem, but $L$ is imprimitive as shown in Example 1.17. This can be seen from the Sykes result (1969) also: g.c.d.$\{2\} = 2$. In addition, the inherent net reproductive number $R_0 = b_2 s_1$.

For the particular case $R_0 = 1$, where

$$L = \begin{pmatrix} 0 & 2 \\ 0.5 & 0 \end{pmatrix},$$

it can be seen that $\lambda_{1,2} = \pm 1$. The eigenvector corresponding to $\lambda_1 = 1$ is found by solving $(L - I)V_1 = \mathbf{0}$ or

$$\begin{pmatrix} -1 & 2 \\ 0.5 & -1 \end{pmatrix} \begin{pmatrix} x_1 \\ x_2 \end{pmatrix} = \begin{pmatrix} 0 \\ 0 \end{pmatrix},$$

where $V_1 = (x_1, x_2)^T$. Since $x_1 = 2x_2$, then $V_1 = (2, 1)^T$ or any constant multiple of $V_1$. Using the formula given by Pielou, $V_1 = (1, 0.5)^T$, which gives the same result.

Suppose the initial population distribution is $X(0) = (10, 5)^T$. Then we compute $X(1)$,

$$X(1) = \begin{pmatrix} 0 & 2 \\ 0.5 & 0 \end{pmatrix} \begin{pmatrix} 10 \\ 5 \end{pmatrix} = \begin{pmatrix} 10 \\ 5 \end{pmatrix}.$$

Since $X(0)$ is an eigenvector associated with $\lambda_1 = 1$, $LX(0) = X(0)$ and $L^t X(0) = X(0)$.

Suppose the initial distribution is $X(0) = (10, 10)^T$. Then $X(1) = (20, 5)^T$, $X(2) = (10, 10)^T$. The population oscillates between two states. The general solution $X(t)$ can be determined for any initial distribution from the fact that there are two linearly independent eigenvectors. The solution has the form

$$X(t) = c_1(1)^t \begin{pmatrix} 2 \\ 1 \end{pmatrix} + c_2(-1)^t \begin{pmatrix} -2 \\ 1 \end{pmatrix} = \begin{pmatrix} 2c_1 - 2c_2(-1)^t \\ c_1 + c_2(-1)^t \end{pmatrix}.$$

Thus, if $X(0) = (10, 10)^T$, then the constants $c_1 = 15/2$ and $c_2 = 5/2$. The solution is

$$X(t) = \begin{pmatrix} 15 + (-1)^{t+1} 5 \\ 15/2 + (-1)^t (5/2) \end{pmatrix}.$$

■

The theory developed in this section was applied to Leslie matrices. However, the theory applies to many different types of structured models involving nonnegative matrices (Caswell, 2001; Cushing, 1998). Cushing gives a general form for a structured matrix $L$. A general structured matrix is the sum of two matrices, $L = F + T$, where $F$ is a fertility matrix and $T$ is a transition matrix. Matrix $F = (f_{ij})$, $f_{ij} \geq 0$ and $T = (t_{ij})$, $0 \leq t_{ij} \leq 1$, $\Sigma_{i=1}^m t_{ij} \leq 1$ for all $j$ (each column sum of $T$ is less than one). The value of $f_{ij}$ is the number of newborns added to newborn class $i$ from stage $j$ and $t_{ij}$ is the probability that an individual in stage $j$ transfers to stage $i$. The column sums of $T$ are less than one, $\Sigma_{i=1}^m t_{ij} \leq 1$, because the probability of transfer out of any state cannot be greater than one.

A stage-structured model for the loggerhead sea turtle is discussed in the next example.

**Example 1.21** We consider a stage-structured model developed by Crouse et al. (1987) for loggerhead sea turtles (*Caretta caretta*). The loggerhead sea turtle is classified as threatened under the U.S. Federal Endangered Species Act. There are seven species of sea turtles worldwide, and currently six of them are either listed as threatened or endangered (Caribbean Conservation Corporation & Sea Turtle Survival League, 2003). Loggerheads are located over a wide geographic range in the Atlantic and Pacific oceans. They are the most abundant species along the U.S. coast. Unfortunately, they are often captured accidentally in fishing nets. In addition, because they lay their eggs on beaches, they are subject to predation by other animals and also disturbance by humans. The modeling objectives of Crouse et al. (1987) were to compare the results of management strategies at the egg and the adult stages. See also Caswell (2001).

The model for the loggerhead sea turtle divides the population into seven stages, Stage 1: eggs or hatchlings (<1 year of age), Stage 2: small juveniles (1–7 years of age), Stage 3: large juveniles (8–15 years of age), Stage 4: subadults

(16–21 years of age), Stage 5: novice breeders (22 years old), Stage 6: first-year remigrants or females returning to nest in the second year (23 years old), and Stage 7: mature breeders (24–54 years of age). The time interval is one year. Therefore, individuals in stages 2, 3, 4, and 7 can remain in these stages for up to 6, 7, 5, and 30 years, respectively, before moving to the next stage. There is only one newborn class, Stage 1. The stage-structured matrix $L$ has the following form:

$$L = \begin{pmatrix} 0 & 0 & 0 & 0 & 127 & 4 & 80 \\ 0.6747 & 0.7370 & 0 & 0 & 0 & 0 & 0 \\ 0 & 0.0486 & 0.6610 & 0 & 0 & 0 & 0 \\ 0 & 0 & 0.0147 & 0.6907 & 0 & 0 & 0 \\ 0 & 0 & 0 & 0.0518 & 0 & 0 & 0 \\ 0 & 0 & 0 & 0 & 0.8091 & 0 & 0 \\ 0 & 0 & 0 & 0 & 0 & 0.8091 & 0.8089 \end{pmatrix}.$$

Matrix $L = F + T$, where the fertility matrix has its first row identical to $L$ and zeros elsewhere and the transition matrix has zeros in the first row, but is identical to $L$ in the remaining six rows. The eigenvalues of matrix $L$ are difficult to calculate by hand. Use of a calculator or software packages such as Maple or MATLAB® can be applied to calculate the eigenvalues and eigenvectors. (See the Maple and MATLAB® commands in the Appendix of Chapter 1.) The strictly dominant eigenvalue is $\lambda_1 \approx 0.94503$ and the normalized eigenvector corresponding to this eigenvalue (stable stage distribution) is

$$\frac{V_1}{v_1} \approx \begin{pmatrix} 0.20651 \\ 0.66975 \\ 0.11460 \\ 0.00662 \\ 0.00036 \\ 0.00031 \\ 0.00185 \end{pmatrix}.$$

If the population distribution is initially close to the distribution $V_1$, the population will decline at a rate of $\lambda_1$ each year. After 25 years, the original population can be reduced to less than 25% of the original population size ($\lambda_1^{25} \approx 0.243$). Increasing the probability of survival during any one of the stages will increase the value of the strictly dominant eigenvalue $\lambda_1$. ■

Stage-structured models have been frequently applied to fish populations. For example, they have been used to estimate the harvest that can be taken without reducing the population size (Usher, 1972).

**Example 1.22** Consider the following simple example of harvesting of a fish population. Let $N$ represent the total population size and let $\lambda_1 > 1$ be the strictly dominant eigenvalue of the Leslie matrix governing the fish population. Then the population size increases from $N$ to $\lambda_1 N$ over one time period when the age distribution is stable. The harvest that can be taken is given by $H = [(\lambda_1 - 1)/\lambda_1]\,100\%$, where $H$ is expressed as a percentage of the total population size and

$$N \to \lambda_1 N - \left(\frac{\lambda_1 - 1}{\lambda_1}\right)\lambda_1 N = N.$$

For example, suppose the Leslie matrix satisfies

$$L = \begin{pmatrix} 0 & 9 & 12 \\ 1/3 & 0 & 0 \\ 0 & 1/2 & 0 \end{pmatrix}.$$

The inherent net reproductive number $R_0 = 5$ and $\lambda_1 = 2$ so that $1 < \lambda_1 < R_0$. The stable age distribution (normalized with last entry equal to one) satisfies $V_1 = (24, 4, 1)^T$. Without any harvesting, the population will eventually approach a constant multiple of the stable age distribution. If the population distribution is close to the distribution of $V_1$, then $H = [(2 - 1)/2]100\% = 50\%$ of the population can be harvested (50% from every age group) and the total population size will remain constant. ■

For many other applications and extensions of the Leslie matrix model, consult the references at the end of this chapter (e.g., Allen, 1989; Caswell, 2001; Cushing, 1998; Cushing et al., 2003; Getz and Haight, 1989; Levin and Goodyear, 1980; and references therein). Some introductory textbooks on difference equations include *An Introduction to Difference Equations* by Elaydi (1999), *Difference Equations Theory and Applications* by Mickens (1990), and *Introduction to Difference Equations* by Goldberg, originally published in 1958 and republished by Dover in 1986.

## 1.8 Exercises for Chapter 1

**1.** Classify the difference equations or systems as to (i) order, (ii) linear or nonlinear, and (iii) autonomous or nonautonomous. If the equation or system is linear, determine if it is (iv) homogeneous or nonhomogeneous.

(a) $x_{t+3} + 4t^3 x_{t+2} + \sin(t)x_t = 3t + 1$

(b) $x_{t+1} = x_t(r + 1 - x_t), r > 0$

(c) $x_t = x_{t+1} + t^5 x_{t+2}$

(d) $x_{t+2} = t x_{t+1}/x_t$

(e) $x_{t+1} = x_t \exp(r[1 - x_t] - y_t),\ y_{t+1} = c x_t(1 - \exp(-y_t)), r, c > 0$

**2.** For the difference equation, $x_{t+1} = a x_t + b = f(x_t)$, where $0 < a < 1$ and $b > 0$, use the solution given in (1.12) to find the following limit: $\lim_{t\to\infty} x_t$. Show that this limit is also a fixed point of the difference equation, that is, it is a solution $\bar{x}$ of $\bar{x} = f(\bar{x})$ (see Figure 1.2).

**3.** Find the general solution to each of the following homogeneous difference equations.

(a) $x_{t+2} - 16x_t = 0$

(b) $x_{t+3} + 5x_{t+2} - x_{t+1} - 5x_t = 0$

(c) $x_{t+4} - 8x_{t+2} + 16x_t = 0$

(d) $3x_{t+2} - 6x_{t+1} + 4x_t = 0$

**4.** (a) For equations (a) and (d) in Exercise 3, find the unique solution satisfying the initial condition $x_0 = 0$ and $x_1 = 1$.

(b) For equation (b) in Exercise 3, find the unique solution satisfying $x_0 = 0 = x_1$ and $x_2 = 5$.

(c) For equation (c) in Exercise 3, find the unique solution satisfying $x_0 = 0 = x_2$ and $x_1 = 2 = x_3$.

**5.** Find the general solutions to the nonhomogeneous, linear difference equations.

(a) $x_{t+2} + x_{t+1} - 6x_t = 5$

(b) $x_{t+2} - 4x_t = 6t - 1$

(c) $x_{t+1} - 5x_t = 2t$

**6.** A second-order difference equation $x_{t+2} + ax_{t+1} + bx_t = 0, a \neq 0$, has a characteristic polynomial with a root of multiplicity two, $\lambda_1 = \lambda_2 \neq 0$.

(a) Show that $t\lambda_1^t$ is a solution of the difference equation. (*Hint:* First show $\lambda_1 = -a/2$.)

(b) Show that the Casoratian $C(\lambda_1^t, t\lambda_1^t) \neq 0$ for any $t$.

**7.** The solution $x_t, t = 0, 1, \ldots$, of the following second-order difference equation is known as the *Fibonacci* sequence:

$$x_{t+2} = x_{t+1} + x_t, \ x_0 = 1 = x_1. \tag{1.27}$$

Leonardo de Pisa (also known as Fibonacci), an Italian mathematician, was one of the first to study the properties of this sequence. Fibonacci related equation (1.27) to reproduction in rabbits. A pair of rabbits reproduce twice, at one month and at two months old. The pair of rabbits produces a pair of baby rabbits each time. The rabbits are not reproductive until one month old. All rabbits are assumed to survive. Let $x_t$ be the number of pairs of rabbits in month $t$. Then beginning with one pair of rabbits in months one and two, the number of pairs of rabbits in month $t + 2$ satisfies the Fibonacci equation (1.27).

(a) Solve the Fibonacci equation (1.27).

(b) The ratio $x_{t+1}/x_t$ converges to a constant $g$ known as the golden ratio:

$$\lim_{t\to\infty} \frac{x_{t+1}}{x_t} = g.$$

Find the golden ratio. [*Hint:* Divide the difference equation (1.27) by $x_{t+1}$; then find the limit.]

**8.** The properties of the Fibonacci equation hold for other second-order difference equations as well. Consider the difference equation, $x_{t+1} = px_t + qx_{t-1}$, where $x_0$, $x_1$, $p$, and $q$ are positive numbers. Show that

$$\lim_{t\to\infty} \frac{x_{t+1}}{x_t} = r,$$

where $r$ is the positive root of the quadratic equation (Falbo, 2005)

$$x^2 - px - q = 0.$$

**9.** Verify that the inequality (1.14) in Example 1.10 is satisfied iff $|\lambda_1| < 1$.

**10.** Consider the following two difference equations:

$$(1) \quad x_{t+4} + ax_t = 0,$$
$$(2) \quad x_{t+4} + ax_{t+2} + bx_t = 0.$$

(a) Find the characteristic equation for the difference equations (1) and (2).

(b) Convert the higher-order difference equations (1) and (2) to first-order systems, $Y(t + 1) = AY(t)$. Then show that the characteristic equation for the matrix $A$ in each case is the same as in part (a).

**11.** Convert each of the four linear difference equations in Exercise 3 to equivalent first-order systems.

**12.** Show that the characteristic equation of the companion matrix $A$ defined in equation (1.16) is the same as the characteristic equation for the homogeneous scalar equation (1.15).

**13.** Assume $B$ is a negative matrix: $B = -A$, where $A$ is a positive matrix. Apply the Perron Theorem to show that matrix $B$ has a negative eigenvalue $\lambda_1$ and an associated positive eigenvector $V_1 > \mathbf{0}$. In addition, show that $\lambda_1$ is a simple root of the characteristic equation whose magnitude exceeds the magnitude all other eigenvalues. (*Hint:* Consider $AV = \lambda V$ and show that $B$ and $A$ have the same eigenvectors but the eigenvalues of $B$ are the negative of those corresponding to $A$.)

**14.** Consider the first-order system

$$X(t+1) = \begin{pmatrix} 0 & a \\ b & 0 \end{pmatrix} X(t), \quad a, b > 0.$$

What condition is required on the constants $a$ and $b$ so that $\lim_{t\to\infty} X(t) = \mathbf{0}$ for any initial condition $X(0) \geq \mathbf{0}$?

**15.** Find the eigenvalues and eigenvectors of matrix $A$. Then find the general solution to $X(t+1) = AX(t)$.

(a) $A = \begin{pmatrix} 1 & 2 & 3 \\ 0 & 2 & 0 \\ 0 & 0 & -3 \end{pmatrix}$ (b) $A = \begin{pmatrix} 1 & 2 \\ 4 & 3 \end{pmatrix}$

**16.** Let $X(t+1) = AX(t)$, where

$$A = \begin{pmatrix} 1 & 1 \\ 0 & 2 \end{pmatrix}.$$

The general solution is $X(t) = A^t X(0)$. Find $A^2$, $A^3$, and $A^4$. Then find a general expression for $A^t$ and write the general solution.

**17.** Suppose the Leslie matrix is given by

$$L = \begin{pmatrix} 0 & 3a^2/2 & 3a^3/2 \\ 1/2 & 0 & 0 \\ 0 & 1/3 & 0 \end{pmatrix}, \quad a > 0.$$

(a) Find the characteristic equation, eigenvalues, and inherent net reproductive number $R_0$ of $L$.

(b) Show that $L$ is primitive.

(c) Find the stable age distribution.

**18.** Suppose the Leslie matrix is

$$L = \begin{pmatrix} 0 & 0 & 6a^3 \\ 1/2 & 0 & 0 \\ 0 & 1/3 & 0 \end{pmatrix}, \quad a > 0.$$

The only reproductive age class is the third age class.

(a) Find the characteristic equation, eigenvalues, and inherent net reproductive number $R_0$ of $L$.

(b) Show that $L$ is imprimitive.

(c) For $a = 1$ and $X(0) = (6, 0, 0)^T$, find $X(1)$, $X(2)$, $X(3)$; then describe the behavior of $X(t)$ as $t \to \infty$.

(d) For $a = 1$ and $X(0) = (12, 2, 1)^T$, find $X(1)$, $X(2)$, $X(3)$; then describe the behavior of $X(t)$ as $t \to \infty$.

**19.** Which of the following matrices are irreducible? primitive? Draw life cycle graphs for each of the matrices. All entries $b_i$ and $s_i$ are assumed to be positive.

(a) $M_1 = \begin{pmatrix} 0 & 0 & b_3 \\ s_1 & 0 & 0 \\ 0 & s_2 & s_3 \end{pmatrix}$ (b) $M_2 = \begin{pmatrix} 0 & b_2 & 0 \\ s_1 & 0 & 0 \\ 0 & s_2 & s_3 \end{pmatrix}$

(c) $M_3 = \begin{pmatrix} 0 & b_2 & 0 & b_4 \\ s_1 & 0 & 0 & 0 \\ 0 & s_2 & 0 & 0 \\ 0 & 0 & s_3 & 0 \end{pmatrix}$

**20.** Consider the two Leslie age-structured models, $X(t+1) = L_1 X(t)$, and $Y(t+1) = L_2 Y(t)$, where

$$L_1 = \begin{pmatrix} 0 & b_2 & b_3 \\ s_1 & 0 & 0 \\ 0 & s_2 & 0 \end{pmatrix} \quad \text{and} \quad L_2 = \begin{pmatrix} f_1 & f_2 \\ p_1 & 0 \end{pmatrix}.$$

All entries $b_i$, $s_i$, $f_i$, and $p_1$ are assumed to be positive.

(a) Show that matrices $L_1$ and $L_2$ are irreducible and primitive.

(b) If $b_2 = 2$, $b_3 = 4$, and $s_1 = 1/2$, use the inherent net reproductive number to determine conditions on $s_2$ so that the dominant eigenvalue $\lambda_1$ of $L_1$ satisfies $\lambda_1 < 1$ or $\lambda_1 > 1$.

(c) If $f_1 = 1$, determine whether the dominant eigenvalue $\lambda_2$ of $L_2$ satisfies $\lambda_2 < 1$ or $\lambda_2 > 1$.

(d) For the parameter values in (b), $s_2 = 1/2$, and $X(0) = (10, 2, 2)^T$, find $X(1)$ and $X(2)$.

**21.** Suppose a nonnegative structured matrix has the following form:

$$M = \begin{pmatrix} b_1 & b_2 & \cdots & b_{m-1} & b_m \\ s_1 & 0 & \cdots & 0 & 0 \\ 0 & s_2 & \cdots & 0 & 0 \\ \vdots & \vdots & \ddots & \vdots & \vdots \\ 0 & 0 & \cdots & s_{m-1} & s_m \end{pmatrix}.$$

Note that matrix $M$ is similar to a Leslie matrix except for the term $s_m$. If the nonnegative matrix $M$ with $s_m = 0$ (a Leslie matrix) is primitive, then show that it is also primitive if $s_m > 0$.

**22.** An annual plant model where the seeds remain viable for three years takes the form of a Leslie matrix model (Edelstein-Keshet, 1988). Let $\alpha_i$ be the fraction of $i$-year-old seeds that germinate to plants. Let $\sigma$ be the probability that seeds overwinter and $\gamma$ be the number of viable seeds produced per plant. If $p_t$ is the number of plants in year $t$ and $s_t^i$ the number of seeds in years $i = 1, 2$, then the model has the following form:

$$\begin{aligned} p_{t+1} &= \alpha_1 \sigma \gamma p_t + \alpha_2 \sigma (1 - \alpha_1) s_t^1 + \alpha_3 \sigma (1 - \alpha_2) s_t^2, \\ s_{t+1}^1 &= \sigma \gamma p_t, \\ s_{t+1}^2 &= \sigma (1 - \alpha_1) s_t^1. \end{aligned}$$

(a) Express the model in the form of a Leslie matrix model: $Y(t+1) = LY(t)$.

(b) What conditions on $L$ will guarantee that $L$ is irreducible? primitive?

(c) Find an expression for the inherent net reproductive number, $R_0$.

**23.** For the loggerhead sea turtle model of Example 1.21, the structured matrix has the form

$$L = \begin{pmatrix} 0 & 0 & 0 & 0 & b_5 & b_6 & b_7 \\ s_1 & p_2 & 0 & 0 & 0 & 0 & 0 \\ 0 & s_2 & p_3 & 0 & 0 & 0 & 0 \\ 0 & 0 & s_3 & p_4 & 0 & 0 & 0 \\ 0 & 0 & 0 & s_4 & 0 & 0 & 0 \\ 0 & 0 & 0 & 0 & s_5 & 0 & 0 \\ 0 & 0 & 0 & 0 & 0 & s_6 & p_7 \end{pmatrix},$$

where the elements $b_i$, $s_i$ and $p_i$ are assumed to be positive.

(a) Draw a life cycle graph of the seven stages. Label the directed paths.

(b) Show that matrix $L$ is irreducible and primitive.

(c) Consider the structured matrix $L$ in Example 1.12. Increase $s_1 = 0.6747$ to one. Is the spectral radius of this new matrix greater than one? Increase $p_7$ to 0.95. Is the spectral radius of this new matrix greater than one? Based on your results, is it better to increase the egg survival or the adult survival?

**24.** The Leslie matrix can be generalized to a structured matrix model as follows. Let $L = F + T$, where matrix $F$ the fertility matrix and $T$ is the transition or survival matrix. The elements of $T$ have the property that their column sums are all less than one (meaning that the probability of a transition or survival from state $i$ to all other states must be less than one). An inherent net reproductive number $R_0$ can be defined for this general structured model (Cushing, 1998; Li and Schneider, 2002). First note that the spectral radius of $T$ is less than one, $\rho(T) < 1$. Then $(I - T)^{-1}$ exists and can be defined as $(I - T)^{-1} = I + T + T^2 + \cdots$. Let

$$Q = F(I - T)^{-1} = F + FT + FT^2 + \cdots.$$

Matrix $Q$ is referred to as the *next generation matrix*. The spectral radius of $Q$, $\rho(Q)$, is defined as the inherent net reproductive number. Express the Leslie matrix $L$, given by equation (1.21), as

$$L = F + T = \begin{pmatrix} b_1 & b_2 & \cdots & b_m \\ 0 & 0 & \cdots & 0 \\ \vdots & \vdots & \cdots & \vdots \\ 0 & 0 & \cdots & 0 \end{pmatrix} + \begin{pmatrix} 0 & 0 & \cdots & 0 & 0 \\ s_1 & 0 & \cdots & 0 & 0 \\ 0 & s_2 & \cdots & 0 & 0 \\ \vdots & \vdots & \ddots & \vdots & \vdots \\ 0 & 0 & \cdots & s_{m-1} & 0 \end{pmatrix}.$$

Then form the matrix $Q$ and show that the inherent net reproductive number $R_0 = \rho(Q)$. (*Hint:* $T^m = \mathbf{0}$, zero matrix.)

**25.** In the circulatory system, red blood cells are constantly being destroyed and replaced. They carry oxygen throughout the body and they must be maintained at a constant level. The spleen filters out and destroys a fraction of the cells daily and the bone marrow produces a number proportional to the number lost on the previous day. The cell count on day $t$ is modeled as follows (see Edelstein-Keshet, 1988, pp. 27, 33):

$R_t$ = number of red blood cells in circulation on day $t$.

$M_t$ = number of red blood cells produced by marrow on day $t$.

$f$ = fraction of red blood cells removed by spleen, $0 < f < 1$.

$\gamma$ = production constant, $\gamma > 0$.

The system of difference equations satisfied by $R_t$ and $M_t$ is

$$R_{t+1} = (1 - f)R_t + M_t,$$
$$M_{t+1} = \gamma f R_t.$$

(a) Express the system as a matrix equation $X_{t+1} = AX_t$. Find the eigenvalues of $A$ and determine their sign.

(b) Let the positive eigenvalue $\lambda_1 = 1$ ($R_t$ is approximately constant); then what does this imply about $\gamma$?

(c) Let $\lambda_1 = 1$. Find $\lambda_2$. Then describe the behavior of $R_t$.

## 1.9 References for Chapter 1


Allen, L. J. S. 1989. A density-dependent Leslie matrix model. *Math. Biosci.* 95: 179–187.

Bernadelli, H. 1941. Population waves. *J. Burma Research Soc.* XXI (Part I): 3–18.

Brauer, F. and C. Castillo-Chávez. 2001. *Mathematical Models in Population Biology and Epidemiology*. Springer-Verlag, New York.

Britton, N. F. 2003. *Essential Mathematical Biology*. Springer Undergraduate Mathematics Series, Springer-Verlag, London, Berlin, Heidelberg.

Caribbean Conservation Corporation & Sea Turtle Survival League. 2003. Web site: http://www.cccturtle.org /species_world.htm

Caswell, H. 2001. *Matrix Population Models: Construction, Analysis and Interpretation.* 2nd ed. Sinauer Assoc. Inc., Sunderland, Mass.

Crouse, D. T., L. B. Crowder, and H. Caswell. 1987. A stage-based population model for loggerhead sea turtles and implications for conservation. *Ecology* 68: 1412–1423.

Cushing, J. 1998. *An Introduction to Structured Population Dynamics*, CBMS-NSF Regional Conference Series in Applied Mathematics # 71, SIAM, Philadelphia.

Cushing, J. M., R. F. Constantino, B. Dennis, R. A. Desharnais, and S. M. Henson. 2003. *Chaos in Ecology Experimental Nonlinear Dynamics*. Academic Press, New York.

Edelstein-Keshet, L. 1988. *Mathematical Models in Biology*. The Random House/Birkhäuser Mathematics Series, New York.

Elaydi, S. N. 1999. *An Introduction to Difference Equations*. 2nd ed. Springer-Verlag, New York.

Elaydi, S. and W. Harris. 1998. On the computation of $A^n$. *SIAM Review*. 40: 965–971.

Falbo, C. 2005. The golden ratio—a contrary viewpoint. *The College Math. Journal* 36: 123–134.

Gantmacher, F. R. 1964. *The Theory of Matrices*, Vol. II. Chelsea Pub. Co., New York.

Getz, W. M. and R. G. Haight. 1989. *Population Harvesting: Demographic Models of Fish, Forest, and Animal Resources*. Princeton Univ. Press, Princeton, N.J.

Goldberg, S. 1986. *Introduction to Difference Equations*. Dover Pub., Inc., New York.

Keener, J. and J. Sneyd. 1998. *Mathematical Physiology*. Springer-Verlag, New York.

Kot, M. 2001. *Elements of Mathematical Ecology*. Cambridge Univ. Press, Cambridge.

Lefkovitch, L. P. 1965. The study of population growth in organisms grouped by stages. *Biometrics* 21: 1–18.

Leslie, P. H. 1945. On the use of matrices in certain population mathematics. *Biometrics*. 21: 1–18.

Levin, S. A. and C. P. Goodyear. 1980. Analysis of an age-structured fishery model. *J. Math. Biol.* 9: 245–274.

Lewis, E. G. 1942. On the generation and growth of a population. *Sankhya: The Indian Journal of Statistics* 6: 93–96.

Li, C.-K. and H. Schneider. 2002. Applications of Perron-Frobenius theory to population dynamics. *J. Math. Biol.* 44: 450–462.

Mickens, R. E. 1990. *Difference Equations Theory and Applications*. 2nd ed. Van Nostrand Reinhold Co., New York.

Murray, J. D. 1993. *Mathematical Biology*. 2nd ed. Springer-Verlag, Berlin, Heidelberg, New York.

Murray, J. D. 2002. *Mathematical Biology: I An Introduction*. 3rd ed. Springer-Verlag, New York.

Murray, J. D. 2003. *Mathematical Biology: II Spatial Models and Biomedical Applications*. 3rd ed. Springer-Verlag, New York.

Ortega, J. M. 1987. *Matrix Theory: A Second Course*. Plenum Press, New York.

Pielou, E. C. 1977. *Mathematical Ecology*. John Wiley & Sons, New York.

Sykes, Z. M. 1969. On discrete stable population theory. *Biometrics* 25: 285–293.

Thieme, H. R. 2003. *Mathematics in Population Biology*. Princeton Univ. Press, Princeton and Oxford.

Usher, M. B. 1972. Developments in the Leslie matrix model. In: *Mathematical Models in Ecology*. J. M. R. Jeffers (Ed.). Blackwell Scientific Publishers, London, pp. 29–60.

Wang, X. 2004. A simple proof of Descartes's Rule of Signs. *Amer. Math. Monthly* 111: 525–526.

## 1.10 Appendix for Chapter 1

### 1.10.1 Maple Program: Turtle Model

The following program uses the linear algebra package in Maple to find the eigenvalues and eigenvectors of the matrix for the loggerhead sea turtle.

```
> with(linalg):
> A:=matrix(7, 7, [0, 0, 0, 0, 127, 4, 80, 0.6747,
0.737, 0, 0, 0, 0, 0, 0, 0.0486, 0.6610, 0, 0, 0, 0,
0, 0, 0.0147, 0.6907, 0, 0, 0, 0, 0, 0, 0.0518, 0, 0,
0, 0, 0, 0, 0, 0.8091, 0, 0, 0, 0, 0, 0, 0, 0.8091,
0.8089]);
> eigenvals(A);
> eigenvects(A);
```

### 1.10.2 MATLAB® Program: Turtle Model

The following program uses MATLAB® to find the eigenvalues and eigenvectors of the matrix for the loggerhead sea turtle.

```
clear
r1=[0, 0, 0, 0, 127, 4, 80];  % First row.
r2=[.6747, .7370, 0, 0, 0, 0, 0];  % Second row.
r3=[0, .0486, .661, 0, 0, 0, 0];  % Third row, etc.
r4=[0, 0, .0147, .6907, 0, 0, 0];
r5=[0, 0, 0, .0518, 0, 0, 0];
r6=[0, 0, 0, 0, .8091, 0, 0];
r7=[0, 0, 0, 0, 0, .8091, .8089];
L=[r1;r2;r3;r4;r5;r6;r7];  % Leslie matrix.
E=eig(L)
lambda1=max(E)  % Dominant eigenvalue.
[V,D]=eig(L);
V1=V(:,i);  % choose i such that it is the eigenvector λ1.
s=sum(V1);
V1=V1/s  % Normalized eigenvector.
```

**Note:** A statement following % explains the MATLAB® command. This statement is not executed. If a semicolon is omitted after an executable command, then the value generated by the command prints to the computer screen.

Chapter 2

# Nonlinear Difference Equations, Theory, and Examples

## 2.1 Introduction

Mathematical models of biological systems are generally much more complex than the linear models studied in Chapter 1. In this chapter, we present the mathematical theory and techniques that are important in the study of nonlinear difference equations and systems. Techniques developed in this chapter will be useful to the study of biological applications in Chapter 3.

Some mathematical problems of interest in nonlinear difference equations include identification of equilibrium and periodic solutions and analyses of the stability of these types of solutions. Equilibrium solutions are biologically interesting because they represent "resting states" or "stationary states" of the system. The zero solution is often an equilibrium solution. If the zero solution is stable, then the system may approach zero. However, if a positive solution is an equilibrium solution and it is stable, then for initial values close to this equilibrium, solutions approach it. In population dynamics, the zero equilibrium represents population extinction and a positive equilibrium represents survival of the population. The zero equilibrium is often not a desired state, unless, for example, the state represents the proportion of the population that is infected or a population of pests.

It is important to distinguish between local and global stability. Local stability of an equilibrium implies that solutions approach the equilibrium only if they are initially close to it. For example, if the initial population size is very small and the zero equilibrium is stable, then extinction of the population may occur. However, if the initial population size is large, then local stability of the zero equilibrium tells nothing about population extinction. Global stability of an equilibrium is much stronger. Global stability implies that regardless of the initial population size, solutions approach the equilibrium. We state conditions for local stability and global stability of an equilibrium in the case of a scalar difference equation, where only one state is modeled such as population size. In addition, we state conditions for local stability of an equilibrium when several states are modeled by first-order difference equations or when one state is modeled by a second-order or higher-order difference equation. These latter conditions are known as the Jury conditions.

An important nonlinear difference equation studied in this chapter is the logistic difference equation,

$$x_{t+1} = rx_t(1 - x_t). \tag{2.1}$$

The first-order nonlinear difference equation (2.1) is very interesting mathematically. It was one of the first equations whose solution behavior was shown to exhibit what is known as "chaotic behavior." Equation (2.1) is discussed in detail in Section 2.6.

We introduce some terminology and techniques associated with nonlinear dynamical systems. These techniques are useful in studying how the dynamics of a system change as a parameter is varied (bifurcation theory). The types of bifurcations that may occur are defined and examples are given. In addition, we state a criterion that helps determine whether solutions are chaotic. This criterion depends on the magnitude of a Liapunov exponent. Liapunov exponents are defined for scalar and systems of difference equations. Finally, in Section 2.10, we give an example of a discrete-time epidemic model. Additional biological examples are studied in Chapter 3.

## 2.2 Basic Definitions and Notation

Recall that if the function $f$ in the $k$th-order difference equation $f(x_{t+k}, \ldots, x_{t+1}, x_t) = 0$ or the function $F$ in the first-order system $X_{t+1} = F(X_t)$ do not depend explicitly on $t$, they are referred to as *autonomous* difference equations. If there is an explicit $t$ dependence in these equations, then they are referred to as *nonautonomous* difference equations. In this chapter, we study autonomous difference equations. Scalar difference equations of order two or more, such as $x_{t+k} = g(x_{t+k-1}, \ldots, x_t)$, can be expressed as a system of first-order equations, $X_{t+1} = F(X_t)$. Therefore, we concentrate on first-order equations and systems.

**Definition 2.1.** For the first-order difference equation,

$$x_{t+1} = f(x_t), \tag{2.2}$$

an *equilibrium solution* or *steady-state solution* is a constant solution $\bar{x}$ to the difference equation, that is, a solution $\bar{x}$ satisfying

$$\bar{x} = f(\bar{x}). \tag{2.3}$$

For the first-order system, $X_{t+1} = F(X_t)$, an *equilibrium solution* or a *steady-state solution* is a constant solution $\bar{X}$ satisfying

$$\bar{X} = F(\bar{X}). \tag{2.4}$$

Solutions $\bar{x}$ satisfying (2.3) or $\bar{X}$ satisfying (2.4) are also called *fixed points* of the function $f$ or $F$, respectively.

The term "equilibrium solution" or "steady-state solution" is often shortened to "equilibrium" or "steady-state." For the two-dimensional, first-order system,

$$x_{t+1} = f(x_t, y_t),$$
$$y_{t+1} = g(x_t, y_t),$$

an *equilibrium solution* is a solution $(\bar{x}, \bar{y})$ such that $\bar{x} = f(\bar{x}, \bar{y})$ and $\bar{y} = g(\bar{x}, \bar{y})$. An equilibrium solution for a higher-order difference equation $f(x_{t+k}, \ldots, x_{t+1}, x_t) = 0$ is a solution $\bar{x}$ satisfying $f(\bar{x}, \ldots, \bar{x}, \bar{x}) = 0$.

For convenience, we introduce an alternate notation for the solution at time $t$, $x_t$ in (2.2). The solution can be expressed in terms of the initial value $x_0$. Denote $f(f(x_0)) = f^2(x_0)$, so that $x_2 = f^2(x_0)$. In general,

$$x_t = f(f(\cdots f(x_0) \cdots)) = f^t(x_0),$$

where the superscript $t$ represents the number of time steps or iterations beginning from the initial value $x_0$.

Solutions to the difference equation (2.2) may exhibit periodic behavior.

**Definition 2.2.** A *periodic solution of period* $m > 1$ of the difference equation (2.2) is a real-valued solution $\bar{x}_k$ satisfying

$$f^m(\bar{x}_k) = \bar{x}_k \quad \text{and} \quad f^i(\bar{x}_k) \neq \bar{x}_k \quad \text{for} \quad i = 1, 2, \ldots, m - 1.$$

An *m-cycle* is a set of points $\{\bar{x}_1, \bar{x}_2, \ldots, \bar{x}_m\}$, where for each $k = 1, \ldots, m$, $\bar{x}_k$ is a periodic solution of period $m$. The set $\{\bar{x}_1, f(\bar{x}_1), \ldots, f^{m-1}(\bar{x}_1)\}$ is called the *periodic orbit of* $\bar{x}_1$. A *periodic solution of period m* of the first-order system $X_{t+1} = F(X_t)$ is a real-valued vector $\bar{X}_k$ satisfying

$$F^m(\bar{X}_k) = \bar{X}_k \quad \text{and} \quad F^i(\bar{X}_k) \neq \bar{X}_k \quad \text{for} \quad i = 1, 2, \ldots, m - 1.$$

An *m-cycle* is a set of vectors $\{\bar{X}_1, \bar{X}_2, \ldots, \bar{X}_m\}$, where each $\bar{X}_k$ is a periodic solution of period $m$; $\{\bar{X}_1, F(\bar{X}_1), \ldots, F^{m-1}(\bar{X}_1)\}$ is called the *periodic orbit* of $\bar{X}_1$.

If $\bar{x}_k, k = 1, \ldots, m - 1$ is a periodic solution, then each $\bar{x}_k$ is a fixed point of the functions $f^m, f^{2m}, f^{3m}$, and so on (or $\bar{X}_k$ is a fixed point of the functions $F^m$, $F^{2m}$, $F^{3m}$, and so on). In addition, Definition 2.2 implies that a solution of period $m$ cannot have period $k < m$. In other words, period $m$ is the smallest value such that $f^m(\bar{x}_k) = \bar{x}_k$ or $F^m(\bar{X}_k) = \bar{X}_k$.

**Example 2.1** Let

$$x_{t+1} = \frac{ax_t}{b + x_t} = f(x_t), \quad a, b > 0. \tag{2.5}$$

Equilibrium solutions $\bar{x}$ of (2.5) satisfy

$$\bar{x} = \frac{a\bar{x}}{b + \bar{x}}.$$

Simplifying, $\bar{x}(\bar{x} + b - a) = 0$. Hence, there are two equilibria, $\bar{x} = 0$ and $\bar{x} = a - b$. Period 2 solutions or 2-cycles are found by solving for $\bar{x}$ in the equation $f^2(\bar{x}) = \bar{x}$. For (2.5),

$$f^2(\bar{x}) = f(f(\bar{x})) = \frac{a(a\bar{x}/[b + \bar{x}])}{b + a\bar{x}/[b + \bar{x}]} = \bar{x}.$$

Simplifying the above equation, we obtain

$$\frac{(a + b)(a - b - \bar{x})\bar{x}}{b^2 + b\bar{x} + a\bar{x}} = 0.$$

The only solutions are $\bar{x} = 0$ and $\bar{x} = a - b$, the two equilibria found earlier. These solutions are not period 2 solutions because they have period 1 (equilibrium solutions). Thus, there are no 2-cycles. ■

Next, we define the local stability of an equilibrium. An equilibrium is called locally asymptotically stable if for any small perturbation away from the equilibrium, the solution returns to the equilibrium value. In mathematical terminology, a more precise definition is given below.

**Definition 2.3.** An equilibrium solution $\bar{x}$ of (2.2) is *locally stable* if, for any $\epsilon > 0$, there exists $\delta > 0$ such that if $|x_0 - \bar{x}| < \delta$, then

$$|x_t - \bar{x}| = |f^t(x_0) - \bar{x}| < \epsilon \text{ for every } t \geq 0.$$

If $\bar{x}$ is not stable it is said to be *unstable*. The equilibrium solution $\bar{x}$ is *locally attracting* if there exists $\gamma > 0$ such that for all $|x_0 - \bar{x}| < \gamma$,

$$\lim_{t\to\infty} x_t = \lim_{t\to\infty} f^t(x_0) = \bar{x}.$$

The equilibrium solution $\bar{x}$ is *locally asymptotically stable* if it is locally stable and locally attracting.

The following example illustrates that an equilibrium can be locally attracting but not locally stable.

**Example 2.2** Suppose $f$ is defined as follows:

$$f(x) = \begin{cases} x + 1, & \text{if } x < 1 \\ 1, & \text{if } x \geq 1. \end{cases}$$

Note that $f$ is not continuous. The map $x_{t+1} = f(x_t)$ has an equilibrium at $\bar{x} = 1$. It can be seen that $\bar{x}$ is locally attracting (in fact for all initial conditions $\lim_{t\to\infty} x_t = 1$) but it is not locally stable (unstable). Let $\epsilon = 1/2$; then for $1/2 < x_0 < 1, x_1 = 1 + x_0 > 3/2$ so that $|x_1 - 1| > 1/2$. ■

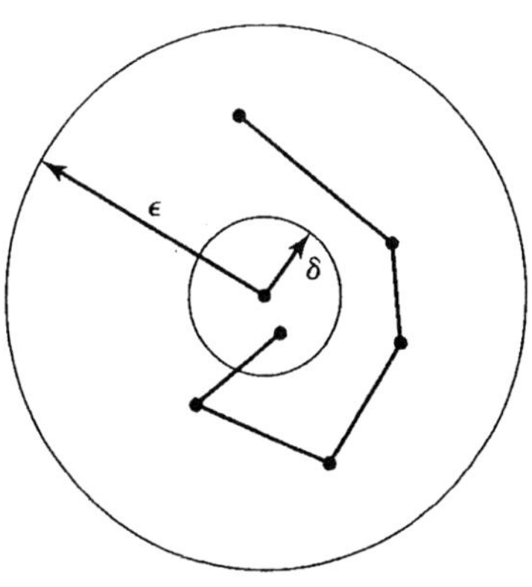


**Figure 2.1** Solutions beginning a distance of $\delta$ away from a stable equilibrium will stay within $\epsilon$ of the equilibrium.

Local asymptotic stability is also referred to as *neighborhood stability*. Solutions that are locally asymptotically stable converge to the stable equilibrium if they begin in a small neighborhood of that equilibrium. The stability and attractivity definitions for the system $X_{t+1} = F(X_t)$ are the same as in Definition 2.3 if the scalars $x$ and $\bar{x}$ are replaced by the vectors $X$ and $\bar{X}$, and the absolute value is replaced by the Euclidean norm, $\|X\|_2 = \|(x_1, x_2, \ldots, x_m)^T\|_2 = \sqrt{x_1^2 + x_2^2 + \cdots + x_m^2}$.

The value of $\delta$ in the definition of local asymptotic stability is dependent on the particular system and the initial condition. If the value of $\delta$ is very small, then initially the solution will have to be close to the equilibrium before it converges to it. Figure 2.1 shows the relationship among $\delta$, $\epsilon$, and a stable solution.

The convergence behavior for a first-order difference equation of the form (2.2) that is locally asymptotically stable may take one of two forms, either *convergent oscillations* or *convergent exponential* solutions. If the solution values tend to amplify themselves and do not converge to the equilibrium no matter how small the value of $\epsilon$, then the equilibrium is unstable. Such instability may appear as *divergent oscillations* or *divergent exponential* solutions. The case where solutions do not converge toward or diverge away from the equilibrium is sometimes referred to as *neutral stability* (stable, but not asymptotically stable).

In the next sections, we give criteria for determining local stability for first-order equations and systems.

## 2.3 Local Stability in First-Order Equations

In a study of local stability, first equilibrium solutions are identified, then linearization techniques are applied to determine the behavior of solutions near the equilibrium. If the equilibrium is stable for any set of initial conditions, then this type of stability is referred to as global stability. Some techniques for determining global stability of first-order difference equations are studied in the next section. In the particular case of *linear* difference equations or *linear* first-order systems, it will be seen that local and global stability are equivalent.

Suppose that the difference equation (2.2) has an equilibrium at $\bar{x}$. The equilibrium is translated to the origin by defining a new variable,

$$u_t = x_t - \bar{x}.$$

Then $u_{t+1}$ satisfies

$$u_{t+1} = x_{t+1} - \bar{x} = f(x_t) - \bar{x} = f(u_t + \bar{x}) - f(\bar{x}) = g(u_t), \tag{2.6}$$

where $g(u) = f(u + \bar{x}) - f(\bar{x})$. The equilibrium $\bar{x}$ in the original system has been translated to zero in the new system. Note that zero is a fixed point of $g$ iff $\bar{x}$ is a fixed point of $f$. In addition, zero is a locally stable (unstable or locally asymptotically stable) fixed point of $g$ iff $\bar{x}$ is a locally stable (unstable or locally asymptotically stable) fixed point of $f$.

To find conditions for local asymptotic stability of $\bar{x}$, we assume $f$ has a continuous second-order derivative in some interval $I$ containing $\bar{x}$. Then Taylor's Theorem with remainder can be applied,

$$f(x) = f(\bar{x}) + f'(\bar{x})(x - \bar{x}) + \frac{f''(\xi)}{2!}(x - \bar{x})^2$$

for some $\xi \in I$. For $(x_t - \bar{x})$ sufficiently small, the following linear approximation is valid: $f(x_t) - \bar{x} \approx f'(\bar{x})(x_t - \bar{x})$ or $u_{t+1} \approx f'(\bar{x})u_t$. We refer to this latter approximation as the *linear approximation* to the difference equation (2.2) at the equilibrium $\bar{x}$:

$$u_{t+1} = f'(\bar{x})u_t. \tag{2.7}$$

If $x_0$ is sufficiently close to $\bar{x}$, then the dynamics of $u_t$ are determined by the linearization (2.7). The value of $f'(\bar{x})$ determines whether $\bar{x}$ is locally asymptotically stable or unstable. If $|f'(\bar{x})| > 1$, then $u_t$ will not approach 0 (and $x_t$ will not approach $\bar{x}$), and if $|f'(\bar{x})| < 1$, then $u_t$ approaches 0 (and $x_t$ approaches $\bar{x}$). There is exponential convergence if $0 < f'(\bar{x}) < 1$ and oscillatory convergence if $-1 < f'(\bar{x}) < 0$. We have the following theorem.

**Theorem 2.1** *Assume $f'$ is continuous on an open interval $I$ containing $\bar{x}$ and $\bar{x}$ is a fixed point of $f$. Then $\bar{x}$ is a locally asymptotically stable equilibrium of $x_{t+1} = f(x_t)$ if*

$$|f'(\bar{x})| < 1$$

*and unstable if*

$$|f'(\bar{x})| > 1.$$

The shorthand notation $f'(\bar{x})$ in the theorem means differentiation of $f$ followed by evaluation at $\bar{x}$, that is,

$$f'(\bar{x}) = \left.\frac{df(x)}{dx}\right|_{x=\bar{x}}.$$

A rigorous proof of Theorem 2.1 is based on the Mean Value Theorem and only requires that $f'$ and *not* $f''$ be continuous.

**Proof** Suppose $|f'(\bar{x})| < 1$. Because $f'$ is continuous on $I$, there exists a sufficiently small subinterval $[\bar{x} - \epsilon, \bar{x} + \epsilon] \subset I$ such that $|f'(x)| < c < 1$ for $x \in [\bar{x} - \epsilon, \bar{x} + \epsilon]$. Now, we apply the Mean Value Theorem to $f$ with $x_0 \in [\bar{x} - \epsilon, \bar{x} + \epsilon]$,

$$|\bar{x} - f(x_0)| = |f(\bar{x}) - f(x_0)| = |f'(\xi_1)||\bar{x} - x_0| \leq c|\bar{x} - x_0|,$$

where $\xi_1$ is between $x_0$ and $\bar{x}$. Thus $f(x_0) \in [\bar{x} - \epsilon, \bar{x} + \epsilon]$. The Mean Value Theorem and the preceding inequality can be applied using $f(x_0)$ instead of $x_0$,

$$\begin{aligned}|\bar{x} - f^2(x_0)| &= |f^2(\bar{x}) - f^2(x_0)| \\ &= |f'(f(\xi_2))f'(\xi_2)||\bar{x} - f(x_0)| \\ &\leq c|\bar{x} - f(x_0)| \\ &\leq c^2|\bar{x} - x_0|,\end{aligned}$$

where $\xi_2$ is between $\bar{x}$ and $f(x_0)$ and $|\bar{x} - f(\xi_2)| < c|x - \xi_2|$. By induction it follows that

$$|\bar{x} - f^t(x_0)| \leq c^t\, |\bar{x} - x_0|,$$

for $t = 1, 2, \ldots$. Hence, $\lim_{t\to\infty} x_t = \bar{x}$; $\bar{x}$ is locally asymptotically stable.

Suppose $|f'(\bar{x})| > 1$. There exists an $\epsilon > 0$ such that for $x \in [\bar{x} - \epsilon, \bar{x} + \epsilon] \subset I$, $|f'(x)| > c > 1$. For $0 < |x_0 - \bar{x}| < \epsilon$, we apply the Mean Value Theorem,

$$|\bar{x} - f(x_0)| = |f'(\xi_1)||\bar{x} - x_0| \geq c|\bar{x} - x_0|,$$

where $\xi_1$ is between $\bar{x}$ and $x_0$. We apply this argument again, if $|\bar{x} - f(x_0)| < \epsilon$, to obtain $|\bar{x} - f^2(x_0)| \geq c^2|\bar{x} - x_0|$. This argument cannot be continued indefinitely because there must exist $t$ such that $c^t|\bar{x} - x_0| > \epsilon$. Hence, there exists $t$ such that $|\bar{x} - f^t(x_0)| > \epsilon$; $\bar{x}$ is unstable. □

The local asymptotic stability results in Theorem 2.1 apply only to the equilibrium $\bar{x}$ where $|f'(\bar{x})| \neq 1$. When $|f'(\bar{x})| \neq 1$, then $\bar{x}$ is referred to as a hyperbolic equilibrium.

**Definition 2.4.** An equilibrium $\bar{x}$ of $x_{t+1} = f(x_t)$ is said to be *hyperbolic* if $|f'(\bar{x})| \neq 1$. Otherwise, it is said to be *nonhyperbolic*.

The criterion for stability in Theorem 2.1 can be applied to periodic solutions. In the case of a periodic solution of period $m$, the function $f^m(x)$ is used instead of $f(x)$.

**Theorem 2.2** *Suppose $f'$ is continuous on an open interval $I$ and the m-cycle,*

$$\{\bar{x}_1, f(\bar{x}_1), \ldots, f^{m-1}(\bar{x}_1)\},$$

*of the difference equation (2.2) is contained in I. Then the m-cycle is locally asymptotically stable if*

$$\left|\frac{d[f^m(\bar{x}_k)]}{dx}\right| < 1 \tag{2.8}$$

*for some k and unstable if*

$$\left|\frac{d[f^m(\bar{x}_k)]}{dx}\right| > 1 \tag{2.9}$$

*for some k.* □

The conditions in Theorem 2.2 need to be verified for only one of the $\bar{x}_k$. The reason it is necessary to check only one equilibrium value is because if condition (2.8) or condition (2.9) hold for some $k$, then they hold for all $k$. Simplification of the conditions in Theorem 2.2 show that all of the values $\bar{x}_j$, $j = 1, \ldots, k$ are used to compute (2.8) and (2.9). Consider a 2-cycle. From the chain rule it follows that

$$\frac{d[f^2(x)]}{dx} = f'(f(x))f'(x).$$

Evaluating at $\bar{x}_1$, then

$$\frac{d[f^2(\bar{x}_1)]}{dx} = f'(f(\bar{x}_1))f'(\bar{x}_1) = f'(\bar{x}_2)f'(\bar{x}_1) = \frac{d[f^2(\bar{x}_2)]}{dx}$$

because for a 2-cycle $f(\bar{x}_1) = \bar{x}_2$ and $\bar{x}_1 = f(\bar{x}_2)$. This latter equality illustrates an alternate method for checking the stability of $m$-cycles.

**Corollary 2.1** *Suppose $\{\bar{x}_1, \bar{x}_2, \ldots, \bar{x}_m\}$ is an m-cycle of the difference equation $x_{t+1} = f(x_t)$. Then the m-cycle is locally asymptotically stable if*

$$|f'(\bar{x}_1)f'(\bar{x}_2)\cdots f'(\bar{x}_m)| < 1.$$ ◄

**Example 2.3** Consider again

$$x_{t+1} = \frac{ax_t}{b + x_t} = f(x_t), \quad a, b > 0.$$

It has already been shown in Example 2.1 that this difference equation has two equilibria, $\bar{x} = 0$ and $\bar{x} = a - b$. We test for local stability by finding the derivative of $f$,

$$f'(x) = \frac{ba}{(b + x)^2}.$$

Evaluating $f'$ at $\bar{x} = 0$ yields $f'(0) = a/b$. If $a/b < 1$ or $a < b$, then zero is locally asymptotically stable and if $a > b$, the zero equilibrium is unstable. Evaluating $f'$ at $\bar{x} = a - b$ yields $f'(a - b) = b/a$. Thus, $\bar{x} = a - b$ is locally asymptotically stable if $a > b$ (when $\bar{x} > 0$) and unstable if $a < b$. ■

**Example 2.4** Let

$$x_{t+1} = r - x_t^2 = f(x_t), \quad r > 0.$$

The equilibria are found by solving $f(x) = x$ or

$$x^2 + x - r = 0. \tag{2.10}$$

There are two equilibria,

$$\bar{x}_{\pm} = \frac{-1 \pm \sqrt{1 + 4r}}{2};$$

one is positive and one is negative. Since $f'(x) = -2x$, it follows that the negative equilibrium, $\bar{x}_-$, is unstable,

$$f'(\bar{x}_-) = 1 + \sqrt{1 + 4r} > 1.$$

The positive equilibrium, $\bar{x}_+$, satisfies

$$f'(\bar{x}_+) = 1 - \sqrt{1 + 4r}.$$

Since $f'(\bar{x}_+) < 0$, the positive equilibrium, $\bar{x}_+$, is locally asymptotically stable if $f'(\bar{x}_+) > -1$ or $-f'(\bar{x}_+) < 1$. Simplifying this latter inequality leads to

$$\sqrt{1 + 4r} - 1 < 1 \quad \text{or} \quad r < \frac{3}{4}.$$

The equilibrium $\bar{x}_+$ is locally asymptotically stable if $r < 3/4$ and unstable if $r > 3/4$.

Next, we shall determine if there are any 2-cycles for this example by solving $f^2(x) = f(f(x)) = x$ for $x$, where $f(f(x)) = r - (r - x^2)^2$. Expanding $r - (r - x^2)^2 - x = 0$, we obtain $r - r^2 + 2rx^2 - x^4 - x = 0$. Factoring yields

$$-(x^2 + x - r)(x^2 - x + 1 - r) = 0. \tag{2.11}$$

Note that the equilibrium solutions from (2.10) are also solutions to the latter equation. However, they are not 2-cycle solutions. The second factor in (2.11) set equal to zero yields two new solutions,

$$\bar{x}_{1,2} = \frac{1 \pm \sqrt{4r - 3}}{2}.$$

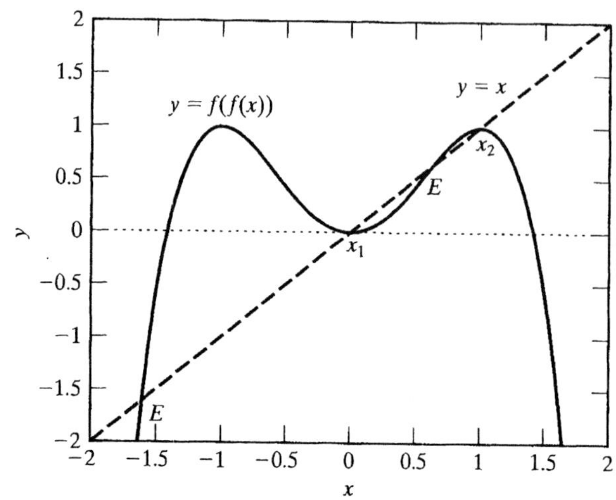


**Figure 2.2** Graphs of $y = f(f(x)) = r - (r - x^2)^2$ and $y = x$ when $r$ is some value satisfying $3/4 < r < 5/4$. The equilibria $\bar{x}$ are denoted by $E$ and the 2-cycle by $\{x_1, x_2\}$.

These two solutions represent a 2-cycle. We assume $r > 3/4$ so that the solutions are real. The stability of the 2-cycle is determined by the magnitude of the derivative of $f(f(x))$ evaluated at $\bar{x}_1$ or $\bar{x}_2$ or by applying Corollary 2.1. We leave it as an exercise to show that the 2-cycle is locally asymptotically stable if $3/4 < r < 5/4$ and unstable if $r > 5/4$ (Exercise 6). Figure 2.2 illustrates $y = f(f(x)) = r - (r - x^2)^2$ and $y = x$ when $3/4 < r < 5/4$. It is easy to verify that the slope of $f(f(x))$ at $\bar{x}_j$ satisfies $|f'(f(\bar{x}_j))| < 1$ for $j = 1, 2$. ■

For nonhyperbolic equilibria, it is clear that the higher-order terms that do not appear in the linear approximation are important in assessing local asymptotic stability. In the cases where $f'(\bar{x}) = 1$ or $f'(\bar{x}) = -1$ there are some results to show local asymptotic stability or instability of $\bar{x}$. They require the third-order derivative and the Schwarzian derivative (see pp. 24–26, Elaydi, 1999). These results are stated in Theorems 2.3 and 2.4. Applications of these theorems can be found in Exercise 5. First, we define the Schwarzian derivative.

**Definition 2.5.** The *Schwarzian derivative* of a function $f$ at $x$ is denoted $(Sf)(x)$ and defined as follows:

$$(Sf)(x) = \frac{f'''(x)}{f'(x)} - \frac{3}{2}\left(\frac{f''(x)}{f'(x)}\right)^2.$$

The Schwarzian derivative is named after Hermann Schwarz (1843–1921), a German mathematician who made many contributions to mathematics, especially in the area of complex function theory.

**Example 2.5** If $f(x) = x^2$, then $(Sf)(x) = -\dfrac{3}{2x^2} < 0$. Even at $x = 0$, the Schwarzian derivative is defined to be $(Sf)(0) = \lim_{x\to 0}\left(-\dfrac{3}{2x^2}\right) = -\infty$. ■

The first theorem gives criteria for local asymptotic stability of $\bar{x}$ for the case $f'(\bar{x}) = 1$ and the second theorem for the case $f'(\bar{x}) = -1$.

**Theorem 2.3** *Suppose $f'(\bar{x}) = 1$, where $\bar{x}$ is an equilibrium point of $x_{t+1} = f(x_t)$ and $f'''$ is continuous on an open interval containing $\bar{x}$.*

(i) *If $f''(\bar{x}) \neq 0$, then $\bar{x}$ is unstable.*
(ii) *If $f''(\bar{x}) = 0$ and $f'''(\bar{x}) > 0$, then $\bar{x}$ is unstable.*
(iii) *If $f''(\bar{x}) = 0$ and $f'''(\bar{x}) < 0$, then $\bar{x}$ is locally asymptotically stable.* □

Verification of Theorem 2.3 can be easily shown geometrically and is left as an exercise (Exercise 7).

**Theorem 2.4** *Suppose $f'(\bar{x}) = -1$, where $\bar{x}$ is an equilibrium point of $x_{t+1} = f(x_t)$ and $f'''$ is continuous on an open interval containing $\bar{x}$.*

(i) *If $(Sf)(\bar{x}) > 0$, then $\bar{x}$ is unstable.*
(ii) *If $(Sf)(\bar{x}) < 0$, then $\bar{x}$ is locally asymptotically stable.*

**Proof** The proof applies Theorem 2.3 to the function $x_{t+1} = f^2(x_t) = g(x_t)$ (Elaydi, 1999). Since $\bar{x}$ is a fixed point of $f$ it is also a fixed point of $f^2$, $\bar{x} = f^2(\bar{x}) = g(\bar{x})$. In addition, $g'(x) = f'(f(x))f'(x)$ and

$$g''(x) = f''(f(x))[f'(x)]^2 + f'(f(x))f''(x)$$

so that $g'(\bar{x}) = f'(f(\bar{x}))f'(\bar{x}) = [f'(\bar{x})]^2 = 1$ and

$$\begin{aligned} g''(\bar{x}) &= f''(\bar{x})[f'(\bar{x})]^2 + f'(\bar{x})f''(\bar{x}) \\ &= f''(\bar{x})f'(\bar{x})[f'(\bar{x})+1] = 0. \end{aligned}$$

After computing $g'''(\bar{x})$, Theorem 2.3 parts (ii) and (iii) can be used to complete the proof of the theorem. (See Exercise 8.) □

**Example 2.6** Let $x_{t+1} = -x_t^3 - x_t = f(x_t)$. The only real equilibrium solution is $\bar{x} = 0$. Then $f'(0) = -1$, $f''(0) = 0$, and $f'''(0) = -6$. Calculation of the Schwarzian derivative, $(Sf)(0) = 6 > 0$, and application of Theorem 2.4 show that $\bar{x} = 0$ is unstable. ■

## 2.4 Cobwebbing Method for First-Order Equations

The cobwebbing method, introduced in Chapter 1, is applied to Examples 2.3 and 2.4. Recall that, in the cobwebbing method, the line $y = x$ and the reproduction curve $y = f(x)$ are graphed in the $x$-$y$ plane.

In Example 2.3, the reproduction curve $f(x) = ax/(b + x)$ is sketched for $x > 0$ in Figure 2.3. Two cases are considered, $a < b$ and $a > b$. Successive iterates $x_t$ converge monotonically to zero if $a < b$ and if $a > b$, then $x_t$ converges monotonically to $\bar{x}$. There does not exist a positive equilibrium when $a < b$.

In Example 2.4, the reproduction curve $f(x) = r - x^2$ is sketched for $x > 0$. Two cases are considered, $0 < r < 3/4$ and $r > 3/4$. Recall that if $0 < r < 3/4$, then the equilibrium $\bar{x}_+ = (-1 + \sqrt{1 + 4r})/2$ is locally asymptotically stable and if $r > 3/4$ it is unstable. The stability behavior is evident in the cobwebbing method graphed in Figure 2.3. This latter example [Figure 2.3(c)] illustrates why the term "cobwebbing" is used to describe this method.

Recall that the difference equation in Example 2.4 has a 2-cycle which is stable if $3/4 < r < 5/4$. Graphs of $y = x$ and $f^2(x) = f(f(x))$ in Figure 2.2,

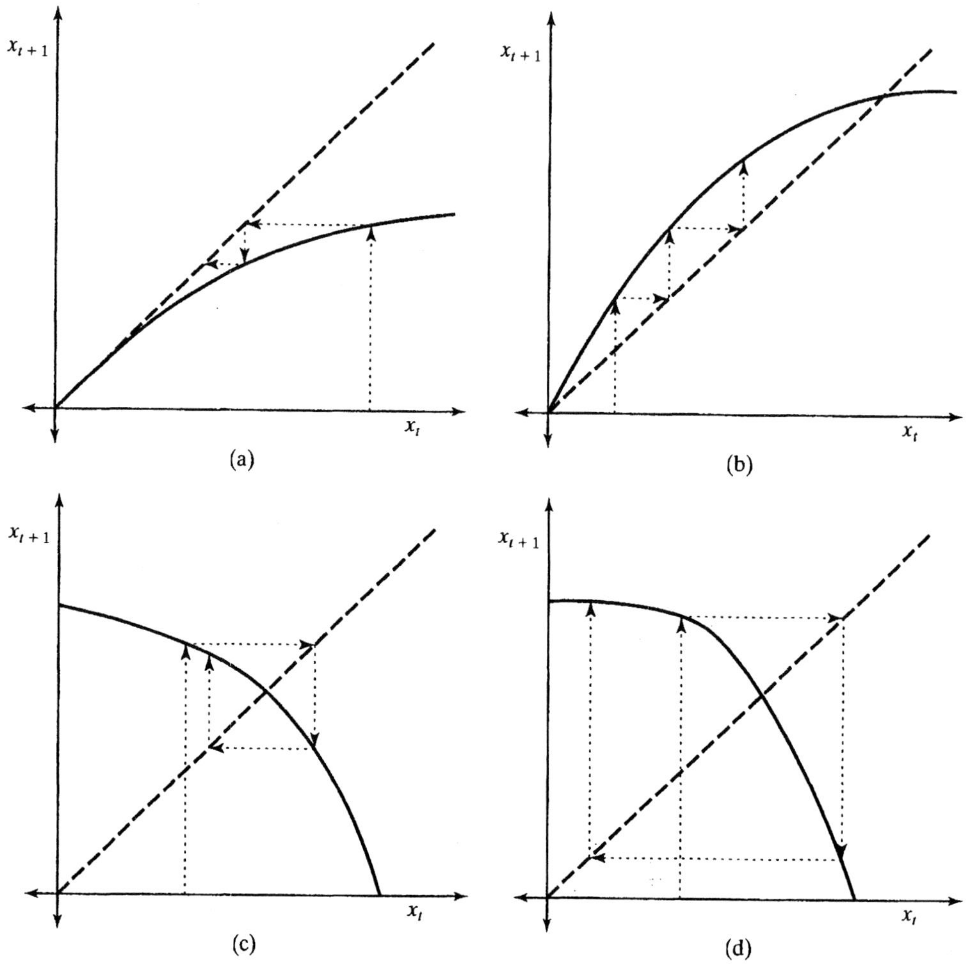


**Figure 2.3** Cobwebbing method for Examples 2.3 when (a) $a < b$ and (b) $a > b$ and for Example 2.4 when (c) $0 < r < 3/4$ and (d) $r > 3/4$.

where $f(x) = r - x^2$ and $3/4 < r < 5/4$, illustrate the 2-cycle and the equilibrium points (points of intersection). Also, note that the slope at the equilibrium points $\bar{x}$ satisfies $|df^2(\bar{x})/dx| > 1$ and at the 2-cycle, $|df^2(\bar{x}_i)/dx| < 1$.

## 2.5 Global Stability in First-Order Equations

Global stability of an equilibrium removes the restrictions on the initial conditions. In global asymptotic stability, solutions approach the equilibrium for all initial conditions. However, because our applications apply to biological systems, we consider only positive initial conditions. In addition, we distinguish between global attractivity and global asymptotic stability.

**Definition 2.6.** Suppose $\bar{x}$ is an equilibrium of the difference equation,

$$x_{t+1} = f(x_t), \tag{2.12}$$

where $f : [0,a) \to [0,a), 0 < a \leq \infty$. Then $\bar{x}$ is said to be *globally attractive* if for all initial conditions $x_0 \in (0,a)$, $\lim_{t\to\infty} x_t = \bar{x}$. The equilibrium $\bar{x}$ is said to be *globally asymptotically stable* if $\bar{x}$ is globally attractive and if $\bar{x}$ is locally stable.

Globally attractive equilibria are locally attractive, and therefore globally asymptotically stable equilibria are locally asymptotically stable. Sedaghat (1997) proved that if the map $f$ is continuous, then a globally attracting equilibrium must be locally asymptotically stable. Thus, for a continuous map $f$, global attractivity is equivalent to global asymptotic stability. However, if $f$ is not continuous, the following example shows that an equilibrium can be globally attractive but not globally asymptotically stable. (Also see Example 2.2.)

**Example 2.7** Define the map $f : [0, \infty) \to [0, \infty)$ as follows:

$$f(x) = \begin{cases} 2, & x \in [0, 2] \\ x - 1 & x \in (2, \infty). \end{cases}$$

Then, for $x_{t+1} = f(x_t)$, it is easy to see for any initial conditions that $\lim_{t\to\infty} x_t = 2$. The equilibrium $\bar{x} = 2$ is globally attractive. Let $\epsilon = 1/2$. Then for $2 < x_0 < 2 + \epsilon, x_1 < 1 + \epsilon$ and $|x_1 - 2| > 1/2$. The equilibrium $\bar{x} = 2$ is not locally stable. ■

In the global definitions given in (2.6), it is assumed that solutions are nonnegative; the initial conditions and $f$ are restricted to the interval $[0,a)$. For biological models, this is a reasonable assumption. It is often the case in biological models that zero is an equilibrium, $f(0) = 0$. If there is an additional positive equilibrium, $f(\bar{x}) = \bar{x}$, a question of interest is whether the zero or positive equilibrium is globally asymptotically stable. An analogous definition for global asymptotic stability of an equilibrium $\bar{X}$ for first-order systems, $X_{t+1} = F(X_t)$, can be stated. We concentrate on first-order scalar difference equations and state conditions on $f$ such that the solutions are nonnegative and the zero equilibrium or a positive equilibrium is globally asymptotically stable.

First, we make some assumptions about the function $f$.

(i) $f$ is a continuous function on $[0, a), 0 < a \leq \infty$.

(ii) $f : [0, a) \to [0, a), 0 < a \leq \infty$.

Because of assumption (i), continuity of $f$, global asymptotic stability and global attractivity are equivalent.

The first result shows global asymptotic stability of the origin; solutions approach zero (extinction).

**Theorem 2.5** *If the function $f$ of (2.12) satisfies (i), (ii), and $0 < f(x) < x$ for all $x \in (0, a)$, then the origin is globally asymptotically stable.*

**Proof** The result follows by noticing that $0 < f^t(x_0) < \cdots < f^2(x_0) < f(x_0) < x_0$ for $x_0 \in (0, a)$. The sequence $\{f^t(x_0)\}_{t=0}^{\infty}$ is monotone decreasing and bounded below by zero. Thus, $\lim_{t\to\infty} f^t(x_0) = \bar{x}$ exists. Note that $\bar{x}$ is a fixed point of $f$ by the continuity of $f$. Because zero is the only nonnegative fixed point of $f$, it follows that $\bar{x} = 0$. See Figure 2.4. □

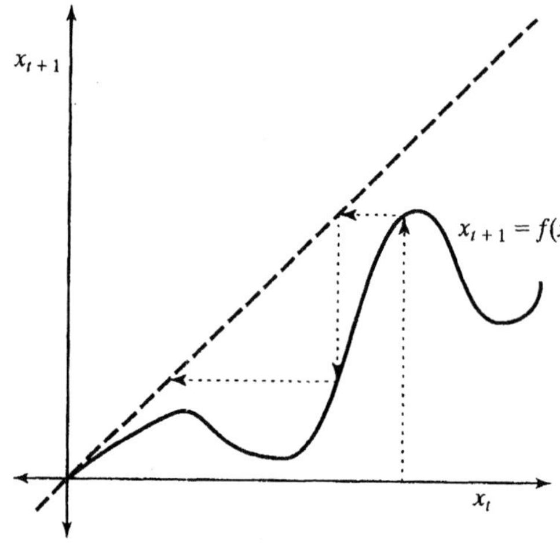


**Figure 2.4** The function $f$ satisfies $0 < f(x) < x$. Solutions to $x_{t+1} = f(x_t)$ approach zero; see Theorem 2.5.

**Example 2.8** Consider again

$$x_{t+1} = \frac{ax_t}{b + x_t} = f(x_t),$$

where $f(x) = ax/(b + x)$ satisfies $f : [0, \infty) \to [0, \infty)$. When $0 < a \le b$, there is only one nonnegative fixed point at $\bar{x} = 0$ and $0 < f(x) < x$ for $x \in (0, \infty)$. It follows directly from Theorem 2.5 that $\lim_{t\to\infty} x_t = \lim_{t\to\infty} f^t(x_0) = 0$ for all positive initial conditions. ■

Some additional assumptions on the reproduction curve $y = f(x)$ are needed to verify global asymptotic stability for a positive equilibrium. Assume

(iii) $f(0) = 0, f(\bar{x}) = \bar{x}$.

(iv) $f(x) > x$ for $0 < x < \bar{x}$.

(v) $f(x) < x$ for $\bar{x} < x < a$.

(vi) If $f$ has a maximum at $x_M$ in $(0, \bar{x})$, then $f$ is decreasing for $x > x_M$.

Note that if $f$ has no maximum in $(0, \bar{x})$, then condition (vi) is automatically satisfied. Assumptions (i)–(vi) imply $f$ has only two fixed points in its domain, $x = 0$ and $x = \bar{x}$. In addition, assumptions (i)–(vi) imply that $f$ has at most one maximum and if there is a maximum at $x_M$, then $f$ is increasing on $(0, x_M)$. Some functions satisfying the assumptions (i)–(vi) include

$$f(x) = \frac{4x}{3 + x}, \quad \text{where} \quad f : [0,\infty) \to [0,\infty),$$
$$f(x) = 2xe^{-x}, \quad \text{where} \quad f : [0,\infty) \to [0,\infty),$$
$$f(x) = 2x(1 - x), \quad \text{where} \quad f : [0,1) \to [0,1).$$

These functions are graphed in Figure 2.5.

The next theorem gives necessary and sufficient conditions for global asymptotic stability of a positive equilibrium $\bar{x}$. The theorem is due to Cull (1981).

**Theorem 2.6** *The difference equation $x_{t+1} = f(x_t)$ satisfying (i)–(vi) has a globally asymptotically stable equilibrium at $\bar{x}$ iff $f$ has no 2-cycles.* □

Global asymptotic stability implies $f$ has no 2-cycles is obvious. To prove the reverse implication is a lengthy argument and is given in the Appendix for Chapter 2.

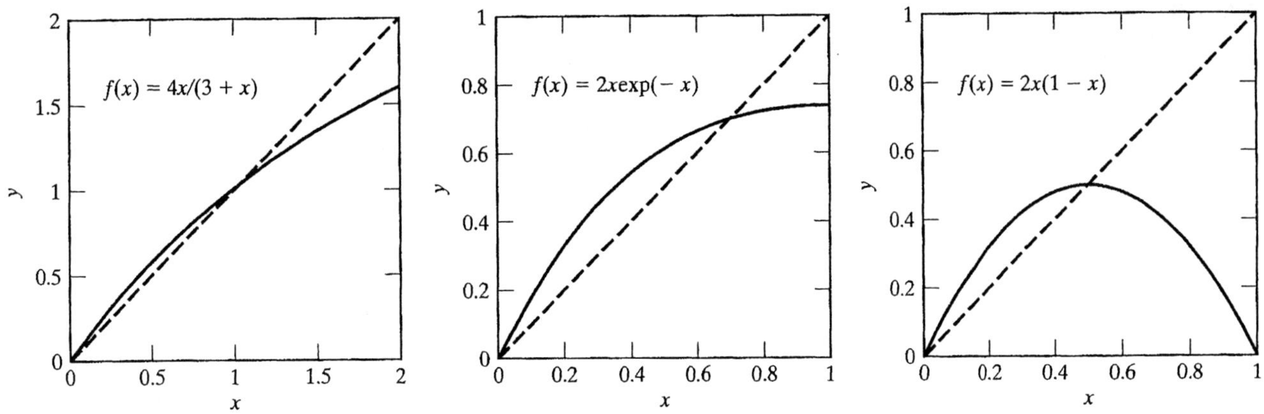


**Figure 2.5** Three functions satisfying assumptions (i)–(vi).

---

**Example 2.9** Theorem 2.6 can be applied to the difference equation $x_{t+1} = ax_t/(b + x_t)$, $a > b > 0$, and $f : [0,\infty) \to [0,\infty)$. There are two fixed points, $\bar{x} = 0$ and $\bar{x} = a - b > 0$, and assumptions (i)–(vi) are satisfied. Also, note that $f'(x) = ab/(b + x)^2 > 0$ implies $f$ has no maximum on $[0,\infty)$. It has already been shown that this model has no 2-cycles. According to Theorem 2.6, $\bar{x} = a - b$ is globally asymptotically stable. ■

A sufficient condition for nonexistence of 2-cycles is given in the following theorem. The theorem has a straightforward proof due to McCluskey and Muldowney (1998) and does not require assumptions (iii)–(vi), but $f$ must have a continuous first derivative.

---

**Theorem 2.7** *Let $f'$ be continuous on an interval $I$ and $f : I \to I$. If $1 + f'(x) \neq 0$ for all $x \in I$, then $x_{t+1} = f(x_t)$ has no 2-cycles in $I$.*

**Proof** Recall for a 2-cycle to exist, $f^2(x_0) = f(x_1) = x_0$ for some $x_0$ and $x_1$ in $I$. We show this cannot happen. Suppose $x_0 \in I$, then $x_1 = f(x_0) \in I$ and

$$0 \neq \int_{x_0}^{x_1} (1 + f'(x))\,dx = (x_1 + f(x_1)) - (x_0 + f(x_0)) = f^2(x_0) - x_0.$$

Thus, there can be no 2-cycles in $I$. □

---

**Example 2.10** Theorem 2.7 can be applied to the difference equation

$$x_{t+1} = \frac{ax_t}{b + (x_t)^k}, \quad a, b, k > 0 \text{ and } x_0 > 0.$$

The expression

$$1 + f'(x) = 1 + \frac{a[b + x^k(1 - k)]}{(b + x^k)^2}.$$

If $k \leq 1$, then $1 + f'(x) > 0$ for all $x > 0$; there are no 2-cycles. For $k = 1$, the model simplifies to the difference equation in Examples 2.1 and 2.9. ■

**Figure 2.6** A function $f$ satisfying Theorem 2.8.

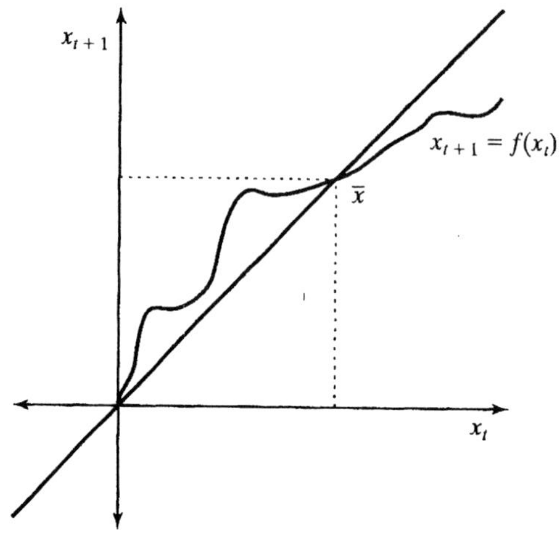


Some additional results concerning global asymptotic stability of $\bar{x}$ are stated in the following theorems. In some cases the following theorems may be easier to apply than Theorem 2.6.

**Theorem 2.8** *If $f$ satisfies (i) and (ii) and $\bar{x} \in (0, a)$ such that $x < f(x) < \bar{x}$ for $0 < x < \bar{x}$ and $\bar{x} < f(x) < x$ for $x > \bar{x}$, then the difference equation $x_{t+1} = f(x_t)$ has a globally asymptotically stable equilibrium at $\bar{x}$.*

**Proof** The proof follows from the fact that $x_t < x_{t+1} = f(x_t) < \bar{x}$ for $x_0 < \bar{x}$ and $\bar{x} < x_{t+1} = f(x_t) < x_t$ for $x_0 > \bar{x}$. In the first case, $x_0 < \bar{x}$, the sequence $\{f^t(x_0)\}_{t=0}^{\infty}$ is monotone increasing and bounded above by $\bar{x}$ and in the second case, $\bar{x} < \tilde{x}_0$, $\{f^t(\tilde{x}_0)\}_{t=0}^{\infty}$ is a monotone decreasing sequence, bounded below by $\bar{x}$. See Figure 2.6. Both sequences must converge to a positive limit $z_1$ and $z_2$: $z_1 = \lim_{t\to\infty} f^t(x_0)$ and $z_2 = \lim_{t\to\infty} f^t(\tilde{x}_0)$. The limits $z_1$ and $z_2$ are fixed points of $f$;

$$z_1 = \lim_{t\to\infty} f(f^{t-1}(x_0)) = f(z_1), \quad z_2 = \lim_{t\to\infty} f(f^{t-1}(\tilde{x}_0)) = f(z_2).$$

The only positive fixed point is $\bar{x}$, so $z_1 = z_2 = \bar{x}$. □

**Example 2.11** The difference equation

$$x_{t+1} = \frac{ax_t}{b + x_t} = f(x_t)$$

satisfies the hypotheses of Theorem 2.8 for $a > b > 0$. This theorem provides another method to show that $\bar{x} = a - b$ is globally asymptotically stable. ■

The next theorem on global asymptotic stability is due to Cull (1986).

**Theorem 2.9** *Let $x_{t+1} = f(x_t)$.*

(a) *Suppose $f$ satisfies assumptions (i)–(v) but $f$ has no maximum in $(0, \bar{x})$. Then $\bar{x}$ is globally asymptotically stable.*

(b) *Suppose $f$ satisfies conditions (i)–(vi) and $f$ has a maximum $x_M$ in $(0, \bar{x})$. Then $\bar{x}$ is globally asymptotically stable iff $f(f(x)) > x$ for all $x \in [x_M, \bar{x})$.*

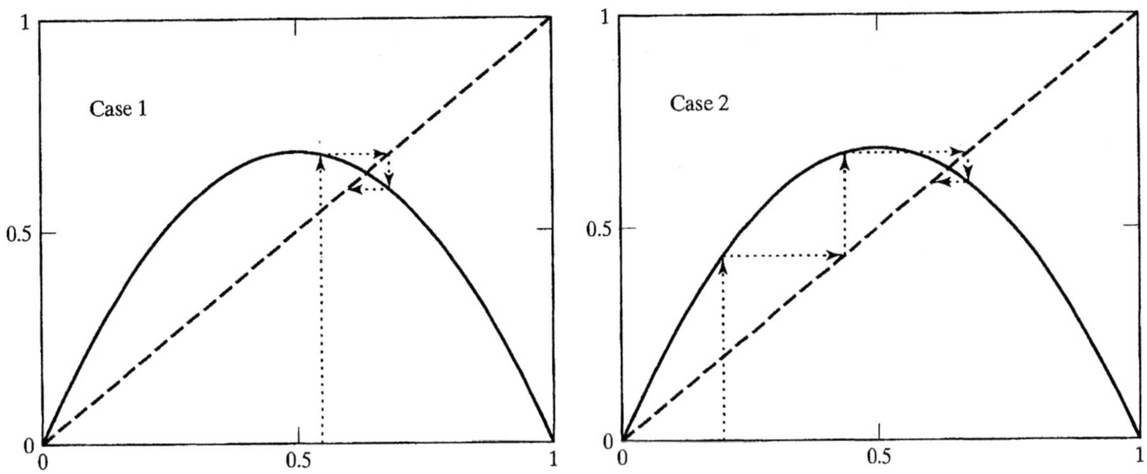


**Figure 2.7** Examples that satisfy the assumptions in Cases 1 and 2 of Theorem 2.9.

**Proof** (a) We show that $f$ has no 2-cycles. Suppose $\{x_1, x_2\}$ is a 2-cycle. Then properties (iv) and (v) imply $x_1 < \bar{x} < x_2$. But $x_1 \in (0, \bar{x})$ implies $x_1 < x_2 = f(x_1) < \bar{x}$ because $f$ has no maximum in $(0, \bar{x})$. Therefore, we have a contradiction. There can be no 2-cycles. Theorem 2.6 implies that $\bar{x}$ is globally asymptotically stable.

(b) First, if $\bar{x}$ is globally asymptotically stable, there can be no 2-cycles. The proof that global asymptotic stability implies $f(f(x)) > x$ on $[x_M, \bar{x})$ is given in the proof of Theorem 2.6 in the Appendix for Chapter 2.

Second, to show the reverse implication, three cases are considered: Case 1: $x_0 \in [x_M, \bar{x})$, Case 2: $x_0 \in (0, x_M)$, and Case 3: $x_0 \in (\bar{x}, a)$. It is assumed that $f(f(x)) > x$ on $[x_M, \bar{x})$.

Case 1 $x_0 \in [x_M, \bar{x})$. Then $x_1 = f(x_0) > \bar{x}$ and $x_0 < x_2 = f(f(x_0)) < \bar{x}$ by hypothesis and property (vi). The sequence of even iterates satisfies $x_0 < f^2(x_0) < f^4(x_0) < \cdots < \bar{x}$ and $\lim_{t\to\infty} f^{2t}(x_0) = \bar{x}$. But the sequence of odd iterates, $f^{2t+1}(x_0)$, satisfies, by the continuity of $f$, $\lim_{t\to\infty} f^{2t+1}(x_0) = \lim_{t\to\infty} f(x_{2t}) = f(\bar{x}) = \bar{x}$. Thus, $\lim_{t\to\infty} f^t(x_0) = \bar{x}$. (See Figure 2.7.)

Case 2 $x_0 \in (0, x_M)$. Since $f(x) > x$ on $(0, x_M]$, eventually $f^t(x_0) > x_M$. (See Figure 2.7.) If for some $t_0$, $f^{t_0}(x_0) \in [x_M, \bar{x}]$, then by Case 1, $f^{t_0+t}(x_0)$ converges to the equilibrium $\bar{x}$. If $f^{t_0}(x_0) > \bar{x}$, then there exists $\hat{x}$ such that $f(\hat{x}) = f^{t_0}(x_0)$ and $\hat{x} \in [x_M, \bar{x}]$, then by Case 1, $f^t(\hat{x}) = f^{t+t_0}(x_0)$ converges to $\bar{x}$.

Case 3 $x_0 \in (\bar{x}, a)$. Then $x_1 = f(x_0) < \bar{x}$, so either Case 1 or 2 apply. □

We apply Theorem 2.9 (a) to the following example.

**Example 2.12** Consider the model

$$x_{t+1} = 2x_t e^{-rx_t} = f(x_t), \quad r > 0. \tag{2.13}$$

This model has two equilibria, one at zero and one at $\bar{x} = \ln(2)/r$. Assumptions (i)–(v) are satisfied. In addition, the first derivative, $f'(x) = 2\exp(-rx)(1 - rx)$, can be used to show that $f$ has a maximum at $x_M = 1/r$. Since

$\bar{x} = \ln(2)/r < 1/r = x_M$, $f$ has no maximum in $(0, \bar{x})$, so condition (vi) is automatically satisfied. It follows by Theorem 2.9(a) that $\bar{x}$ is globally asymptotically stable. ■

Model (2.13) is a special case of a more general model known as the Ricker model:

$$x_{t+1} = ax_t e^{-rx_t} = f(x_t), \quad a, r > 0.$$

This equation has been applied extensively in population models but it was originally developed for fish populations based on stock and recruitment (Ricker, 1954). We discuss this model in Chapter 3.

## 2.6 The Approximate Logistic Equation

The logistic model is one of the most well-known population models. Logistic growth is often referred to as sigmoid growth, due to its S-shaped solution curve. A population that grows logistically, initially increases exponentially; then the growth slows down and eventually approaches an upper bound or limit. The most well-known form of the model is the logistic differential equation. Let $y(t)$ represent the size or density of a population at time $t$; then the growth rate in a logistic differential equation satisfies

$$\frac{dy}{dt} = ay\left(1 - \frac{y}{K}\right), \tag{2.14}$$

where $a > 0$ is known as the intrinsic growth rate and $K > 0$ is the carrying capacity. For positive initial conditions, $y(0) > 0$, solutions approach the carrying capacity, $\lim_{t\to\infty} y(t) = K$. This equation will be discussed in more detail when continuous time models are introduced in Chapter 5. One approximation of the logistic differential equation leads to a difference equation often referred to as the *discrete logistic equation*. (An exact discrete logistic equation is discussed in the Exercises.)

To derive the discrete logistic equation, the derivative $dy/dt$ is approximated by a difference quotient,

$$\frac{dy}{dt} \approx \frac{y(t + \Delta t) - y(t)}{\Delta t}.$$

Approximating the differential equation by a difference equation leads to

$$y(t + \Delta t) = y(t) + a\Delta t y(t)(1 - y(t)/K).$$

(This approximation is known as Euler's method in numerical solutions of differential equations.) We can simplify the difference equation by assuming $\Delta t = 1$, meaning one time unit. However, keep in mind that the magnitude of $a\Delta t$ or the time step $\Delta t$ plays an important role in the dynamics of this simple equation. After simplification, the discrete logistic equation is obtained:

$$y_{t+1} = y_t + ay_t\left(1 - \frac{y_t}{K}\right) = (1 + a)y_t - \frac{a}{K}y_t^2,$$

where $y(t) = y_t$. The parameter $a$ determines exponential growth. The expression $-ay_t^2/K$ limits population growth (density-dependent factor).

The discrete logistic equation is put in a simpler form prior to analysis. This simpler form is referred to as a *dimensionless form* because the new variable and parameters are dimensionless. Make the change of variable, $x_t = ay_t/(K[1 + a])$. Recall that $a$ is actually $a\Delta t$, and since $a$ has units $1/\text{time}$, $a\Delta t$ is dimensionless. The parameter $K$ has the same units as $y$; therefore, the new variable $x$ is dimensionless. The new variable $x$ satisfies

$$\begin{aligned} x_{t+1} &= \frac{a}{K(1 + a)} y_{t+1} = \frac{a}{K} y_t - \frac{a^2}{K^2(1 + a)} y_t^2 \\ &= \frac{ay_t}{K}(1 - x_t) = (1 + a)x_t(1 - x_t). \end{aligned}$$

In the dimensionless form, $(1 + a)$ appears as a factor. Denote this factor as $r$, $r = 1 + a \equiv 1 + a\Delta t$. Then the following dimensionless difference equation is obtained:

$$x_{t+1} = rx_t(1 - x_t) = f(x_t). \tag{2.15}$$

Equation (2.15) is the normalized or dimensionless *discrete logistic equation*. The simplified equation (2.15) is analyzed in this section. Note that there are two parameters in the difference equation for $y$ but only one in the difference equation for $x$. Thus, it is much easier to analyze the dimensionless discrete logistic equation (2.15). As we shall see, the behavior of this equation is much different from that of the original logistic differential equation (2.14). The behavior of (2.15) depends on the magnitude of the parameter $r = 1 + a \equiv 1 + a\Delta t$. Therefore, it may be more appropriate to refer to equation (2.15) as an approximate logistic difference equation rather than the discrete logistic equation.

The parameters and initial conditions in (2.15) are restricted so that solutions are nonnegative. Assume

$$0 < r < 4, \quad \text{and} \quad 0 \le x_0 < 1.$$

The maximum of $f(x) = rx(1 - x)$ occurs at $x = 1/2$ and $f(1/2) = r/4$. When $r > 4$, then $f(1/2) > 1$ and $f(f(1/2)) < 0$. Thus, if $0 < r < 4$ and $0 \le x_0 < 1$, then $0 \le f(x_t) < 1$. Hence, $f : [0, 1) \to [0, 1)$. Also, note that $f(0) = f(1) = 0$.

The equilibrium solutions of (2.15) satisfy $\bar{x} = r\bar{x}(1 - \bar{x})$ or $\bar{x}(r\bar{x} + (1 - r)) = 0$. Thus, $\bar{x} = 0$ and $\bar{x} = (r - 1)/r$.

If $0 < r \le 1$, then zero is the only nonnegative fixed point. In this case, $0 < f(x) = rx(1 - x) < x$ for $x \in (0, 1)$. By Theorem 2.5, it follows that $\lim_{t\to\infty} f^t(x_0) = 0$, $x_0 \in [0, 1]$.

For $1 < r < 4$, we test the stability of the positive equilibrium $\bar{x} = (r - 1)/r$. Evaluating $f'(x) = r - 2rx$ at $(r - 1)/r$ yields

$$f'((r - 1)/r) = -r + 2.$$

Local asymptotic stability of $\bar{x} = (r - 1)/r$ requires, by Theorem 2.1,

$$|2 - r| < 1 \quad \text{or} \quad 1 < r < 3.$$

In addition, $\bar{x} = (r - 1)/r$ is unstable if $r > 3$.

At $r = 1$ and $r = 3$ there are changes in the stability of the equilibria. The values where these changes occur are referred to as *bifurcation values*. At $r = 1$, the bifurcation is called a *transcritical bifurcation*. When the parameter $r = 1$, the zero equilibrium is no longer locally asymptotically stable; the derivative satisfies

$$f'(0) = r = 1.$$

The change in behavior that occurs when the parameter $r$ reaches 3 is called a *period-doubling bifurcation*. When $r = 3$, the equilibrium $\bar{x} = (r - 1)/r$ is no longer locally asymptotically stable; the derivative satisfies

$$f'(\bar{x}) = -r + 2 = -1.$$

The parameter $r$ is referred to as a *bifurcation parameter*, because as this parameter varies the dynamics of $x_{t+1} = f(x_t)$ change.

When $r$ increases slightly beyond 3, stable oscillations of period 2 appear, and when it increases slightly more, stable oscillations of period 4, period 8, and period $2^n$ appear. This behavior has been called *period doubling*. To examine this solution behavior for $r > 3$, we calculate the 2-cycles. Two-cycles are solutions of the equation

$$f^2(x) = f(f(x)) = f(rx(1 - x)) = r^2x(1 - x)(1 - rx(1 - x)) = x.$$

Simplifying,

$$x[1 - r^2(1 - x)(1 - rx(1 - x))]$$
$$= x[rx - (r - 1)][r^2x^2 - r(r + 1)x + r + 1] = 0$$

The first two factors set equal to zero give the following equilibria: $x = 0$ and $x = (r - 1)/r$. The other two solutions represent the 2-cycle, $\{\bar{x}_1, \bar{x}_2\}$:

$$\bar{x}_{1,2} = \frac{r + 1 \pm \sqrt{(r + 1)^2 - 4(r + 1)}}{2r}$$
$$= \frac{r + 1 \pm \sqrt{(r - 3)(r + 1)}}{2r}. \tag{2.16}$$

The 2-cycle exists if $r > 3$. The graphs of $y = f^2(x)$ and $y = x$ in Figure 2.8 show the four intersection points when $r$ is slightly larger than 3.

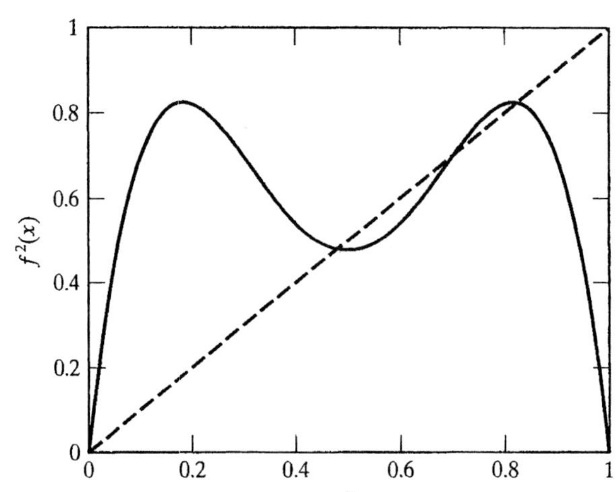


**Figure 2.8** Graphs of $y = x$ and $y = f^2(x)$ for the discrete logistic equation for a fixed value of $r$ satisfying $3 < r < 1 + \sqrt{6}$.

We apply Corollary 2.1 to determine where the 2-cycle is locally asymptotically stable:

$$|f'(\bar{x}_2)f'(\bar{x}_1)| < 1.$$

Using this criterion, $f'(x) = r - 2rx$, and substituting $\bar{x}_1$ and $\bar{x}_2$ from equation (2.16) yields

$$\left|\left(r - (r+1) + \sqrt{(r-3)(r+1)}\right)\left(r - (r+1) - \sqrt{(r-3)(r+1)}\right)\right| < 1.$$

The inequality simplifies to

$$|1 - (r-3)(r+1)| < 1 \quad \text{or} \quad 0 < (r-3)(r+1) < 2.$$

Let $q(r) = (r-3)(r+1) = r^2 - 2r - 3$. Then $q(r) = 0$ for $r = -1, 3$. The minimum value of $q(r)$ occurs at $r = 1$, where $q(1) < 0$. In addition, $q(r) = 2$ for $r = 1 \pm \sqrt{6}$. Thus, the inequalities $0 < q(r) < 2$ are satisfied if

$$3 < r < 1 + \sqrt{6}.$$

The 2-cycle is unstable if $r > 1 + \sqrt{6}$. The behavior changes at $r = 1 + \sqrt{6}$, another bifurcation value. The derivative of $f^2(x)$ at $\bar{x}_i$ satisfies $df^2(\bar{x}_i)/dx = -1$; a period-doubling bifurcation occurs. The graph of $f^2(x)$ when $r = 1 + \sqrt{6}$ has a slope of $-1$ at $\bar{x}_i$, $i = 1, 2$. See Figure 2.8.

It can be shown that when the parameter $r$ satisfies $1 < r < 3$, the positive equilibrium $\bar{x} = (r-1)/r$ is globally asymptotically stable. The expression for $f(f(x))$ and the conditions on $r$ for existence of the 2-cycle imply $f(f(x)) > x$ if $0 < x < \bar{x} = (r-1)/r$ and $1 < r < 3$. Hence, by Theorem 2.9, the equilibrium $\bar{x}$ is globally asymptotically stable for $1 < r < 3$.

The analysis of the discrete logistic equation can be continued to find bifurcation values at $r \approx 3.5441, 3.5644, 3.5688$, and so on. For the parameter region $1 + \sqrt{6} < r < 3.5441$, there is a stable 4-cycle; for $3.5441 < r < 3.5644$, there is a stable 8-cycle; for $3.5644 < r < 3.5688$, there is a stable 16-cycle, and so on. As $r$ gets closer to the critical value $r_c = 3.570$, there are stable cycles of increasing period $2^n$ and the range of their stability gets smaller. For $r > r_c = 3.570$, a cycle of period 3 exists and there exists what is referred to as *aperiodic solutions* or *chaotic solutions*. (See the Appendix for Chapter 2.)

Solutions to the discrete logistic equation for various values of $r$ are graphed in Figure 2.9. When $r = 2.75$, there is a stable equilibrium at $\bar{x} = 0.\overline{63}$; when $r = 3.1$, there is a 2-cycle. When $r = 3.8$, solutions are *chaotic* and when $r = 3.828$, there is a 3-cycle.

## 2.7 Bifurcation Theory

In the next two sections, we discuss various types of changes in behavior that can occur at bifurcation values. The types of bifurcations depend on how the dynamics of $x_{t+1} = f(x_t)$ change as a single parameter $r$ is varied. In addition, we define what is known as a Liapunov exponent, a real number dependent on the initial condition $x_0$ and the parameter $r$. We shall see that a positive Liapunov exponent is associated with chaotic behavior.

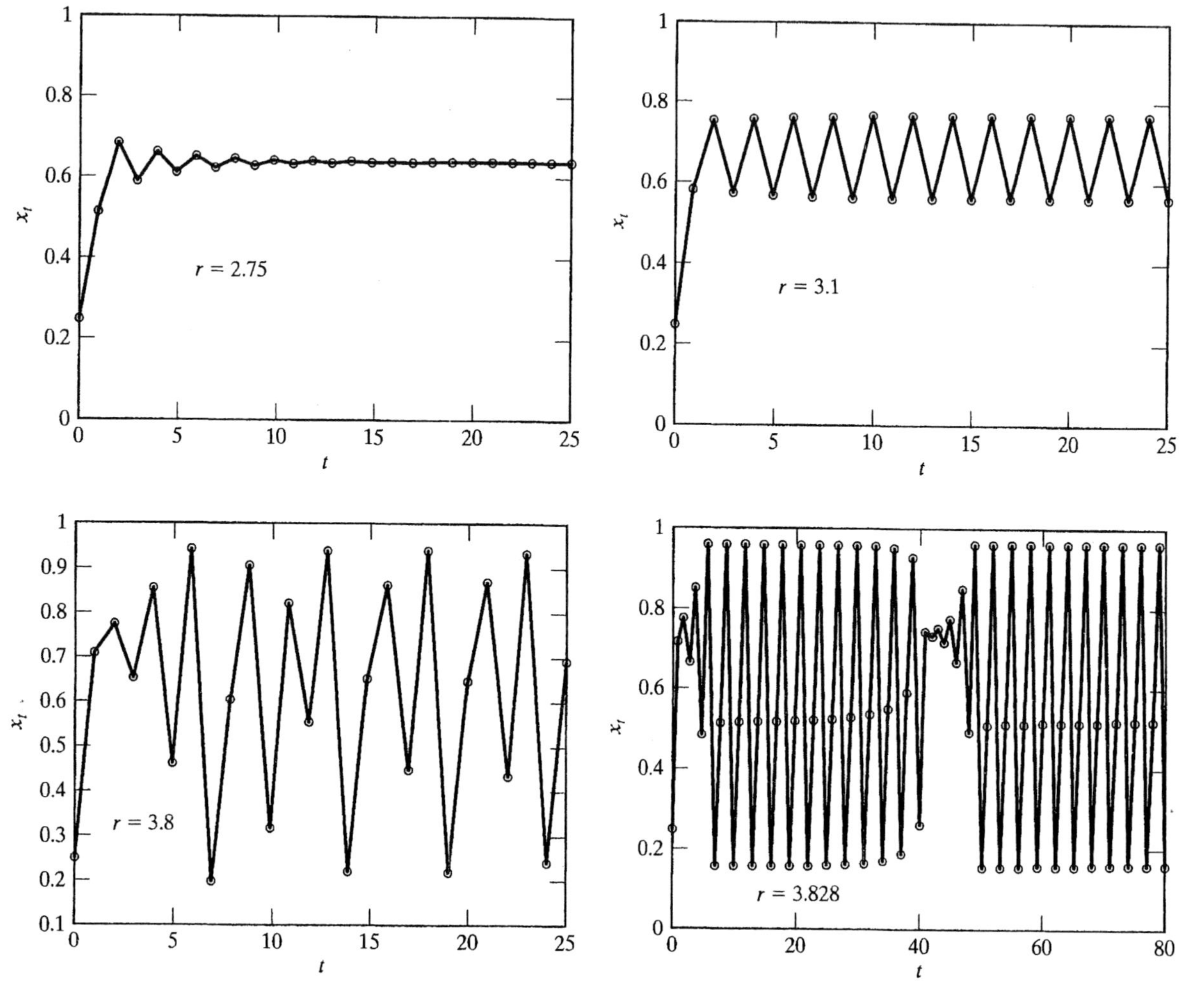


**Figure 2.9** The dynamics of the discrete logistic equation for $r = 2.75$, 3.1, 3.8, and 3.828.

## 2.7.1 Types of Bifurcations

Denote the dependence of the first-order difference equation, $x_{t+1} = f(x_t)$ on the parameter $r$ by setting $f(x_t) \equiv f(x_t, r)$. We study the dynamics of the following first-order nonlinear difference equation:

$$x_{t+1} = f(x_t, r). \tag{2.17}$$

To show the dependence of equilibrium solutions on the parameter $r$, we denote the equilibria of the difference equation as $\bar{x}(r)$. As demonstrated in the case of the discrete logistic equation, the behavior of the difference equation changes as $r$ changes. The values of $r$ where the behavior changes are known as the *bifurcation values* and the points $(r, \bar{x}(r))$ are referred to as the *bifurcation points*. A change in the solution behavior occurs when an equilibrium or $m$-cycle changes stability, that is, when the derivative of $f$ or $f^m$ equals one or negative one. The types of bifurcations that can occur when $f'(\bar{x}(\bar{r})) = \pm 1$ have been classified into four different types for a difference equation of the form (2.17).

I. saddle node (or tangent)
II. pitchfork
III. transcritical
IV. period doubling (flip)

At the first three types of bifurcations (I, II, and III),

$$f'(\bar{x}(\bar{r})) = 1.$$

At a period-doubling bifurcation (IV),

$$f'(\bar{x}(\bar{r})) = -1.$$

Additional criteria on $f$ and its derivatives that define the type of bifurcation can be found, for example, in Elaydi (2000) or Rasband (1990).

The four types of bifurcations can be illustrated in a bifurcation diagram. A *bifurcation diagram* is a graph of the stable and unstable equilibria or cycles as a function of the parameter $r$, $\bar{x}(r)$. The horizontal axis represents values of the bifurcation parameter $r$ and the vertical axis represents the values of the equilibria or cycles. An unstable equilibrium is denoted by a dashed curve and a stable one by a solid curve. The four types of bifurcations are illustrated in Figure 2.10. In all of the graphs, the origin is the bifurcation point, $(0, \bar{x}(0)) = (0, 0)$. At the bifurcation point, the dynamics change.

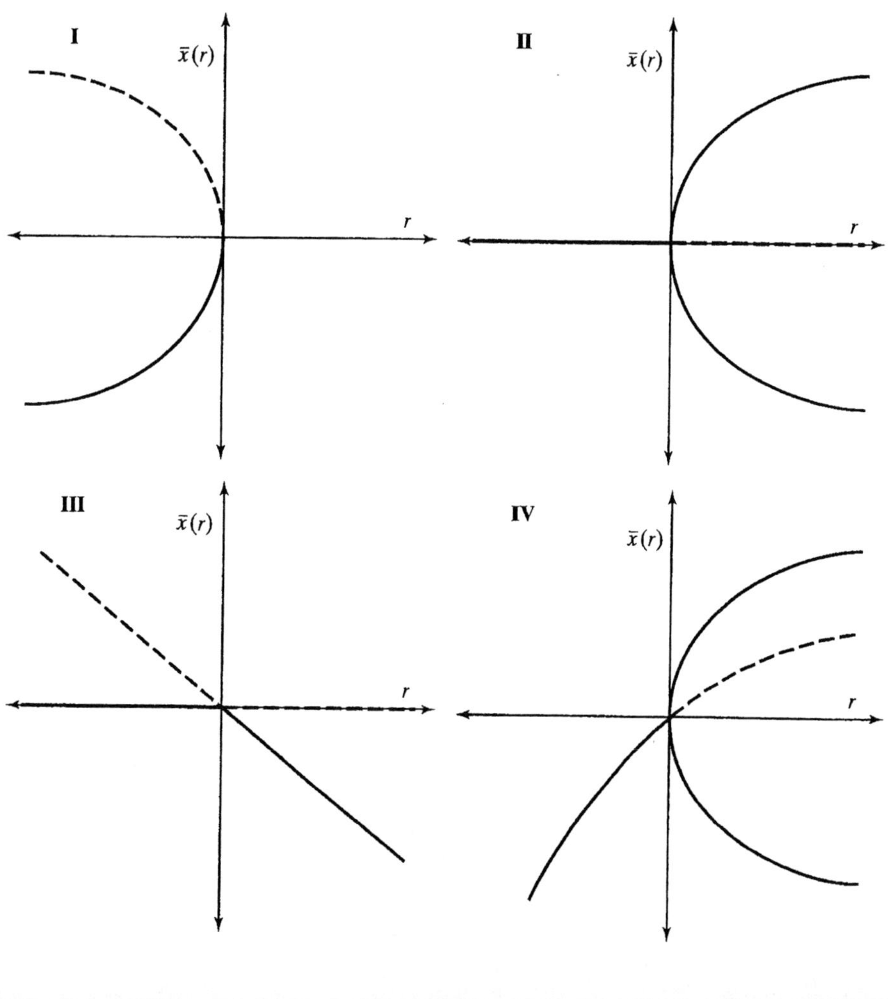


**Figure 2.10** Bifurcation diagrams corresponding to the four bifurcation types I, II, III, and IV. Dashed curves denote unstable equilibria or cycles and solid curves denote stable ones.

We describe briefly the bifurcation dynamics as $r$ passes through the bifurcation point (illustrated in Figure 2.10). In a saddle node bifurcation, as the bifurcation point is passed, two equilibria, one stable and one unstable, disappear. The saddle node bifurcation is sometimes referred to as a *blue sky bifurcation* (Strogatz, 2000) because equilibria appear suddenly, out of the "clear blue sky." In a pitchfork bifurcation, as the bifurcation point is passed three equilibria appear, two stable ones separated by an unstable one. In Figure 2.10 II, the bifurcation is referred to as a *supercritical pitchfork bifurcation*. A system with two stable equilibria is said to have the property of *bistability*. If the stability of the equilibria in Figure 2.10 II is reversed, that is, if there are two unstable equilibria separated by a stable one, then it is referred to as a *subcritical pitchfork bifurcation*. (See Exercise 10.) In a transcritical bifurcation (Figure 2.10 III), two equilibria, one stable and one unstable, exchange stability as the bifurcation point is passed. Finally, in a *period-doubling* or *flip bifurcation* (Figure 2.10 IV), as the bifurcation point is passed, a stable equilibrium becomes unstable and a stable 2-cycle appears. This type of period-doubling bifurcation is referred to as *supercritical*. If the stability is reversed so that the 2-cycle is unstable, it is referred to as a *subcritical bifurcation*.

The following difference equations are canonical examples of the four types of bifurcations that are illustrated in Figure 2.10. (See Exercises 16, 17, and 18.)

I. $x_{t+1} = r + x_t + x_t^2$
II. $x_{t+1} = (r + 1)x_t - x_t^3$
III. $x_{t+1} = (r + 1)x_t + x_t^2$
IV. $x_{t+1} = r - x_t - x_t^2$

**Example 2.13** The difference equation $x_{t+1} = r + x_t + x_t^2 = f(x_t)$ has a saddle node bifurcation at $r = 0$. The equilibria satisfy $\bar{x} = r + \bar{x} + \bar{x}^2$ or $\bar{x} = \pm\sqrt{-r}$ or $r = -\bar{x}^2$. Therefore, real-valued equilibria exist only for $r < 0$. Since $f'(x) = 1 + 2x$, it follows that when $\bar{x}(r) = \sqrt{-r}$, $f'(\sqrt{-r}) = 1 + 2\sqrt{-r} > 1$ so that $\bar{x}(r) = \sqrt{-r}$ is unstable. However, for $\bar{x}(r) = -\sqrt{-r}, f'(-\sqrt{-r}) = 1 - 2\sqrt{-r}$ so that for $-1 < r < 0$, $|f'(-\sqrt{-r})| < 1$; equilibrium $\bar{x}(r) = -\sqrt{-r}$ is locally asymptotically stable. Hence, the bifurcation diagram has the form given in Figure 2.10 I. ∎

The period-doubling or flip bifurcation is unique to difference equations. Although the bifurcation diagrams corresponding to pitchfork and period-doubling bifurcations look alike, they represent different types of bifurcations. In the pitchfork bifurcation, there are two stable fixed points as $r$ passes through the bifurcation value, whereas in the period-doubling bifurcation, there is a stable 2-cycle.

For the discrete logistic equation, it was noted that there exists a transcritical bifurcation and period-doubling bifurcations. When $r = 1$, there is a transcritical bifurcation, and at $r = 3$, there is a period-doubling bifurcation. In the second iteration map of the discrete logistic equation, $x_{t+1} = f^2(x_t, r)$, where $f(x, r) = rx(1 - x)$, there is a period-doubling bifurcation at $r = 1 + \sqrt{6}$. A saddle node or tangent bifurcation occurs in the discrete logistic equation as well. It has been shown that when $r = 1 + 2\sqrt{2}$, there is a saddle node bifurcation (Saha and Strogatz, 1995). When $r = 1 + 2\sqrt{2}$, 3-cycles are created.

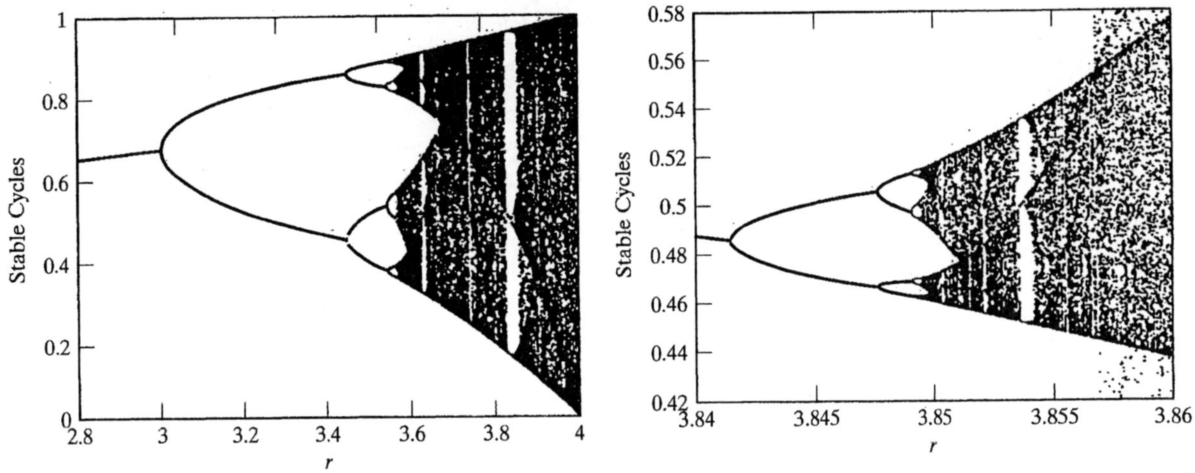


**Figure 2.11** The stable cycles $\bar{x}(r)$ in the discrete logistic equation, $x_{t+1} = rx_t(1 - x_t)$ are illustrated when the bifurcation parameter $r \in [2.8, 4]$ and when $r \in [3.84, 3.86]$.

The periodic behavior of the discrete logistic equation can be seen in the bifurcation diagram of Figure 2.11, which graphs only the stable equilibria or stable cycles as a function of $r$. Two bifurcation diagrams for the discrete logistic equation are given in Figure 2.11 for different ranges in the parameter value $r$. The bifurcations at $r = 3$ and $r = 1 + \sqrt{6}$ are period-doubling bifurcations.

There is an interesting relationship between the range of values for each successive periodic cycle. Let $r_1 < r_2 < \cdots < r_n$ be the values of $r$ at which the successive bifurcations occur as $r$ increases, that is, $r_1 = 1, r_2 = 3, r_3 = 1 + \sqrt{6}$, $r_4 = 3.5441$, and so on. Then it can be shown that the following ratio converges:

$$\lim_{n\to\infty} \frac{r_{n+1} - r_n}{r_{n+2} - r_{n+1}} = \delta \approx 4.66920.$$

For example,

$$\frac{r_4 - r_3}{r_5 - r_4} \approx \frac{3.5441 - 3.4495}{3.5644 - 3.5441} \approx 4.66$$

(Devaney, 1989, 1992; Elaydi, 2000; Rasband, 1990). The number $\delta$ was discovered by Feigenbaum and is appropriately called *Feigenbaum's number* (Feigenbaum, 1978). The popular book by Gleick (1987) discusses Feigenbaum's discovery. Feigenbaum's number is found in many other difference equations of the form $x_{t+1} = f(x_t)$, where $f$ has a form similar to the logistic map and exhibits period-doubling behavior. For example, another equation that exhibits period-doubling is $x_{t+1} = x_t \exp[r(1 - x_t)]$, where the bifurcation parameter is again $r$. This latter equation has frequently been used to model population dynamics. A function of the type $f(x) = x\exp[r(1 - x)]$ is known as a *Ricker curve*.

Three classical papers that discuss some of the mathematical properties of one-dimensional maps, including the equation $x_{t+1} = rx_t(1 - x_t)$, should be mentioned. These papers are by Sharkovskii (1964, in Russian), Li and Yorke (1975), and May (1975). A. N. Sharkovskii was a Russian mathematician who defined an important ordering of the integers. Sharkovskii's ordering lies at the heart of the classical paper by James Yorke and Tien-Yien Li. Discussions about Sharkovskii's ordering can be found in many textbooks (see, e.g., Alligood et al., 1996; Elaydi, 2000; Gulick, 1992). The first instance of the term "chaos" in connection with dynamical systems was in the paper by Li and Yorke (1975). Their paper is entitled "Period three implies chaos." The papers by Robert May (1975, 1976) were the first publications to discuss the period-doubling behavior and chaotic dynamics in discrete models of population growth.

### 2.7.2 Liapunov Exponents

Sensitive dependence on initial conditions is one of the criteria necessary for showing a solution to a difference equation exhibits chaotic behavior. The discrete logistic equation, solutions graphed in Figure 2.9, exhibits chaotic dynamics (sensitive dependence on initial conditions) when $r = 3.8$. The map $x_{t+1} = f(x_t)$ exhibits sensitive dependence on initial conditions when solutions become separated even if they are initially close together. This property is measured by the sign of what is known as a Liapunov exponent. The name "Liapunov" refers to the Russian mathematician A. M. Liapunov (also spelled Lyapunov). One definition of *chaos* for $x_{t+1} = f(x_t)$, where $f : I \to I$, is that solutions are bounded and there exists a positive Liapunov exponent for each point $x_0 \in I$, which is not eventually periodic (Alligood et al., 1996). An *eventually periodic point* is a point $x_0$ having the property that there exist positive integers $p$ and $N$ such that $f^{t+p}(x_0) = f^t(x_0)$ for all $t \geq N$, where $p$ is the period. Other definitions of chaos require additional conditions on $f$ (Elaydi, 2000).

First, we give a precise definition of sensitive dependence on initial conditions. Then we define what is meant by a Liapunov exponent.

**Definition 2.7.** Let $f: I \to I$. Then the map $f$ has *sensitive dependence on initial condition* $x_0$ if there exists $\delta > 0$ such that for any $\epsilon > 0$, there exists $y_0 \in I$ and an integer $k$ such that

$$|x_0 - y_0| < \epsilon \quad \text{and} \quad |f^k(x_0) - f^k(y_0)| > \delta.$$

The function $f$ has sensitive dependence on initial conditions on its domain $I$ if $f$ has sensitive dependence on initial conditions for each $x_0 \in I$ (referred to as simply *sensitive dependence on initial conditions*). Next we define what is meant by a Liapunov exponent at a point $x_0$. The definition of a Liapunov exponent provides a method for calculating sensitive dependence on initial conditions without directly applying the definition.

Denote the Liapunov exponent at a point $x_0$ as $\lambda(x_0)$. The definition of a Liapunov exponent depends on the initial condition $x_0$. The Liapunov exponent measures the exponential stretching of nearby points during successive iterations. For two points $x_0$ and $x_0 + \epsilon$ and $t$ iterates of the function $f$, the Liapunov exponent $\lambda(x_0)$ is approximately

$$\epsilon e^{t\lambda(x_0)} \approx f^t(x_0 + \epsilon) - f^t(x_0),$$

when $\epsilon$ is small and $t$ is large. Dividing by $\epsilon$ and letting $\epsilon \to 0$,

$$e^{t\lambda(x_0)} \approx \lim_{\epsilon \to 0} \frac{f^t(x_0 + \epsilon) - f^t(x_0)}{\epsilon} = \left. \frac{df^t(x)}{dx} \right|_{x=x_0}$$

for large $t$. Then taking the logarithm of both sides, dividing by $t$, and letting $t \to \infty$, the Liapunov exponent of $f$ at $x_0$ is defined as follows:

$$\lambda(x_0) = \lim_{t\to\infty} \frac{1}{t} \ln \left| \frac{df^t(x)}{dx} \bigg|_{x=x_0} \right|.$$

Hence, it is clear from the previous derivation that when $\lambda(x_0) > 0$, points in the domain do not stay close to $x_0$. Because $df^t(x_0)/dx = f'(x_0)f'(x_1)\cdots f'(x_{t-1})$ and

$$\ln |f'(x_0)f'(x_1)\cdots f'(x_{t-1})| = \sum_{k=0}^{t-1} \ln |f'(x_k)|,$$

the Liapunov exponent can be defined more simply in terms of $f'$.

**Definition 2.8.** The *Liapunov exponent* at $x_0$ of the difference equation $x_{t+1} = f(x_t)$ is denoted as $\lambda(x_0)$ and is defined as

$$\lambda(x_0) = \lim_{t\to\infty} \frac{1}{t} \sum_{k=0}^{t-1} \ln |f'(x_k)|.$$

If $\lambda(x_0)$ is independent of $x_0$, then $\lambda(x_0)$ is denoted simply as $\lambda$ and referred to as the *Liapunov exponent* of $f$. There exists sensitive dependence on initial conditions if $\lambda(x_0) > 0$ for all initial conditions $x_0$ in the domain.

**Example 2.14** Suppose $x_{t+1} = (1/2)(x_t + x_t^2) = f(x_t)$. Then the equilibria are $\bar{x} = 0$ and $\bar{x} = 1$. We compute the Liapunov exponents at the equilibria. First $f'(x) = 1/2 + x$ so that $f'(0) = 1/2$ and $f'(1) = 3/2$. If $x_0 = \bar{x}$, then $x_k = \bar{x}$ so that

$$\lambda(0) = \lim_{t\to\infty} \frac{1}{t} \sum_{k=0}^{t-1} \ln |f'(0)| = -\ln(2) < 0$$

and

$$\lambda(1) = \lim_{t\to\infty} \frac{1}{t} \sum_{k=0}^{t-1} \ln |f'(1)| = \ln(3/2) > 0.$$

The stable equilibrium $\bar{x} = 0$ has a negative Liapunov exponent and the unstable equilibrium has a positive Liapunov exponent (sensitive dependence on initial condition $x_0 = 1$). This property of negative and positive Liapunov exponents applies to all stable and unstable equilibria (Exercise 20). ■

As mentioned previously, sensitive dependence on initial conditions (a positive Liapunov exponent) is indicative of chaotic behavor. A good discussion about Liapunov exponents in the context of biological applications is Olsen and Degn (1985). Figure 2.12 graphs the Liapunov exponents of the discrete logistic equation for values of $r$ on the interval $[3, 3.99]$ when $x_0 = 1/2$. It can be seen that the Liapunov exponents are positive for many values of $r > 3.57$. Some calculated values of the Liapunov exponent are given in Table 2.1 for $x_0 = 1/2$. Although the Liapunov exponents are shown only for the initial condition $x_0 = 1/2$, it can be shown that $\lambda(x_0) = \lambda(x_j)$ for $x_j = f^j(x_0)$. (See Exercise 19.)

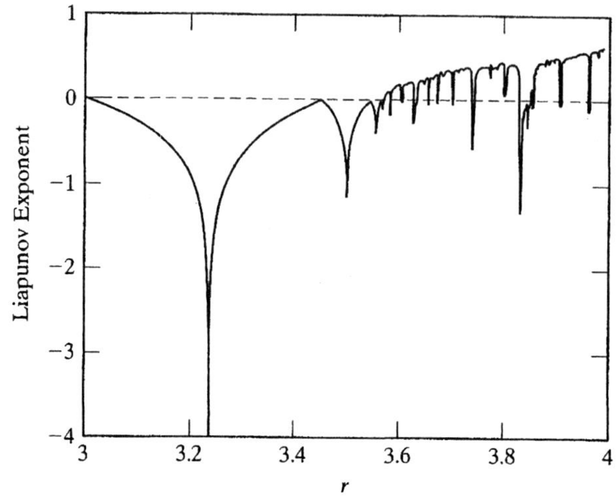


**Figure 2.12** Liapunov exponents $\lambda(1/2)$ for the discrete logistic equation $x_{t+1} = rx_t(1 - x_t)$ for $r \in [3, 3.99]$.

We used the method given in Elaydi (2000) to calculate the Liapunov exponents when $x_0 = 1/2$. Because $\lambda(x_j) = \lambda(x_0)$, we calculate $\lambda(x_{401})$. The iterations for $t = 0, \ldots, 400$ are calculated. Then the next 500 iterations are used to estimate the Liapunov exponent, that is,

$$\lambda(1/2) = \lambda(x_{401}) \approx \frac{1}{500} \sum_{k=401}^{900} \ln |r - 2rx_k|,$$

for $r \in [3, 3.99]$.

Liapunov exponents can be calculated for first-order systems as well. A brief description of how to calculate Liapunov exponents for systems is presented in the Appendix to this chapter.

## 2.8 Stability in First-Order Systems

We provide a simple criterion for verifying local asymptotic stability of an equilibrium solution to a first-order system of difference equations. We derive criteria for the equilibrium $(\bar{x}, \bar{y})$ of the following two-dimensional first-order system to be locally asymptotically stable:

$$\begin{aligned} x_{t+1} &= f(x_t, y_t), \\ y_{t+1} &= g(x_t, y_t). \end{aligned} \tag{2.18}$$

**Table 2.1** Liapunov exponents $\lambda(1/2)$, $x_0 = 1/2$, of the discrete logistic equation $x_{t+1} = rx_t(1 - x_t)$.

| $r$ | $\lambda(1/2)$ | $r$ | $\lambda(1/2)$ |
|---|---|---|---|
| 3.3 | −0.6189 | 3.82 | 0.4148 |
| 3.5 | −0.8726 | 3.83 | −0.3729 |
| 3.6 | 0.1790 | 3.84 | −0.0470 |
| 3.7 | 0.3724 | 3.86 | 0.3693 |
| 3.8 | 0.4118 | 3.9 | 0.4998 |

Before we state the stability result for system (2.18), we linearize the system about the equilibrium.

We assume that $f$ and $g$ have continuous second-order partial derivatives in an open set containing the equilibrium $(\bar{x}, \bar{y})$. Then applying a Taylor series expansion about the equilibrium for the function $f(x, y)$ yields

$$\begin{aligned} f(x, y) &= f(\bar{x}, \bar{y}) + \frac{\partial f(\bar{x}, \bar{y})}{\partial x}(x - \bar{x}) + \frac{\partial f(\bar{x}, \bar{y})}{\partial y}(y - \bar{y}) \\ &\quad + \frac{\partial^2 f(\bar{x}, \bar{y})}{\partial x^2}\frac{(x - \bar{x})^2}{2!} + \frac{\partial^2 f(\bar{x}, \bar{y})}{\partial y^2}\frac{(y - \bar{y})^2}{2!} + \cdots, \end{aligned}$$

where the notation

$$\frac{\partial f(\bar{x}, \bar{y})}{\partial x} = \left.\frac{\partial f(x,y)}{\partial x}\right|_{(x,y)=(\bar{x},\bar{y})}$$

has been used.

Denote $u = x - \bar{x}$ and $v = y - \bar{y}$. Then $f - \bar{x}$ is approximately linear:

$$\begin{aligned} f(x, y) &\approx f(\bar{x}, \bar{y}) + \frac{\partial f(\bar{x}, \bar{y})}{\partial x}(x - \bar{x}) + \frac{\partial f(\bar{x}, \bar{y})}{\partial y}(y - \bar{y}) \\ &= \bar{x} + \frac{\partial f(\bar{x}, \bar{y})}{\partial x}u + \frac{\partial f(\bar{x}, \bar{y})}{\partial y}v. \end{aligned}$$

A similar approximation for $g$ yields

$$g(x, y) \approx \bar{y} + \frac{\partial g(\bar{x}, \bar{y})}{\partial x}u + \frac{\partial g(\bar{x}, \bar{y})}{\partial y}v.$$

The *linearization of the system (2.18) about the equilibrium* $(\bar{x}, \bar{y})$, where $u_t = x_t - \bar{x}$ and $v_t = y_t - \bar{y}$, is given by

$$X_{t+1} = JX_t,$$

where $X_t = (u_t, v_t)^T$ and $J$ is the Jacobian matrix of $(f, g)^T$ evaluated at the equilibrium $(\bar{x}, \bar{y})$,

$$J = \begin{pmatrix} \dfrac{\partial f(\bar{x}, \bar{y})}{\partial x} & \dfrac{\partial f(\bar{x}, \bar{y})}{\partial y} \\ \dfrac{\partial g(\bar{x}, \bar{y})}{\partial x} & \dfrac{\partial g(\bar{x}, \bar{y})}{\partial y} \end{pmatrix}.$$

The criterion for determining local asymptotic stability of the equilibrium only requires that the functions $f$ and $g$ have continuous first-order partial derivatives in $x$ and $y$ (Hale and Koçak, 1991). The eigenvalues of the Jacobian matrix $J$ determine the local stability of the nonlinear system. If the eigenvalues satisfy $|\lambda_i| < 1$ so that the spectral radius $\rho(J) < 1$, then according to Theorem 1.1, $\lim_{t\to\infty} J^t = \mathbf{0}$.

Let the Jacobian matrix $J$ be denoted by $(a_{ij})$, where the elements $a_{ij}$ represent the partial derivatives of $f$ or $g$ evaluated at the equilibrium point. Recall that the eigenvalues of the Jacobian matrix $J = (a_{ij})$ are found by solving $\det(J - \lambda I) = 0$ or

$$\det\begin{pmatrix} a_{11} - \lambda & a_{12} \\ a_{21} & a_{22} - \lambda \end{pmatrix} = 0.$$

In a simplified form, the characteristic equation is

$$\lambda^2 - (a_{11} + a_{22})\lambda + a_{11}a_{22} - a_{21}a_{12} = 0 \quad \text{or} \quad \lambda^2 - \text{Tr}(J)\lambda + \det(J) = 0.$$

There is an easy check for local stability in the two-dimensional case. Local stability depends on the values of the trace and the determinant of the Jacobian matrix. The proof follows Edelstein-Keshet (1988).

**Theorem 2.10** *Assume the functions $f(x, y)$ and $g(x, y)$ have continuous first-order partial derivatives in $x$ and $y$ on some open set in $\mathbf{R}^2$ that contains the point $(\bar{x}, \bar{y})$. Then the equilibrium point $(\bar{x}, \bar{y})$ of the nonlinear system*

$$x_{t+1} = f(x_t, y_t), \quad y_{t+1} = g(x_t, y_t),$$

*is locally asymptotically stable if the eigenvalues of the Jacobian matrix $J$ evaluated at the equilibrium satisfy $|\lambda_i| < 1$ iff*

$$|\text{Tr}(J)| < 1 + \det(J) < 2. \tag{2.19}$$

*The equilibrium is unstable if some $|\lambda_i| > 1$, that is, if any one of three inequalities is satisfied,*

$$\text{Tr}(J) > 1 + \det(J), \quad \text{Tr}(J) < -1 - \det(J), \quad \text{or } \det(J) > 1. \tag{2.20}$$

**Proof** We show that condition (2.19) holds iff the eigenvalues of $J$ have magnitude less than one.

Denote $\tau = \text{Tr}(J)$ and $\delta = \det(J)$. Then the characteristic equation of the Jacobian matrix $J$ is

$$p(\lambda) = \lambda^2 - \tau\lambda + \delta = 0.$$

The eigenvalues are the zeros of $p(\lambda)$, that is,

$$\lambda_{1,2} = \frac{\tau \pm \sqrt{\tau^2 - 4\delta}}{2}.$$

We show that $|\lambda_i| < 1$ iff $|\tau| < 1 + \delta < 2$.

First, we prove that $|\lambda_i| < 1$ implies $|\tau| < 1 + \delta < 2$. Two cases are considered, when the eigenvalues are real and when they are complex conjugates.

**Case I** Suppose $|\lambda_i| < 1$ and the eigenvalues are real. Then $\tau^2 \geq 4\delta$. The parabola $\lambda^2 - \tau\lambda + \delta$ crosses the $\lambda$ axis in two places $\lambda_1$ and $\lambda_2$. See Figure 2.13. Suppose $\lambda_2 \leq \lambda_1$. Since the vertex of the parabola occurs at $\tau/2$, it follows from the assumption $|\lambda_i| < 1$ that

$$-1 < \lambda_2 \leq \tau/2 \leq \lambda_1 < 1.$$

Thus, $|\tau|/2 < 1$. Then $4 > \tau^2 \geq 4\delta$ implies $\delta < 1$. Also, the distances $|\tau/2 - 1| > |\tau/2 - \lambda_1|$ and $|\tau/2 + 1| > |\tau/2 - \lambda_2|$. Since $|\tau/2 - \lambda_i| = \sqrt{(\tau^2 - 4\delta)}/2$, it follows that

$$1 - \frac{|\tau|}{2} > \frac{\sqrt{\tau^2 - 4\delta}}{2}.$$

Squaring both sides, $1 - |\tau| + \tau^2/4 > (\tau^2 - 4\delta)/4$. Then simplifying leads to $1 + \delta > |\tau|$. Thus, $|\lambda_i| < 1$ implies $|\tau| < 1 + \delta < 2$.

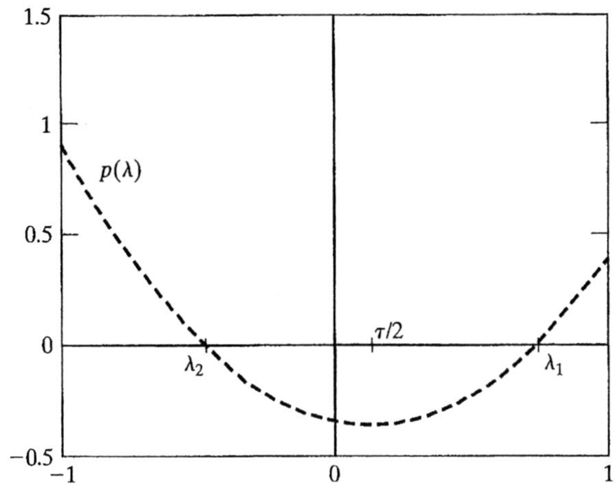


**Figure 2.13** The characteristic polynomial $p(\lambda)$ when the eigenvalues are real and satisfy $|\lambda_i| < 1$.

Case 2 Suppose $|\lambda_i| < 1$ and the eigenvalues are complex conjugates. Then $\tau^2 < 4\delta$. The eigenvalues satisfy

$$\lambda_{1,2} = \frac{\tau}{2} \pm i\frac{\sqrt{4\delta - \tau^2}}{2}$$

and

$$|\lambda_i|^2 = \frac{\tau^2}{4} + \delta - \frac{\tau^2}{4} = \delta.$$

Thus, $0 < \delta < 1$. But $\tau^2 < 4\delta$ implies $|\tau| < 2\sqrt{\delta}$ and $2\sqrt{\delta} < 1 + \delta$. Hence, $|\lambda_i| < 1$ implies $|\tau| < 1 + \delta < 2$.

Next, we show that $|\tau| < 1 + \delta < 2$ implies $|\lambda_i| < 1$. Again, we consider two cases, when the eigenvalues are real and when they are complex.

Case 1 Suppose $|\tau| < 1 + \delta < 2$ and the eigenvalues $\lambda_i$ are real, so that $\tau^2 \geq 4\delta$. Denote $\lambda_1 = \tau/2 + (1/2)\sqrt{\tau^2 - 4\delta}$ and $\lambda_2 = \tau/2 - (1/2)\sqrt{\tau^2 - 4\delta}$ so that $\lambda_2 \leq \lambda_1$. We need to show $\lambda_1 < 1$ and $\lambda_2 > -1$. Rearranging the inequality $|\tau| < 1 + \delta$ leads to $1 - |\tau| > -\delta$. Add $\tau^2/4$ to both sides of the last inequality,

$$(1 - |\tau|/2)^2 > \tau^2/4 - \delta \geq 0,$$

and take the square root of both sides,

$$1 - \frac{|\tau|}{2} > \frac{\sqrt{\tau^2 - 4\delta}}{2}.$$

Thus, $1 > |\tau|/2 + \sqrt{\tau^2 - 4\delta}/2 \geq \lambda_1$ and $-1 < -|\tau|/2 - \sqrt{\tau^2 - 4\delta}/2 \leq \lambda_2$. The first case is proved.

Case 2 Suppose $|\tau| < 1 + \delta < 2$ and the eigenvalues $\lambda_i$ are complex. Thus, $|\lambda_i|^2 = \delta < 1$. All cases have been verified.

The equilibrium $(\bar{x}, \bar{y})$ is unstable if $|\lambda_i| > 1$. This occurs if any of three inequalities in (2.20) are reversed. As can be seen in Figure 2.14, the condition $|\text{Tr}(J)| > 2$ (or $|\tau| > 2$) is implied by the three inequalities in (2.20). □

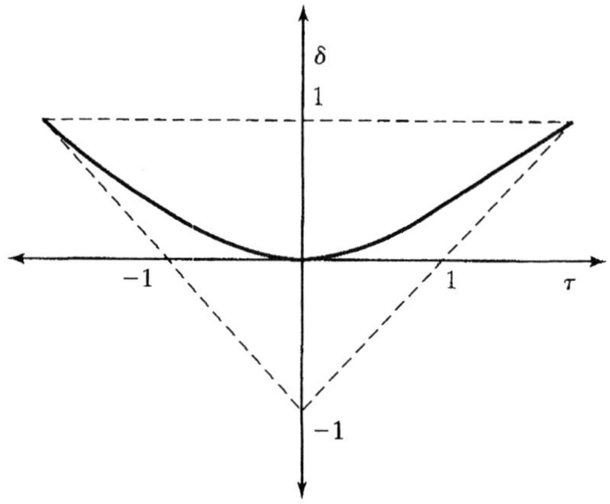


**Figure 2.14** The triangular region inside the dashed lines is the region of local asymptotic stability ($|\lambda_i| < 1$) for the system of difference equations in the $\tau$-$\delta$ plane, $\tau = \text{Tr}(J)$ and $\delta = \det(J)$. The solid curve represents $\tau^2 = 4\delta$. Below this curve, the eigenvalues are real and above this curve they are complex. If the parameters $\tau$ and $\delta$ lie outside of this triangular region, then at least one eigenvalue satisfies $|\lambda_i| > 1$.

When one of the inequalities in the local stability criteria does not hold, at least one eigenvalue either lies on the unit circle in the complex plane or outside of the unit circle, $|\lambda_i| \geq 1$. If $\text{Tr}(J) > 1 + \det(J)$, it can be seen from the previous proof that the eigenvalues are real and $\lambda_1 > 1$. If $\text{Tr}(J) < -1-\det(J)$, then the eigenvalues are real and $\lambda_2 < -1$. Finally, if the eigenvalues are complex conjugates and $\det(J) > 1$, then $|\lambda_i| > 1$ for $i = 1, 2$. The eigenvalues are complex conjugates (with nonzero imaginary part) only if $\det(J) > [\text{Tr}(J)]^2/4$. The region of local asymptotic stability is graphed in Figure 2.14 in terms of $\tau = \text{Tr}(J)$ and $\delta = \det(J)$. In summary, crossing one of three boundaries from inside to outside of the parameter region in Figure 2.14 results in either a positive eigenvalue greater than one (boundary $\tau = 1 + \delta$) or a negative eigenvalue less than one (boundary $\tau = -1 - \delta$) or a complex conjugate pair satisfying $|\lambda_i| > 1$ (boundary $\delta = 1$).

**Example 2.15** Suppose a predator-prey system satisfies the following system of difference equations. Variable $x$ represents the prey and variable $y$ represents the predator,

$$x_{t+1} = x_t(a - x_t - y_t), \quad a > 0,$$
$$y_{t+1} = y_t(b + x_t), \quad 0 < b < 1.$$

The prey population grows logistically in the absence of the predator, $x_{t+1} = x_t(a - x_t)$. When the predator is present, the prey population size decreases. The predator population decreases exponentially in the absence of the prey, $y_{t+1} = by_t$, but in the presence of the prey, the predator population increases. Note that it is possible for solutions for the prey population to become negative if $x_t$ or $y_t$ are large. Nonnegative solutions are very important for a realistic model of prey and predator. Here, we use this simple model only to illustrate the local stability criteria. There are three equilibria: (0, 0), $(a - 1, 0)$, and $(\bar{x}, \bar{y}) = (1 - b, a + b - 2)$. The Jacobian matrix for this system has the form

$$J = \begin{pmatrix} a - 2x - y & -x \\ y & b + x \end{pmatrix}.$$

The Jacobian matrices, evaluated at each of the equilibria, are

$$J(0,0) = \begin{pmatrix} a & 0 \\ 0 & b \end{pmatrix}, \quad J(a-1,0) = \begin{pmatrix} 2 - a & 1 - a \\ 0 & a + b - 1 \end{pmatrix}.$$

and

$$J(\bar{x}, \bar{y}) = \begin{pmatrix} b & -\bar{x} \\ \bar{y} & 1 \end{pmatrix}.$$

Equilibrium (0, 0) is locally asymptotically stable if $a < 1$, both the prey and predator become extinct. When $a < 1$, even in the absence of the predator, it is impossible for the prey to persist (see the discrete logistic equation). Equilibrium $(a - 1, 0)$ is nonnegative and locally asymptotically stable if $1 < a$, $|2 - a| < 1$, and $|a + b - 1| < 1$. These inequalities simplify to $1 < a < 2 - b$. When these latter conditions hold, the prey survives but not the predator. Finally, equilibrium $(\bar{x}, \bar{y})$ is positive and locally asymptotically stable if $\bar{x}, \bar{y} > 0$ and

$$|1 + b| < 1 + b + \bar{x}\bar{y} < 2.$$

The first inequality is automatically satisfied if $\bar{x}, \bar{y} > 0$. The second inequality is satisfied if $\bar{x}, \bar{y} > 0$ and $a + b < 3$. The equilibrium $(\bar{x}, \bar{y})$ is positive if $b < 1$ and $a + b > 2$. Combining these inequalities leads to the conditions needed for local asymptotic stability of $(\bar{x}, \bar{y})$,

$$2 < a + b < 3;$$

both prey and predator survive. ■

## 2.9 Jury Conditions

Local stability criteria for first-order systems or for higher order difference equations depend on the behavior of the linearized system. Consider a first-order system consisting of $n$ equations, $X(t) = (x_1(t), x_2(t), \ldots, x_n(t))^T$,

$$X(t + 1) = F(X(t)), \tag{2.21}$$

where $F = (f_1, f_2, \ldots, f_n)^T$ and $f_i \equiv f_i(x_1, x_2, \ldots, x_n)$, $i = 1, 2, \ldots, n$. Suppose system (2.21) has an equilibrium at $\bar{X}$. Then if $U_t = X_t - \bar{X}$, linearization of system (2.21) about $\bar{X}$ leads to the system

$$U_{t+1} = JU_t,$$

where $J$ is the Jacobian matrix evaluated at $\bar{X}$,

$$J = \begin{pmatrix} \dfrac{\partial f_1(\bar{X})}{\partial x_1} & \dfrac{\partial f_1(\bar{X})}{\partial x_2} & \cdots & \dfrac{\partial f_1(\bar{X})}{\partial x_n} \\ \dfrac{\partial f_2(\bar{X})}{\partial x_1} & \dfrac{\partial f_2(\bar{X})}{\partial x_2} & \cdots & \dfrac{\partial f_2(\bar{X})}{\partial x_n} \\ \vdots & \vdots & \cdots & \vdots \\ \dfrac{\partial f_n(\bar{X})}{\partial x_1} & \dfrac{\partial f_n(\bar{X})}{\partial x_2} & \cdots & \dfrac{\partial f_n(\bar{X})}{\partial x_n} \end{pmatrix}.$$

Local asymptotic stability of $\bar{X}$ depends on the eigenvalues of the Jacobian matrix, which in turn depend on the existence of the partial derivatives in a region containing $\bar{X}$. Therefore, for local asymptotic stability of $\bar{X}$, we require that the partial derivatives of $f_i$ be continuous in an open set containing $\bar{X}$.

The eigenvalues of the Jacobian matrix are the solutions of the characteristic equation

$$\det(J - \lambda I) = 0.$$

The eigenvalues are the zeros of the following $n$th-degree characteristic polynomial:

$$p(\lambda) = \lambda^n + a_1\lambda^{n-1} + a_2\lambda^{n-2} + \cdots + a_n. \tag{2.22}$$

The coefficients in (2.22) are real because the functions $f_i$ are real valued. For $n = 2, a_1 = -\text{Tr}(J)$ and $a_2 = \det(J)$.

The conditions that must be satisfied for local asymptotic stability are known as the *Jury conditions* or the *Jury test*. These conditions ensure that the roots of the characteristic polynomial (2.22) satisfy $|\lambda_i| < 1$. These conditions are also referred to as the *Schur-Cohn criteria* (Elaydi, 1999). Schur and Cohn derived necessary and sufficient conditions so that the solutions of (2.22) satisfy $|\lambda_i| < 1$ when the coefficients $a_j$ are complex. Jury (1964, 1971, 1974) simplified some of the conditions derived by Schur and Cohn when the coefficients $a_j$ are real. The conditions are complicated for large $n$. However, the Jury conditions for a third-degree characteristic polynomial,

$$p(\lambda) = \lambda^3 + a_1\lambda^2 + a_2\lambda + a_3, \tag{2.23}$$

are easy to state.

**Theorem 2.11** **(Jury conditions, Schur-Cohn criteria, $n = 3$).** *Suppose the characteristic polynomial $p(\lambda)$ is given by (2.23). The solutions $\lambda_i$, $i = 1, 2, 3$, of $p(\lambda) = 0$ satisfy $|\lambda_i| < 1$ iff the following three conditions hold:*

*(i)* $p(1) = 1 + a_1 + a_2 + a_3 > 0,$
*(ii)* $(-1)^3 p(-1) = 1 - a_1 + a_2 - a_3 > 0,$
*(iii)* $1 - (a_3)^2 > |a_2 - a_3 a_1|.$ □

The Jury conditions (or Schur-Cohn criteria) for the general case are stated in the Appendix for Chapter 2 (see also Edelstein-Keshet, 1988; Elaydi, 1999; Jury, 1964; Murray, 1993, 2002). The Jury conditions for a second-degree polynomial, $p(\lambda) = \lambda^2 + a_1\lambda + a_2$, are the conditions in Theorem 2.10, that is,

$$|a_1| < 1 + a_2 < 2.$$

Some necessary (but not sufficient) conditions for $|\lambda_i| < 1$ are easy to check for a general system with $n$ equations.

**Theorem 2.12** *If the solutions $\lambda_i$, $i = 1, 2, \ldots, n$, of (2.22), $p(\lambda) = 0$, satisfy $|\lambda_i| < 1$, then*

*(a)* $p(1) = 1 + a_1 + a_2 + \cdots + a_n > 0,$
*(b)* $(-1)^n p(-1) = 1 - a_1 + a_2 - \cdots + (-1)^n a_n > 0$ *(alternate in sign),*
*(c)* $|a_n| < 1.$ □

Condition (c) in Theorem 2.12 follows from the fact that $a_n$ is the product of the eigenvalues, $a_n = \lambda_1 \cdots \lambda_n$. If one of the preceding conditions is *not* satisfied, then there exists a root $|\lambda_i| \geq 1$. However, if all of the preceding conditions are satisfied, then the other conditions in the Jury test must be verified to determine local asymptotic stability.

**Definition 2.9.** Let $\bar{X}$ denote an equilibrium of the first-order system $X_{t+1} = F(X_t)$ and $J$ denote the Jacobian matrix evaluated at the equilibrium $\bar{X}$. The equilibrium $\bar{X}$ is referred to as a *hyperbolic equilibrium* if none of the eigenvalues of the Jacobian matrix has magnitude equal to one, $|\lambda_i| \neq 1$. Otherwise, it is referred to as a *nonhyperbolic equilibrium*.

In the case of a nonhyperbolic equilibrium, the local stability criteria are indeterminate.

**Example 2.16** Suppose the characteristic polynomial has the form

$$p(\lambda) = \lambda^3 - 0.25\lambda^2 - 0.25\lambda + a.$$

A check of the Jury conditions shows

(i) $p(1) = 1 - 0.25 - 0.25 + a > 0$ if $a > -0.5$.
(ii) $(-1)^3 p(-1) = 1 + 0.25 - 0.25 - a > 0$ if $a < 1$.
(iii) $1 - a^2 > |0.25 - 0.25a|$ if $(1 - a)(1 + a) > 0.25|1 - a|$.

The combined inequalities in (i)–(iii) yield a bound on $a$. If $-0.5 < a < 1$, then $p(\lambda)$ will have roots satisfying $|\lambda_i| < 1$. ■

The Jury conditions are used to study stability in the epidemic model introduced in the next section. The stability criteria in Theorem 2.10 are applied.

## 2.10 An Example: Epidemic Model

The following system represents a simple discrete-time epidemic model. The population is subdivided into three groups: susceptible, infected, and immune (or removed) groups. The variable $S_t$ is the number of susceptible individuals at time $t$, $I_t$ is the number of infected individuals at time $t$, and $R_t$ is the number of immune or recovered individuals at time $t$. The total population size is assumed to be constant, $N$. The model is referred to as an SIR epidemic model. The relationship among the various states, $S$, $I$, and $R$, is illustrated in the compartmental diagram in Figure 2.15. The model is applicable to infectious diseases such as measles, chickenpox, or mumps, where infection confers immunity. Early contributions to epidemic modeling were made by Hamer (1906), Ross (1911), and Kermack and McKendrick (1927). Sir Ronald Ross received the Nobel Prize for Medicine in 1902 for his work on malaria.

The parameter $\beta$ is the contact number, the average number of successful contacts (resulting in infection) made by one infected (and infectious) individual during the time $t$ to $t + 1$. Thus, $\beta S/N$ is the proportion of contacts by one infected individual that result in an infection of a susceptible individual and $\beta SI/N$ is the total number of contacts by the infected class that result in infection. The probability of a birth equals the probability of a death, which is given by the parameter $b$. Also, individuals are born susceptible; there is no vertical transmission (from mother to offspring) of the disease. The parameter $\gamma$ is the probability of recovery. The ratio $1/\gamma$ is the average length of the infectious

**Figure 2.15** A compartmental diagram for the SIR epidemic model.

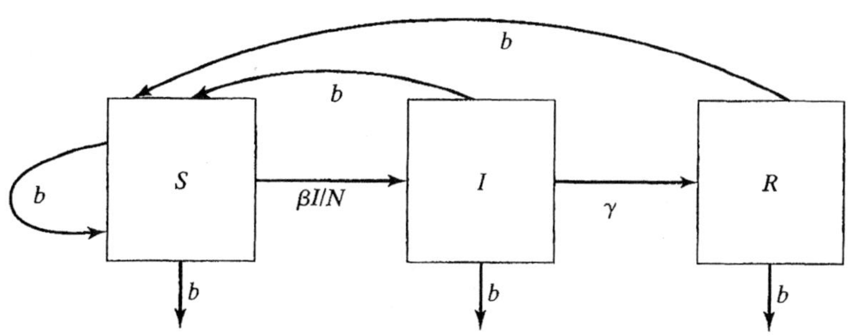

period when there are no deaths (we are assuming an infected individual is also infectious, i.e., can transmit the disease). The length of the infectious period may be shortened because of death. Therefore, the ratio $1/(b + \gamma)$ is the average length of the infectious period when deaths are included. Based on these assumptions, the SIR epidemic model has the form

$$
\begin{aligned}
S_{t+1} &= S_t - \frac{\beta}{N} I_t S_t + b(I_t + R_t),\\
I_{t+1} &= I_t(1 - \gamma - b) + \frac{\beta}{N} I_t S_t,\\
R_{t+1} &= R_t(1 - b) + \gamma I_t.
\end{aligned}
$$

The initial conditions are $S_0 + I_0 + R_0 = N$, where $S_0, I_0, R_0 > 0$. Restrictions are imposed on the parameters to ensure that solutions are nonnegative. Sufficient conditions for nonnegative solutions are $b, \gamma > 0$,

$$0 < b + \gamma < 1, \quad \text{and} \quad 0 < \beta < 1.$$

These conditions put a restriction on the length of the time interval, $t$ to $t + 1$. The length must be short enough (hours or days) so that the conditions on the parameters are satisfied. If we add the three equations, we obtain $S_{t+1} + I_{t+1} + R_{t+1} = S_t + I_t + R_t$. Because $S_0 + I_0 + R_0 = N$, it follows by induction that the total population size equals $N$ for all time, $S_t + I_t + R_t = N$ (an assumption we made earlier). Now, to show that solutions are nonnegative, note that $S_0, I_0, R_0 > 0$. The assumptions on the parameters imply $\beta I_0/N \leq I_0/N \leq 1$ so that $S_1, I_1, R_1 > 0$. By induction, it follows that solutions are nonnegative for all time.

Because of the relationship $S_t + I_t + R_t = N$, one of the variables is extraneous and can be eliminated. For example, the variable $R_t$ can be replaced by $N - I_t - S_t$. Then the three difference equations simplify to two equations in $S$ and $I$,

$$
\begin{aligned}
S_{t+1} &= S_t - \frac{\beta}{N} I_t S_t + b(N - S_t) = f(S_t, I_t),\\
I_{t+1} &= I_t(1 - \gamma - b) + \frac{\beta}{N} I_t S_t = g(S_t, I_t).
\end{aligned}
$$

There are two equilibria of the preceding equations in $S$ and $I$. These equilibria can be found by solving the following equations for $\bar{S}$ and $\bar{I}$:

$$
\begin{aligned}
\bar{S} &= \bar{S} - \frac{\beta}{N} \bar{I}\bar{S} + b(N - \bar{S})\\
\bar{I} &= \bar{I}(1 - \gamma - b) + \frac{\beta}{N} \bar{I}\bar{S}.
\end{aligned}
$$

Simplifying and solving the equations simultaneously, the following two equilibrium solutions are obtained:

$$(1)\ \bar{S} = N, \bar{I} = 0 \quad \text{and} \quad (2)\ \bar{S} = \frac{N(\gamma + b)}{\beta}, \quad \bar{I} = bN\left[\frac{\beta - (\gamma + b)}{\beta(\gamma + b)}\right].$$

The first equilibrium is referred to as the *disease-free equilibrium* and the second one is referred to as the *endemic equilibrium*. Note that the second one is positive iff $\beta > \gamma + b$.

The Jacobian matrix for the $S$-$I$ system has the following form:

$$J(S,I) = \begin{pmatrix} 1 - b - \dfrac{\beta}{N}I & -\dfrac{\beta}{N}S \\ \dfrac{\beta}{N}I & 1 - \gamma - b + \dfrac{\beta}{N}S \end{pmatrix}.$$

At the disease-free equilibrium, the Jacobian matrix satisfies

$$J(N,0) = \begin{pmatrix} 1 - b & -\beta \\ 0 & 1 - \gamma - b + \beta \end{pmatrix}.$$

Because the Jacobian matrix is upper triangular, the eigenvalues are along the diagonal,

$$\lambda_1 = 1 - b \quad \text{and} \quad \lambda_2 = 1 - \gamma - b + \beta.$$

The disease-free equilibrium is locally asymptotically stable if $|\lambda_{1,2}| < 1$. From the assumptions on the parameters it follows that $0 < \lambda_1 < 1$. The second eigenvalue satisfies $0 < \lambda_2 < 1$ if $\beta/(\gamma + b) < 1$. This latter ratio defines a threshold parameter for the *SIR* epidemic model denoted as $\mathcal{R}_0$ and is referred to as the *basic reproduction number* or the *basic reproduction ratio*,

$$\mathcal{R}_0 = \frac{\beta}{\gamma + b}.$$

This ratio also has a biological interpretation. The value of $\mathcal{R}_0$ is the number of secondary infections caused by one infectious individual during the individual's infectious period (Anderson and May, 1991). If $\mathcal{R}_0 < 1$, then there exists only one equilibrium, the disease-free equilibrium, and it is locally asymptotically stable.

Estimates of $\mathcal{R}_0$ have been obtained for diseases such as measles, chickenpox, and smallpox (May, 1983). Table 2.2 provides some estimates for $\mathcal{R}_0$ for

**Table 2.2** Basic reproduction number as reported by May (1983).

| Infection | Location and Time | $\mathcal{R}_0$ |
|---|---|---|
| Smallpox | Developing Countries Before Global Campaign | 3–5 |
| Measles | England & Wales, 1956–1968 | 13 |
| | U.S., various locations, 1910–1930 | 12–13 |
| Whooping Cough | England & Wales, 1942–1950 | 17 |
| | Maryland, U.S., 1908–1917 | 13 |
| German Measles | England & Wales, 1979 | 6 |
| | West Germany, 1972 | |
| Chickenpox | U.S. various locations, 1913–1921, 1943 | 9–10 |
| Diphtheria | U.S. various locations, 1910–1947 | 4–6 |
| Scarlet Fever | U.S. various locations, 1910–1920 | 5–7 |
| Mumps | U.S. various locations, 1912–1916, 1943 | 4–7 |
| Poliomyelitis | Holland, 1960; U.S., 1955 | 6 |

these infectious diseases and for others which resulted in large-scale epidemics. Note that the estimates depend on the location of the epidemic and the time at which the epidemic occurred. Contact rates and level of immunity in a population vary with geographic location and time period of the epidemic. The estimate for $\mathcal{R}_0$ for smallpox ($\approx 3-5$) was prior to its eradication. (The World Health Assembly declared the world free of smallpox in 1980.) Eradication of smallpox was achieved through an extensive vaccination campaign. (The value of $\mathcal{R}_0$ was reduced to a value less than one.) The magnitude of the basic reproduction number gives an indication of the difficulty in controlling an epidemic or eradicating the disease (the larger the value of $\mathcal{R}_0$, the harder it is to control). A worldwide vaccination campaign is currently underway to eradicate poliomyelitis.

Vaccination can be included in an epidemic model by assuming a proportion of susceptible individuals, $pS_t$, are vaccinated during each time interval. These individuals become immune and are added to the compartment for immune individuals, $R_{t+1}$. The effects of vaccination are investigated in a measles epidemic model, which is formulated and analyzed in Chapter 3.

To continue the analysis of the SIR epidemic model, the local asymptotic stability is investigated for the endemic equilibrium. It is assumed that the endemic equilibrium is positive, $\mathcal{R}_0 > 1$. The Jacobian matrix evaluated at the endemic equilibrium has the form

$$J(\bar{S},\bar{I}) = \begin{pmatrix} 1 - b\mathcal{R}_0 & -\dfrac{\beta}{\mathcal{R}_0} \\ b(\mathcal{R}_0 - 1) & 1 \end{pmatrix}.$$

A special case is considered. Assume

$$2 - b\mathcal{R}_0 \geq 0.$$

This assumption implies the trace of $J(\bar{S},\bar{I})$ is nonnegative. Theorem 2.10 can be applied to determine whether the equilibrium is locally asymptotically stable,

$$\text{Tr}(J) = 2 - b\mathcal{R}_0 \geq 0 \quad \text{and} \quad \det(J) = 1 - b\mathcal{R}_0 + \beta b(1 - \frac{1}{\mathcal{R}_0}).$$

The conditions for local asymptotic stability given in Theorem 2.10 are

$$2 - b\mathcal{R}_0 < 2 - b\mathcal{R}_0 + \beta b\left(1 - \frac{1}{\mathcal{R}_0}\right) < 2,$$

which simplify to

$$0 < \beta b\left(1 - \frac{1}{\mathcal{R}_0}\right) < b\mathcal{R}_0.$$

The preceding inequalities are satisfied because $\mathcal{R}_0 > 1 > \beta(1 - 1/\mathcal{R}_0)$. Thus, the endemic equilibrium exists and is locally asymptotically stable if

$$2 - b\mathcal{R}_0 \geq 0 \quad \text{and} \quad \mathcal{R}_0 > 1.$$

Equivalently, when $1 < \mathcal{R}_0 \leq 2/b$, the endemic equilibrium is locally asymptotically stable. When $\mathcal{R}_0 > 1$ and the above conditions are not satisfied, the endemic equilibrium may be stable or unstable; also, periodic or chaotic solutions may exist. In the latter case, the behavior of the epidemic cannot be predicted. Two examples illustrating the solution to the SIR epidemic model

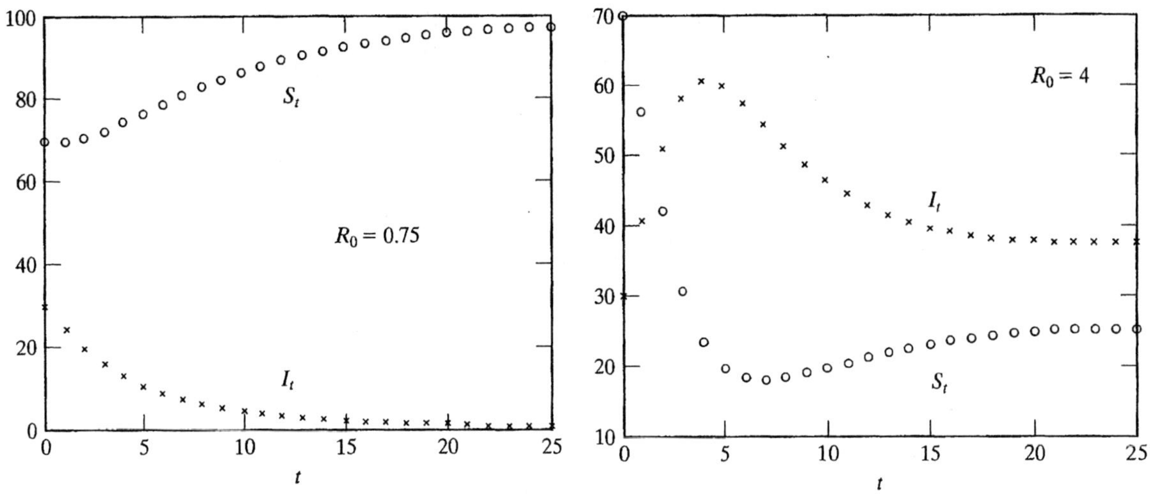


**Figure 2.16** Solutions to the SIR epidemic model when $\mathcal{R}_0 = 0.75$ ($\beta = 0.3, b = 0.2 = \gamma$) and $\mathcal{R}_0 = 4$ ($\beta = 0.8, b = 0.1 = \gamma$). In both cases, $S_0 = 70$, $I_0 = 30$, and $\mathcal{R}_0 = 0$.

are given in Figure 2.16 for $\mathcal{R}_0 < 1$ and for $\mathcal{R}_0 > 1$. A MATLAB program for the simulation with $\mathcal{R}_0 < 1$, illustrated in Figure 2.16, is given in the Appendix for Chapter 2.

Many other types of difference equation models for epidemics can be formulated. For example, there are models with no recovery or temporary recovery (SIS epidemic model), a growing population, age classes as in the Leslie model, and several host or pathogen populations (see, e.g., Allen, 1994; Allen and Burgin, 2000; Allen et al., 2004; Castillo-Chávez and Yakubu, 2001; Martin et al., 1996). An in-depth analysis of a gonorrhea SIS epidemic model is discussed in the monograph by Hethcote and Yorke (1984).

## 2.11 Delay Difference Equations

Many of the models considered thus far have been first-order equations or systems, where the state in generation $t + 1$ depends on the previous generation $t$. As a first approximation this assumption may be adequate. However, populations where there is delayed sexual maturation before reproduction may be affected by many previous generations. A simple model, where the growth rate of a single population is affected by two previous generations, takes the form of the following difference equation:

$$x_{t+1} = f(x_t, x_{t-T}). \tag{2.24}$$

The effect of generation $t - T$ on the current generation $t + 1$ may be considered a "delayed" effect. Equation (2.24) has order $T + 1$. The current generation is affected by generations from 1 and $T + 1$ generations ago. For example, the discrete logistic model of Section 2.6 with a delay in the density-dependent effects has the form

$$x_{t+1} = rx_t(1 - x_{t-T}). \tag{2.25}$$

Here, we consider a general model such as (2.25), but where the delay is 2 generations ($T = 1$). We examine how the delay affects the stability of the population, that is, we compare a model with $T = 0$ to a model with $T = 1$. Since a higher-order difference equation can be converted to a first-order system, we can analyze the local stability behavior using techniques from Sections 2.8 and 2.9. Thus, no new techniques will be introduced here but the stability behavior will be interpreted in terms of the effect of delays.

Consider the following two models:

$$(1)\ x_{t+1} = g(x_t)x_t \quad \text{and} \quad (2)\ x_{t+1} = g(x_{t-1})x_t.$$

In the first model, the per capita rate of growth $g$ depends only on the population one generation in the past but, in the second model, it depends on two generations in the past; there is a delay of $T = 1$. It is assumed that $g$ is continuously differentiable.

In the first model,

$$x_{t+1} = g(x_t)x_t = f(x_t),$$

there exists at least one equilibrium, the zero equilibrium. Other equilibria are solutions to $g(x) = 1$. We assume that there is a unique positive solution to $g(x) = 1$ and denote it as $\bar{x}$. The criterion for local asymptotic stability of the positive equilibrium $\bar{x}$ depends on $f'(\bar{x})$ (Theorem 2.1). Since

$$f'(x)|_{x=\bar{x}} = [g'(x)x + g(x)]|_{x=\bar{x}} = g'(\bar{x})\bar{x} + 1,$$

the equilibrium $\bar{x}$ is locally asymptotically stable if $-1 < 1 + g'(\bar{x})\bar{x} < 1$ or

$$0 < -g'(\bar{x})\bar{x} < 2.$$

In the second model, $x_{t+1} = g(x_{t-1})x_t$, there are the same two equilibria as in the first model, zero and $\bar{x} > 0$ ($g(\bar{x}) = 1$). To determine the local asymptotic stability of $\bar{x}$ we need to determine the linear approximation. We can either linearize the higher-order equation directly or convert the equation to a first-order system and find the Jacobian matrix. We illustrate both linearization techniques.

In the first method, the second-order equation is linearized directly. Let $u_t = x_t - \bar{x}$. Then

$$\begin{aligned} u_{t+1} &= x_{t+1} - \bar{x} \\ &= g(x_{t-1})x_t - \bar{x} \\ &\approx [g(\bar{x}) + g'(\bar{x})(x_{t-1} - \bar{x})]x_t - \bar{x} \\ &= [1 + g'(\bar{x})u_{t-1}]x_t - \bar{x} \\ &= x_t - \bar{x} + g'(\bar{x})u_{t-1}(u_t + \bar{x}) \\ &\approx u_t + g'(\bar{x})\bar{x}u_{t-1}. \end{aligned}$$

The *linearized equation about* $\bar{x}$ is

$$u_{t+1} = u_t + g'(\bar{x})\bar{x}u_{t-1}.$$

The characteristic equation for this linearized equation is given by

$$\lambda^2 - \lambda - g'(\bar{x})\bar{x} = 0. \tag{2.26}$$

In the second method, the second-order equation is converted to a first-order system. Let $y_{t+1} = x_t$. Then the system can be written as

$$\begin{aligned} x_{t+1} &= g(y_t)x_t = F(x_t, y_t) \\ y_{t+1} &= x_t = G(x_t, y_t). \end{aligned}$$

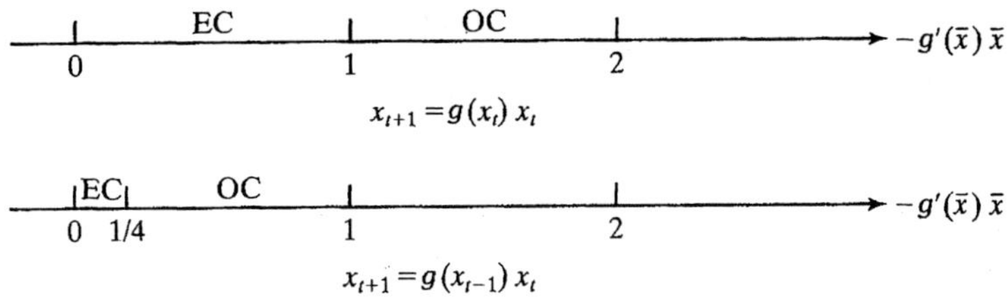


**Figure 2.17** The region of local asymptotic stability for the difference equations, $x_{t+1} = g(x_t)x_t$ and $x_{t+1} = g(x_{t-1})x_t$; EC = exponential convergence and OC = oscillating convergence.

The Jacobian matrix is

$$J(x, y) = \begin{pmatrix} g(y) & g'(y)x \\ 1 & 0 \end{pmatrix}.$$

Evaluating at the equilibrium, $(\bar{x}, \bar{y})$, where $\bar{y} = \bar{x}$ and $g(\bar{y}) = 1$, the Jacobian matrix is

$$J(\bar{x}, \bar{y}\,) = \begin{pmatrix} 1 & g'(\bar{x})\bar{x} \\ 1 & 0 \end{pmatrix}.$$

It is easy to see that the characteristic equation, $\det(J - \lambda I) = 0$, is the same as (2.26), where $\mathrm{Tr}(J) = \tau = 1$ and $\det(J) = \gamma = -g'(\bar{x})\bar{x}$.

Theorem 2.10 can be applied to the characteristic equation (2.26) to determine local asymptotic stability of the equilibrium. The eigenvalues have modulus less than one if $1 < 1 - g'(\bar{x})\bar{x} < 2$ or, equivalently,

$$0 < -\, g'(\bar{x})\bar{x} < 1.$$

The stability regions for the two models are compared in Figure 2.17.

Calculation of the regions for exponential and oscillating convergence is left as an exercise. The region of asymptotic stability for the second model is much smaller than for the first model. Time lags generally have a destabilizing effect and the longer the delay, the greater the destabilizing effect.

**Example 2.17** Consider the two models,

$$x_{t+1} = x_t \exp(r(1 - x_t)),$$

$$x_{t+1} = x_t \exp(r(1 - x_{t-1})).$$

In both models, $g(x) = \exp(r(1 - x))$ and the positive equilibrium is $\bar{x} = 1$. Also, $g'(\bar{x}) = -r$. The first model, without delay, has a locally asymptotically stable equilibrium $\bar{x} = 1$ if $0 < r < 2$ and the second model, with a delay of one generation, has a locally asymptotically stable equilibrium if $0 < r < 1$. ■

**Example 2.18** Murray (1993, pp. 43–54) presents a model of interest to the International Whaling Commission (IWC). We discuss this model briefly. The goal of the IWC is to manage populations of baleen whales in order to prevent their extinction and to obtain a sustained yield. The following model was developed by Clark (1976):

$$x_{t+1} = (1 - \mu)x_t + R(x_{t-T}),$$

where $x_t$ is the sexually mature whales, $(1 - \mu)x_t$ is the population of whales surviving each year, $0 < \mu < 1$, and $R(x_{t-T})$ is the recruitment of whales into the adult population through whales born $T$ years earlier. The delay $T$ is the time to sexual maturity which ranges from 5 to 10 years. A formula for $R$ is given by

$$R(x) = \mu x\left(1 + q\left[1 - \left(\frac{x}{K}\right)^z\right]\right),$$

where $K > 0$ is the equilibrium density, $q > 0$ is related to the fecundity of females, and $z > 0$ is a measure of the strength of the density-dependent interactions.

It is easy to see that there exists an equilibrium at $\bar{x} = K$. We rescale the model so that the equilibrium is at one. Making the substitution $y_t = x_t/K$ yields

$$y_{t+1} = (1 - \mu)y_t + \mu y_{t-T}[1 + q(1 - y_{t-T}^z)].$$

Let $u_t = y_t - 1$ and substitute $u_t$ into the preceding equation,

$$\begin{aligned} y_{t+1} = u_{t+1} + 1 &\approx (1 - \mu)(u_t + 1) + \mu(u_{t-T} + 1)[1 + q(1 - [1 + zu_{t-T}])], \\ &= (1 - \mu)(u_t + 1) + \mu(u_{t-T} + 1)[1 - qzu_{t-T}]. \end{aligned}$$

The linearization of $y^z$ about one is $y^z \approx 1 + z(y - 1) = 1 + zu$. Thus,

$$u_{t+1} \approx (1 - \mu)u_t + \mu u_{t-T}(1 - qzu_{t-T}) - \mu qzu_{t-T}$$

or dropping terms of higher order in $u$ leads to the linearization,

$$u_{t+1} = (1 - \mu)u_t + \mu(1 - qz)u_{t-T}.$$

The characteristic equation is found by assuming $u_t = \lambda^t$:

$$p(\lambda) = \lambda^{T+1} - (1 - \mu)\lambda^T - \mu(1 - qz) = 0.$$

If the eigenvalues $\lambda$ of the characteristic equation have magnitude less than one, then $u_t \to 0$, $y_t \to 1$, and $x_t \to K$ (if initially the population is close to $K$). Since $T > 1$, the Jury conditions must be applied to determine the local asymptotic stability of the equilibrium $K$. However, we can check the necessary conditions stated in Theorem 2.12:

(i) $p(1) = 1 - (1 - \mu) - \mu(1 - qz) = \mu qz > 0$,

(ii) $(-1)^{T+1}p(-1) = (-1)^{T+1}[(-1)^{T+1} - (1 - \mu)(-1)^T - \mu(1 - qz)]$
$= 2 - \mu + \mu(-1)^T(1 - qz) > 0$,

(iii) $|\mu(1 - qz)| < 1$.

Condition (i) is satisfied. Conditions (ii) and (iii) are satisfied if, for example, $qz < 1 + 1/\mu$. This latter inequality puts a bound on the fertility coefficient $q$ and the strength of the density-dependent interaction, measured by $z$. If harvesting of the whale population is included, then it is important to maintain a stable population from year to year. ■

## 2.12 Exercises for Chapter 2

1. Find all of the equilibria for the following difference equations.
   (a) $x_{t+1} = ax_t\exp(-rx_t)$, $a, r > 0$
   (b) $x_{t+1} = ax_t/(b + x_t)^k$, $a, b, k > 0$ (nonnegative equilibria)
   (c) $x_{t+1} = ax_t^3$, $a > 0$
   (d) $x_{t+1} = ax_t\exp(-x_t) + bx_t$, $a > 0$ and $0 < b < 1$
   (e) $x_{t+1} = \begin{cases} ax_t, & |x_t| < 1, \\ 0, & |x_t| \geq 1, \end{cases} \quad a > 1$

**2.** For the difference equations in Exercise 1(a)–(d), determine the range of parameter values so that each equilibrium is either (i) locally asymptotically stable or (ii) unstable. Apply Theorem 2.1.

**3.** For the difference equation in Exercise 1(e), show that the origin is globally attracting (for all initial conditions) but it is not locally asymptotically stable.

**4.** Find all of the equilibria for the system of difference equations

$$x_{t+1} = \frac{ax_t y_t}{1 + x_t}, \quad a > 0,$$

$$y_{t+1} = \frac{bx_t y_t}{1 + y_t}, \quad b > 0.$$

**5.** Find all of the equilibria for the following difference equations. Then determine whether they are (i) locally asymptotically stable or (ii) unstable. Apply Theorems 2.1, 2.3, or 2.4.

(a) $x_{t+1} = ax_t^3 + x_t, a \neq 0$

(b) $x_{t+1} = \dfrac{1}{2}x_t^2 + 3x_t$

(c) $x_{t+1} = \dfrac{1}{2}(x_t^3 + x_t)$

**6.** Show that the 2-cycle

$$\bar{x}_{1,2} = \frac{1 \pm \sqrt{4r - 3}}{2}$$

of the difference equation $x_{t+1} = r - x_t^2$ is locally asymptotically stable if $3/4 < r < 5/4$ and unstable if $r > 5/4$.

**7.** The proof of Theorem 2.3 depends on the shape of $f(x)$ at $\bar{x}$. Use the cobwebbing method to show that the conditions in parts (i) and (ii) imply that $\bar{x}$ is unstable and the conditions in part (iii) imply $\bar{x}$ is locally asymptotically stable. Graphs of $f(x)$ near $\bar{x}$ are given in Figure 2.18.

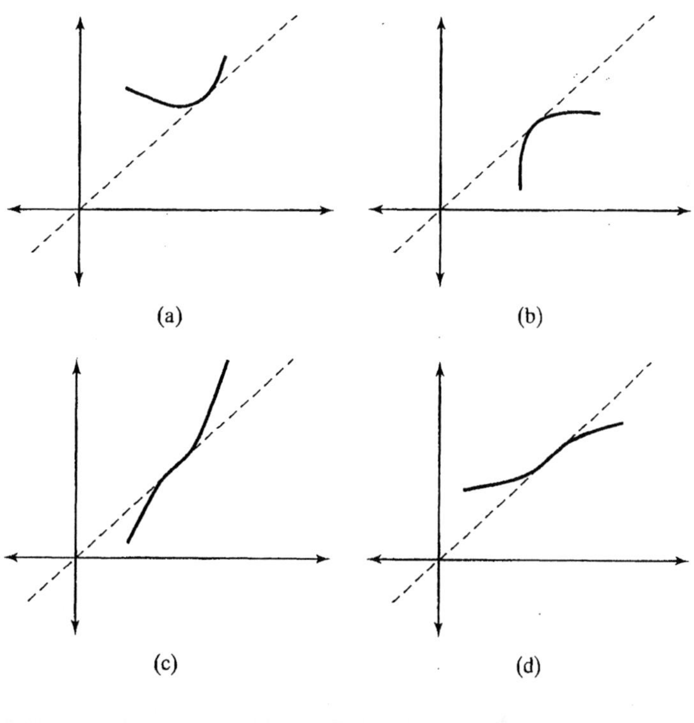


**Figure 2.18** The solid curve is $x_{t+1} = f(x_t)$ near $\bar{x}$ and the dotted curve is the line $x_{t+1} = x_t$. In (a), $f''(\bar{x}) > 0$; in (b), $f''(\bar{x}) < 0$. In (c) and (d), $f''(\bar{x}) = 0$ but in (c), $f'''(\bar{x}) > 0$ and in (d), $f'''(\bar{x}) < 0$.

**8.** Complete the proof of Theorem 2.4.

**9.** The differential equation

$$\frac{dx}{dt} = rx\left(1 - \frac{x}{K}\right)$$

is the well-known logistic equation, $r$ is the intrinsic growth rate, and $K$ is the carrying capacity. For $r, K > 0$ and $x(0) > 0$, it can be shown that $\lim_{t\to\infty} x(t) = K$. In this exercise, the exact difference equation version of logistic growth is derived.

(a) Denote $x(t)$ by $x_t$. Separate variables and integrate the differential equation, using partial fraction decomposition,

$$\int_{x_t}^{x_{t+1}} \frac{K}{x(K - x)}\, dx = \int_t^{t+1} r\, d\tau,$$

to show that

$$\ln\left|\frac{K - x_{t+1}}{Kx_{t+1}} \Big/ \frac{K - x_t}{Kx_t}\right| = -r.$$

Assume either $x_t < x_{t+1} < K$ or $x_t > x_{t+1} > K$; solutions are either increasing or decreasing.

(b) Show that the equation in (a) can be expressed in the following form:

$$x_{t+1} = \frac{\lambda K\, x_t}{K + (\lambda - 1)x_t},$$

where $\lambda = e^r > 1$. This last equation is the *exact* discrete version of logistic growth. The positive equilibrium $\bar{x} = K$ of the difference equation is globally asymptotically stable.

**10.** (a) For the logistic difference equation in Exercise 9 (b), let $u_t = 1/x_t$ and show that the difference equation in the variable $u$ is linear,

$$u_{t+1} = \frac{1}{\lambda}u_t + \frac{\lambda - 1}{\lambda K} = au_t + b, \quad u_0 = 1/x_0.$$

(b) Solve the linear difference equation for $u_t$.

(c) Show that

$$x_t = \frac{x_0 K \lambda^t}{K + x_0(\lambda^t - 1)}.$$

Then find $\lim_{t\to\infty} x_t$.

**11.** Show that the difference equation $x_{t+1} = ax_t^3 = f(x_t)$, $a > 0$, $f : [0,\infty) \to [0,\infty)$ has no 2-cycles on the interval $[0,\infty)$. (*Hint:* Apply a theorem.)

**12.** Show that if $x_{t+1} = f(x_t)$, $f(0) = 0$, $f : [0,\infty) \to [0,\infty)$, and $0 < f'(x) < 1$ on $[0,\infty)$, then for $x_0 > 0$,

$$\lim_{t\to\infty} f^t(x_0) = 0.$$

**13.** Find the 2-cycle of $x_{t+1} = x_t^2 - 1$ and determine if it is locally asymptotically stable. If $x_0 = 0.5$, find $x_1, \ldots, x_{20}$.

**14.** Suppose $x_t$ is the size of a population in generation $t$ and in generation $t + 1$,

$$x_{t+1} = f(x_t),$$

where $f'(x) > 0$ for $x \geq 0$. Suppose there are exactly three hyperbolic equilibria given by $\bar{x} = 0, 2, 4$.

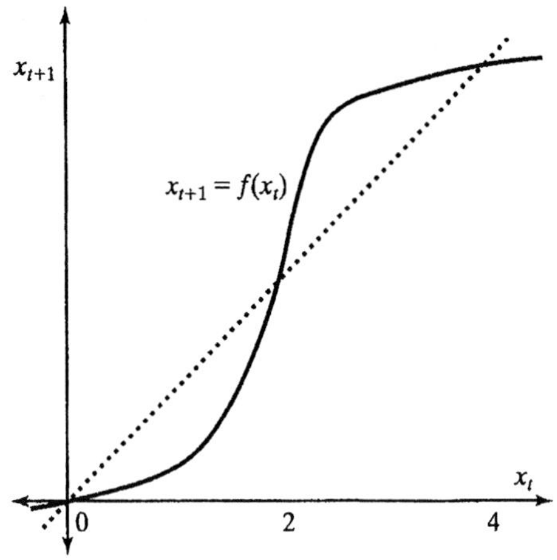


**Figure 2.19** Reproduction curve.

(a) The graph in Figure 2.19 is the reproduction curve $y = f(x)$ or $x_{t+1} = f(x_t)$ and the line $y = x$ or $x_{t+1} = x_t$. Determine the local stability of each of the three equilibria by noting the slope at each of the equilibria.

(b) Prove that $x_{t+1} = f(x_t)$ has no 2-cycles on $[0,\infty)$. (*Hint:* What happens if $x_0 < x_1$? $x_0 > x_1$?)

**15.** Suppose $x_{t+1} = f(x_t), f : [0,\infty) \to [0,\infty)$ and $f'(x) > 0$ for $x \in [0, \infty)$. In addition, assume there are exactly two hyperbolic equilibria, $\bar{x} > 0$ and zero. Prove that if $0 < x_0 < \bar{x}$, then either $\lim_{t\to\infty} x_t = 0$ or $\lim_{t\to\infty} x_t = \bar{x}$.

**16.** For the following nonlinear difference equations, find the equilibria as a function of $r$. Then draw the bifurcation diagrams near $r = 0$. Show that the bifurcation diagrams have a form corresponding to Figure 2.10 I, II, and III, respectively.

(a) $x_{t+1} = r + x_t + x_t^2$

(b) $x_{t+1} = (r + 1)x_t - x_t^3$

(c) $x_{t+1} = (r + 1)x_t + x_t^2$

**17.** For the nonlinear difference equation, $x_{t+1} = r - x_t - x_t^2$, show that there is a stable 2-cycle for $0 < r < 1/2$. Then show that the bifurcation diagram near $r = 0$ is given by Figure 2.10 IV.

**18.** Express the equilibria for the following difference equations as a function of $r$. Then draw the bifurcation diagrams near $r = 0$. Note the differences in signs between these equations and the equations given in Exercise 16.

(a) $x_{t+1} = r + x_t - x_t^2$ (saddle node)

(b) $x_{t+1} = (r + 1)x_t + x_t^3$ (pitchfork)

(c) $x_{t+1} = (r + 1)x_t - x_t^2$ (transcritical)

**19.** Show that the Liapunov exponent $\lambda(x_0) = \lambda(x_j)$ for $j = 1, 2, \ldots$, where $x_j = f(x_{j-1})$. *Hint:* Note that $\lambda(x_j) = \lim_{t\to\infty} \frac{1}{t} \sum_{i=j}^{t+j-1} \ln|f'(x_i)|$. Then show that

$$\lambda(x_0) = \lim_{t\to\infty} \frac{1}{t + j}\left[\sum_{i=0}^{j-1} \ln|f'(x_i)| + \sum_{i=j}^{t+j-1} \ln|f'(x_i)|\right].$$

**20.** Show that if $\bar{x}$ is an unstable equilibrium of $x_{t+1} = f(x_t)$ (as defined in Theorem 2.1), then the Liapunov exponent $\lambda(\bar{x}) > 0$, and if $\bar{x}$ is a locally asymptotically stable equilibrium (as defined in Theorem 2.1), then the Liapunov exponent satisfies $\lambda(\bar{x}) < 0$.

**21.** For the system of difference equations in Exercise 4, show that the zero equilibrium is always locally asymptotically stable. Then find conditions on the parameters so that the nonzero equilibrium is locally asymptotically stable.

**22.** The following system of difference equations represents two species $x$ and $y$ competing for a common resource (Leslie 1959; Pielou, 1977). Note that increases in the population size of $x$ or $y$ decrease the population size for the other species.

$$x_{t+1} = \frac{(a_1 + 1)x_t}{1 + x_t + b_1 y_t}, \quad a_1,\ b_1 > 0$$

$$y_{t+1} = \frac{(a_2 + 1)y_t}{1 + b_2 x_t + y_t}, \quad a_2,\ b_2 > 0$$

(a) There are three equilibria of the form $(0, 0)$, $(x^*, 0)$, $(0, y^*)$. Find $x^*$ and $y^*$.

(b) Determine conditions on the parameters so that the equilibria in part (a) are locally asymptotically stable.

(c) There is a fourth equilibrium $(\bar{x}, \bar{y})$. Find conditions on the parameters so that $\bar{x}$ and $\bar{y}$ are positive.

(d) Assume $a_2/b_2 > a_1$ and $a_1/b_1 > a_2$. What can you say about the stability of the equilibrium $(\bar{x}, \bar{y})$? If one of the inequalities is reversed, what can you say about the stability of $(\bar{x}, \bar{y})$?

**23.** The following epidemic model is referred to as an SIS epidemic model. Infected individuals recover but do not become immune. They become immediately susceptible again.

$$S_{t+1} = S_t - \frac{\beta}{N} I_t S_t + (\gamma + b) I_t,$$

$$I_{t+1} = I_t(1 - \gamma - b) + \frac{\beta}{N} I_t S_t$$

Assume that $0 < \beta < 1$, $0 < b + \gamma < 1$, $S_0 + I_0 = N$ and $S_0, I_0 > 0$.

(a) Show that $S_t + I_t = N$ for $t = 1, 2, \ldots$.

(b) Show that there exist two equilibria and they are both nonnegative if $\mathcal{R}_0 = \beta/(b + \gamma) \geq 1$.

(c) Reduce the system to a single difference equation in $I$, [e.g., $I_{t+1} = f(I_t)$]. Show that the zero equilibrium ($\bar{I} = 0$) is locally asymptotically stable if $\mathcal{R}_0 < 1$ and the positive equilibrium ($\bar{I} > 0$) is locally asymptotically stable if $\mathcal{R}_0 > 1$.

(d) Write a computer program and perform some numerical simulations for the model when $\mathcal{R}_0 < 1$ and $\mathcal{R}_0 > 1$. (See the MATLAB program in the Appendix to this chapter.)

**24.** Let $\mathcal{R}_0$ be the bifurcation parameter in the SIS model of Exercise 23. In particular, let $b + \gamma = 1/2$ and $\beta = \mathcal{R}_0/2$. Then make a change of variable $x_t = I_t/N$ to simplify the difference equation in $I_t$ to

$$x_{t+1} = \frac{1}{2} x_t (1 + \mathcal{R}_0[1 - x_t]). \tag{2.27}$$

(a) Find all of the equilibria for model (2.27).

(b) Show that there exists a transcritical bifurcation at $\mathcal{R}_0 = 1$ for model (2.27). Sketch a bifurcation diagram near $\mathcal{R}_0 = 1$.

**25.** Consider the Ricker model for population growth,

$$y_{t+1} = y_t \exp(r[1 - y_t/K]),\ r, K > 0. \tag{2.28}$$

This equation exhibits period-doubling behavior similar to the discrete logistic equation (May, 1975).

(a) Let $x_t = y_t/K$ and express the difference equation in the dimensionless form: $x_{t+1} = x_t \exp(r[1 - x_t])$. Show that $x_t > 0$ if $x_0 > 0$.

(b) For what values of $r$ is the positive equilibrium $\bar{x} = 1$ locally asymptotically stable?

(c) Show by computing 2-cycles $\{\bar{x}_1, \bar{x}_2\}$ and checking stability for values of $r$ between $2 < r < 2.526$ that the 2-cycles are stable (use $r = 2.1,\ 2.2,\ 2.3,\ 2.4$, and 2.5). What happens when $r = 2.6$?

**26.** Show that the Ricker model (2.28) does not have any 2-cycles for $0 < r < 2$,

$$x_{t+1} = x_t \exp(r[1 - x_t/K) = f(x_t),$$

$r, K > 0$. In particular, show $1 + f'(x) \neq 0$ for $x \in (0,\infty)$. By applying Theorems 2.6 and 2.7, conclude that the positive equilibrium $K$ is globally asymptotically stable.

**27.** The following difference equation is an extension of the Ricker model. An additional term $cx_t^2$ is included to represent negative feedback on population growth due to predation and other processes. The model takes the form

$$x_{t+1} = x_t \exp(r - x_t - cx_t^2), \quad r, c > 0.$$

(a) Show that there exists a unique positive equilibrium $\bar{x} = (\sqrt{1 + 4rc} - 1)/2c$.

(b) Show that the positive equilibrium is locally asymptotically stable if $0 < r + c\bar{x}^2 < 2$.

(c) Show that $f'(\bar{x}) = -1$ when $r + c\bar{x}^2 = 2$ (property of a period-doubling bifurcation).

(d) The following bifurcation diagram shows the period-doubling behavior when $r = 1.9$. Find the value of $c$ where a 2-cycle first appears. The following graph is a bifurcation diagram for $c \in [0,0.8]$ and $r = 1.9$.

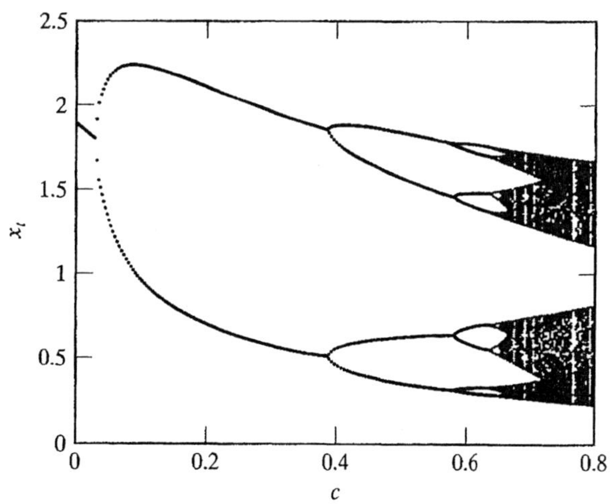


**28.** The Jacobian matrix for the system of difference equations, $X_{t+1} = F(X_t)$ evaluated at a positive equilibrium, $\bar{X} = (\bar{x}, \bar{y})^T$, satisfies

$$J = \begin{pmatrix} a & 1 + a \\ a & 1 \end{pmatrix},$$

where $a$ is a real number. Give conditions on the parameter $a$ so that the positive equilibrium is locally asymptotically stable.

**29.** In the delay difference equation, $x_{t+1} = x_t g(x_{t-1})$, use the formula for the eigenvalues,

$$\lambda_{1,2} = \frac{1 \pm \sqrt{1 + 4g'(\bar{x})\bar{x}}}{2},$$

to show directly that

(a) $0 < \lambda_{1,2} < 1$ iff $0 < -g'(\bar{x})\bar{x} < 1/4$ and

(b) the eigenvalues are complex conjugates satisfying $|\lambda_{1,2}| < 1$ iff $1/4 < -g'(\bar{x})\bar{x} < 1$.

Thus, examining Figure 2.17, we can see condition (a) implies solutions to the difference equations have exponential convergence (EC) to the equilibrium $\bar{x}$ and condition (b) implies solutions have oscillating convergence (OC) to the equilibrium $\bar{x}$.

**30.** Consider the following nonlinear difference equation of order 2:

$$x_{t+1} = rx_t(1 - x_{t-1}), \quad r > 1. \tag{2.29}$$

(a) Convert this difference equation to a first-order system. Then determine the Jacobian matrix of the system.

(b) Determine conditions on $r$ that ensure the equilibrium $\bar{x} = (r-1)/r$ is locally asymptotically stable.

**31.** Suppose the delay is in the intrinsic growth rate rather than the density-dependent term, that is,

$$x_{t+1} = rx_{t-1}(1 - x_t), \quad r > 0.$$

(a) Convert this difference equation to a first-order system. Then determine the Jacobian matrix of the system.

(b) Determine conditions on $r$ such that the equilibrium $\bar{x} = 0$ is locally asymptotically stable.

(c) Show for $r > 1$ that the equilibrium $\bar{x} = (r - 1)/r$ is never locally asymptotically stable. Compare the stability results for this model to the model (2.29) and to the model without delay, $x_{t+1} = rx_t(1 - x_t)$ (discrete logistic equation). These examples illustrate the importance of the term containing the delay in the model.

**32.** Let matrix $A = \begin{pmatrix} a_{11} & a_{12} \\ 0 & a_{22} \end{pmatrix}$.

(a) Find $\rho(A)$ and the matrix norms $\|A\|_1$, and $\|A\|_\infty$. (See definitions of matrix norms in the section on Liapunov exponents in the Appendix to this Chapter.)

(b) Give examples of matrices $B$, $C$, and $D$ satisfying the following relationships:

(i) $\rho(B) < \|B\|_1 < \|B\|_\infty$

(ii) $\rho(C) < \|C\|_\infty < \|C\|_1$

(iii) $\rho(D) = \|D\|_1 = \|D\|_\infty$

## 2.13 References for Chapter 2


Allen, L. J. S. 1994. Some discrete-time *SI*, *SIR*, and *SIS* epidemic models. *Math. Biosci.* 124: 83–105.

Allen, L. J. S. and A. M. Burgin. 2000. Comparison of deterministic and stochastic SIS and SIR models in discrete time. *Math. Biosci.* 163: 1–33.

Allen, L. J. S., N. Kirupaharan, and S. M. Wilson. 2004. SIS epidemic models with multiple pathogen strains. *J. Difference Eqns. and Appl.* 10: 53–75.

Alligood, K. T., T. D. Sauer, and J. A. Yorke. 1996. *Chaos: An Introduction to Dynamical Systems*. Springer-Verlag, New York.

Anderson, R. M. and R. M. May. 1991. *Infectious Diseases of Humans, Dynamics and Control*. Oxford University Press, Oxford.

Castillo-Chávez, C. and A.-A. Yakubu. 2001. Dispersal, disease and life-history evolution, *Math. Biosci.* 173: 35–53.

Clark, C. W. 1976. A delayed-recruitment model of population dynamics with an application to baleen whale populations. *J. Math. Biol.* 3: 381–391.

Cull, P. 1981. Global stability of population models. *Bull. Math. Biol.* 43: 47–58.

Cull, P. 1986. Local and global stability for population models. *Biological Cybernetics* 54: 141–149.

Cull, P. 1988. Stability of discrete one-dimensional population models. *Bull. Math. Biol.* 50: 67–75.

Devaney, R. L. 1989. *An Introduction to Chaotic Dynamical Systems*. 2nd ed. Addison-Wesley, Reading, Mass.

Devaney, R. L. 1992. *A First Course in Chaotic Dynamical Systems: Theory and Experiments*. Addison-Wesley, Reading, Mass.

Edelstein-Keshet, L. 1988. *Mathematical Models in Biology*. The Random House/Birkhäuser Mathematics Series, New York.

Elaydi, S. N. 1999. *An Introduction to Difference Equations*. 2nd ed. Springer-Verlag, New York.

Elaydi, S. N. 2000. *Discrete Chaos*. Chapman & Hall/CRC, Boca Raton, London, New York, Washington, D.C.

Feigenbaum, M. J. 1978. Quantitative universality for a class of nonlinear transformations. *J. Stat. Phys.* 19: 25–52.

Gleick, J. 1987. *Chaos: The Making of a New Science*. Viking, New York.

Gulick, D. 1992. *Encounters with Chaos*. McGraw-Hill, Inc., New York.

Hale, J. and H. Koçak. 1991. *Dynamics and Bifurcations*. Springer-Verlag, New York, Berlin, Heidelberg.

Hamer, W. H. 1906. Epidemic disease in England. *Lancet* 1: 733–739.

Hethcote, H. W. and J. A. Yorke. 1984. *Gonorrhea Transmission Dynamics and Control*. Springer-Verlag, Berlin and Heidelberg.

Jury, E. I. 1964. *Theory and Applications of the Z Transform*. John Wiley & Sons, New York.

Jury, E. I. 1971. Inners' approach to some problems of system theory. *IEEE Trans. Automat. Control* 16: 233–240.

Jury, E. I. 1974. *Inners and Stability of Dynamic Systems*. John Wiley & Sons, New York.

Kermack, W. O. and A. G. McKendrick. 1927. Contributions to the mathematical theory of epidemics, part 1. *Proc. Roy. Soc. London Series A* 115: 700–721.

Leslie, P. H. 1959. The properties of a certain lag type of population growth and the influence of an external random factor on a number of such populations. *Physiol. Zool.* 32: 151–159.

Li, T.-Y., M. Misiurewicz, G. Pianigiani, and J. A. Yorke. 1982. Odd chaos. *Phys. Letters*. 87A: 271–273.

Li, T.,-Y. and J. A. Yorke. 1975. Period three implies chaos. *Amer. Math. Monthly*. 82: 985–992.

Martin, C. F., L. J. S. Allen, and M. Stamp. 1996. An analysis of the transmission of chlamydia in a closed population. *J. Difference Equations and Appl.* 2: 1–29.

May, R. M. 1975. Biological populations obeying difference equations: stable points, stable cycles, and chaos. *J. Theor. Biol.* 51: 511–524.

May, R. M. 1976. Simple mathematical models with very complicated dynamical behavior. *Nature* 261: 459–467.

May, R. M. 1983. Parasitic infections as regulators of animal populations. *Amer. Scientist* 71: 36–45.

McCluskey C. C. and J. S. Muldowney. 1998. Bendixson-Dulac criteria for difference equations. *J. Dynamics and Differential Equations* 10: 567–576.

Murray, J. D. 1993. *Mathematical Biology*. 2nd ed. Springer-Verlag, Berlin, Heidelberg, New York.

Murray, J. D. 2002. *Mathematical Biology: I An Introduction*. 3rd ed. Springer-Verlag, New York.

Olsen, L. F. and H. Degn. 1985. Chaos in biological systems. *Quarterly Rev. Biophysics*. 18: 165–225.

Ortega, J. M. 1987. *Matrix Theory A Second Course*. Plenum Press, New York.

Pielou, E. C. 1977. *Mathematical Ecology*. John Wiley & Sons, New York.

Rasband, S. N. 1990. *Chaotic Dynamics of Nonlinear Systems*. John Wiley & Sons, New York.

Ricker, W. E. 1954. Stock and recruitment. *Journal of the Fisheries Research Board of Canada* 11: 559–623.

Ross, R. 1911. *The Prevention of Malaria*. 2nd ed. John Murray, London.

Saha, P. and S. H. Strogatz. 1995. The birth of period 3. *Math. Mag.* 68: 42–47.

Sedaghat, H. 1997. The impossibility of unstable, globally attracting fixed points for continuous mappings of the line. *Amer. Math. Monthly.* 104: 356–359.

Sharkovskii, A. N. 1964. Coexistence of cycles of a continuous map of a line into itself (in Russian). *Ukr. Mat. Z.* 16: 61–71.

Strogatz, S. H. 2000. *Nonlinear Dynamics and Chaos*. Perseus Pub., Cambridge, Mass.

## 2.14 Appendix for Chapter 2

### 2.14.1 Proof of Theorem 2.6

The proof of Theorem 2.6 on global asymptotic stability of the fixed point $\bar{x}$ of $x_{t+1} = f(x_t)$ iff there are no 2-cycles is verified. The proof is due to Cull (1981). See also Cull (1986, 1988).

**Proof of Theorem 2.6** If $f$ has a globally asymptotically stable equilibrium at $\bar{x}$, then $f$ cannot have any 2-cycles because for a 2-cycle $\{\bar{x}_1, \bar{x}_2\}$, $\lim_{t\to\infty} f^t(\bar{x}_1)$ will not exist.

The reverse implication takes more work. We assume $f$ has no 2-cycles, then show global asymptotic stability of $\bar{x}$. The proof is divided into three cases.

Case 1 We show that $f(f(x)) > x$ for $x \in (0, \bar{x})$. First note that $f(f(x)) \neq x$ for $x \in (0, \bar{x})$ or $x \in (\bar{x}, a)$. The function $f(f(x)) - x$ can only have one sign on the interval $(0, \bar{x})$. We show that this sign must be positive. The proof of Case 1 is divided into two parts, (a) and (b).

(a) Suppose $f(x) < \bar{x}$ for all $x \in (0, \bar{x})$. Then by assumption (iv), $x < f(x)$ and $y = f(x) > \bar{x}$ implies $f(f(x)) = f(y) > y = f(x) > x$. Thus, $f(f(x)) > x$ on $(0, \bar{x})$.

(b) Suppose there exists some $\hat{x} \in (0, \bar{x})$ such that $f(\hat{x}) = \bar{x}$. Then $f(f(\hat{x})) = f(\bar{x}) = \bar{x} > \hat{x}$. So $f(f(\hat{x})) - \hat{x} > 0$. Since $f(f(x)) - x$ can only have one sign on $(0, \bar{x})$ and $f(f(\hat{x})) - \hat{x} > 0$, it follows that $f(f(x)) > x$ for all $x \in (0, \bar{x})$.

Case 2 We show that if $x_0 \in (0, \bar{x})$, then $\lim_{t\to\infty} f^t(x_0) = \bar{x}$. The proof of Case 2 is divided into three parts, (a), (b), and (c).

(a) Suppose $x_0 < f(x_0) \leq \bar{x}$ for all $x_0 \in (0, \bar{x})$. Since $y_0 = f(x_0) \in (0, \bar{x})$, we can apply the inequality again. $f(x_0) = y_0 < f(y_0) = f^2(x_0) \leq \bar{x}$. By continued application of the inequality, we obtain a monotone increasing sequence $\{f^t(x_0)\}_{t=0}^{\infty}$ bounded above by $\bar{x}$. Thus, the sequence converges to a point $0 < z \leq \bar{x}$. But since $f$ is continuous,

$$z = \lim_{t\to\infty} f^t(x_0) = \lim_{t\to\infty} f^{t+1}(x_0) = f(z),$$

which shows $z$ is a positive fixed point. The only positive fixed point is $\bar{x}$, so that $z = \bar{x}$. Thus, $\lim_{t\to\infty} f^t(x_0) = \bar{x}$ for $x_0 \in (0, \bar{x})$.

(b) Suppose $f(x_0) > \bar{x}$ for some $x_0 \in (0, \bar{x})$. Then $f(f(x_0)) < f(x_0)$ and $f(f(x_0)) < \bar{x}$ by properties (v) and (vi). We have from Case 1, $x_0 < f^2(x_0) < \bar{x}$ for all $x_0 \in (0, \bar{x})$ for which $f(x_0) > \bar{x}$. Case 2 (a) applies to $f^2$: $\lim_{t\to\infty} f^{2t}(x_0) = \bar{x}$. Since $f$ is continuous, $\lim_{t\to\infty} f^{2t+1}(x_0) = f(\bar{x})$ so that $\lim_{t\to\infty} f^t(x_0) = \bar{x}$ for $x_0 \in (0, \bar{x})$.

(c) Suppose there exists $t_0 > 0$ such that $f^{t_0}(x_0) > \bar{x}$ for some $x_0 \in (0, \bar{x})$ and $f^t(x_0) < \bar{x}$ for $t = 0, \ldots, t_0 - 1$. The finite sequence $\{x_0, f(x_0), \ldots, f^{t_0-1}(x_0)\}$ must be monotone by property (iv). By Case 2 (a), $x_0 < f^{t_0-1}(x_0) < \bar{x}$, from which it follows that $\lim_{t\to\infty} f^{(t_0-1)t}(x_0) = \bar{x}$. Since $f$ is continuous,

$$\lim_{t\to\infty} f^{(t_0-1)t+k}(x_0) = f^k(\bar{x}) = \bar{x}$$

for $k = 0, 1, 2, \ldots, t_0 - 2$, so that $\lim_{t\to\infty} f^t(x_0) = \bar{x}$.

Case 3 We show if $x_0 \in (\bar{x}, a)$, then $\lim_{t\to\infty} f^t(x_0) = \bar{x}$. The proof of Case 3 is divided into two parts, (a) and (b).

(a) Suppose $f^t(x_0) > \bar{x}$ for all $t > 0$, where $x_0 \in (\bar{x}, a)$. Then $\bar{x} < f(x_0) < x_0$ by property (v). Again by property (v) it follows that

$$\bar{x} < f^2(x_0) < f(x_0) < x_0.$$

The sequence $\{f^t(x_0)\}_{t=0}^{\infty}$ is monotone decreasing and bounded below by $\bar{x}$. It follows that the sequence must converge to the fixed point $\bar{x}$, $\lim_{t\to\infty} f^t(x_0) = \bar{x}$.

(b) Suppose there exists $t_0$ such that $0 < f^{t_0}(x_0) < \bar{x}$ for some $x_0 \in (\bar{x}, a)$. Then for $y_0 = f^{t_0}(x_0)$ it follows as in Case 2 that $\lim_{t\to\infty} f^t(y_0) = \bar{x}$. Hence,

$$\lim_{t\to\infty} f^t(x_0) = \lim_{s\to\infty} f^s(y_0) = \bar{x},$$

where $t = s + t_0$.

### 2.14.2 A Definition of Chaos

The following definition for chaos is given by Li et al. (1982).

**Definition 2.10.** The equation $x_{t+1} = f(x_t)$ is *chaotic* if there are infinitely many periodic points of $f$ with different periods, and there is some uncountable set $C$ of critical values which satisfies the following conditions

(i) For every $p, q \in C$ with $p \neq q$, there is a sequence $n_1, n_2, \ldots$ (tending to $\infty$) such that

$$\lim_{i\to\infty} |f^{n_i}(p) - f^{n_i}(q)| > \epsilon,$$

for some $\epsilon > 0$.

(ii) In addition, there is another sequence $m_1, m_2, m_3, \ldots$ (tending to $\infty$) such that

$$\lim_{i\to\infty} |f^{m_i}(p) - f^{m_i}(q)| = 0.$$

The definition states that there exists an uncountable number of points that are at some times close together and at other times not close together. The approximate logistic difference equation $x_{t+1} = rx_t(1 - x_t)$ satisfies this definition of chaos for some values of $r > r_c = 3.570$. There are other criteria that have been used to define chaos (see, e.g., Alligood et al., 1996; Elaydi, 2000; Gulick, 1992).

### 2.14.3 Jury Conditions (Schur-Cohn Criteria)

The conditions for an $n$th-degree polynomial with real coefficients, $p(\lambda)$, to have solutions $\lambda$ satisfying $|\lambda| < 1$ are known as the *Jury conditions* or the *Schur-Cohn criteria* (Edelstein-Keshet, 1988; Elaydi, 1999; Murray, 1993, 2002). Before we state the conditions, we need to define the "inner matrices" of a square matrix.

**Definition 2.11.** Let $B_n$ be an $n \times n$ square matrix, $n \geq 3$. Delete the first row and column and the $n$th row and column of $B_n$ to obtain an $(n-2)\times(n-2)$ matrix $B_{n-2}$. Matrix $B_{n-2}$ is called an *inner matrix of* $B_n$. If $n-2 \geq 3$, delete the first row and column and last row and column of $B_{n-2}$ to obtain another *inner matrix of* $B_n$ of size $(n-4)\times(n-4)$, $B_{n-4}$. This process can be continued until the last inner matrix is either a $1 \times 1$ or a $2 \times 2$ matrix. The *inner matrices of* $B_n$ are $\{B_{n-2}, \ldots, B_2\}$, if $n$ is even, or $\{B_{n-2}, \ldots, B_1\}$, if $n$ is odd.

The inner matrices of a $5 \times 5$ matrix are the $3 \times 3$ submatrix $B_3$ and the $1 \times 1$ submatrix $B_1$:

$$B_5 = \begin{pmatrix} b_{11} & b_{12} & b_{13} & b_{14} & b_{15} \\ b_{21} & b_{22} & b_{23} & b_{24} & b_{25} \\ b_{31} & b_{32} & b_{33} & b_{34} & b_{35} \\ b_{41} & b_{42} & b_{43} & b_{44} & b_{45} \\ b_{51} & b_{52} & b_{53} & b_{54} & b_{55} \end{pmatrix}, \quad B_3 = \begin{pmatrix} b_{22} & b_{23} & b_{24} \\ b_{32} & \boxed{b_{33}} & b_{34} \\ b_{42} & b_{43} & b_{44} \end{pmatrix}, \quad B_1 = (b_{33}).$$

The following theorem was proved by Jury (1971) using the results of Schur and Cohn.

**Theorem 2.13 (Jury Conditions, Schur-Cohn Criteria).** *Suppose the characteristic polynomial,*

$$p(\lambda) = \lambda^n + a_1\lambda^{n-1} + a_2\lambda^{n-2} + \cdots + a_{n-1}\lambda + a_n = 0, \tag{2.31}$$

*has real coefficients. Define two $(n-1)\times(n-1)$ matrices $B^{\pm}_{n-1}$ as follows:*

$$B^{\pm}_{n-1} = \begin{pmatrix} 1 & a_1 & a_2 & \cdots & a_{n-1} \\ 0 & 1 & a_2 & \cdots & a_{n-2} \\ 0 & 0 & 1 & \cdots & a_{n-3} \\ \vdots & \vdots & \vdots & \ddots & \vdots \\ 0 & 0 & 0 & \cdots & 1 \end{pmatrix} \pm \begin{pmatrix} 0 & 0 & 0 & \cdots & a_n \\ \vdots & \vdots & \vdots & \cdots & \vdots \\ 0 & 0 & a_n & \cdots & a_4 \\ 0 & a_n & a_{n-1} & \cdots & a_3 \\ a_n & a_{n-1} & a_{n-2} & \cdots & a_2 \end{pmatrix}.$$

*Then the solutions $\lambda$ of (2.31) satisfy $|\lambda| < 1$ iff the following three conditions hold:*

*(i)* $p(1) > 0$,

*(ii)* $(-1)^n p(-1) > 0$, *and*

*(iii) the determinant of each of the inner matrices of $B^{\pm}_{n-1}$ is positive.*

Another equivalent Jury test is discussed by Edelstein-Keshet (1988) and Murray (1993, 2002).

### 2.14.4 Liapunov Exponents for Systems of Difference Equations

Liapunov exponents can be calculated for systems of difference equations in a manner similar to scalar difference equations. Recall that positive Liapunov exponents for all initial conditions imply that the difference equation exhibits sensitive dependence on initial conditions, an indication of chaos. For a scalar difference equation, $x_{t+1} = f(x_t)$, the Liapunov exponent at $x_0$ is defined as

$$\lambda(x_0) = \lim_{t\to\infty} \frac{1}{t} \sum_{k=0}^{t-1} \ln|f'(x_k)|.$$

A similar definition can be applied to systems of difference equations. However, the derivative is replaced by the Jacobian matrix and the absolute value is replaced by the magnitude of the eigenvalues of the Jacobian matrix. In addition, there are $n$ Liapunov exponents for a system of $n$ difference equations. The magnitude of the largest eigenvalue, the spectral radius of the matrix, determines whether there is sensitive dependence on initial conditions. Recall that the spectral radius of $J$ is $\rho(J) = \max_{1\le i\le} |\lambda_i|$, where $\lambda_i, i = 1, \dots, n$ are the eigenvalues of $J$.

We define the Liapunov exponent at $(x_0, y_0)$ for a system of two difference equations, $x_{t+1} = f(x_t, y_t)$ and $y_{t+1} = g(x_t, y_t)$. Let $J(x_k, y_k)$ denote the Jacobian matrix of this system evaluated at the point $(x_k, y_k)$. To determine whether this system has sensitive dependence on initial condition $(x_0, y_0)$, the spectral radius of the product of matrices of the form $J(x_k, y_k)$ is calculated (Olsen and Degn, 1985). The Liapunov exponent for initial condition $(x_0, y_0)$ is denoted $\lambda(x_0, y_0)$ and defined as follows:

$$\lambda(x_0, y_0) = \lim_{t\to\infty} \frac{1}{t} \ln \rho\Big(\prod_{k=0}^{t-1} J(x_k, y_k)\Big). \tag{2.32}$$

If $\lambda(x_0, y_0) > 0$, then the solution to the system of difference equation exhibits sensitive dependence on the initial condition $(x_0, y_0)$. In addition, according to Alligood, et al. (1996), if the solution is not asymptotically periodic and no Liapunov exponent is exactly zero, then the solution is *chaotic*. An efficient numerical method for calculating Liapunov exponents is described by Alligood, et al. (1996).

**Example 2.19.** Suppose $(x_0, y_0)$ is an equilibrium solution of a system of difference equations. If this equilibrium is locally asymptotically stable, then $\rho(J(x_0, y_0)) < 1$. In addition, $\rho(J^t(x_0, y_0)) = \rho^t(J(x_0, y_0))$. Thus, it is easy to see that $\lambda(x_0, y_0) < 0$. If $(x_0, y_0)$ is unstable $(\rho(x_0, y_0) > 1)$, then it clear that $\lambda(x_0, y_0) > 0$. (See also Example 2.14.)

It can be shown that any matrix norm, induced by a vector norm, $\|A\| = \sup_{\|x\|=1} \|Ax\|$, is an upper bound for the spectral radius,

$$\rho(A) \le \|A\|$$

(Ortega, 1987). In addition, any such matrix norm satisfies $\|AB\| \le \|A\|\,\|B\|$ (Ortega, 1987). For example, the $L_1$ and $L_\infty$ norms are induced matrix norms. They are defined by

$$\|A\|_1 = \max_{1 \le j \le n} \sum_{i=1}^{n} |a_{ij}|, \quad \text{and} \quad \|A\|_\infty = \max_{1 \le j \le n} \sum_{j=1}^{n} |a_{ij}|,$$

respectively. The $L_1$ norm is the maximum absolute column sum and the $L_\infty$ norm is the maximum absolute row sum.

**Example 2.20.** Let

$$A = \begin{pmatrix} -1 & 2 \\ -1 & -2 \end{pmatrix}.$$

Then $\|A\|_1 = \max\{1 + 1, 2 + 2\} = 4$ and $\|A\|_\infty = \max\{1 + 2, 1 + 2\} = 3$. The eigenvalues of $A$ equal $-3/2 \pm i\sqrt{7}/2$. Thus, $\rho(A) = 2$.

**Example 2.21.** Suppose $a < 0 < b < |a|$ and matrix

$$A = \begin{pmatrix} a & 0 \\ b & b \end{pmatrix}.$$

Then $\rho(A) = |a|, \|A\|_1 = |a| + b$, and $\|A\|_\infty = \max\{|a|, 2b\}$.

If, for an induced matrix norm, $\|\cdot\|$,

$$\lim_{t \to \infty} \frac{1}{t} \sum_{k=0}^{t-1} \ln \|J(xk, yk))\| < 0,$$

then the Liapunov exponent $\lambda(x_0, y_0)$ given in (2.32) will be negative. The system will not exhibit sensitive dependence on the initial conditions. Negative Liapunov exponents also mean that the solution will not exhibit chaotic behavior. Hence, matrix norms, which are easier to compute than eigenvalues, can be used to show nonexistence of chaotic solutions.

### 2.14.5 MATLAB Program: SIR Epidemic Model

The following MATLAB program numerically solves the SIR epidemic model given specific initial conditions for $S$ and $I$.

```
clear % Clears variables and functions.
set(0,'DefaultAxesFontSize',18) % Increases axes labels
time=25;
beta=.3;
b=.2;
gama=b;
N=100;
s(1)=70;
i(1)=30;
for t=1:time
  s(t+1)=s(t)-beta*i(t)*s(t)/N+b*(N-s(t));
  i(t+1)=i(t)*(1-gama-b)+beta*i(t)*s(t)/N;
end
plot([0:1:time],s,'bo',[0:1:time],i,'rx','LineWidth',2)
xlabel('t');
```

**Note:** A statement following % explains the MATLAB command. This statement is not executed. If a semicolon is left off an executable command, then the value generated by the command prints to the computer screen.

Chapter 3

# Biological Applications of Difference Equations

## 3.1 Introduction

In this chapter, we discuss a variety of biological models that are expressed in terms of difference equations. First, we discuss some well-known single-species models, Beverton-Holt and Ricker models. In these models, the population size is limited due to a density-dependent relationship that results in a population decline when densities become too large. Then we discuss models with several interacting populations, host-parasitoid and predator-prey models. One of the most well-known host-parasitoid models is the Nicholson-Bailey model. This model has been applied to insect populations. An insect parasitoid is a parasite that is free living as an adult but lays eggs in an insect host. The Nicholson-Bailey model is discussed in detail in Section 3.3. An application to population genetics is studied in Section 3.7, where it is shown that under certain conditions gene frequencies remain the same from generation to generation. This property is known as the *Hardy-Weinberg law*. Some nonlinear, age-, stage-, and sex-structured models are developed and analyzed in Section 3.8. These models are generalizations of the Leslie matrix model and the structured models in Chapter 1. The age- and stage-structured models assume a density-dependent relationship in the birth or death rates. The sex-structured model assumes the birth rate is frequency dependent. A stage-structured model that has been the subject of much current research is known as the LPA model, where $L$ = larvae, $P$ = pupae, and $A$ = adults. The LPA model follows the dynamics of flour beetle populations. Theoretical studies of the LPA model combined with data collected from laboratory experiments on flour beetles and statistical analyses have demonstrated the existence of periodic and chaotic solution behavior. We discuss the LPA model in Section 3.8.2. The chapter ends with a discussion of some epidemic models with vaccination. These epidemic models represent an early attempt to understand the impact that vaccination programs have on populations of varying sizes.

## 3.2 Population Models

The simplest population model is the exponential growth model, but this model does not put a limit on the population size. Two well-known population models in which the population size is limited have been applied to a variety of populations. They are known as the Beverton-Holt model and the Ricker model. The names refer to the investigators who developed and applied these models primarily to fish populations (Beverton and Holt, 1957; Ricker, 1954). A nice derivation of these two models is presented by Thieme (2003), where the Ricker and Beverton-Holt equations are derived based on the assumption that juveniles are cannibalized by adults.

The Beverton-Holt model has the following form:

$$N_{t+1} = \frac{\lambda K N_t}{K + (\lambda - 1)N_t} = f(N_t), \lambda > 1, K > 0, \quad \text{and} \quad N_0 > 0$$

(Beverton and Holt, 1957; Kot et al., 1996; Pielou, 1977). The parameter $\lambda = e^r$, where $r$ is the intrinsic growth rate. The parameter $K$ is the carrying capacity. From the Exercises in Chapter 2, we know that the Beverton-Holt model is the exact discrete logistic model.

For the Beverton-Holt model, $f: [0, \infty) \to [0, \infty)$. The equilibria are found by solving for $\bar{N}$ in the following equation:

$$\bar{N} = f(\bar{N}) = \frac{\lambda K \bar{N}}{K + (\lambda - 1)\bar{N}}.$$

The fixed points or equilibria satisfy $\bar{N} = 0$ and $K + (\lambda - 1)\bar{N} = \lambda K$, which implies $\bar{N} = K$. Thus, $f(0) = 0$ and $f(K) = K$. Also,

$$f'(N) = \frac{\lambda K^2}{[K + (\lambda - 1)N]^2} > 0$$

and $f'(0) = \lambda > 1$. The function $f$ satisfies $N < f(N) < K$ for $N \in (0, K)$ and $K < f(N) < N$ for $N \in (K, \infty)$. (See Figure 3.1.) From Theorem 2.8, it follows that

$$\lim_{t\to\infty} N_t = K.$$

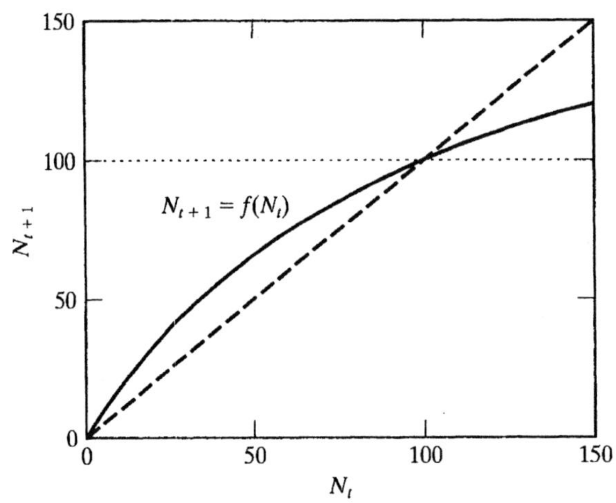


**Figure 3.1** Graph of $N_{t+1} = f(N_t)$ in the Beverton-Holt model, $K = 100$ and $\lambda = 2$.

The Ricker model has the following form:

$$N_{t+1} = N_t \exp\left[r\left(1 - \frac{N_t}{K}\right)\right] = f(N_t), \quad r, K > 0$$

(Caswell, 2001; Kot et al., 1996; Ricker, 1954). The parameter $r$ is the intrinsic growth rate and $K$ is the carrying capacity. Caswell (2001) distinguishes the Beverton-Holt and Ricker models by noting the differences in the growth rate $f(N)$ and per capita growth rate $g(N) = f(N)/N$. The Beverton-Holt function, $f(N) = \dfrac{\lambda KN}{K + (\lambda - 1)N}$, is referred to as *compensatory*, whereas the Ricker function, $f(N) = N\exp[r(1 - N/K)]$, is referred to as *overcompensatory*. Both functions have the properties that $f$ is increasing and $g$ is decreasing in $N$. But the Beverton-Holt function satisfies $\lim_{N\to\infty} f(N) > 0$, whereas the Ricker function satisfies $\lim_{N\to\infty} f(N) = 0$. Caswell (2001) states that, in the Beverton-Holt model, the decrease in the per capita growth rate with population size $N_t$ is exactly compensated for by the increase in population size, $N_{t+1}$, $\lim_{N\to\infty} f(N) > 0$. However, this is not the case in the Ricker model. In the Ricker model, the decrease in the per capita growth rate $g(N_t)$ and subsequent increase in $N_{t+1}$ is insufficient to compensate for the decrease in $g(N_t)$. At high densities $\lim_{N\to\infty} f(N) = 0$. This difference in the two models accounts for differences in their behavior when the intrinsic growth rate $r$ is large. Their behavior is explored in the Exercises.

Returning to the Ricker model, we show that there exist two equilibria for $0 < r < 2$, $\bar{N} = 0$ and $\bar{N} = K$, and that the equilibrium $K$ is globally asymptotically stable. For the Ricker model, $f : [0, \infty) \to [0, \infty)$. The equilibria are found by solving

$$\bar{N} = \bar{N} \exp\left[r\left(1 - \frac{\bar{N}}{K}\right)\right].$$

It is easy to see that $\bar{N} = 0$ and $\bar{N} = K$ are the two equilibria. Next, $f(x) = x\exp[r(1 - x/K)]$ implies

$$f'(x) = \exp\left[r\left(1 - \frac{x}{K}\right)\right]\left(1 - r\frac{x}{K}\right)$$

and $f'(K) = 1 - r$. The equilibrium $K$ is locally asymptotically stable if

$$-1 < r - 1 < 1 \text{ or } 0 < r < 2.$$

In addition, it can be shown for the case $0 < r < 2$ that the positive equilibrium $K$ is globally asymptotically stable. Cull (1986) shows that when $0 < r < 2$, $f(f(x)) - x > 0$ for all $x \in [x_M, K)$, where $x_M$ is the value of $x$ where the maximum of $f$ occurs. Thus, according to Theorem 2.9, for the parameter region $0 < r < 2$, the equilibrium $K$ is globally asymptotically stable. The global stability result can be verified also by showing that there do not exist any 2-cycles (see Theorem 2.6). Recall in Theorem 2.7 that $x_{t+1} = f(x_t)$ has no 2-cycles iff $1 + f'(x) \neq 0$. See Exercise 26 in Chapter 2.

**Example 3.1** The whooping crane population (*Grus americana*) severely declined at the beginning of the twentieth century due to loss of wetland habitat. It was listed as an endangered species in 1970. In 1938 the total number of whooping cranes in North America was estimated at 29 (U.S. Fish & Wildlife Service, 2005). Conservation efforts have resulted in an increase in the population size. The largest population of

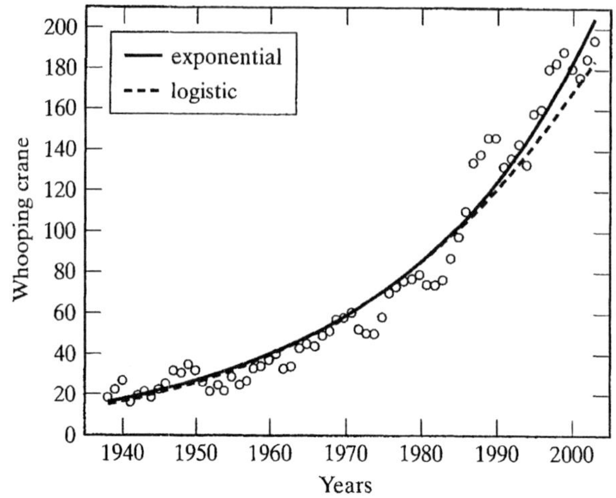


**Figure 3.2** Whooping crane population data from 1938 to 2003 fit to the exponential curve and the Beverton-Holt curve.

whooping cranes in North America breeds in Wood Buffalo National Park in northern Canada and spends the winter in Aransas National Wildlife Refuge on the Texas coast. Data from the Wood Buffalo/Aransas (WBA) winter population are graphed in Figure 3.2 (U.S. Fish & Wildlife Service, 2005). (See the Appendix to Chapter 3.) An exponential curve, $N_t = N_0\lambda^t$, and the Beverton-Holt curve are fit to the whooping crane population data $(t, N_t)$ using a least squares approximation. (The least squares approximation minimizes the sum of the squares of the errors, where an error is the distance from the data value at $t$ to the predicted value at $t$, $N_t$.)

For the exponential curve, let $y_t = \ln(N_t) = \ln(N_0) + t\ln(\lambda)$. The command *polyfit* in MATLAB finds the least squares estimates for the parameters. The estimate for $N_0 \approx 15.88$ and for $\lambda \approx 1.040$.

The Beverton-Holt difference equation has an explicit solution which is given in Exercise 10 in Chapter 2 and equals

$$N_t = \frac{N_0 K \lambda^t}{K + N_0(\lambda^t - 1)}.$$

If we assume $K$ is known, then we can make a change of variable and fit a linear curve. Making the change of variable

$$y_t = \ln\left(\frac{1}{N_t} - \frac{1}{K}\right)$$

in the Beverton-Holt solution leads to

$$y_t = \ln\left(\frac{1}{N_0} - \frac{1}{\mathrm{K}}\right) - t\ln\lambda.$$

A least squares approximation is used to fit the whooping crane data to the normalized logistic curve $y_t$ with $K = 500$. Estimates for the other parameters are $N_0 \approx 14.71$ and $\lambda = 1.046$. The estimate for $K$ is used only for illustration purposes; an estimate for $K$ is not known. For both models, the parameter estimates for $\lambda$ show that the whooping crane population is increasing on the average about 4% per year. See Figure 3.2. ■

## 3.3 Nicholson-Bailey Model

One of the earliest applications of a discrete-time model to a biological system involved two insects, a parasitoid and its host. The model is named after the two researchers, Nicholson and Bailey, who developed the model and applied it to the parasitoid, *Encarsia formosa*, and the host, *Trialeurodes vaporariorum* (1935). The term "parasitoid" means a parasite which is free living as an adult but lays eggs in the larvae or pupae of the host. Hosts that are not parasitized give rise to their own progeny. Hosts that are successfully parasitized die but the eggs laid by the parasitoid may survive to be the next generation of parasitoids.

Parasitoids are frequently used as biological control agents to control insect pests. Introduction of a parasitoid led to a successful biological control program involving the California citrus industry. In the late 1800s citrus crops in California were being ruined by the insect pest, cottony cushion scale insect. A parasitoid fly and a predatory insect, the vedalia beetle, were introduced from Australia (Hoffmann and Frodsham, 1993). Within a few years, the pest was under control.

The following derivation for the host-parasitoid model of Nicholson-Bailey is based on the discussion presented by Edelstein-Keshet (1988). The basic variables and parameters are defined as follows:

$N_t$ = density of host species in generation $t$.

$P_t$ = density of parasitoid species in generation $t$.

$f(N_t, P_t)$ = fraction of hosts not parasitized.

$r$ = number of eggs laid by a host that survive through the larvae, pupae and adult stages.

$e$ = number of eggs laid by a parasitoid on a single host that survive through larvae, pupae, and adult stages.

The parameters $r$ and $e$ are positive. Applying these definitions, the general host-parasitoid model has the following form:

$$N_{t+1} = rN_t f(N_t, P_t),$$
$$P_{t+1} = eN_t(1 - f(N_t, P_t)).$$

Note that if $N_t = 0$, then $P_{t+1} = 0$. The parasitoid cannot survive without the host. This is one reason why parasitoids are good biological control agents.

The Nicholson-Bailey model assumes a simple functional form for $f(N_t, P_t)$. The function $f$ depends on the searching behavior of the parasitoid. The number of encounters of the parasitoids, $P_t$, with the hosts, $N_t$, is in direct proportion to host density $N_t$, that is, it follows the law of mass action,

$$aN_tP_t,$$

where the constant $a$ is referred to as the *searching efficiency*, the probability that a given parasitoid will encounter a given host during its searching lifetime. Nicholson and Bailey assumed that $a$ was a constant. The number of encounters are distributed randomly among the available hosts and assumed to follow the Poisson distribution,

$$p(n) = \frac{\exp(-\mu)\mu^n}{n!}, \quad n = 0, 1, 2, \ldots,$$

where $n$ is the number of encounters and $\mu$ is the average number of encounters per host in one generation. Once a host is parasitized, it cannot be parasitized again. Therefore, only the first encounter between parasitoid and host is important. Hosts with no encounters $p(0)$ are separated from those with more than one encounter, $1 - p(0)$.

Now, we estimate the fraction of hosts not parasitized, $f(N_t, P_t)$. The probability of no encounters of the host by the parasitoid represents the fraction of hosts that are not parasitized, that is,

$$p(0) = \frac{\exp(-\mu)\mu^0}{0!} = \exp(-\mu).$$

The parameter $\mu$ can be estimated from the number of encounters:

$$\mu = \frac{\text{\# encounters}}{N_t} = \frac{aN_tP_t}{N_t} = aP_t.$$

Thus,

$$p(0) = \exp(-aP_t) = f(N_t, P_t).$$

It follows from the preceding discussion that the Nicholson-Bailey model has the following form:

$$\begin{aligned} N_{t+1} &= rN_t\exp(-aP_t) = F(N_t, P_t), \qquad (3.1)\\ P_{t+1} &= eN_t(1 - \exp(-aP_t)) = G(N_t, P_t). \end{aligned}$$

The equilibrium solutions and the local asymptotic stability of model (3.1) are analyzed. The equilibrium solutions of the Nicholson-Bailey model are found by solving the following equations simultaneously for $\bar{N}$ and $\bar{P}$,

$$\begin{aligned} \bar{N} &= r\bar{N}\exp(-a\bar{P}),\\ \bar{P} &= e\bar{N}(1 - \exp(-a\bar{P})). \end{aligned}$$

There are two equilibria. One equilibrium is the zero equilibrium, where $\bar{N} = 0$ and $\bar{P} = 0$. The other equilibrium is

$$\bar{P} = \frac{\ln r}{a}, \quad \text{and} \quad \bar{N} = \frac{r \ln r}{(r-1)ae}.$$

Note that the latter equilibrium is positive if $r > 1$.

To analyze the local asymptotic stability, the Jacobian matrix is calculated and evaluated at an equilibrium. Computer algebra systems can be used to simplify calculations. In the Appendix to Chapter 3, commands in Maple are given for setting up and evaluating the Jacobian matrix. The partial derivatives of $F(N, P) = rN\exp(-aP)$ and $G(N, P) = eN(1 - \exp(-aP))$ are calculated below for the Jacobian matrix,

$$J(N, P) = \begin{pmatrix} r\exp(-aP) & -arN\exp(-aP) \\ e(1 - \exp(-aP)) & aeN\exp(-aP) \end{pmatrix}.$$

At the zero equilibrium $J$ is given by

$$J(0, 0) = \begin{pmatrix} r & 0 \\ 0 & 0 \end{pmatrix}.$$

Since the eigenvalues are $\lambda_{1,2} = r, 0$, the zero equilibrium is locally asymptotically stable if $|r| < 1$ and unstable if $|r| > 1$. Since $r > 0$, the zero equilibrium is locally asymptotically stable if $0 < r < 1$ and unstable if $r > 1$.

The Jacobian matrix at the positive equilibrium, $\bar{P} = \ln(r)/a$ and $\bar{N} = r\ln r/[(r-1)ae]$, $r > 1$, can be simplified by noting $r\exp(-a\bar{P}) = 1$. Thus,

$$J(\bar{N}, \bar{P}) = \begin{pmatrix} 1 & \dfrac{-r \ln r}{(r-1)e} \\ \dfrac{(r-1)e}{r} & \dfrac{\ln r}{(r-1)} \end{pmatrix}.$$

The equilibrium is locally asymptotically stable if

$$\left| 1 + \frac{\ln r}{(r-1)} \right| < 1 + \frac{\ln r}{(r-1)} + \ln r < 2. \tag{3.2}$$

The first inequality is satisfied when $r > 1$. It will be shown by the following theorem that the second inequality is *not* satisfied, which implies that the positive equilibrium is unstable. There exist eigenvalues of the Jacobian matrix satisfying $|\lambda_i| > 1$.

**Theorem 3.1** *The positive equilibrium in the Nicholson-Bailey model (3.1), where*

$$\bar{P} = \frac{\ln r}{a}, \quad \text{and} \quad \bar{N} = \frac{r \ln r}{(r-1)ae}$$

*for $r > 1$, is unstable.*

**Proof** Consider the second inequality in (3.2) and define $s(r) = (r-1)[1 + \ln(r)/(r-1) + \ln(r) - 2] = 1 - r + r\ln(r)$. If $s(r) < 0$, then the second inequality is satisfied. However, it will be shown that $s(r) > 0$, which implies the equilibrium is unstable.

First, note that $s(1) = 1 - 1 - \ln 1 = 0$ and $s'(r) = -1 + \ln r + 1 = \ln r$. Thus, $s'(r) > 0$ for $r > 1$; $s$ is strictly increasing for $r > 1$. Thus, $s(r) > 0$ for $r > 1$. The equilibrium is unstable. □

Figure 3.3 illustrates the dynamics of the Nicholson-Bailey model (3.1) when $r < 1$ and $r > 1$. When $r > 1$, the oscillations increase in amplitude, but they also get very close to zero so that numerically the values are set to zero; population extinction occurs even in this case.

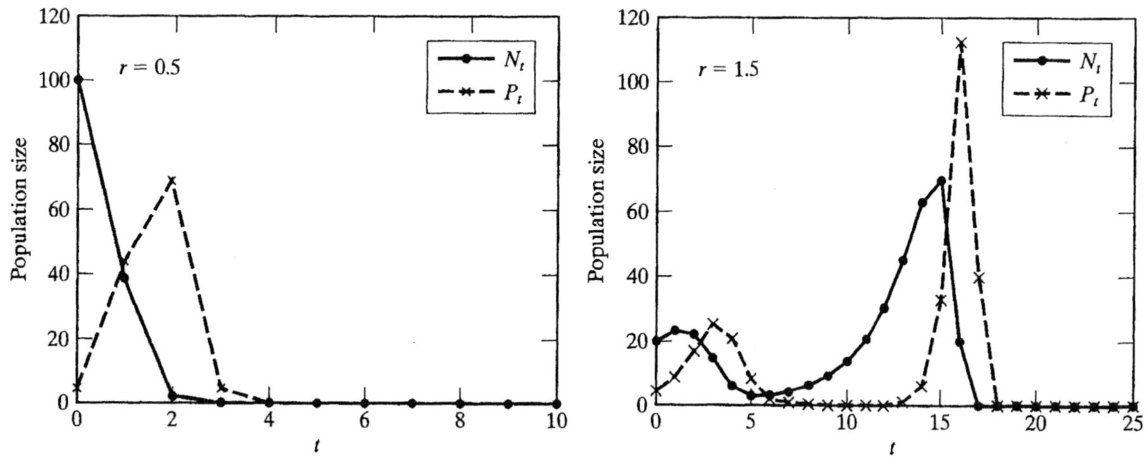


**Figure 3.3** Some simulations of the Nicholson-Bailey model when $r < 1$ and $r > 1$ ($a = 0.05$ and $e = 2$).

## 3.4 Other Host-Parasitoid Models

There are a number of unrealistic assumptions in the Nicholson-Bailey model. For example, a constant reproductive rate of the host, a constant searching efficiency, and a homogeneous environment are unrealistic assumptions. With additional realistic assumptions, the positive equilibrium can be locally asymptotically stable. Some of these more realistic assumptions are discussed next.

Suppose the reproductive rate $r$ in the host equation is replaced by a density-dependent factor, $g(N_t)$,

$$\begin{aligned} N_{t+1} &= N_t g(N_t)\exp(-aP_t), \\ P_{t+1} &= eN_t(1 - \exp(-aP_t)). \end{aligned}$$

Some forms for the density-dependent factor $g$ that have been studied include

$$\frac{r}{N_t^b}, \quad \exp\left(r\left(1 - \frac{N_t}{K}\right)\right) \quad \text{and} \quad \frac{r}{(1 + cN_t)^b}.$$

For some parameter values the density-dependent factor has a stabilizing influence. If the host population does not grow too fast (the value of $r$ in the density-dependent factor should not be too large), then there exists a locally asymptotically stable equilibrium. See Figure 3.4 when $g(N_t) = \exp(1.5(1 - N_t/25))$.

Suppose the environment is not homogeneous. Suppose the environment is patchy, so that a proportion of the host population may find a refuge and be safe from attack by parasitoids. Let $\gamma$ be the proportion of hosts that are not safe from attack by parasitoids and $1 - \gamma$ be the proportion of hosts that are safe within a refuge. In this case, the model has the form

$$\begin{aligned} N_{t+1} &= r(1 - \gamma)N_t + r\gamma N_t\exp(-aP_t), \\ P_{t+1} &= \gamma eN_t[1 - \exp(-aP_t)]. \end{aligned}$$

Another way to model the effect of a refuge is to assume there is a constant number of hosts $N_0$ in a refuge. The hosts, $N_0$, are safe from attack by parasitoids. The model has the form

$$\begin{aligned} N_{t+1} &= rN_0 + r(N_t - N_0)\exp(-aP_t), \\ P_{t+1} &= e(N_t - N_0)[1 - \exp(-aP_t)]. \end{aligned}$$

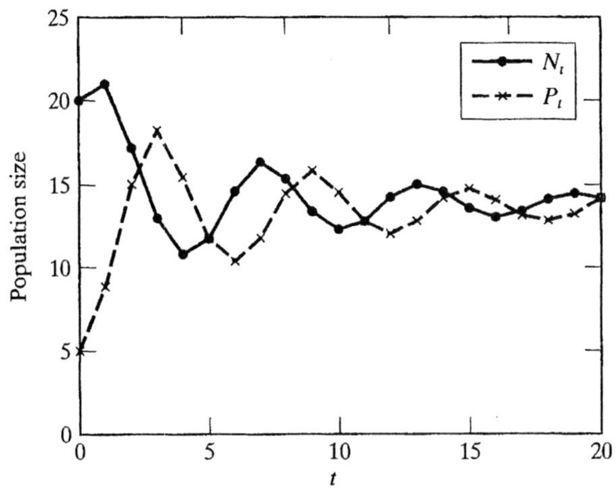


**Figure 3.4** Modified Nicholson-Bailey model with density-dependent factor $g(N_t) = \exp(1.5(1 - N_t/25))$ ($a = 0.05$ and $e = 2$).

Another generalization of the Nicholson-Bailey host-parasitoid model is to include more than one parasitoid. Suppose there is one host and two parasitoids which parasitize the same host. Suppose the host density is denoted by $N_t$ and the two parasitoid densities are denoted by $P_t$ and $Q_t$. In addition, suppose the parasitoid $P$ acts first, followed by $Q$ that acts on the surviving hosts. For example, the host species may be attacked in different developmental stages by a range of parasitoids. Hassell (1978) discusses a winter moth that is parasitized by egg, larval, and pupal parasitoids. A model with this form is

$$\begin{aligned} N_{t+1} &= rN_t f_1(P_t) f_2(Q_t), \\ P_{t+1} &= e_1 N_t[1 - f_1(P_t)], \\ Q_{t+1} &= e_2 N_t f_1(P_t)[1 - f_2(Q_t)]. \end{aligned}$$

If the proportion of hosts not parasitized by the parasitoids takes the same form as in the Nicholson-Bailey model [i.e., $f_i(x) \equiv \exp(-a_i x)$], then the positive equilibrium will be unstable. However, for some other forms, there are regions in parameter space that give rise to a stable positive equilibrium.

A model in which a parasitoid parasitizes another parasitoid is referred to as a host-parasitoid-hyperparasitoid system. In this case, $N$ is the host, $P$ is the primary parasitoid, and $Q$ is the hyperparasitoid which only parasitizes hosts first parasitized by $P$. The function $f_1$ is the fraction of hosts $N$ not parasitized by $P$ and $f_2$ is the fraction of parasitoids $P$ not parasitized by $Q$. The model has the form

$$\begin{aligned} N_{t+1} &= rN_t f_1(P_t), \\ P_{t+1} &= e_1 N_t[1 - f_1(P_t)] f_2(Q_t), \\ Q_{t+1} &= e_2 N_t[1 - f_1(P_t)][1 - f_2(Q_t)]. \end{aligned}$$

Hassell (1978) studies this host-parasitoid-hyperparasitoid model when

$$f_i(x) = \left(1 + \frac{a_i x}{k_i}\right)^{-k_i}, \quad a_i, k_i > 0, \quad i = 1, 2.$$

These particular forms are derived from a negative binomial distribution rather than a Poisson distribution. The search for hosts by parasitoids is not random. For example, parasitoids may aggregate in patches where there is high prey density. The parameter $k_i$ is related to parasitoid aggregation. Aggregation is strongest when $k_i \to 0$,

$$\lim_{k_i \to 0} \left(1 + \frac{a_i x}{k_i}\right)^{-k_i} = 1,$$

and weakest when $k_i \to \infty$,

$$\lim_{k_i \to \infty} \left(1 + \frac{a_i x}{k_i}\right)^{-k_i} = \exp(-a_i x).$$

In this latter case, the form of $f$ is the one in Nicholson-Bailey's model (3.1). Some extensions and applications to biological control in host-parasitoid systems can be found in the references (Barlow et al., 2004; Hassell, et al., 1991; King and Hastings, 2003; Lynch et al., 2002; Várkonyi et al., 2002).

## 3.5 Host-Parasite Models

Host-parasite models are similar to host-parasitoid models, except that the parasite does not necessarily kill the host. A host-parasite model was formulated by Leslie and Gower in 1960 and has a particularly simple form. Let $H_t$ denote the host and $P_t$ the parasite populations at time $t$. Then the host-parasite model is defined as follows:

$$H_{t+1} = \frac{\alpha_1 H_t}{1 + \gamma_1 P_t}, \tag{3.3}$$
$$P_{t+1} = \frac{\alpha_2 P_t}{1 + \gamma_2 P_t/H_t},$$

where $H_0 > 0$, $P_0 > 0$, $\alpha_i > 0$, and $\gamma_i > 0$, $i = 1, 2$. The parameters $\alpha_i$ are growth rates of the host and parasite populations in the absence of the other population. The larger the quantity $\gamma_1 P_t$, the greater the reduction of the host population. The larger the ratio $P_t/H_t$, the smaller the number of hosts per parasite resulting in a reduction in the parasite population (Pielou, 1977).

It is easy to see that solutions to (3.3) are positive for $t > 0$. In addition, if $\alpha_1 < 1$, then $\lim_{t\to\infty} H_t = 0$ and if $\alpha_2 < 1$, then $\lim_{t\to\infty} P_t = 0$. In Exercise 6, the dynamics when $\alpha_i < 1$ for $i = 1$ or $i = 2$ are summarized. In the following analysis, the parameters $\alpha_i$ are assumed to be greater than one, $\alpha_i > 1$, $i = 1, 2$.

Model (3.3) has a unique positive equilibrium given by

$$\bar{P} = \frac{\alpha_1 - 1}{\gamma_1} \quad \text{and} \quad \bar{H} = \frac{\gamma_2}{\alpha_2 - 1}\bar{P} = \frac{\gamma_2(\alpha_1 - 1)}{\gamma_1(\alpha_2 - 1)}.$$

The Jacobian matrix evaluated at the positive equilibrium has a simple form,

$$J = \begin{pmatrix} 1 & -\dfrac{\gamma_2(\alpha_1 - 1)}{\alpha_1(\alpha_2 - 1)} \\ \dfrac{(\alpha_2 - 1)^2}{\alpha_2\gamma_2} & \dfrac{1}{\alpha_2} \end{pmatrix}.$$

The conditions in Theorem 2.10 can be easily checked to see whether the equilibrium is locally asymptotically stable. It is straightforward to see that the trace is positive and the trace is also less than one plus the determinant. Hence, the only condition to check for local stability is that the determinant is less than one,

$$\frac{\alpha_1\alpha_2 - \alpha_2 + 1}{\alpha_1\alpha_2} < 1.$$

This latter condition reduces to $\alpha_2 > 1$. The stability conditions in Theorem 2.10 are always satisfied under the conditions $\alpha_i > 1, i = 1, 2$.

If the host population is a pest, then according to the Leslie-Gower model, a fast-growing parasite population with a growth rate larger than that of the host that significantly reduces the host population, $\alpha_2 > \alpha_1$ and $\gamma_1 > \gamma_2$, would help in reducing the pest population. A numerical example in Figure 3.5 illustrates the dynamics of model (3.3).

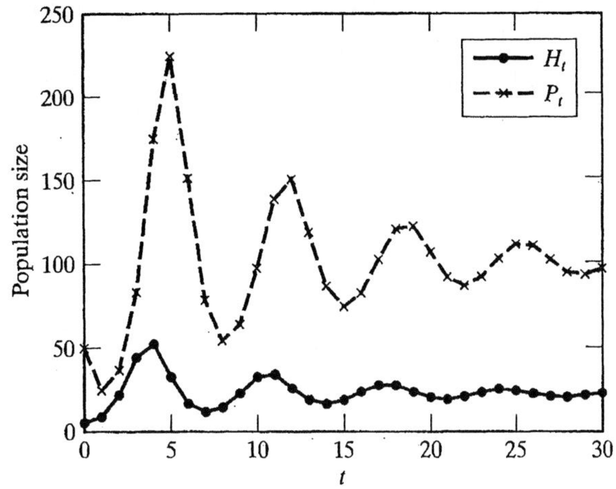


**Figure 3.5** Leslie-Gower model (3.5) of a host-parasite system, ($\alpha_1 = 5$, $\alpha_2 = 10$, $\gamma_1 = 0.04$, and $\gamma_2 = 2$). The stable equilibrium is $\bar{P} = 100$ and $\bar{H} = 22.22$.

## 3.6 Predator-Prey Models

Predator-prey models are similar to both host-parasite and host-parasitoid models. However, unlike the latter two systems, the predator does not live on the host. The prey serves as a food source for the predator. In this section, we discuss a discrete-time predator-prey model studied by Neubert and Kot (1992). This model is related to the well-known continuous-time, Lotka-Volterra predator-prey model, named after Alfred Lotka (1880–1949) and Vito Volterra (1860–1940), who contributed to the analysis of the continuous-time predator-prey model. The continuous-time predator-prey model is studied in Chapter 6, Section 6.3. The dynamics of this model are much more complex than its continuous analogue or the Leslie-Gower model, although it appears to be much simpler. The Neubert-Kot model exhibits some new dynamics that are not present in scalar difference equations.

In the model of Neubert and Kot (1992), the per capita growth rates are linear. The model has the following form:

$$N_{t+1} = N_t + rN_t\left(1 - \frac{N_t}{K}\right) - eN_tP_t, \tag{3.4}$$

$$P_{t+1} = bN_tP_t + (1 - d)P_t,$$

where $N_t$ is prey density at time $t$ and $P_t$ is predator density at time $t$. The parameters $r$, $e$, $b$, $d$, and $K$ are positive. The term $rN_t(1 - N_t/K)$ represents logistic growth, $r$ is the intrinsic growth rate, and $K$ is the carrying capacity. The term $eP_t$ is the per capita prey reduction due to consumption by the predator (assumes the law of mass action). The term $bN_t$ is per capita predator increase due to prey consumption and $d$ is the death rate of the predator. Note that the original system (3.4) has five parameters. A simpler version of the predator-prey model (3.4) is formulated by reducing the number of parameters and assuming that predators only live one generation, $d = 1$.

A change of variable reduces the model with $d = 1$ containing four parameters to one with only two parameters. This technique was used in Chapter 2, when the approximate logistic equation was reduced to a simpler form containing only one parameter. The new variables are chosen to be

$$x_t = \frac{N_t}{K}, \quad y_t = \frac{eP_t}{bK}, \quad \text{and} \quad c = bK.$$

Note that the variables $x_t$ and $y_t$ are dimensionless (e.g., $K$ has the same units as $N_t$ and $eP_t$ has the same units as $bK$). There are other choices for the change of variables to put the system in dimensionless form, but this is a good choice for our purposes. With this change of variables, the new system has only two parameters, $r$ and $c$:

$$\begin{aligned} x_{t+1} &= (r+1)x_t - rx_t^2 - cx_t y_t, \\ y_{t+1} &= cx_t y_t. \end{aligned} \tag{3.5}$$

System (3.5) is much simpler to analyze than the original system because now there are only two parameters.

Before we begin the analysis of (3.5), it must be noted that the question of nonnegative solutions has not been addressed. If the initial conditions, $x_0$ and $y_0$, are sufficiently large, it is easy to see that $x_1$ could become negative. The variable $x_{t+1}$ could be redefined to alleviate this problem, for example,

$$x_{t+1} = \max\{0, (r+1)x_t - rx_t^2 - cx_t y_t\}. \tag{3.6}$$

If $x_{t+1} = 0$, then at the next time step, both predator and prey are zero, $(x_{t+2}, y_{t+2}) = (0, 0)$; complete extinction occurs. In all of the numerical simulations performed here, solutions remained nonnegative, even without the restriction given in (3.6). Therefore, in the following analysis, it is assumed that $x_t \geq 0$ and $y_t \geq 0$ for all time. Note if the predator is absent, $y_0 = 0$, then $x_t$ satisfies the approximate logistic equation. In this case, for $x_t$ to be positive for all time requires $0 < x_0 < (r+1)/r$ and $r < 3$ (no restrictions on $c$). But the presence of the predator, $y_t$, reduces the size of $x_{t+1}$. Thus, for solutions of (3.5) to be positive, the restrictions on $r$ can be relaxed somewhat; new restrictions on $r$ will depend on $c$.

First, the equilibria of system (3.5) are identified, then their local stability is analyzed. The equilibria satisfy

$$\begin{aligned} \bar{x} &= \bar{x}(r + 1 - r\bar{x} - c\bar{y}), \\ \bar{y} &= c\bar{x}\bar{y}. \end{aligned}$$

It can be seen that there exist equilibria at $\bar{x} = 0 = \bar{y}$ and at $\bar{x} = 1$ and $\bar{y} = 0$. Also, there exists a positive equilibrium if $c > 1$,

$$\bar{x} = \frac{1}{c} \quad \text{and} \quad \bar{y} = \frac{r(c-1)}{c^2}.$$

The Jacobian matrix for system (3.5) is

$$J(x, y) = \begin{pmatrix} r + 1 - 2rx - cy & -cx \\ cy & cx \end{pmatrix}.$$

The Jacobian matrix evaluated at the zero equilibrium is

$$J(0,0) = \begin{pmatrix} r+1 & 0 \\ 0 & 0 \end{pmatrix}.$$

This matrix has a positive eigenvalue $r + 1 > 1$. The equilibrium with both species extinct is unstable.

The Jacobian matrix evaluated at $(1, 0)$ is

$$J(1,0) = \begin{pmatrix} 1-r & -c \\ 0 & c \end{pmatrix}.$$

This matrix has the eigenvalues $\lambda_{1,2} = 1 - r, c$. The equilibrium with only the prey present is locally asymptotically stable if $0 < r < 2$ and $0 < c < 1$. The growth rates of the prey and predator must be within a certain range, but not too large.

Finally, the Jacobian matrix at the positive equilibrium $(\bar{x}, \bar{y})$, where $c > 1$, is

$$J(\bar{x}, \bar{y}) = \begin{pmatrix} 1 - r/c & -1 \\ r - r/c & 1 \end{pmatrix}.$$

The stability criterion from Theorem 2.10 is

$$|2 - r/c| < 2 + r - 2r/c < 2.$$

Each of these inequalities can be stated separately,

$$2 - r/c < 2 + r - 2r/c,$$
$$2 - r/c > -2 - r + 2r/c,$$
$$2 + r - 2r/c < 2.$$

If the first condition fails there is a real eigenvalue, $\lambda_i \geq 1$. If the second condition fails there is a real eigenvalue, $\lambda_i \leq -1$. If the third condition fails and the eigenvalue is complex, its magnitude $|\lambda_i| \geq 1$. The three stability conditions simplify as follows:

$$c > 1,$$
$$r < \frac{4c}{3 - c} \quad \text{if} \quad c < 3,$$
$$c < 2.$$

The three conditions together imply

$$1 < c < 2 \quad \text{and} \quad r < \frac{4c}{3 - c}.$$

These conditions are graphed in the $r$-$c$ parameter space (Figure 3.6). The parameter $c$ must be large to guarantee that the predator persists, but not too large. In addition, the parameter $r$ must not be too large and its magnitude depends on the magnitude of $c$.

As a stability boundary is crossed in Figure 3.6, at least one eigenvalue has magnitude equal to one. The type of eigenvalue, positive or negative, real or complex determines the type of bifurcation. There exists a trans-critical bifurcation when $\lambda = 1$, a flip bifurcation when $\lambda = -1$, and a Hopf bifurcation when the eigenvalues are complex and satisfy $|\lambda| = 1$ (Neubert and Kot, 1992). The terminology "Hopf bifurcation" is more often used

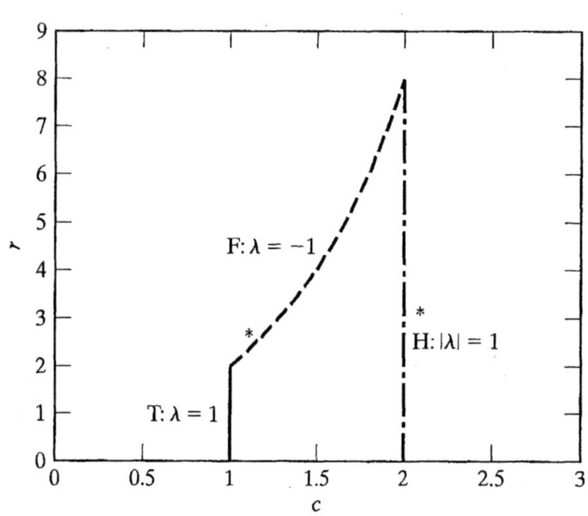


**Figure 3.6** Stability region in the $r$-$c$ parameter space. The labels $T$, $F$, and $H$ mean either a trans-critical, flip (period-doubling), or Hopf bifurcation occurs as the particular boundary is crossed.

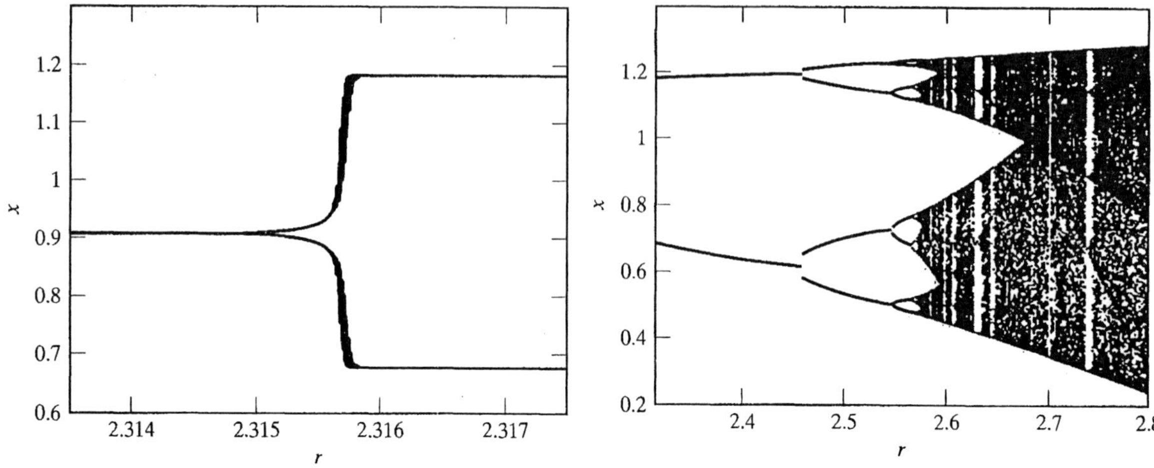


**Figure 3.7** Bifurcation diagram for the normalized predator-prey system when $c = 1.1$ and $r \in [2.3135, 2.3175]$ and $r \in [2.317, 2.8]$ (flip or period-doubling bifurcations). The cycles of the prey $x$ are graphed.

in connection with differential equations (see Chapter 5). In difference equations, such types of bifurcations are referred to as *Neimark-Sacker bifurcations* (Elaydi, 2000; Hale and Koçak, 1991). Figures 3.7 and 3.8 are bifurcation diagrams for a period-doubling bifurcation (flip) and a Neimark-Sacker bifurcation, respectively. When a flip bifurcation occurs, the predator becomes extinct and the prey exhibits period-doubling behavior (Figure 3.7).

A Neimark-Sacker bifurcation represents a new type of bifurcation that occurs in systems of difference equations, but not in scalar difference equations. When this bifurcation occurs, there is a pair of purely imaginary eigenvalues. In this case, there exists a periodic solution, but the period may not be integer valued (Figure 3.8).

Solutions to the normalized predator-prey system are graphed in Figure 3.9 when $r = 3.1$ and $c = 2.1$ and when $r = 2.5$ and $c = 1.1$ (when the flip and Hopf bifurcation boundaries are crossed in Figure 3.6). They illustrate a periodic solution for the prey and predator and a 2-cycle for the prey, respectively.

**Figure 3.8** Bifurcation diagram for the normalized predator-prey system when $r = 3.1$ and $c \in [1.9, 2.4]$. The cycles of the predator $y$ are graphed.

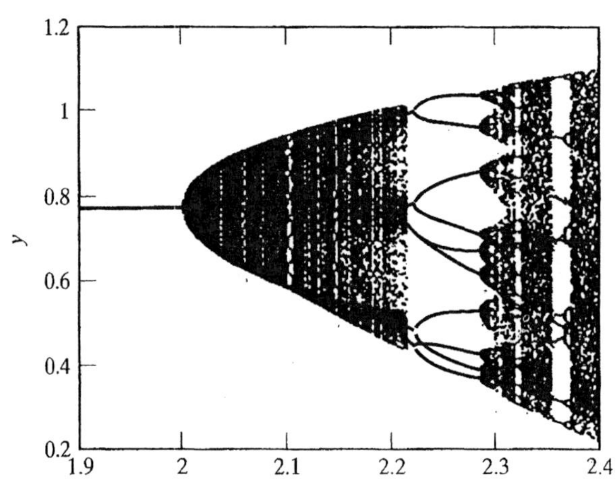

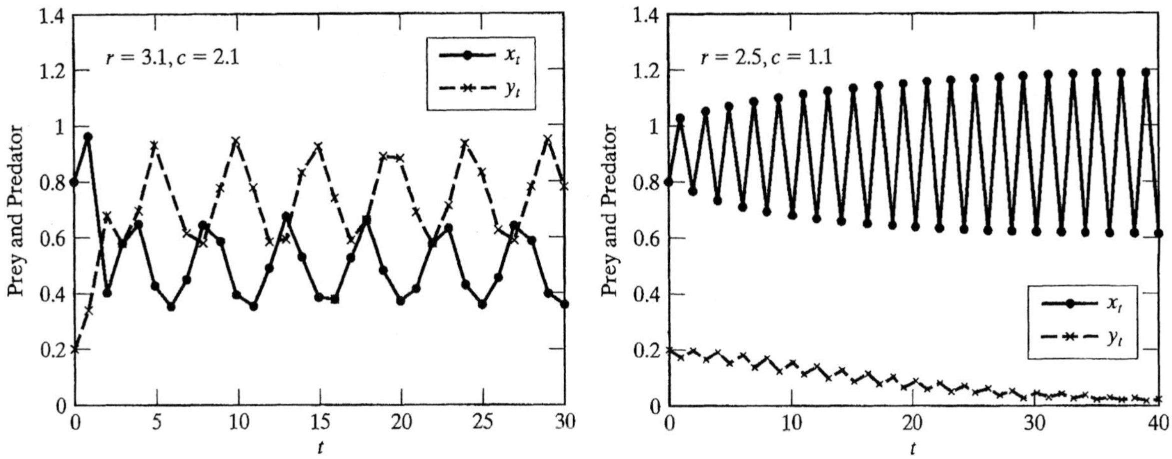


**Figure 3.9** Solutions to the normalized predator-prey system when $r = 3.1$ and $c = 2.1$ (periodic solution) and when $r = 2.5$ and $c = 1.1$ (period 2 solutions for the prey and the predator goes extinct).

## 3.7 Population Genetics Models

Inheritance depends on the information contained in the chromosomes that are passed down from generation to generation. Humans have two sets of 23 chromosomes (diploid), making a total of 46 chromosomes; one set is obtained from each parent. Certain locations along the chromosomes contain the instructions for some characteristic, such as eye or hair color. The locations along the chromosomes are called the *loci* (a single location is called a *locus*). The instructions within the chromosomes are referred to as the *genes*. Each gene gives a unique instruction (for color of eyes, color of hair, etc.) and each human has two genes per locus because there are two sets of chromosomes. The physical characteristics (eye or hair color) unique to each individual are determined by that individual's genes. In simple organisms, such as bacteria, there are 2000 to 3000 genes, whereas in higher organisms such as plants and animals there are 50,000 to 100,000 genes (Clark and Russell, 1997). Each gene has different variant forms (the gene for eye color can be green, blue, brown, etc.). These different variant forms of the genes are referred to as *alleles*. Here, we shall consider the simplest possible case, the case where there are only two different alleles associated with a particular gene.

Suppose there are two alleles for a given gene. The two alleles are denoted $a$ and $A$. A human with two sets of chromosomes could then have one of three different combinations on his or her chromosome: $AA$, $Aa$, or $aa$. The combinations $AA$ and $aa$ are *homozygous*, whereas the combination $Aa$ is *heterozygous*. The three combinations, $AA$, $Aa$, and $aa$, are called the *genotypes* of the locus.

One of the two alleles may be *dominant*. For example, if $A$ is the dominant allele, then $a$ is referred to as the *recessive allele*. Then genotypes $AA$ and $Aa$ correspond to the same physical trait, but different from that of $aa$. This is also described as saying that genotypes $AA$ and $Aa$ have *phenotype A* and $aa$ has *phenotype a* (Hoppensteadt, 1975).

We explore the question of whether the allele frequencies (associated with a particular gene) change in a given population over time as individuals within that population mate and reproduce. Our population genetics model is a simple one-locus, two-allele model. We assume that during each time step, the population in generation $t$ is replaced by the population in generation $t + 1$.

An important principle in population genetics is known as the *Hardy-Weinberg law*. First, the following assumptions must hold: (1) Mating is random, (2) there is no variation in the number of progeny from parents of different genotypes, (3) all genotypes are equally fit, and (4) there are no mutations. Then the Hardy-Weinberg law asserts that gene frequencies and allele frequencies do not change from one generation to the next. Our population genetics model is based on these assumptions. We will show that our model follows the Hardy-Weinberg law. The name Hardy-Weinberg recognizes the work of G. H. Hardy, a famous English mathematician (number theorist) and Wilhelm Weinberg, a German physician and human geneticist, who independently discovered this result in 1908 (Felsenstein, 2003).

The following definitions and assumptions are needed. Let $N$ be the total population size. Since each individual has two alleles per locus, there are a total of $2N$ alleles in the population. Let

$$\begin{aligned} p &= \text{frequency of allele } A = (\text{total number of} A \text{ alleles})/(2N), \\ q &= \text{frequency of allele } a = (\text{total number of } a \text{ alleles})/(2N), \end{aligned}$$

then $p + q = 1$. Let

$$\begin{aligned} p_{AA} &= \text{frequency of } AA \text{ genotype}, \\ p_{Aa} &= \text{frequency of } Aa \text{ genotype}, \\ p_{aa} &= \text{frequency of } aa \text{ genotype}. \end{aligned}$$

Thus, the frequency of $A$ alleles is

$$p = \frac{2Np_{AA} + Np_{Aa}}{2N} = p_{AA} + \frac{p_{Aa}}{2}.$$

The frequency of $a$ alleles is

$$q = 1 - p = \frac{p_{Aa}}{2} + p_{aa}.$$

To determine what happens after one generation of mating, it is necessary to consider all possible matings, their frequency, and all possible offspring and their frequency. The possible matings are obtained by considering all possible pairings of $AA$, $Aa$, and $aa$ (representing a genotype for each parent) for which there are $3(2) = 6$:

$$AA \times AA,\ AA \times Aa,\ AA \times aa,\ Aa \times Aa,\ Aa \times aa,\ aa \times aa.$$

Each of these occur with the corresponding frequencies:

$$p_{AA}^2,\ 2p_{AA}p_{Aa},\ 2p_{AA}p_{aa},\ p_{Aa}^2,\ 2p_{Aa}p_{aa},\ p_{aa}^2.$$

This information plus the information needed for the offspring frequencies are given in the following mating and offspring table, Table 3.1 (Hastings, 1998). Note that the sum of the mating frequencies equals $(p_{AA} + p_{Aa} + p_{aa})^2$.

Let $p'_{AA}$, $p'_{Aa}$, and $p'_{aa}$ denote the genotypic frequencies in the next generation. Then, applying the results from Table 3.1,

$$\begin{aligned} p'_{AA} &= p_{AA}^2 + p_{AA}p_{Aa} + p_{Aa}^2/4 \\ &= (p_{AA} + p_{Aa}/2)^2 \\ &= p^2, \end{aligned}$$

**Table 3.1** Mating and offspring table.

| Mating | Mating Frequency | Offspring Fraction $AA$ | $Aa$ | $aa$ | Next Generation $AA$ | $Aa$ | $aa$ |
|---|---|---|---|---|---|---|---|
| $AA \times AA$ | $p_{AA}^2$ | 1 | 0 | 0 | $p_{AA}^2$ | 0 | 0 |
| $AA \times Aa$ | $2p_{AA}p_{Aa}$ | 1/2 | 1/2 | 0 | $p_{AA}p_{Aa}$ | $p_{AA}p_{Aa}$ | 0 |
| $AA \times aa$ | $2p_{AA}p_{aa}$ | 0 | 1 | 0 | 0 | $2p_{AA}p_{aa}$ | 0 |
| $Aa \times Aa$ | $p_{Aa}^2$ | 1/4 | 1/2 | 1/4 | $p_{Aa}^2/4$ | $p_{Aa}^2/2$ | $p_{Aa}^2/4$ |
| $Aa \times aa$ | $2p_{Aa}p_{aa}$ | 0 | 1/2 | 1/2 | 0 | $p_{Aa}p_{aa}$ | $p_{Aa}p_{aa}$ |
| $aa \times aa$ | $p_{aa}^2$ | 0 | 0 | 1 | 0 | 0 | $p_{aa}^2$ |

$$\begin{aligned} p'_{aa} &= p_{Aa}^2/4 + p_{Aa}p_{aa} + p_{aa}^2 \\ &= (p_{aa} + p_{Aa}/2)^2 \\ &= q^2, \end{aligned}$$

and

$$\begin{aligned} p'_{Aa} &= p_{AA}p_{Aa} + 2p_{AA}p_{aa} + p_{Aa}^2/2 + p_{Aa}p_{aa} \\ &= 2(p_{AA} + p_{Aa}/2)(p_{aa} + p_{Aa}/2) \\ &= 2pq. \end{aligned}$$

These results can be used to find the allele frequencies in the next generation, $p'$ and $q'$,

$$\begin{aligned} p' &= p'_{AA} + p'_{Aa}/2 \\ &= p^2 + pq \\ &= p(p + q) = p \end{aligned}$$

and

$$\begin{aligned} q' &= p'_{aa} + p'_{Aa}/2 \\ &= q^2 + pq \\ &= q(p + q) = q. \end{aligned}$$

The frequencies remain constant from generation to generation, that is, from generation $t$ to $t + 1$, $p_{t+1} = p_t$ and $q_{t+1} = q_t$. Hence,

$$p_t = p_0 \quad \text{and} \quad q_t = q_0.$$

The Hardy-Weinberg law has been verified.

**Theorem 3.2** **(Hardy-Weinberg Law).** *Assume in a parent population, a particular gene has two alleles A and a, and the initial proportion of allele A is $p_0$ and the initial proportion of allele a is $q_0$. In addition, assume (i) mating is random, (ii) there is no variation in the number of progeny from parents of different genotypes, (iii) all genotypes have equal survival probability, (iv) there is no immigration nor* *emigration, (v) there are no mutations, and (vi) generations are nonoverlapping. Then, in generation t, the allele frequencies do not change,*

$$p_t = p_0 \quad \text{and} \quad q_t = q_0.$$

*In addition, the genotypic frequencies do not change from the second generation onwards,*

$$p_{AA} = p_0^2, \quad p_{Aa} = 2p_0q_0, \quad \text{and} \quad p_{aa} = q_0^2.$$ □

According to the Hardy-Weinberg law, the recessive trait will not die out but remain in the population at a fixed proportion. If the assumptions in Theorem 3.2 are violated, then the Hardy-Weinberg proportions change. We consider a violation of assumption (iii).

Suppose the survival rates depend on genotype. In this case, different genotypes have different fitnesses. The frequencies of allele $A$ and the proportion $p$ are modeled over time. Let $p_t$ be the frequency of allele $A$ in generation $t$ and $q_t$ be the frequency of allele $a$ ($p_t + q_t = 1$). Let $w_{AA}$ and $w_{aa}$ be the constant survival rates of genotypes $AA$ and $aa$ relative to the heterozygote genotype $Aa$, which is assumed to satisfy $w_{Aa} = 1$. Note that $w_{AA}$ and $w_{aa}$ can be less than or greater than one, but must be nonnegative. Let the *mean fitness* be denoted as

$$w = p^2w_{AA} + 2pqw_{Aa} + q^2w_{aa}.$$

Suppose initially the genotypic frequencies $AA$, $Aa$, and $aa$ are in the proportions $p^2$, $2pq$, and $q^2$, respectively. Then, it follows from the following genotypic frequency table, Table 3.2, that the next generation satisfies

$$\begin{aligned} p_{t+1} &= p_{AA} + p_{Aa}/2 \\ &= p_t^2w_{AA}/w_t + (1/2)2p_tq_tw_{Aa}/w_t \\ &= p_t(p_tw_{AA} + q_tw_{Aa})/w_t \\ &= p_t(p_tw_{AA} + (1 - p_t)w_{Aa})/w_t, \end{aligned}$$

where

$$w_t = p_t^2w_{AA} + 2p_tq_tw_{Aa} + q_t^2w_{aa}$$

is the mean fitness in generation $t$ (see Hastings, 1998). The following difference equation models the change in the allele frequency $A$ from generation $t$ to generation $t + 1$,

$$p_{t+1} = \frac{p_t^2w_{AA} + p_t(1 - p_t)w_{Aa}}{w_t}. \tag{3.7}$$

**Table 3.2** Genotypic frequency table, where the mean fitness is given by $w = p^2w_{AA} + 2pq + q^2w_{aa}$.

| | Genotype | | |
|---|---|---|---|
| | $AA$ | $Aa$ | $aa$ |
| Juvenile frequencies | $p^2$ | $2pq$ | $q^2$ |
| Relative survival rates | $w_{AA}$ | $w_{Aa}$ | $w_{aa}$ |
| Relative adult frequencies | $p^2w_{AA}$ | $2pqw_{Aa}$ | $q^2w_{aa}$ |
| Adult frequencies | $p^2w_{AA}/w$ | $2pqw_{Aa}/w$ | $q^2w_{aa}/w$ |

Note that if $w_{AA} = w_{Aa} = w_{aa} = 1$, then $w_t = 1$ and $p_{t+1} = p_t$.

Suppose that the relative survival rates satisfy

$$w_{AA} = 1 - s, \quad w_{Aa} = 1 \quad \text{and} \quad w_{aa} = 1 - r.$$

Then $s$ and $r$ can be positive or negative but $w_{AA}$ and $w_{aa}$ must be nonnegative so that $r, s < 1$ (but not both zero). Then

$$w_t = p_t^2(1 - s) + 2p_tq_t + q_t^2(1 - r) = 1 - p_t^2 s - (1 - p_t)^2 r.$$

The difference equation in $p$ satisfies

$$p_{t+1} = \frac{p_t[p_t(1 - s) + (1 - p_t)]}{1 - p_t^2 s - (1 - p_t)^2 r} = \frac{p_t(1 - p_t s)}{1 - p_t^2 s - (1 - p_t)^2 r} = f(p_t). \quad (3.8)$$

Next the equilibria are determined for the difference equation (3.8) and their local stability assessed.

There are three equilibria for the difference equation (3.8). They are $\bar{p} = 0$ and the solutions to

$$\bar{p}^2 s + (1 - \bar{p})^2 r = \bar{p}s.$$

Solutions to the latter equation satisfy $(1 - \bar{p})^2 r = s\bar{p}(1 - \bar{p})$. Therefore, $\bar{p} = 1$ is another equilibrium. Dividing by $1 - \bar{p}$ leads to $(1 - \bar{p})r = s\bar{p}$ so that the third equilibrium is

$$\bar{p} = \frac{r}{r + s}.$$

When $\bar{p} = 0$, only the allele $a$ is present in the population. When $\bar{p} = 1$, only allele $A$ is present, and when $\bar{p} = r/(r + s)$ both alleles are present. The local stability of these equilibria are determined next.

The derivative of $f(p) = p(1 - ps)/(1 - p^2 s - (1 - p)^2 r)$ in simplified form is

$$f'(p) = \frac{1 + p^2 s - r + rp^2 - 2ps + 2psr - 2p^2 sr}{(1 - p^2 s - r + 2rp - rp^2)^2}$$

$$= \frac{(1 - s)p^2 + 2(1 - s)(1 - r)p(1 - p) + (1 - r)(1 - p)^2}{(1 - p^2 s - r + 2rp - rp^2)^2}. \quad (3.9)$$

Note that $f'(p) > 0$ for $0 \le p \le 1$ and $r, s < 1$.

The equilibrium $\bar{p}$ is locally asymptotic stable if $-1 < f'(\bar{p}) < 1$. However, since $f'(p) > 0$ for $0 \le p \le 1$ and $r, s < 1$, we only need to show for local stability that $f'(\bar{p}) < 1$. At $\bar{p} = 0$,

$$f'(0) = \frac{1}{1 - r}.$$

Thus, $\bar{p} = 0$ is locally asymptotically stable if $r < 0$ (i.e., the relative survival rate of $aa$ is $w_{aa} = 1 - r > 1$ or the fitness value of the homozygote $aa$ is greater than that of the heterozygote, $w_{Aa} = 1$).

At $\bar{p} = 1$,

$$f'(1) = \frac{1}{1 - s}.$$

Thus, $\bar{p} = 1$ is locally asymptotically stable if $s < 0$ (i.e., the relative survival rate of $AA$ is $w_{AA} = 1 - s > 1$ or the fitness value of the homozygote $AA$ is greater than that of the heterozygote, $w_{Aa} = 1$).

Finally, at the equilibrium $\bar{p} = r/(r+s)$,

$$f'\left(\frac{r}{r+s}\right) = \frac{2rs - r - s}{rs - r - s}. \tag{3.10}$$

In order for $\bar{p}$ to be positive and less than one, either both $r$ and $s$ are positive or both are negative, $rs > 0$. Suppose $rs - r - s > 0$; then $f'(r/(r+s)) > 1$, which means for stability $rs - r - s \leq 0$. This means $r$ and $s$ must both be positive for stability.

We will show that if $r, s \in (0,1)$, then the equilibrium $\bar{p} = r/(r+s)$, where $0 < \bar{p} < 1$, is locally asymptotically stable. First, if $r, s \in (0,1)$, then $rs < r + s$. Second, it follows from (3.10) that $2rs < r + s$ [because $f'(\bar{p}) > 0$]. Hence $f'(\bar{p}) = (r + s - 2rs)/(r + s - rs) < 1$. Therefore, if $r, s \in (0,1)$, the equilibrium $\bar{p} = r/(r+s)$ is locally asymptotically stable. This result can be interpreted biologically. When $r, s \in (0,1)$, the heterozygous genotype has the largest survival rate, $w_{Aa} > \max\{w_{AA}, w_{aa}\}$. Because the heterozygote has an advantage, both alleles persist in the population.

It is interesting to note that in all cases, the mean fitness, $w = w_t$, increases over time until an equilibrium is reached, either $\bar{p} = 0$, $\bar{p} = 1$ or $\bar{p} = r/(r+s)$. This result can be verified mathematically (see Exercise 8). That is, for $t = 0, 1, \ldots,$

$$w_{t+1} \geq w_t.$$

An alternate method to verify stability of the equilibria is to consider the graph of $p_{t+1} = f(p_t)$ and use the cobwebbing method or apply the theorems in Chapter 2. For example, there are four possible configurations for $p_{t+1} = f(p_t)$ in the $p_t$-$p_{t+1}$ plane. They are graphed in Figure 3.10. In Exercise 9, the stability conditions derived for the positive equilibrium are shown to be global asymptotic stability conditions.

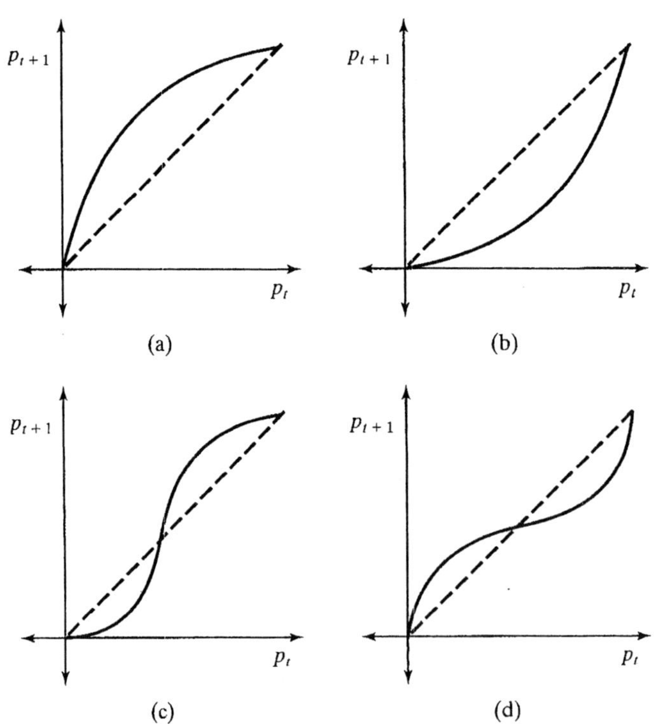


**Figure 3.10** Graphs of the function $p_{t+1} = f(p_t)$ in $p_t$-$p_{t+1}$ plane. The solid curve is the function $p_{t+1} = f(p_t)$ and the dotted curve is the line, $p_{t+1} = p_t$. The intersection points of these two curves represent the equilibria: 0, 1, and $\bar{p}$. In (a), $s < 0$ and $0 < r < 1$, $\lim_{t\to\infty} p_t = 1$. In (b), $r < 0$ and $0 < s < 1$, $\lim_{t\to\infty} p_t = 0$. In (c), $r, s < 0$, so that the limit depends on initial conditions. Solutions approach one of the two equilibria, 0 or 1. In (d), $0 < r, s < 1$, $\lim_{t\to\infty} p_t = \bar{p}$, the positive polymorphic equilibrium.

Selection depends on many different factors, and the importance of these factors differs significantly between animal and plant populations. Hedrick (2000) classifies the types of selection into several categories based on the underlying biological principles: viability selection (dependent on survival such as in the previous model), fecundity selection (differential production of offspring), sexual selection (males or females have preferential mating), and gametic selection (for example, equal proportion of alleles may not be produced). In addition, population size may affect selection for mates. For example, when the fitnesses $w_{AA}$ and $w_{Aa}$ depend on the frequency $p$ of allele $A$, this is referred to as *frequency-dependent selection* (Nagylaki, 1992). According to Nagylaki (1992), polymorphism can be maintained by frequency-dependent selection which favors rare genotypes.

**Example 3.2** Assume $w_{Aa} = 1$ and $w_{AA}(p)w_{aa}(p) = 1$ so that the heterozygote has fitness equal to the geometric mean of the homozygotes (Elaydi, 2000). In addition, assume that the frequency-dependent fitnesses are symmetric, $w_{AA}(p) = w_{aa}(1-p) = w_{aa}(q)$ and $w_{AA}(p) = f(p) = 1/w_{aa}(p)$. The function $f(p)$ is a positive, strictly decreasing function with $f(0) > 1$ and $f'$ continuous. Thus, when $p$ is small or $A$ is rare, the fitness $w_{AA}(p)$ is large and when $p$ is large, $1 - p$ is small, $a$ is rare, the fitness $w_{aa}(p)$ is large. The population genetics model (3.7) has the form

$$p_{t+1} = \frac{p_t f(p_t)[p_t + (1-p_t)/f(p_t)]}{(p_t f(p_t) + 1 - p_t)[p_t + (1-p_t)/f(p_t)]} = \frac{p_t f(p_t)}{p_t f(p_t) + 1 - p_t} = F(p_t).$$

This model has three equilibria, $\bar{p} = 0, \bar{p} = 1$, and the polymorphic equilibrium $p^*$ satisfying $f(p^*) = 1$. The polymorphic equilibrium is unique because of the assumptions on $f$. Now,

$$F'(p) = \frac{f(p) + p(1-p)f'(p)}{[pf(p) + 1 - p]^2},$$

so that $F'(0) = f(0) > 1$ and $F'(1) = 1/f(1) = w_{aa}(1) = w_{AA}(0) = f(0) > 1$. The equilibria $\bar{p} = 0$ and $\bar{p} = 1$ are unstable. Since the derivative $F'(p^*) = 1 + p^*(1-p^*)f'(p^*) < 1$, for stability of $p^*$, it is only necessary to show that $F'(p^*) > -1$ or $p^*(1-p^*)f'(p^*) > -2$. For example, if $f(p) = \exp(1-2p)$, then $p^* = 1/2$ and $p^*(1-p^*)f'(p^*) = -1/2 > -2$ so that the polymorphic equilibrium is locally asymptotically stable. Elaydi (2000) has shown that for suitably chosen $f(p)$, the model can exhibit period-doubling behavior (see Exercise 10). ■

**Example 3.3** We formulate a population genetics model for two populations. Assume the two populations are diploid. We model one gene in each population and assume there are only two alleles. In the first population, the two alleles are $V$ and $v$ and in the second population, the two alleles are $R$ and $r$. We model the frequency of alleles $V$ and $R$ for the first and second population, respectively. Let the proportion of allele $V$ in the first population be denoted as $n$, and the proportion of allele $R$ in the second population be denoted as $p$. Then the model takes the form

$$n_{t+1} = \frac{n_t^2 w_{VV} + n_t(1-n_t)w_{Vv}}{n_t^2 w_{VV} + 2n_t(1-n_t)w_{Vv} + (1-n_t)^2 w_{vv}} = f(n_t, p_t),$$

$$p_{t+1} = \frac{p_t^2 w_{RR} + p_t(1-p_t)w_{Rr}}{p_t^2 w_{RR} + 2p_t(1-p_t)w_{Rr} + (1-p_t)^2 w_{rr}} = g(n_t, p_t). \quad (3.11)$$

Population genetics models of this form have been studied in relation to plant pathogens. The first population represents a pathogen, whereas the second population represents a plant that is attacked by the pathogen. Allele $V$ represents a virulent allele and $v$ an avirulent allele in the pathogen population and $R$ represents a resistant allele and $r$ a susceptible allele in the host plant. A virulent gene in the pathogen population is matched by a resistant gene in the plant population. Such types of gene relationships are referred to as *gene-for-gene* systems and have been studied by Leonard (1977, 1994) and many others (see, e.g., Sasaki, 2002; Kesinger and Allen, 2002 and references therein). The gene-for-gene hypothesis states that for each gene determining resistance in the host there is a corresponding gene for avirulence in the parasite with which it interacts (Thompson and Burdon, 1992). This hypothesis was originally applied to flax and flax rust (Flor, 1956) but has been applied to variety of plant pathogens including wheat stem rust and potato late blight (Vanderplank, 1984).

The fitnesses of the various pathogen genotypes, $w_{VV}$, $w_{Vv}$, and $w_{vv}$, depend on the frequency of the plant resistance allele $p$. The fitnesses of the plant genotypes, $w_{RR}$, $w_{Rr}$, and $w_{rr}$, depend on the frequency of the pathogen virulence gene $n$. Model (3.11) is studied in more detail in Exercise 12. ■

The subject of inheritance and population genetics is much more complicated than the short introduction we have given here. For example, selection, mutation, nonrandom mating, migration, recombination, and gene linkage affect the outcome of the genetic makeup of a population. Please consult population genetics textbooks Hartl and Clark (1997) or Hedrick (2000) for a wealth of biological examples.

## 3.8 Nonlinear Structured Models

Two theoretical nonlinear Leslie matrix models and two structured models applied to specific populations are studied. The nonlinear Leslie matrix models are presented in the next subsection. Then the two structured models are presented in the next two subsections. The structured models are applied to a flour beetle population and to the northern spotted owl. The final example in this section is a generalization of the Leslie matrix model to a two-sex model.

### 3.8.1 Density-Dependent Leslie Matrix Models

Assume that the population size or density affects the survival and/or fecundity of each age class. Assume that as the total population size or density increases, food resources are depleted resulting in a decrease in survival and/or fecundity. Competition, cannibalism, and predation also tend to increase with population density, which ultimately leads to decreased survival and fecundity.

Recall that the Leslie matrix model has the form $X(t+1) = LX(t)$, where

$$L = \begin{pmatrix} b_1 & b_2 & \cdots & b_{m-1} & b_m \\ s_1 & 0 & \cdots & 0 & 0 \\ 0 & s_2 & \cdots & 0 & 0 \\ \vdots & \vdots & \ddots & \vdots & \vdots \\ 0 & 0 & \cdots & s_{m-1} & 0 \end{pmatrix},$$

$X(t) = (x_1(t), x_2(t), \ldots, x_m(t))^T$, and $x_i(t)$ is the $i$th age class in year $t$. In the first nonlinear model, we shall assume that survival and fecundity are decreased by the same proportion when population size increases. Let $x(t) = \|X(t)\|_1 = \sum_{i=1}^m x_i(t)$ denote the total population size at time $t$; the notation $\|\cdot\|_1$ represents the $L_1$ norm. Assume each term in the Leslie matrix model, $b_i$ and $s_i$, is replaced by a function of the total population size, that is, $b_i q(x(t))$ and $s_i q(x(t))$, respectively. The function $q : [0, \infty) \to (0, 1]$ is nonnegative and decreasing with the property $q(0) = 1$. The model can be expressed as the matrix equation,

$$X(t + 1) = q(x(t))LX(t).$$

Suppose that $L$ has a dominant eigenvalue $\lambda_1 > 1$; $L$ is irreducible but may be primitive or imprimitive (the dominant eigenvalue is not strict). We consider the case of logistic density dependence. Suppose $q$ satisfies

$$q(x(t)) = \frac{K}{K + (\lambda_1 - 1)x(t)}, \quad K > 0,$$

where $K$ is the carrying capacity and $\lambda_1 = e^r > 1$. Recall that the scalar difference equation $x(t + 1) = \lambda_1 q(x(t))x(t) = f(x(t))$ has a globally asymptotically stable equilibrium at $\bar{x} = K$. This follows because $f$ is monotonically increasing with a single positive equilibrium at $\bar{x}$ (Theorem 2.8). A generalization of this particular model can be found in Cushing (1998).

The following ideas are used in the proof of Theorem 3.3 (see Allen, 1989). Matrix $L$ can be expressed in a Jordan canonical form. There exists a nonsingular matrix $P$ such that $L = PJP^{-1}$, where $J = \text{diag}(J_1, J_2, \ldots, J_s)$ and $J_i$ is a Jordan block (see Ortega, 1987). The Jordan blocks are upper triangular of the form

$$J_i = \begin{pmatrix} \lambda_i & 1 & 0 & \cdots & 0 \\ 0 & \lambda_i & 1 & \cdots & 0 \\ \vdots & \vdots & \vdots & \ddots & 0 \\ 0 & 0 & 0 & \ddots & 1 \\ 0 & 0 & 0 & \cdots & \lambda_i \end{pmatrix}.$$

The number $s$ is equal to the number of linearly independent eigenvectors of $L$. If there are $m$ linearly independent eigenvectors, then $J_i = (\lambda_i)$ and $J$ is diagonal. The same eigenvalues may occur in different Jordan blocks but the number of distinct blocks corresponding to a given eigenvalue is equal to the number of independent eigenvectors corresponding to that eigenvalue (Noble, 1969). It can be shown that the density-dependent Leslie matrix behaves in a manner similar to the scalar difference equation.

**Theorem 3.3**

*(i) If the matrix $L$ is primitive, then the asymptotic distribution $\lim_{t\to\infty} X(t) = N$, where $N$ is the stable age distribution corresponding to $LN = \lambda_1 N$ and $\|N\|_1 = K$.*

*(ii) If the matrix $L$ is imprimitive, then the asymptotic distribution $\lim_{t\to\infty} X(t) = N(t)$, where $N(t)$ is periodic, $N(t + h) = N(t)$, and $h$ is the index of imprimitivity of $L$.* □

Recall that $h$, the index of imprimitivity, is the number of eigenvalues of $L$ whose magnitude equals the spectral radius of $L$ (Definition 1.12). A proof of Theorem 3.3 can be found in Allen (1989).

**Example 3.4** Let

$$L_1 = \begin{pmatrix} 0 & \frac{3}{2}a^2 & \frac{3}{2}a^3 \\ \frac{1}{2} & 0 & 0 \\ 0 & \frac{1}{3} & 0 \end{pmatrix}.$$

Matrix $L_1$ is primitive. This follows from the result of Sykes (1969), g.c.d.$\{i \mid b_i > 0\} = 1$. The dominant eigenvalue is $a$ with stable age distribution $(6a^2, 3a, 1)^T$. The other eigenvalues are $-a/2$ with algebraic multiplicity two. Assume the density-dependent Leslie matrix has the form

$$X(t+1) = \frac{100}{100 + (a-1)x(t)} L_1 X(t),$$

where $K = 100$ is the carrying capacity, and $x(t) = \Sigma_{i=1}^{3} x_i(t)$. For example, if $a = 2$, then $\lim_{t\to\infty} X(t) = (100/31)(24, 6, 1)^T$ and if $a = 3$, then $\lim_{t\to\infty} X(t) = (100/64)(54, 9, 1)^T$. ■

**Example 3.5** Let

$$L_2 = \begin{pmatrix} 0 & 0 & 6a^3 \\ \frac{1}{2} & 0 & 0 \\ 0 & \frac{1}{3} & 0 \end{pmatrix}.$$

The model is the same as that given in Example 3.4 but with $L_2$ replacing $L_1$. Matrix $L_2$ is imprimitive (g.c.d.$\{i \mid b_i > 0\} = 3$). The eigenvalues are $a$ and $a(-1/2 \pm i\sqrt{3}/2)$ and $|\lambda| = a$ for all eigenvalues $\lambda$. The stable age distribution is the same as that given in Example 3.4. For the density-dependent model, where all of the terms in $L_2$ are multiplied by

$$q(x(t)) = \frac{100}{100 + (a-1)x(t)},$$

the asymptotic distribution is periodic with period 3 because $L_2$ is imprimitive with index of imprimitivity $h = 3$. Two numerical examples are given in Figure 3.11 with $a = 2$ and either $X(0) = (25, 5, 1)^T$ or $X(0) = (10, 10, 10)^T$. Note that the periodic solutions are not unique; they depend on the initial distribution, $X(0)$. ■

The second density-dependent Leslie matrix model we study is one where only the fertilities are density dependent (Silva and Hallam, 1992, 1993). Structured models of this type have often been applied to fish populations. In this model, the density dependence is referred to as *density-dependent recruitment*. Let $s_i$ be the survival probability from age class $i$ to age class $i+1$, $i = 1, \ldots, m-1$. Define $\tilde{b}_i$ as the fecundity of an age class $i$ individual (e.g., number of eggs produced per female). Let $s_0$ be the fraction surviving to recruitment, meaning survival through the egg and larval stages. Assume $s_0$ is of the form

$$s_0 = ag(w(t)),$$

where $a$ is the density-independent survival probability and $g(w(t))$ is the density-dependent survival probability. The quantity $w(t)$ denotes the weighted population size as it affects density-dependent mortality,

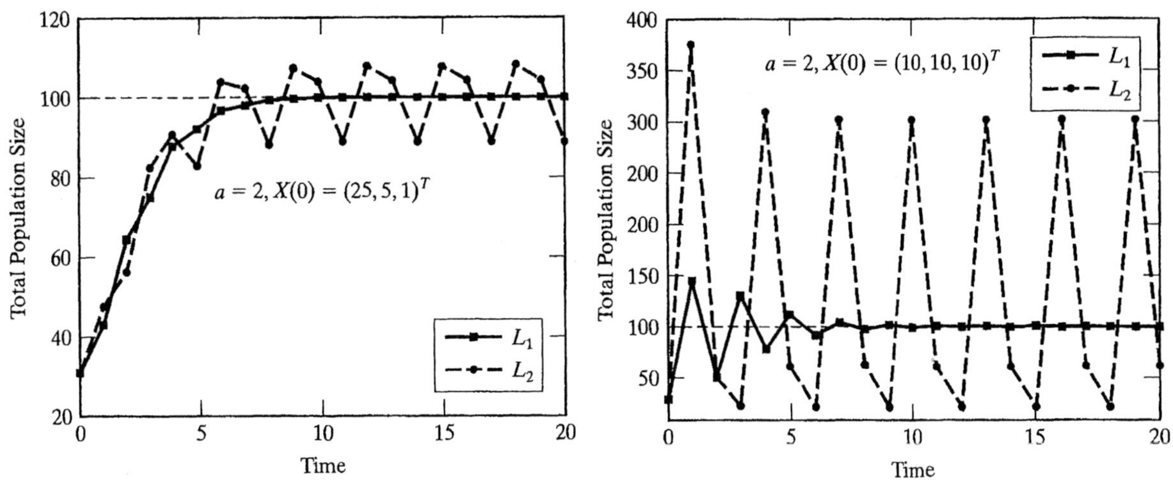


**Figure 3.11** The density-dependent Leslie matrix models with matrices $qL_1$ and $qL_2$ with $a = 2$.

$$w(t) = \sum_{i=1}^{m} \alpha_i x_i(t).$$

Assume the sex ratio is $1:1$. The model has the form

$$\begin{aligned} x_1(t+1) &= ag(w(t))\sum_{i=1}^{m} b_i x_i(t), \\ x_2(t+1) &= s_1 x_1(t), \\ x_3(t+1) &= s_2 x_2(t), \\ &\vdots \\ x_m(t+1) &= s_{m-1}x_{m-1}(t). \end{aligned}$$

The fecundity is halved, $b_i = \tilde{b}_i/2$, because half of the population is females. Parameters $b_i$ and $\alpha_i$ are nonnegative and the parameters $a$ and $s_i$ are positive. The above model shall be referred to as the *Leslie matrix model with density-dependent recruitment* (LMMDDR). The LMMDDR has been applied to fish populations because density dependence has the greatest impact on the young-of-the-year (see DeAngelis et al., 1980; Levin and Goodyear, 1980; Getz and Haight, 1989). Let $E(t) = \sum_{i=1}^{m} b_i x_i(t)$ denote the egg production at time $t$ and $R(t) = x_1(t)$, the number of new recruits at time $t$. Then the new recruits in the population at time $t+1$ are $x_1(t+1)$ or

$$R(t+1) = ag(w(t))E(t).$$

The form of the density-dependent $g$ may be a Ricker or Beverton-Holt or some other function, for example,

$$g(x) = \exp(-rx) \;\text{ or }\; g(x) = 1/(1+ax)^c,$$

($c = 1$ is Beverton-Holt). In general, it shall be assumed that $g$ is a positive, strictly decreasing function satisfying the following two assumptions:

(i) $g : [0, \infty) \to (0, 1]$, $g \in C^2(0, \infty)$ and $g$ is strictly decreasing.

(ii) $g(0) = 1$ and $\lim_{x\to\infty} g(x) = 0$.

If the weighted population average equals the egg production, $w(t) = E(t)$, then recruitment satisfies

$$R(t+1) = ag(w(t))w(t) = ah(w(t)).$$

In this case, the function $h$ can be interpreted as the recruitment function.

Define the inherent net reproductive number as follows:

$$R_0 = b_1 + s_1 b_2 + s_1 s_2 b_3 + \cdots + s_1 s_2 \cdots s_{m-1} b_m.$$

This is the same definition used for the basic Leslie matrix model. With the assumptions (i) and (ii) it will be shown that if $aR_0 < 1$, there is only the zero equilibrium and if $aR_0 > 1$, there exists a unique positive equilibrium.

**Theorem 3.4**

*(i) If $aR_0 < 1$, then for the LMMDDR satisfying assumptions (i)–(ii) there exists only one equilibrium solution, the zero solution, $\bar{X} = (0, 0, \ldots, 0)^T$.*

*(ii) If $aR_0 > 1$, then for the LMMDDR satisfying assumptions (i)–(ii) there exists a unique positive equilibrium solution $\bar{X} = (\bar{x}_1, \bar{x}_2, \ldots, \bar{x}_n)^T$, where $\bar{x}_i = s_1 \ldots s_{i-1}\bar{w}/K$, $i = 1, \ldots, m$, and $g(\bar{w}) = 1/(aR_0)$,*

$$K = \alpha_1 + s_1\alpha_2 + s_1 s_2 \alpha_3 + \cdots + s_1 s_2 \cdots s_{m-1}\alpha_m,$$

*and*

$$\bar{w} = \sum_{i=1}^{m} \alpha_i \bar{x}_i.$$

*Moreover,* $\lim_{aR_0 \to 1^+} \bar{x}_i = 0$ for $i = 1, \ldots, m$.

**Proof** The equilibrium solutions satisfy

$$\bar{x}_2 = s_1\bar{x}_1, \bar{x}_3 = s_1 s_2 \bar{x}_1, \ldots, \bar{x}_m = s_1 s_2 \ldots s_{m-1}\bar{x}_1.$$

Now, $\bar{w}$ can be expressed in terms of $\bar{x}_1$:

$$\bar{w} = [\alpha_1 + \alpha_2 s_1 + \alpha_3 s_1 s_2 + \cdots + \alpha_m s_1 \cdots s_{m-1}]\bar{x}_1 = K\bar{x}_1.$$

Thus, we can find the equilibria if we can solve for $\bar{x}_1$:

$$\bar{x}_1 = ag(\bar{w})R_0\bar{x}_1.$$

Either $\bar{x}_1 = 0$ or $g(\bar{w}) = 1/(aR_0)$. The latter expression determines $\bar{x}_1$ implicitly.

For part (i) of the theorem, $aR_0 < 1$ implies that the equilibrium solution satisfies $1 < g(\bar{w}) = 1/(aR_0)$. Since $0 < g(x) \le 1$ for $x > 0$, there is no positive solution $\bar{w}$ when $g(\bar{w}) > 1$. Thus, the only equilibrium solution is $\bar{x}_1 = 0$. But $\bar{x}_1 = 0$ implies $\bar{x}_i = 0$ for $i = 2, \ldots, m$.

For part (ii) of the theorem, $aR_0 > 1$ implies that the equilibrium solution satisfies $1 > g(\bar{w}) = 1/(aR_0)$. Since $g$ is a strictly decreasing function and $\lim_{x\to\infty} g(x) = 0$, there exists a unique positive solution $\bar{w}$ such that $g(\bar{w}) = 1/(aR_0)$. Since $\bar{w} = K\bar{x}_1$, it follows that $\bar{x}_1 = \bar{w}/K > 0$ is the unique positive solution. The other components are uniquely determined by $\bar{x}_1$:

$$\bar{x}_2 = s_1\bar{w}/K, \ldots, \bar{x}_m = s_1 \ldots s_{m-1}\bar{w}/K.$$

As $aR_0 \to 1^+$, $g(\bar{w}) \to 1$. But when $g(\bar{w}) \to 1$, $\bar{w} \to 0$ because $g$ is continuous and strictly monotonically decreasing (one-to-one). Consequently, $\bar{x}_1 \to 0$ and $\bar{x}_i \to 0$ for $i = 1, \ldots, m$. □

Next, we consider the LMMDDR where there are only two age classes, juveniles and adults. The simpler model has the form

$$x_1(t+1) = ag(w(t))\sum_{i=1}^{2} b_i x_i(t),$$

$$x_2(t+1) = sx_1(t), \tag{3.12}$$

where $w(t) = \alpha_1 x_1(t) + \alpha_2 x_2(t)$. Silva and Hallam (1992) considered several different variations of the model. One variation they referred to as a stock-recruitment model, where $\alpha_i = b_i$ for $i = 1, 2$. The egg production or stock $E(t) = b_1 x_1(t) + b_2 x_2(t) = w(t)$. The stock $E(t)$ is related to the recruitment function $R(t)$ by the relation

$$R(t+1) = ag(w(t))w(t).$$

We shall study this case in more detail.

Let the positive steady state for the stock-recruitment model (3.12) be given by $(\bar{x}_1, \bar{x}_2)$ and $\bar{w} = b_1\bar{x}_1 + b_2\bar{x}_2$. Next, we calculate the Jacobian matrix of the two-age class LMMDDR and evaluate the Jacobian matrix at the positive equilibrium:

$$J = \begin{pmatrix} ag(\bar{w})b_1 + ag'(\bar{w})b_1\bar{w} & ag(\bar{w})b_2 + ag'(\bar{w})b_2\bar{w} \\ s & 0 \end{pmatrix}.$$

Now, the trace and the determinant of the Jacobian matrix are given by

$$\operatorname{Tr} J = ab_1[g(\bar{w}) + g'(\bar{w})\bar{w}],$$

$$\det J = -sab_2[g(\bar{w}) + g'(\bar{w})\bar{w}].$$

Each of the conditions for asymptotic stability can be considered separately:

$$\operatorname{Tr} J < 1 + \det J, \quad \operatorname{Tr} J > -1 - \det J, \quad \text{and} \det J < 1.$$

When one of these conditions fails, a bifurcation occurs (i.e., a change in equilibrium stability). When the first condition fails, $\lambda \geq 1$. When the second condition fails, $\lambda \leq -1$, and when the last condition fails, $|\lambda| \geq 1$. Silva and Hallam (1992) made some additional assumptions concerning the function $g$ and were able to obtain conditions for local asymptotic stability. Some examples and numerical simulations of the stock-recruitment model (3.12) are discussed next.

**Example 3.6** Let $g(x) = \exp(-x)$, $b_1 = 0 = \alpha_1$, $b_2 = 10 = \alpha_2$, $s = 0.5$, and $a = 1$. Then $R_0 = 5$, $\bar{w} = \ln(5)$, $\bar{x}_1 = \ln(5)/5$, $\bar{x}_2 = \ln(5)/10$, and $g(\bar{w}) = 1/5$. The trace and determinant are

$$\operatorname{Tr} J = 0 \quad \text{and} \quad \det J = \ln 5 - 1.$$

Thus, $\lambda_{1,2} = \pm i\sqrt{\ln 5 - 1}$, $|\lambda_{1,2}| \approx 0.609$; the positive equilibrium is locally asymptotically stable.

Now suppose all of the parameters are the same, except $b_2 = 20$. Note the equilibrium changes also, $\bar{x}_1 = \ln(10)/10$, $\bar{x}_2 = \ln(10)/20$, $\bar{w} = \ln(10)$, and $g(\bar{w}) = 1/10$. Since

$$\operatorname{Tr} J = 0 \quad \text{and} \quad \det J = \ln 10 - 1,$$

$\lambda_{1,2} = \pm i\sqrt{\ln 10 - 1}$, $|\lambda_{1,2}| \approx 1.303$; the equilibrium is unstable. A bifurcation occurs when $b_2$ is increased. The system appears to have a period 4 solution. See Figure 3.12. ■

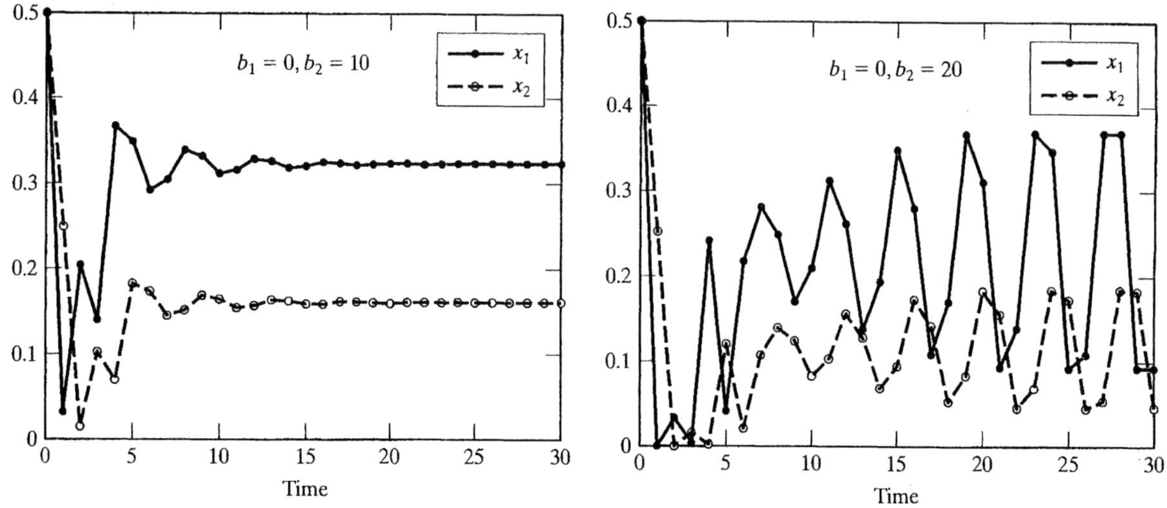


**Figure 3.12** The LMMDDR model with two ages, $g(x) = e^{-x}$ and $R(t+1) = g(w(t))w(t), a = 1$. (a) $b_1 = 0, b_2 = 10$, and $s = 0.5$ (b) $b_1 = 0, b_2 = 20$, and $s = 0.5$.

### 3.8.2 Structured Model for Flour Beetle Populations

The structured model for the flour beetle has received much attention because it is one of the few mathematical models that has not only been investigated theoretically but has been tested against data collected from many laboratory experiments. The population of flour beetles from the species *Tribolium* is modeled. In the structured model, the population is subdivided into three developmental stages, larval, pupal, and adult stages, denoted as $L$, $P$ and $A$, respectively (e.g., Costantino et al., 1997, 1998; Cushing et al., 1998, 2003; Henson and Cushing, 1997; Henson et al., 1998). Deterministic and stochastic formulations of this model have been analyzed mathematically and statistically by Cushing, Dennis, and Henson and experiments have been set up and conducted in the laboratory by Costantino and Desharnais. A nicely written book describing their results is *Chaos in Ecology* (Cushing et al., 2003). You may consult a Web page for a current list of publications by this group of mathematicians, statistician, and biologists: **http://caldera.calstatela.edu/nonlin/lpamodel.html**. The results from the laboratory studies and the model agree very well.

The model is a system of difference equations satisfying

$$\begin{aligned} L_{t+1} &= bA_t \exp(-c_{ea}A_t - c_{el}L_t), \\ P_{t+1} &= L_t(1 - \mu_l), \\ A_{t+1} &= P_t \exp(-c_{pa}A_t) + A_t(1 - \mu_a), \end{aligned}$$

where all of the parameters $b$, $c_{ea}$, $c_{el}$, $c_{pa}$, $\mu_l$, and $\mu_a$ are positive and, in addition, $\mu_l$ and $\mu_a$ are less than one. We shall refer to this model as the LPA model. The time unit, $t$ to $t+1$, is two weeks, which is the average amount of time spent in the larval stage and is also the time unit for the duration of the pupal stage. The exponential terms are Ricker type density dependence and represent the effects of cannibalism. In particular, the coefficients $c_{ea}$, $c_{el}$, and $c_{pa}$ are rates of cannibalism of eggs by adults, eggs by larvae, and pupae by adults, respectively (see also Caswell, 2001; Cipra, 1999; Cushing, 1998). The fractions $\exp(-c_{el}L_t)$ and $\exp(-c_{ea}A_t)$ are the probabilities that an egg is not eaten in the presence of $L_t$ larvae and $A_t$ adults in one time unit. The fraction $\exp(-c_{pa}A_t)$ is the survival probability of a pupa in the presence of $A_t$ adults in one time unit. The coefficient $b$ represents the average number of larvae produced per adult, and $\mu_a$ and $\mu_l$ denote the mortality fractions of adults and larvae.

In the laboratory experiments, the beetles are kept in a bottle containing 20 grams of flour at a constant incubator temperature and humidity (Cushing, 1998). The flour is sifted every two weeks and the number of larvae, pupae, and adults counted, then returned to a fresh bottle of flour.

We examine a simple case for this model: the conditions for stability of the zero equilibrium or extinction equilibrium. The Jacobian matrix of the LPA model satisfies

$$J = \begin{pmatrix} -bc_{el}Ae^{-c_{ea}A-c_{el}L} & 0 & be^{-c_{ea}A-c_{el}L}(1-c_{ea}A) \\ 1-\mu_l & 0 & 0 \\ 0 & e^{-c_{pa}A} & 1-\mu_a-c_{pa}Pe^{-c_{pa}A} \end{pmatrix}.$$

At the extinction equilibrium, the Jacobian matrix has the following form:

$$J(0,0,0) = \begin{pmatrix} 0 & 0 & b \\ 1-\mu_l & 0 & 0 \\ 0 & 1 & 1-\mu_a \end{pmatrix}. \tag{3.13}$$

The characteristic polynomial is

$$p(\lambda) = \lambda^3 - (1-\mu_a)\lambda^2 - b(1-\mu_l).$$

The local asymptotic stability of the extinction equilibrium can be determined from the Jury conditions:

$$\begin{aligned} p(1) &= \mu_a - b(1-\mu_l) > 0, \\ p(-1) &= -2 + \mu_a - b(1-\mu_l) < 0, \\ 1 - b^2(1-\mu_l)^2 &> b(1-\mu_l)(1-\mu_a). \end{aligned}$$

The first condition is satisfied if $b(1-\mu_l) < \mu_a$. The second condition is always satisfied. Finally, the third condition follows from the first condition. That is,

$$b^2(1-\mu_l)^2 + b(1-\mu_l)(1-\mu_a) < \mu_a^2 + \mu_a(1-\mu_a) = \mu_a < 1.$$

Hence, the extinction equilibrium is locally asymptotically stable if

$$b(1-\mu_l) < \mu_a. \tag{3.14}$$

In fact, it can be shown that the extinction equilibrium is globally asymptotically stable if condition (3.14) holds. Denote $X_t = (L_t, P_t, A_t)^T$ and note that $X_{t+1} \le JX_t$, where $J \equiv J(0,0,0)$ is defined in (3.13). Since $J$ is a nonnegative matrix, it easily follows that $X_t \le J^t X_0$. The magnitude of the eigenvalues of $J$ are less than one iff inequality (3.14) holds. Thus, if (3.14) holds, then $\lim_{t\to\infty} X_t = (0,0,0)^T$.

Condition (3.14) can be interpreted biologically. Extinction is possible if, in the absence of cannibalism, the number of new larvae that survive to the pupal stage during the two-week interval is less than the fraction of adults who die during that same period. When $b(1-\mu_l) = \mu_a$ there is a change in behavior; a transcritical bifurcation occurs (see Exercise 14).

It is interesting to note that a positive equilibrium $(\bar{L}, \bar{P}, \bar{A})$ requires

$$1 = b\frac{(1-\mu_l)}{\mu_a}\exp(-c_{ea}\bar{A} - c_{ea}\bar{L} - c_{pa}\bar{A}).$$

At a positive equilibrium, the exponential term is less than one. Hence, if $b(1-\mu_l) \leq \mu_a$, then there cannot exist a positive equilibrium. Existence of a positive equilibrium requires that $b(1-\mu_l) > \mu_a$. When this condition holds, the LPA model exhibits a wide array of behaviors as different parameters are varied—from periodic behavior to chaos. Please consult the references for more information about this interesting model and the experiments that have been conducted to test this model.

### 3.8.3 Structured Model for the Northern Spotted Owl

The northern spotted owl, *Strix occidentalis caurina*, is located in the Pacific Northwest of the United States and Canada. It is a monogamous, territorial bird requiring large tracts of mature, coniferous trees for its survival (Lande, 1988). Due to logging of old-growth forests in the Northwest, researchers have predicted extinction of the spotted owl if suitable habitat is not maintained (Lamberson et al., 1992; Lande, 1988). The species was given threatened status in 1990 (McKelvey et al., 1992). A number of models have been developed for the spotted owl, including a simple Leslie matrix model (Lamberson et al., 1992) and a spatially explicit, stage-structured, stochastic metapopulation model (Akçakaya and Raphael, 1998). We discuss a model first reported by Thomas et al. (1990) and later analyzed by Lamberson et al. (1992). This particular model is a density-dependent, structured model; it is also discussed by Caswell (2001), Cushing (1998), and Haefner (1996). The discussion follows that of Allen et al. (2005).

Suppose the landscape is fixed; only a fraction of the landscape is suitable for spotted owl occupation. The suitable area is made up of sites, $T$ = total number of sites and $U$ = number of available sites, $U < T$. Single females find a single male to become paired with or are eliminated from the population. Juvenile birds that survive disperse at the end of their first year; males seek an unoccupied site and females seek a site occupied by a solitary male. Let $P_t$ be the number of paired owls in year $t$, $S_{m,t}$ be the number of single males, and $S_{f,t}$ be the number of single females. The sex ratio between males and females is one, so that $S_{m,t} = S_{f,t}$. The number of occupied sites equals the number of paired owls plus the number of male owls, $O_t = P_t + S_{m,t}$. The number of available and unoccupied sites is $A_t = U - O_t$. To ensure that available sites remain nonnegative, we modify the definition of occupied sites, $O_t = \min\{U, P_t + S_{m,t}\}$.

We model the number of paired owls $P_t$ and number of single male owls $S_{m,t}$. The model takes the following form:

$$P_{t+1} = P_t p_s + S_{m,t} s_S M_t = F(P_t, S_{m,t}), \tag{3.15}$$

$$S_{m,t+1} = \frac{1}{2} f s_J D_t P_t + S_{m,t} s_S (1 - M_t) + p_b P_t = G(P_t, S_{m,t}). \tag{3.16}$$

The model parameters are defined as follows:

$D_t$ = probability of juveniles surviving dispersal,

$M_t$ = probability of female finding a male,

$s_S$ = fraction of single owls surviving one year,

$s_J$ = fraction of juveniles surviving to single adults in one year,

$p_s$ = probability of both individuals in a pair surviving one year and not splitting (becoming single),

$p_b$ = probability that a pair survives one year and splits (becomes single),

$f$ = number of offspring per breeding pair in one year,

$m$ = unoccupied site search efficiency,

$n$ = unmated male search efficiency.

The probabilities $M_t$ and $D_t$ satisfy

$$M_t = 1 - \left(1 - \frac{\min\{T, 2S_{m,t}\}}{T}\right)^n, \quad D_t = 1 - \left(1 - \frac{A_t}{T}\right)^m,$$

where $M_t, D_t \in [0, 1]$.

Equation (3.15) for paired adults in year $t + 1$ consists of those paired adults that survive and do not split up to form singles and those single male owls that survive and find a mate. Equation (3.16) for single male owls in year $t + 1$ consists of those males from pairs that survive but then split up, those single males from the previous year that did not find a mate, and those male offspring from adult pairs that survived the first year and also survived dispersal. It is easy to see that solutions to (3.15) and (3.16) are nonnegative.

The structured owl model has several equilibria. One equilibrium is the extinction equilibrium, $\bar{P} = 0$ and $\bar{S}_m = 0$. Let $E_0 = (0, 0)$ denote the extinction equilibrium. At the equilibrium $E_0$, $2\bar{S}_m < T$. Therefore, in the linearization at $E_0$, the minimum function in $M_t$ is replaced by $2S_{m,t}$. The Jacobian matrix evaluated at $E_0$ satisfies

$$J(E_0) = \begin{pmatrix} p_s & 0 \\ (1/2)fs_J[1 - (1 - U/T)^m] + p_b & s_S \end{pmatrix}.$$

Since $0 < p_s, s_S < 1$, the extinction equilibrium is always locally asymptotically stable. For small initial values, the population approaches the extinction equilibrium. The other equilibria are found by solving the following equations:

$$\bar{P} = \frac{\bar{S}_m s_S \bar{M}}{1 - p_s}, \tag{3.17}$$

$$\bar{S}_m = \left(\frac{1}{2}fs_J\bar{D} + p_b\right)\bar{P} + \bar{S}_m s_S(1 - \bar{M}), \tag{3.18}$$

where

$$\bar{M} = 1 - \left(1 - \frac{2\bar{S}_m}{T}\right)^n \quad \text{and} \quad \bar{D} = 1 - \left(1 - \frac{U - \bar{P} - \bar{S}_m}{T}\right)^m.$$

Here, we have assumed that $2\bar{S}_m < T$ and $\bar{P} + \bar{S}_m < U$.

Equations (3.17) and (3.18) can be expressed as a polynomial in $\bar{S}_m$,

$$p(\bar{S}_m) = -\bar{S}_m + \left(\frac{1}{2}fs_J\bar{D} + p_b\right)\frac{\bar{S}_m s_S \bar{M}}{1 - p_s} + \bar{S}_m s_S(1 - \bar{M}). \tag{3.19}$$

The roots of this polynomial are $\bar{S}_m$. Then $\bar{P}$ is given by (3.17).

Estimates for some of the parameters are

$$s_S = 0.71, \quad s_J = 0.60, \quad p_s = 0.88, \quad p_b = 0.056, \quad f = 0.66$$

(Haefner, 1996; Lamberson et al., 1992; Thomas et al., 1990). In addition, let $n = m = 20$ and $T = 1000$. Parameter $U$, the number of available sites,

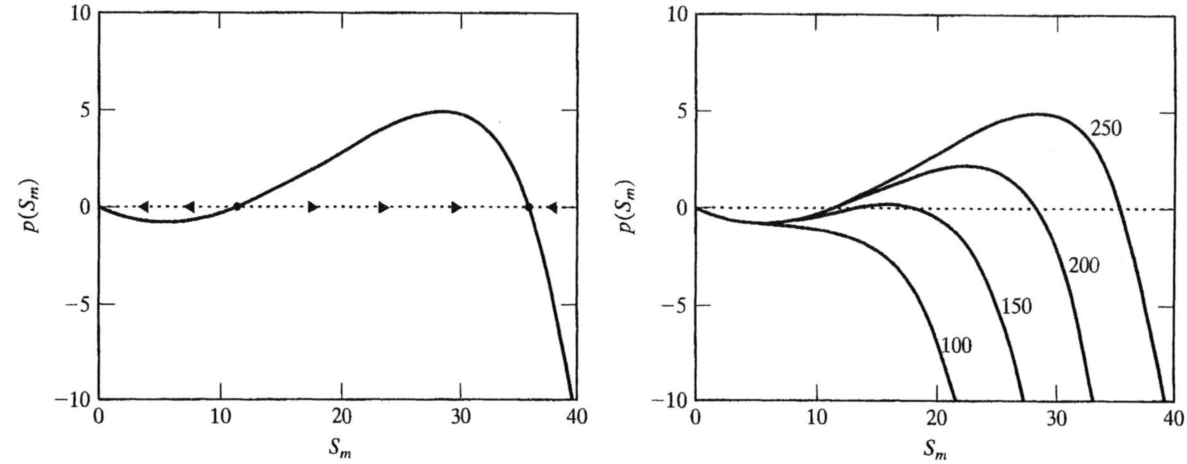


**Figure 3.13** Figure on the left is the polynomial $p(S_m)$ for $U = 250$. The arrows along the line $S_m = 0$ indicate the stability of the three equilibria. Figure on the right is the polynomial $p(S_m)$ for U = 100, 150, 200 and 250.

depends on logging. For different values of $U$, the equilibrium values for male owls can be found by solving (3.19), $p(\bar{S}_m) = 0$ (see Figure 3.13). It can be seen for $U > 149$ that there are three equilibria [three solutions to $p(S_m) = 0$] and for $U < 149$ there is only one solution, the extinction equilibrium.

Suppose $U = 250$. Then there exist three equilibria, $E_0$, $E_1$, and $E_2$, where $E_0 < E_1 < E_2$. The Jacobian matrix at the equilibrium $E_1 = (25.04, 11.43)$ has two positive eigenvalues; one is less than one and the other is greater than one; $E_1$ is unstable. The Jacobian matrix at $E_2 = (163.97, 35.81)$ has two real eigenvalues, both of whose magnitude is less than one; $E_2$ is locally asymptotically stable (Exercise 15). Solutions to the model are graphed in Figure 3.14 for various initial conditions. When the initial conditions are less than $E_1$, solutions tend to zero, but if they are greater than $E_1$, solutions tend to the positive equilibrium. Equilibrium $E_1$ is known as the Allee threshold. The Allee effect is named for Warder C. Allee because of his extensive research on the social behavior and

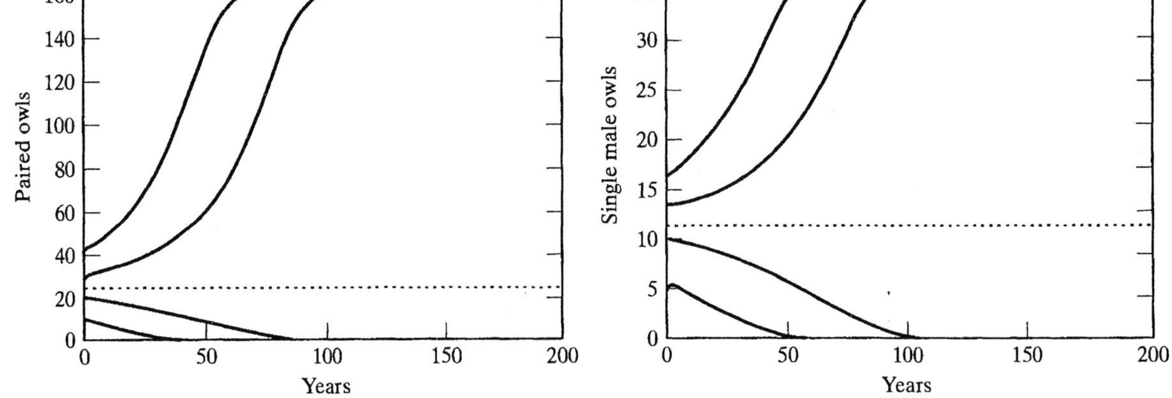


**Figure 3.14** Adult paired owls, $P_t$, and single male owls, $S_{m,t}$ for initial values, $(P_0, S_{m,0}) = (10, 5), (20, 10), (30, 15)$, and $(40, 20)$. The dotted line represents the equilibrium value of $E_1$.

aggregation of animals (Allee, 1931). The Allee effect refers to reduced fitness or decline in population growth at low population sizes or densities. When population sizes or densities are low, it is difficult to find mates and birth rates decrease, which can lead to population extinction. The Allee effect occurs in models when there is a threshold level below which there is population extinction. In the spotted owl model, $E_1$ is the threshold.

The spotted owl model illustrates the importance of providing sufficient habitat for mating and reproduction. If the number of available sites is too small ($U < 149$), then population extinction is certain. However, if the number of available sites is increased ($U > 149$) and if population sizes do not drop below $E_1$, a stable population of spotted owls can be maintained.

### 3.8.4 Two-Sex Model

We describe a model structured by sex of the individual. Two-sex models are needed, if, for example, the life expectancy differs between males and females, the sex ratio is not constant, or behavioral differences between males and females affect the population dynamics. Caswell (2001) gives several examples where life expectancy differs between males and females. In humans, male mortality is generally higher than female mortality; in killer whales, the life expectancy of females is almost twice that of males (Olesiuk et al., 1990); in black widow spiders, the life expectancy of females is about 2.7 times that of males (Deevey and Deevey, 1945).

To formulate a model that distinguishes between males and females, let $m_t$ and $f_t$ denote the number of males and females, respectively, in generation $t$. Assume that offspring are produced by mating of one male and one female. The number of offspring produced per mating couple depends on the form of the birth function. The birth function $B \equiv B(m, f)$ is also referred to as the *marriage function* in human demography. There is no agreement on the functional form taken by the birth function, but certain functional forms are preferred over others. For example, some properties that are reasonable to assume for the birth function include

(i) $B : [0, \infty) \times [0, \infty) \to [0, \infty)$.
(ii) $B(0, f) = 0$ and $B(m, 0) = 0$.
(iii) $\partial B/\partial m \geq 0$ and $\partial B/\partial f \geq 0$.
(iv) $B(km, kf) = kB(m, f)$ for $k \geq 0$.

A function $B$ having property (iv) is said to be a *homogeneous function of degree one*.

Various types of birth functions have been used in the literature. Some common birth functions include male dominant: $B(m, f) = m$, female dominant: $B(m, f) = f$, geometric mean: $B(m, f) = \sqrt{mf}$, and harmonic mean: $B(m, f) = 2mf/(m + f)$ (Caswell, 2001; Kot, 2001). Recall that the geometric mean of a sequence $x_1, \ldots, x_n$ is defined as $(x_1\, x_2 \cdots x_n)^{1/n}$ and the harmonic mean is defined as $n(\sum_{i=1}^{n} 1/x_i)^{-1}$. Thus, the geometric mean of $m$ and $f$ is $\sqrt{mf}$ and the harmonic mean of $m$ and $f$ is $2mf/(m + f)$. Among these four examples, only the geometric mean and the harmonic mean birth functions satisfy conditions (i)–(iv). The harmonic mean birth function is the one most often used by scientists. Keyfitz (1972) used the harmonic mean birth function to study marriage rates of human populations.

**Example 3.7** Assume that the total number of births per generation takes the form of the harmonic mean,

$$B(m_t, f_t) = \frac{2(r_m + r_f)m_t f_t}{m_t + f_t}, \tag{3.20}$$

where $r_m > 0$ and $r_f > 0$ are the average number of male or female offspring, respectively, per mating couple. Note that if there is one male and one female, then $B(1, 1) = r_m + r_f$. This function is maximized when the number of males and females is equal. For animal populations that are dominated by one sex or the other, the birth function can reflect this fact. For example, if males form a harem with $w$ females, the birth function can be written in the form

$$B(m_t, f_t) = \frac{2w(r_m + r_f)m_t f_t}{wm_t + f_t},$$

where $wr_m$ is the average number of male offspring and $wr_f$ is the average number of female offspring produced by the entire harem. In this case, if there is one male and $w$ females, then $B(1, w) = w(r_m + r_f)$. ■

Now, we formulate a difference equation model with two sexes using the harmonic mean birth function (3.20). Suppose males and females die after reproduction and are replaced by their progeny. Then the model has the form

$$m_{t+1} = \frac{2r_m m_t f_t}{m_t + f_t}, \quad m_0 > 0,$$

$$f_{t+1} = \frac{2r_f m_t f_t}{m_t + f_t}, \quad f_0 > 0$$

for $t = 0, 1, 2\ldots.$ It is easy to see that sex ratio is constant after the first generation: $m_t/f_t = r_m/r_f, t = 1, 2, \ldots.$ Using this fact, the model can be rewritten as follows:

$$m_{t+1} = \frac{2r_m r_f}{r_m + r_f} m_t = \lambda m_t,$$

$$f_{t+1} = \frac{2r_m r_f}{r_m + r_f} f_t = \lambda f_t,$$

where $\lambda = 2r_m r_f/(r_m + r_f), t = 1, 2, \ldots.$ Hence, if $\lambda < 1$, then $\lim_{t\to\infty}(m_t, f_t) = (0,0)$, and if $\lambda > 1$, then $\lim_{t\to\infty} m_t = \infty = \lim_{t\to\infty} f_t$. Note that if the average number of male and female offspring each equal one, $r_m = 1 = r_f$, then $\lambda = 1$.

Other two-sex models are described in Exercises 17 and 18. In Exercise 18, adult males and females produce sexually undifferentiated zygotes $z_t$; then the model consists of three difference equations, one each for $z_t$, $m_t$ and $f_t$. Lindström and Kokko (1998) discuss a two-sex model with males, females, and juveniles (where the sex of juveniles is not known). In their model, a Ricker density-dependence survival term is included. See Chapter 17 of the book *Matrix Population Models* by Caswell (2001) and the book *Gender-Structured Population Modeling* by Iannelli et al. (2005) for other types of two-sex models.

## 3.9 Measles Model with Vaccination

A system of difference equations for a measles epidemic with vaccination is derived based on a model of Anderson and May (1982). The states include the number of measles cases, the number of susceptible individuals, and the number of immune cases. It is assumed that each week there is a constant number of births and deaths given by $B$, where the number of births equals the number of deaths. Individuals recover from measles in one week. Thus, there are only new measles cases each week. Newborns are born susceptible. In addition, it is assumed that each week a proportion $p$ of susceptible individuals are vaccinated. The model is of SIR type similar to the one considered in Section 2.10.

Define the following state variables:

$I_t$ = number of new cases of measles per week,
$S_t$ = number of susceptible individuals per week,
$R_t$ = number of cases that become immune per week,
$p$ = proportion of susceptible individuals vaccinated per week, $0 \leq p < 1$,
$N$ = total population size,
$\alpha N$ = average number of secondary infections caused by one infective.

The system of difference equations has the following form:

$$\begin{aligned} S_{t+1} &= (1-p)S_t - \alpha I_t S_t + B, \qquad (3.21) \\ I_{t+1} &= \alpha I_t S_t, \\ R_{t+1} &= R_t + I_t - B + pS_t. \end{aligned}$$

The differences between this model and the epidemic model in Section 2.10 are the form of the births and deaths (constant rather than proportional to the population size), the length of the infectious period (which is the same length as the time step), and the inclusion of vaccination. To ensure solutions are nonnegative, the value of $B$ must be sufficiently small. Note that each variable should have death rates, $D_1$, $D_2$, and $D_3$ for each class, where $B = D_1 + D_2 + D_3$. But here it is assumed that $D_1 = D_2 = 0$ and $D_3 = B$. A more realistic model where births and deaths are proportional to population size is described in the Exercises for this chapter.

Note that the total population size $N_t = I_t + S_t + R_t$ remains constant over time, $N_t = N$. Also, $I_{t+1}$ and $S_{t+1}$ do not depend on $R_t$. The analysis of the system of difference equations can be reduced to the two equations in $I$ and $S$. The equilibria are solutions of the following two equations:

$$p\bar{S} = B - \alpha\bar{I}\bar{S}, \quad \text{and} \quad \bar{I} = \alpha\bar{I}\bar{S}.$$

There exist two equilibria, the disease-free equilibrium,

$$\bar{I} = 0 \quad \text{and} \quad \bar{S} = B/p, p > 0,$$

and the endemic equilibrium,

$$\bar{S} = 1/\alpha \quad \text{and} \quad \bar{I} = B - p/\alpha,$$

where $B > p/\alpha$.

If there is no vaccination in model (3.21), $p = 0$, then there does not exist a disease-free equilibrium. With no vaccination, there exists only the endemic equilibrium, $\bar{I} = B$ and $\bar{S} = 1/\alpha$. The Jacobian matrix satisfies

$$J_p(S, I) = \begin{pmatrix} 1 - p - \alpha I & -\alpha S \\ \alpha I & \alpha S \end{pmatrix} \quad \text{for } p > 0,$$

with vaccination, and

$$J_0(S, I) = \begin{pmatrix} 1 - \alpha I & -\alpha S \\ \alpha I & \alpha S \end{pmatrix} \quad \text{for } p = 0,$$

without vaccination.

If there is no vaccination, then the Jacobian matrix evaluated at the endemic equilibrium is

$$J_0(1/\alpha, B) = \begin{pmatrix} 1 - \alpha B & -1 \\ \alpha B & 1 \end{pmatrix}.$$

The endemic equilibrium is locally asymptotically stable if $|\text{Tr}(J)| < 1 + \det(J) < 2$ or, equivalently,

$$|2 - \alpha B| < 2 < 2.$$

Hence, this equilibrium is not locally asymptotically stable. The eigenvalues can be calculated directly, $\lambda_{1,2} = [\text{Tr}(J_0) \pm \sqrt{\text{Tr}(J_0)^2 - 4\det(J_0)}]/2$,

$$\lambda_{1,2} = \frac{2 - \alpha B \pm \sqrt{\alpha B(\alpha B - 4)}}{2}.$$

If $\alpha B < 4$, the eigenvalues are complex conjugates satisfying $|\lambda_i| = 1$, and solutions oscillate about the equilibrium.

With vaccination, at the disease-free equilibrium,

$$J_p(B/p, 0) = \begin{pmatrix} 1 - p & -\alpha B/p \\ 0 & \alpha B/p \end{pmatrix}.$$

The eigenvalues are $\alpha B/p$ and $1 - p$. Since both of these eigenvalues are positive, the disease-free equilibrium is locally asymptotically stable if $\alpha B < p$. If the vaccination proportion $p$ is sufficiently small, then the disease-free equilibrium is unstable. The expression

$$\mathcal{R}_0 = \alpha B/p$$

represents the basic reproduction number for this model.

With vaccination, at the endemic equilibrium,

$$J_p(\bar{S}, \bar{I}) = \begin{pmatrix} 1 - \alpha B & -1 \\ \alpha B - p & 1 \end{pmatrix}.$$

The endemic equilibrium is locally asymptotically stable if $|2 - \alpha B| < 2 - p < 2$. The second part of this inequality is always satisfied. In addition, note that the endemic equilibrium is positive if $\mathcal{R}_0 > 1$. If $\alpha B < 2$, then the stability criteria can be expressed as $1 < \mathcal{R}_0 < 2/p$.

Two cases were simulated by Anderson and May (1982). In these two cases $B$ represents the birth rate in a developed versus a developing country (e.g., United Kingdom versus Nigeria). Suppose the annual birth rate is 12 per 1000 people for a developed country such as the United Kingdom (UK) and Nigeria's birth rate is three times that rate. Thus, the number of births per week in a developed country whose size is one half a million is

$$B = \frac{12}{1000 \text{ years}} \frac{1 \text{ year}}{52 \text{ weeks}} 500{,}000 \approx 115.$$

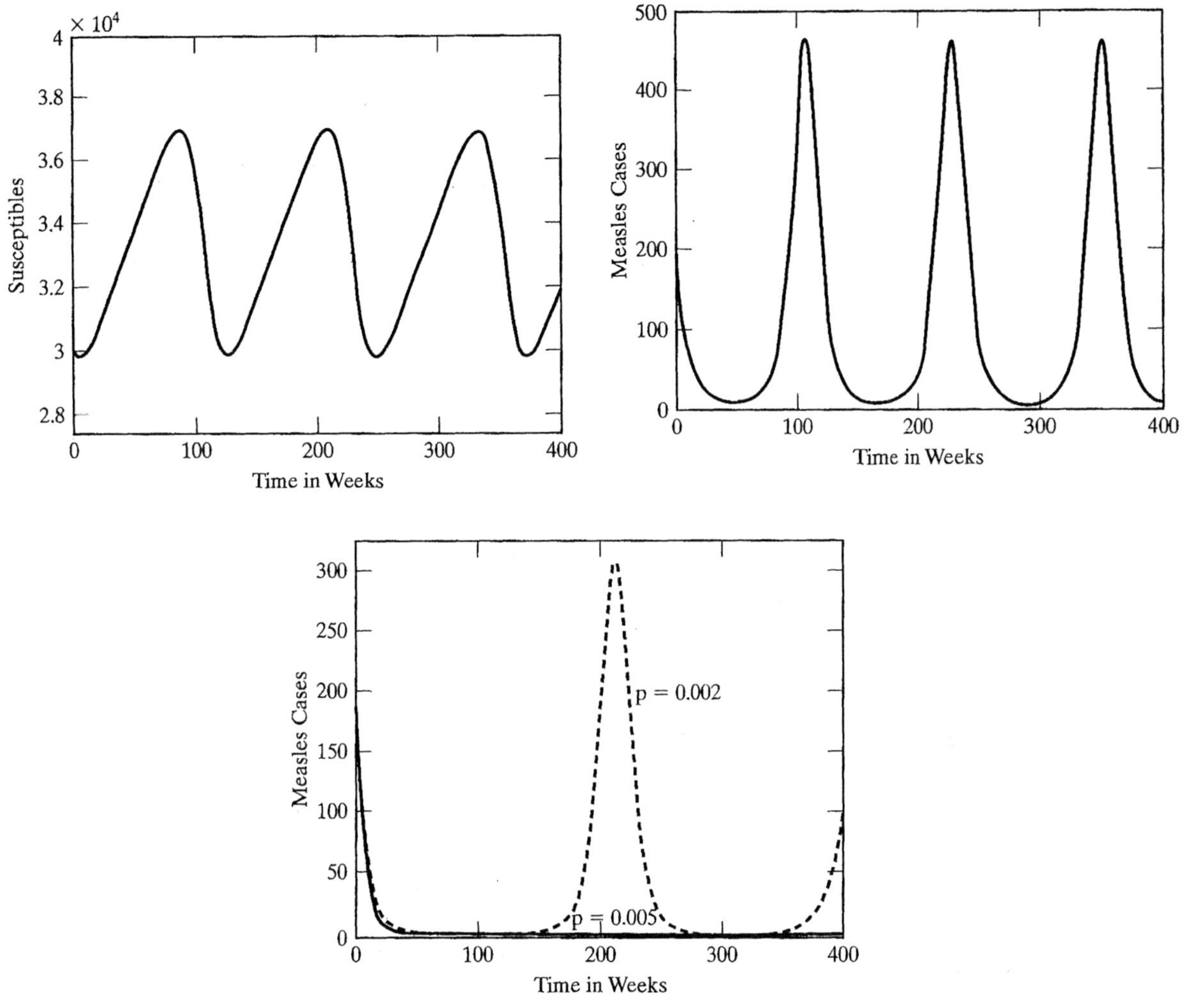


**Figure 3.15** Measles model of Anderson and May (1982) with parameters $B = 115$, $\alpha = 0.3 \times 10^{-4}$. In the top two figures $p = 0$ and in the bottom figure $p = 0.002$ and $p = 0.005$.

In Nigeria, the weekly birth rate is three times that rate, so $B = 345$. Suppose $\alpha = 0.3 \times 10^{-4}$. For the given parameter values, $\alpha B < 4$, so that when there is no vaccination, solutions oscillate about the endemic equilibrium. Measles infection is not eliminated from the population. Figure 3.15 is for the hypothetical UK data when there is no vaccination. The periodicity of measles cases in Figure 3.15 ($p = 0$) is between two and three years.

Now, we suppose vaccination is included. When $p > B\alpha$, the positive equilibrium does not exist and the disease-free equilibrium given by $\bar{I} = 0$ and $\bar{S} = B/p$ is locally asymptotically stable. In the measles model with the hypothetical UK data and a vaccination proportion of $p > B\alpha = 115(0.3 \times 10^{-4}) = 0.00345$, solutions are graphed in Figure 3.15. Solutions satisfy $I_t \to 0$ and $S_t \to \bar{S} = B/p = 1150$. This model shows the significance of the birth rate in controlling measles. Notice that the vaccination proportion for the hypothetical Nigeria data must be three times that for the UK to eliminate measles. If the vaccination proportion is reduced so that $0 < p < B\alpha = 0.00345$ for the hypothetical UK data, then there are still epidemics. However, vaccination increases the length of time between outbreaks.

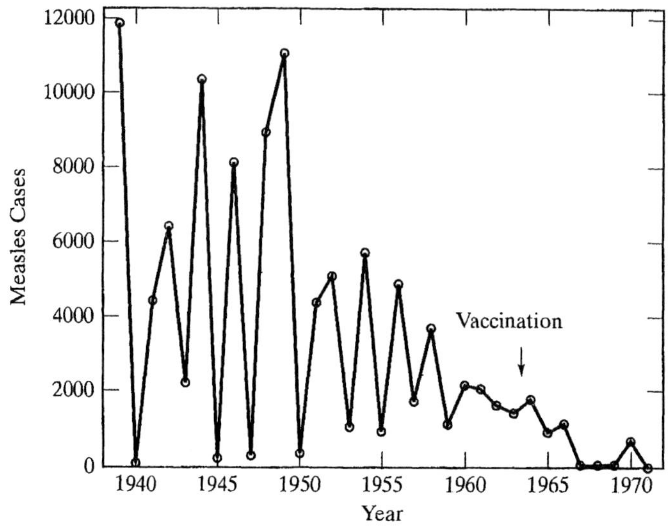


**Figure 3.16** Annual measles cases in Baltimore, Maryland from 1939 to 1971 (Hipel and McLeod, 1994).

In the United States, the measles vaccine was licensed in 1963. Prior to that time epidemic cycles occurred every two to three years. Annual data on measles cases in Baltimore, Maryland show a two- to three-year cycle before vaccination (Figure 3.16). Prior to 1963, almost everyone in the United States acquired measles. The number of measles cases was close to the birth cohort (3.5 to 4 million cases per year) (Prober, 1999). Recall that the basic reproduction number for measles is very high, $\mathcal{R}_0 \approx 12 - 13$ (Table 2.2); measles is a highly contagious disease. The one-dose strategy was not sufficient to keep measles under control. A widespread measles outbreak in 1989–1991 among unvaccinated preschool and vaccinated school-age children prompted a change in the measles vaccination strategy. Instead of a one-dose strategy, a two-dose strategy was recommended in 1989 (Prober, 1999). In the two-dose strategy, the first dose is given at 15 months and the second dose is given prior to entry into middle school or junior high school. This two-dose strategy has been successful in controlling measles in the United States.

A more realistic SIR epidemic model with vaccination assumes the birth and death rates are proportional to the size of the class. A model of this form is given in Section 2.10. If vaccination is included, the model has the form

$$S_{t+1} = (1 - p)S_t - \frac{\beta}{N} I_t S_t + b(R_t + I_t),$$

$$I_{t+1} = \frac{\beta}{N} I_t S_t + (1 - b - \gamma) I_t,$$

$$R_{t+1} = (1 - b)R_t + \gamma I_t + p S_t,$$

where $p, b, \gamma, \beta > 0$, $\beta = \alpha N$. The parameter $b$ is the per capita number of births, $\gamma$ is the probability of recovery, $p$ is the proportion vaccinated, and $\beta$ is the number of successful contacts in time $t$ to $t + 1$. Also, $N$ is the total population size; the population size is constant. If $0 < p + \beta < 1$ and $0 < b + \gamma < 1$, then it can be shown that solutions are nonnegative. The basic reproduction number for this model is

$$\mathcal{R}_0 = \frac{\beta b}{(b + \gamma)(b + p)}.$$

This model is studied in more detail in the Exercises for this chapter. Also, consult Allen and Burgin (2000), Allen et al. (2004), Brauer and Castillo-Chávez (2001), and Castillo-Chávez and Yakubu (2001) for additional examples of discrete-time epidemic models.

## 3.10 Exercises for Chapter 3

**1.** The Beverton-Holt difference equation has an explicit solution given in Exercise 10 in Chapter 2. The explicit solution for $N_t$ is

$$N_t = \frac{N_0 K \lambda^t}{K + N_0(\lambda^t - 1)}.$$

Make the change of variable in the Beverton-Holt solution

$$y_t = \ln\left(\frac{1}{N_t} - \frac{1}{K}\right)$$

so that

$$y_t = \ln\left(\frac{1}{N_0} - \frac{1}{K}\right) - t\ln\lambda.$$

Use a least squares approximation to fit the whooping crane data to the curve $y_t$ when $K = 400$ and when $K = 1000$. (Data are given in the Appendix to Chapter 3.) Find estimates for the parameters $N_0$ and $\lambda$. Graph the whooping crane data and the two Beverton-Holt curves. Compare the least squares approximations for the two Beverton-Holt curves. Note for large values of $K$, the initial part of the logistic curve is close to the exponential approximation.

**2.** Data on the breeding population of ducks from 1968 to 2004 in Minnesota are given in Table 3.4 in the Appendix to Chapter 3 (U.S. Fish and Wildlife Service, 2004). The population data are highly variable due to annual variations in temperature and rainfall. However, it is clear that the size of the breeding population is increasing. Apply the command *polyfit* in MATLAB to find the least squares fit of the data to the curve $N_t = a + bt$. What is the annual rate of increase (in thousands of ducks)? See Figure 3.17.

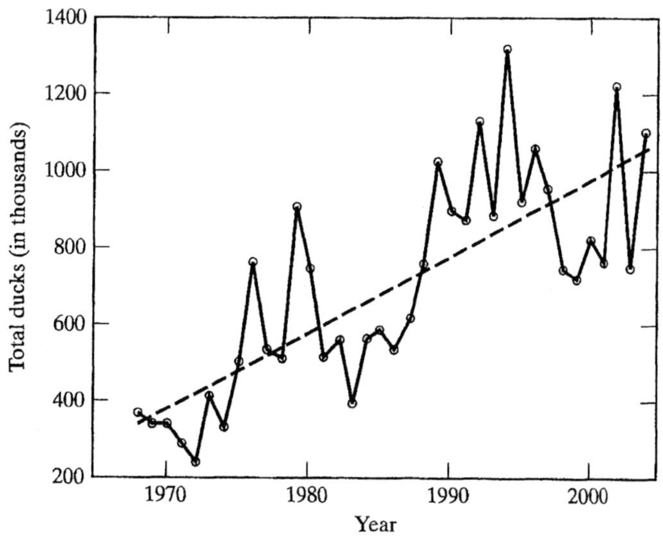


**Figure 3.17** Total duck breeding population in Minnesota from 1968 to 2004 fit to the linear curve $N_t = a + bt$.

**3.** Consider the Nicholson-Bailey model with density dependence in the host population

$$\begin{aligned} N_{t+1} &= N_t \exp(r(1 - N_t/K) - aP_t), \\ P_{t+1} &= eN_t(1 - \exp(-aP_t)). \end{aligned}$$

(a) Make a change of variable to simplify the above system. Let $x_t = N_t/K$ and $y_t = aP_t$ and write the simplified system as follows:

$$\begin{aligned} x_{t+1} &= x_t \exp(r(1 - x_t) - y_t), \\ y_{t+1} &= cx_t(1 - \exp(-y_t)). \end{aligned}$$

The parameter $c$ depends on $e$, $a$, and $K$. What is the value of $c$?

(b) Find implicit equations satisfied by the nonzero equilibrium $(\bar{x}, \bar{y})$ of the simplified system.

(c) Find the Jacobian matrix of the simplified system evaluated at $(\bar{x}, \bar{y})$. Express the Jacobian matrix in a simple form [e.g., use identities such as $\bar{y}/\bar{x} = c(1 - \exp(-\bar{y}))$ and $1 = \exp(r - r\bar{x} - \bar{y})$]. Then find the local asymptotic stability conditions.

(d) Is the equilibrium $(\bar{x}, \bar{y})$ locally asymptotically stable if $r = 1.5$ and $c = 2.5$? (Calculate the equilibrium values to three decimal places and check whether the stability conditions are satisfied.)

**4.** A model for annual plant species competition is discussed by Pakes and Maller (1990). The simplified model has the following form:

$$\begin{aligned} p_{t+1} &= (1 - \alpha)\frac{k_{12}p_t}{k_{12}p_t + q_t} + \alpha p_t = f(p_t, q_t), \\ q_{t+1} &= (1 - \beta)\rho\frac{k_{21}q_t}{p_t + k_{21}q_t} + \beta q_t = g(p_t, q_t), \end{aligned}$$

where $0 < \alpha, \beta < 1$. The expressions $\alpha p_t$ and $\beta q_t$ are the plants that emerge from seeds in the seedbank. The other expressions in the preceding model are plants that emerge from new seeds, where there is competition between the two species. The coefficients $k_{ij}$ are inversely related to the amount of competition; if both are large the competition is weak. If $k_{12} \gg k_{21}$, and $k_{21} \ll 1$, then species 1 ($p$) has a large impact on species 2 ($q$) but species 2 has little impact on species 1 (where $\gg$ means "much greater than"). The parameters $\rho$, $k_{ij}$, $\alpha$, and $\beta$ are positive.

(a) Show that there exist three equilibria, $(\bar{p}, \bar{q})$, given by

$$(1, 0), \quad (0, \rho), \quad \text{and} \quad \left(\frac{k_{21}(k_{12} - \rho)}{k_{12}k_{21} - 1}, \frac{k_{12}(k_{21}\rho - 1)}{k_{12}k_{21} - 1}\right).$$

(b) Calculate the Jacobian matrix of the system, $J(p, q)$.

(c) Evaluate the Jacobian matrix at the equilibria $(1, 0)$ and $(0, \rho)$. Then show the following:

1. Equilibrium $(1, 0)$ is locally asymptotically stable if $\rho k_{21} < 1$.
2. Equilibrium $(0, \rho)$ is locally asymptotically stable if $k_{12} < \rho$.

In Case 1, species 1 wins the competition. In Case 2, species 2 wins the competition. It can also be shown that the two-species equilibrium is locally asymptotically stable if $k_{12} > \rho > 1/k_{21}$. These results are typical of competition models. Generally, only one species wins the competition unless the competition is weak. Competition is weak in the plant model if both $k_{ij}$ are large. The condition $k_{12}k_{21} > 1$ (weak competition) is needed for local asymptotic stability of the two-species equilibrium.

**5.** In the plant species competition model of Exercise 4, assume the parameter $\rho = 1$. Let the positive equilibrium be denoted as $(\bar{p}, \bar{q})$. The following steps show that if $\min\{k_{12}, k_{21}\} > 1$, then the positive equilibrium is locally asymptotically stable.

(a) Find the Jacobian matrix evaluated at the positive equilibrium, $J(\bar{p}, \bar{q})$.

(b) Show that the trace of $J(\bar{p}, \bar{q})$ is positive.

(c) Show that the $\mathrm{Tr}(J) - 1 - \det(J)$ can be expressed as

$$\frac{(k_{12} - 1)(1 - \alpha)(k_{21} - 1)(1 - \beta)}{k_{12}k_{21} - 1}.$$

(d) Express the $\det(J) - 1$ as

$$\frac{(\alpha\beta - 1)(k_{12}k_{21} - 1) + \alpha(1 - \beta)(k_{12} - 1) + \beta(1 - \alpha)(k_{21} - 1)}{k_{12}k_{21} - 1}.$$

Finally, show that this latter expression is negative.

**6.** The dynamics of the Leslie-Gower model are simple in the case $\alpha_i < 1$ for $i = 1$ or $i = 2$. Recall that the model takes the form

$$H_{t+1} = \frac{\alpha_1 H_t}{1 + \gamma_1 P_t}$$

$$P_{t+1} = \frac{\alpha_2 P_t}{1 + \gamma_2 P_t / H_t}.$$

(a) Show that if $\alpha_1 < 1$, then $\lim_{t\to\infty}(H_t, P_t) = (0, 0)$.

(b) Show that if $\alpha_1 > 1$ and $\alpha_2 < 1$, then $\lim_{t\to\infty} P_t = 0$ and $\lim_{t\to\infty} H_t = \infty$.

**7.** Consider the predator-prey model discussed in Section 3.6, where the probability of death $d$ in the interval $t$ to $t + 1$ satisfies $d \neq 1$. In this case, the normalized system has the form

$$x_{t+1} = (r + 1)x_t - rx_t^2 - cx_t y_t,$$
$$y_{t+1} = cx_t y_t + (1 - d)y_t,$$

where $r, c > 0$ and $0 < d < 1$.

(a) Show that there exist three equilibria given by (0, 0), (1, 0), and $(\bar{x}, \bar{y})$, where $\bar{x} > 0$ and $\bar{y} > 0$. Find the values of $\bar{x}$ and $\bar{y}$.

(b) Determine conditions for local stability of the three equilibria. How do the conditions compare with the case $d = 1$?

**8.** Recall that the mean fitness of the population genetics model is given by $w_t = p_t^2 w_{AA} + 2p_t(1 - p_t) + (1 - p_t)^2 w_{aa}$, where $p_t$ is the proportion of the population carrying allele $A$. Selection is governed by the survival rates $w_{AA} = 1 - s$ and $w_{aa} = 1 - r$, where $r, s < 1$. It can be shown that

$$w_{t+1} - w_t = \frac{p_t(-1 + p_t)[p_t^2(r + s) + p_t(s - 3r) + 2r - 2](p_t(s + r) - r)^2}{w_t^2}.$$

Show that $w_{t+1} - w_t \geq 0$ for $p_t \in [0, 1]$, and that $w_{t+1} = w_t$ only if $p_t = 0, 1$, or $r/(r + s)$ (one of the three equilibrium values). Hence, selection leads to an increase in the mean fitness of the population over time. (*Hint:* Consider $g(p) = p^2(r + s) + p(s - 3r) + 2r - 2$. Show $g(0) < 0, g(1) < 0$, and $g(p) = 0$ has no solutions $p \in (0, 1)$.)

**9.** For the population genetics model with selection, $p_{t+1} = f(p_t)$, where $f(p) = p(1 - ps)/(1 - p^2 s - (1 - p)^2 r)$, it was shown in equation (3.9) that

$$f'(p) = \frac{(1 - s)p^2 + 2(1 - s)(1 - r)p(1 - p) + (1 - r)(1 - p)^2}{(1 - p^2 s - r + 2rp - rp^2)^2} > 0$$

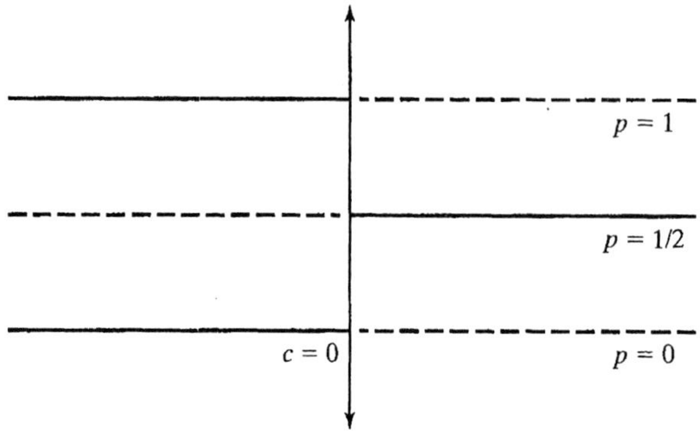


**Figure 3.18** Bifurcation diagram as a function of $c$ for the population genetics model in Exercise 10.

for $0 \le p \le 1$ and $r, s < 1$. Also, $f : [0, 1) \to [0, 1)$ and $f$ has three fixed points, $0, 1$, and $\bar{p} = r/(r + s)$.

Show that the equilibrium $\bar{p}$ is globally asymptotically stable for all initial conditions $p_0 \in (0, 1)$ if $0 < r, s < 1$. (*Hint:* Apply a theorem from Chapter 2.)

**10.** In the population genetics model in Example 3.2,

$$p_{t+1} = \frac{p_t f(p_t)}{p_t f(p_t) + 1 - p_t} = F(p_t).$$

Assume $f(p) = \exp(c[1 - 2p])$, where $c$ is a constant.

(a) Find the positive polymorphic equilibrium, $0 < \bar{p} < 1$, and determine conditions on $c$ such that it is locally asymptotically stable.

(b) Verify that a pitchfork bifurcation occurs at $c = 0$. See Figure 3.18.

(c) Let $c = 5$ and $0 < p_0 < 0.5$. Show by computing $p_t$ for $t = 1, \ldots, 20$, that the solution converges to a 2-cycle.

**11.** In the population genetics model in Example 3.2,

$$p_{t+1} = \frac{p_t f(p_t)}{p_t f(p_t) + 1 - p_t} = F(p_t), \quad 0 < p_0 < 1.$$

Assume $f(p) = \exp(c - p)$, where $c$ is a constant.

(a) Find the positive polymorphic equilibrium, $0 < \bar{p} < 1$, and determine conditions on $c$ such that $\bar{p}$ is locally asymptotically stable.

(b) Verify that a transcritical bifurcation occurs at $c = 0$ and $c = 1$. See Figure 3.19.

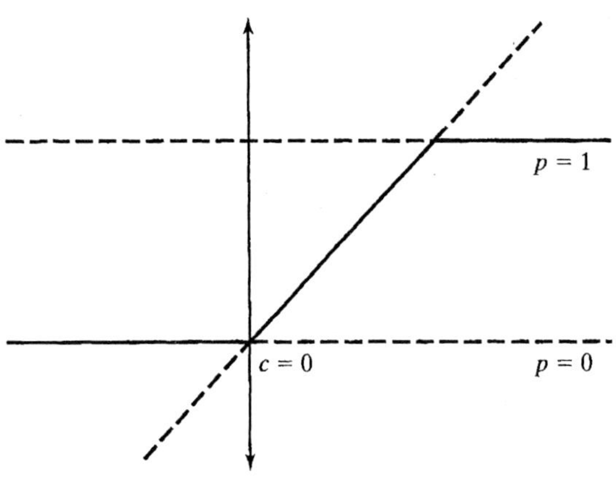


**Figure 3.19** Bifurcation diagram as a function of $c$ for the population genetics model in Exercise 11.

(c) Show that $F'(p) > 0$ for all values of $c$ and $p \in (0, 1)$. Then prove that local asymptotic stability of an equilibrium for this model implies global asymptotic stability.

**12.** In a gene-for-gene system of a plant and a pathogen, we model the frequency of a virulent pathogen allele $n$ (allele $V$) and a resistant plant allele $p$ (allele $R$). Survival of a particular genotype $VV$, $Vv$, $vv$, $RR$, $Rr$, or $rr$, depends on the fitness values $w_{VV}$, $w_{Vv}$, and so on. The model takes the form of the following system of difference equations:

$$n_{t+1} = \frac{n_t^2 w_{VV} + n_t(1 - n_t)w_{Vv}}{n_t^2 w_{VV} + 2n_t(1 - n_t)w_{Vv} + (1 - n_t)^2 w_{vv}} = f(n_t, p_t),$$

$$p_{t+1} = \frac{p_t^2 w_{RR} + p_t(1 - p_t)w_{Rr}}{p_t^2 w_{RR} + 2p_t(1 - p_t)w_{Rr} + (1 - p_t)^2 w_{rr}} = g(n_t, p_t).$$

All of the fitness values are frequency dependent, that is, $w_{VV} \equiv w_{VV}(n, p)$, $w_{RR} \equiv w_{RR}(n, p)$, and so on.

(a) Show that there exist four equilibria $(\bar{n}, \bar{p})$ given by (0,0), (1,0), (0,1), and (1,1).

(b) Verify the following:

Equilibrium (0,0) is locally asymptotically stable (las) if $w_{Vv}(0, 0) < w_{vv}(0, 0)$ and $w_{Rr}(0, 0) < w_{rr}(0, 0)$.

Equilibrium (0,1) is las if $w_{Vv}(0, 1) < w_{vv}(0, 1)$ and $w_{Rr}(0, 1) < w_{RR}(0, 1)$.

Equilibrium (1,0) is las if $w_{Vv}(1, 0) > w_{VV}(1, 0)$ and $w_{Rr}(1, 0) < w_{rr}(1, 0)$.

Equilibrium (1,1) is las if $w_{Vv}(1, 1) > w_{VV}(1, 1)$ and $w_{Rr}(1, 1) < w_{RR}(1, 1)$.

(*Hint:* Maple may be useful in computing and evaluating the Jacobian matrix for this system.)

**13.** In the nonlinear structured model of Silva and Hallam (1992, 1993) assume that there are two age classes. Let $g(x) = \exp(-x)$ and the recruitment satisfy $R(t + 1) = ag(w(t))w(t)$, where $w(t) = E(t) = b_1x_1(t) + b_2x_2(t)$. In addition, assume $a = 1$ and $R_0 > 1$.

(a) Show that $\bar{w} = \ln R_0$.

(b) If the fecundities satisfy $b_1 = 0$ and $b_2 > 0$, show that the eigenvalues of the Jacobian matrix are purely imaginary and the equilibrium is locally asymptotically stable if $(sb_2/R_0)[\ln(R_0) - 1] < 1$.

(c) If the fecundities satisfy $b_2 = 0$ and $b_1 > 0$, show that the eigenvalues are real and the equilibrium is locally asymptotically stable if $(b_1/R_0)[\ln(R_0) - 1] < 1$.

**14.** Recall the flour beetle LPA model discussed in Section 3.8.2,

$$L_{t+1} = bA_t\exp(-c_{ea}A_t - c_{el}L_t),$$
$$P_{t+1} = L_t(1 - \mu_l),$$
$$A_{t+1} = P_t\exp(-c_{pa}A_t) + A_t(1 - \mu_a).$$

In a set of experiments conducted by the beetle group, the adult death rate $\mu_a$ was manipulated (see, e.g., Cushing et al., 1998, 2003). Determine the positive equilibrium and its stability for the following sets of parameter values.

(a) Let $b = 7$, $c_{ea} = 0.01$, $c_{el} = 0.015$, $\mu_l = 0.15$, $\mu_a = 0.005$, and $c_{pa} = 0.005$. Use a calculator or computer to approximate the positive equilibrium and show that it is locally asymptotically stable.

(b) For the same parameter values as in (a), with the exception of $\mu_a = 0.05$ and $\mu_a = 0.2$, find the positive equilibrium. In each case, determine whether the equilibrium is stable or unstable. As the adult death rate increases, the system becomes unstable. (See the Appendix for calculation of the Jacobian matrix and evaluation at an equilibrium using Maple.)

**15.** (a) Suppose the parameter values in the model for the northern spotted owl model satisfy $s_S = 0.71$, $s_J = 0.60$, $p_s = 0.88$, $p_b = 0.056$, $f = 0.66$, $n = m = 20$, $T = 1000$, and $U = 250$. Solve the polynomial $p(\bar{S}_m) = 0$ for $\bar{S}_m$ and apply (3.17) to show that the equilibrium values $E_1$ and $E_2$ satisfy

$$E_1 = (25.04, 11.43) \quad \text{and} \quad E_2 = (163.97, 35.81).$$

(b) Find the Jacobian matrices $J(E_1)$ and $J(E_2)$ and show that $E_1$ is unstable and $E_2$ is locally asymptotically stable.

**16.** In a simplified model for the northern spotted owl, paired owls, $x_t$, and single male owls, $y_t$, are modeled as follows:

$$\begin{aligned} x_{t+1} &= s_p x_t + a(y_t)^2 = F(x_t, y_t), \\ y_{t+1} &= bx_t - c(x_t)^2 - dx_t y_t + s_s y_t - e(y_t)^2 = G(x_t, y_t). \end{aligned}$$

The parameters $a$, $b$ $c$, $d$, and $e$ are positive and depend on environmental conditions, fertility, and survival probabilities. For example, the parameters $s_p$ = probability both individuals in a pair survive one year and do not become single and $s_s$ = probability single male survives one year and does not mate, $0 < s_p, s_s < 1$.

(a) Find the Jacobian matrix for this system.

(b) Show that the extinction equilibrium $(0, 0)$ is locally asymptotically stable. What does this imply about small population sizes?

**17.** Let $M_t$ and $F_t$ denote the number of males and females in generation $t$. Assume that the total number of births per generation has the form of a harmonic mean,

$$B(M_t, F_t) = \frac{2(r_m + r_f)M_t F_t}{M_t + F_t},$$

where $r_m > 0$ and $r_f > 0$ are the average number of male and female offspring, respectively, per mating couple. If the probability of survival for males and females is $s_m$ and $s_f$, respectively, then the model has the form

$$\begin{aligned} M_{t+1} &= \frac{2r_m M_t F_t}{M_t + F_t} + s_m M_t, \quad M_0 > 0, \\ F_{t+1} &= \frac{2r_f M_t F_t}{M_t + F_t} + s_f F_t, \quad F_0 > 0. \end{aligned}$$

(a) For a positive equilibrium $(\bar{M}, \bar{F})$, show that the sex ratio satisfies

$$\frac{\bar{M}}{\bar{F}} = \frac{r_m(1 - s_f)}{r_f(1 - s_m)}.$$

(b) Suppose $r_f = r_m = r$, $s_f = s_m = s$ and $M_0 = F_0$. Then $M_t = F_t = N_t/2$, where $N_t = M_t + F_t$ is the total population size. Find the difference equation satisfied by the total population size $N_t$ and show that $\lim_{t\to\infty} N_t = 0$ if $r + s < 1$ and $\lim_{t\to\infty} N_t = \infty$ if $r + s > 1$.

**18.** Let $X_t = (z_t, m_t, f_t)^T$ denote a population vector of males $m_t$, females $f_t$, and undifferentiated zygotes $z_t$ (the sex is not yet determined). The following model is discussed by Caswell (2001). Suppose the birth function $B$ has the form of a harmonic mean,

$$B(m, f) = \frac{2rmf}{m + f},$$

where $r > 0$ is the average number of offspring produced per mating couple. Then $z_{t+1} = B(m_t, f_t)$. Assume the probability of survival per generation for males and females is $s_m$ and $s_f$, respectively, and differentiation of zygotes into males and females occurs in the proportion $a : 1 - a$. Thus,

$$m_{t+1} = az_t + s_m m_t$$
$$f_{t+1} = (1 - a)z_t + s_f f_t.$$

Denote the per capita fecundities for males and females as $b_m$ and $b_f$, respectively, so that $B(m, f) = b_m m + b_f f$. Since birth requires one male and one female, birth contributions by males should equal birth contributions by females, that is, $b_m m = b_f f$. This implies $b_m = B(m, f)/(2m)$ and $b_f = B(m, f)/(2f)$, or, equivalently,

$$b_m \equiv b_m(m, f) = \frac{rf}{m + f}$$

and

$$b_f \equiv b_f(m, f) = \frac{rm}{m + f}.$$

Now, the two-sex model can be written in matrix form

$$X_{t+1} = \begin{pmatrix} 0 & b_m & b_f \\ a & s_m & 0 \\ 1 - a & 0 & s_f \end{pmatrix} X_t = A(m_t, f_t)X_t.$$

We seek solutions $\bar{X} = (\bar{z}, \bar{m}, \bar{f})^T$ satisfying $A(\bar{m}, \bar{f})\bar{X} = \lambda\bar{X}$, where $\lambda$ is the growth rate. Note that if $\lambda = 1$, then the solutions are equilibria, but if not, then for $X_0 = \bar{X}$, solutions will increase if $\lambda > 1$ or decrease if $0 < \lambda < 1$.

(a) Show that for a solution of the form $\bar{X}$, the sex ratio $\bar{m}/\bar{f}$ equals

$$\frac{\lambda - s_f}{\lambda - s_m}\,\frac{a}{1 - a}.$$

How does this compare with the model in Exercise 17?

(b) Show that the growth rate $\lambda$ satisfies

$$\lambda^2 - [as_f + (1 - a)s_m]\lambda - 2a(1 - a)r = 0.$$

(c) Show that $\lambda > 1$ iff $as_f + (1 - a)s_m + 2a(1 - a)r > 1$. What does this inequality imply about birth and survival rates?

**19.** Suppose the birth function for mating couple is $B(m, f), m \geq 0$ and $f \geq 0$. The maximum value of the per capita birth function $B(m, f)/(m + f) = F(f/m)$ depends on the sex ratio.

(a) Let the birth function be male dominated, $B(m, f) = rm$. Show that the per capita birth function is a maximum when $f/m = 0$.

(b) Let the birth function be based on the geometric mean, $B(m, f) = r\sqrt{mf}$. Show that the per capita birth function is a maximum when $f/m = 1$.

(c) Let the birth function be based on the harmonic mean $B(m, f) = 2rmf/(m + f)$. Show that the per capita birth function is a maximum when $f/m = 1$.

(d) Let the birth function be based on the harmonic mean, modified for a male harem size of $w$, $B(m, f) = 2rwmf/(wm + f)$. Show that the per harem birth function, $B(m, f)/(wm + f)$, is a maximum when $f/m = w$.

**20.** Consider the SIR epidemic model with vaccination

$$S_{t+1} = (1-p)S_t - \frac{\beta}{N} I_t S_t + b(R_t + I_t),$$

$$I_{t+1} = \frac{\beta}{N} I_t S_t + (1 - b - \gamma)I_t,$$

$$R_{t+1} = (1-b)R_t + \gamma I_t + pS_t,$$

where $0 < p + \beta < 1$ and $0 < b + \gamma < 1$. The parameter $b$ is the probability of a birth, $\gamma$ is the probability of recovery, $p$ is the proportion vaccinated, $\beta$ is the contact rate, and $N$ is the total population size. For this model, the basic reproduction number satisfies

$$\mathcal{R}_0 = \frac{\beta b}{(b+\gamma)(p+b)}.$$

Note that if $p = 0$, then $\mathcal{R}_0 = \beta/(b+\gamma)$, the value given in Section 2.10.

(a) Show that if $S_0 + I_0 + R_0 = N$, then $S_t + I_t + R_t = N$.

(b) Show that if $S_0, I_0, R_0 > 0$ and $S_0 + I_0 + R_0 = N$, then solutions are positive for all time, $S_t, I_t, R_t > 0$.

(c) Express the three-dimensional SIR model as a two-dimensional SI model by making the substitution, $R_t = N - I_t - S_t$:

$$S_{t+1} = (1-p)S_t - \frac{\beta}{N} I_t S_t + b(N - S_t),$$

$$I_{t+1} = \frac{\alpha}{N} I_t S_t + (1 - b - \gamma)I_t.$$

Show that this system has two equilibria $(\bar{S}, \bar{I})$, where either $\bar{I} = 0$ or $\bar{I} > 0$. Show that the positive equilibrium exists iff $\mathcal{R}_0 > 1$.

(d) Show that the disease-free equilibrium $\bar{I} = 0$ is locally asymptotically stable if $\mathcal{R}_0 < 1$.

(e) Suppose $\beta = 0.5$, $b = 0.05 = \gamma$. Find the minimum vaccination proportion $p_{\min}$ such that $\mathcal{R}_0 \leq 1$. If $p = (1/2)p_{\min}$, $\beta = 0.5$, and $b = 0.05 = \gamma$, is the positive equilibrium stable?

## 3.11 References for Chapter 3


Akçakaya, H. R. and M. G. Raphael. 1998. Assessing human impact despite uncertainty: viability of the northern spotted owl metapopulation in the northwestern USA. *Biodiversity and Conservation* 7: 875–894.

Allee, W. C. 1931. *Animal Aggregations, a Study in General Sociology*. Univ. Chicago Press, Chicago.

Allen, L. J. S. 1989. A density-dependent Leslie matrix model. *Math. Biosci.* 95: 179–187.

Allen, L. J. S. and A. M. Burgin. 2000. Comparison of deterministic and stochastic SIS and SIR models in discrete time. *Math. Biosci.* 163: 1–33.

Allen, L. J. S., J. F. Fagan, G. Högnäs, and H. Fagerholm. 2005. Population extinction in discrete-time stochastic population models with an Allee effect. *J. Difference Eqns. and Appl.* 11: 273–293.

Allen, L. J. S., N. Kirupaharan, and S. M. Wilson. 2004. SIS epidemic models with multiple pathogen strains. *J. Difference Eqns. and Appl.* 10: 53–75.

Anderson, R. M. and R. M. May. 1982. The logic of vaccination. *New Scientist* November 1982, pp. 410–415.

Barlow, N. D., B. I. P. Barratt, C. M. Ferguson, and M. C. Barron. 2004. Using models to estimate parasitoid impacts on nontarget host abundance. *Environ. Entomol.* 33: 941–948.

Beverton, R. J. H. and S. J. Holt. 1957. On the dynamics of exploited fish populations. *Fisheries Investigations Series 2(19)*. Ministry of Agriculture, Fisheries, and Food, London.

Brauer, F. and C. Castillo-Chávez. 2001. *Mathematical Models in Population Biology and Epidemiology*. Springer-Verlag, New York.

Caswell, H. 2001. *Matrix Population Models: Construction, Analysis and Interpretation*. 2nd ed. Sinauer Assoc. Inc., Sunderland, Mass.

Castillo-Chavez, C. and A.-A. Yakubu. 2001. Dispersal, disease and life-history evolution. *Math. Biosci.* 173: 35–53.

Cipra, B. 1999. *What's Happening in the Mathematical Sciences 1998–1999, Vol. 4.* P. Zorn (Ed.). American Mathematical Society, Providence, Rhode Island.

Clark, D. P. and L. D. Russell. 1997. *Molecular Biology Made Simple and Fun*. Cache River Press, Vienna, Ill.

Costantino, R. F., R. A. Desharnais, J. M. Cushing, and B. Dennis. 1997. Chaotic dynamics in an insect population. *Science* 275: 389–391.

Costantino, R. F., J. M. Cushing, B. Dennis, R. A. Desharnais, and S. M. Henson. 1998. Resonant population cycles in temporally fluctuating habitats. *Bull. Math. Biol.* 50: 247–273.

Cull, P. 1986. Local and global stability for population models. *Biological Cybernetics* 54: 141–149.

Cushing, J. M. 1998. *An Introduction to Structured Population Dynamics*, CBMS-NSF Regional Conference Series in Applied Mathematics # 71, SIAM, Philadelphia.

Cushing, J. M., R. F. Costantino, B. Dennis, R. A. Desharnais, and S. M. Henson. 1998. Nonlinear population dynamics: models experiments and data. *J. Theor. Biol.* 194: 1–9.

Cushing, J. M., R. F. Costantino, B. Dennis, R. A. Desharnais, and S. M. Henson. 2003. *Chaos in Ecology: Experimental Nonlinear Dynamics*. Academic Press, New York.

DeAngelis, D. L., L. J. Svoboda, S. W. Christensen, and D. S. Vaughan. 1980. Stability and return times of Leslie matrices with density-dependent survival: Applications to fish populations. *Ecological Modelling*, 8: 149–163.

Deevey, G. B. and E. S. Deevey, Jr. 1945. A life table for the black widow. *Trans. Connecticut Acad. of Arts and Sciences.* 36: 115–134.

Edelstein-Keshet, L. 1988. *Mathematical Models in Biology*. The Random House/ Birkhäuser Mathematics Series, New York.

Elaydi, S. N. 2000. *Discrete Chaos*. Chapman & Hall/CRC, Boca Raton, London, New York, Washington, D.C.

Felsenstein, J. 2003. *Theoretical Evolutionary Genetics*. Online Notes: http:// evolution.gs. washington.edu/pgbook/pgbook.html.

Flor, H. H. 1956. The complementary genic system in flax and flax rust. *Adv. in Genetics* 8: 29–54.

Getz, W. M. and R. G. Haight. 1989. *Population Harvesting Demographic Models of Fish, Forest, and Animal Resources*. Princeton University Press, Princeton, N.J.

Haefner, J. W. 1996. *Modeling Biological Systems: Principles and Applications*, Chapman and Hall, New York.

Hale, J. and H. Koçak. 1991. *Dynamics and Bifurcations*. Springer-Verlag, New York, Berlin, Heidelberg.

Hassell, M P. 1978. *The Dynamics of Arthropod Predator-Prey Systems*. Princeton Univ. Press.

Hassell, M. P., R. M. May, S. W. Pacala, and P. L. Chesson. 1991. The persistence of host-parasitoid associations in patchy environents. I a general criterion. *Amer. Nat.* 138: 568–583.

Hartl, D. and A. Clark. 1997. *Principles of Population Genetics*. 3rd ed. Sinauer Assoc., Inc., Sunderland, Mass.

Hastings, A. 1998. *Population Biology Concepts and Models*. Springer-Verlag, New York.

Hedrick, P. 2000. *Genetics of Populations*. 2nd ed. Jones and Bartlett Pub., Sudbury, Mass.

Henson, S. M. and J. M. Cushing. 1997. The effect of periodic habitat fluctuations on a nonlinear insect population model. *J. Math. Biol.* 36: 201–226.

Henson, S. M., J. M. Cushing, R. F. Costantino, B. Dennis, and R. A. Desharnais. 1998. Phase switching in population cycles. *Proc. Roy. Soc. Lond. B* 265: 2229–2234.

Hipel, K. W. and A. I. McLeod, 1994. *Time Series Modelling of Water Resources and Environmental Systems*. Elsevier, Amsterdam.

Hoffmann, M. P. and A. C. Frodsham. 1993. *Natural Enemies of Vegetable Insect Pests*. Cooperative Extension, Cornell Univ., Ithaca, N.Y.

Hoppensteadt, F. 1975. *Mathematical Methods of Population Biology*. Cambridge Univ. Press, Cambridge.

Iannelli, M., M. Martcheva, and F. A. Milner. 2005. *Gender-Structured Population Modeling Mathematical methods, Numerics, and Simulations*. SIAM Frontiers in Applied Mathematics, Philadelphia, Pa.

Keyfitz, N. 1972. On the mathematics of sex and marriage. *Proceedings of the 6th Berkeley Symposium of Mathematical Statistics and Probability*. 4: 89–108.

Kesinger, J. C. and L. J. S. Allen. 2002. Genetic models for plant pathosystems. *Math. Biosci.* 177 & 178: 247–269.

King, A. A. and A. Hastings. 2003. Spatial mechanisms for coexistence of species sharing a common natural enemy. *Theor. Pop. Biol.* 64: 431–438.

Kot, M. 2001. *Elements of Mathematical Ecology*. Cambridge Univ. Press, Cambridge.

Kot, M., M. A. Lewis, and P. van den Driessche. 1996. Dispersal data and the spread of invading organisms. *Ecology* 77: 2027–2042.

Lamberson, R. H., R. McKelvey, B. R. Noon, and C. Voss. 1992. A dynamic analysis of northern spotted owl viability in a fragmented forest landscape. *Conservation Biol.* 6(4): 505–512.

Lande, R. 1988. Demographic models of the northern spotted owl, *Strix occidentalis caurina*. *Oecologia* 75: 601–607.

Leonard, K. J. 1977. Selection pressures and plant pathogens. *Ann. N.Y. Acad. Sci.* 287: 207–222.

Leonard, K. J. 1994. Stability of equilibria in gene-for-gene coevolution model of host-parasite interactions. *Phytopathology* 84: 70–77.

Leslie, P. H. and J. C. Gower. 1960. The properties of a stochastic model for the predator-prey type of interaction between two species. *Biometrika* 47: 219–234.

Levin, S. A. and C. P. Goodyear. 1980. Analysis of an age-structured fishery model. *J. Math. Biol.* 9: 245–274.

Lindström, J. and H. Kokko. 1998. Sexual reproduction and population dynamics: the role of polygyny and demographic sex differences. *Proc. Roy. Soc. London B* 265: 483–488.

Lynch, L. D., A. R. Ives, J. K. Waage, M. E. Hochberg, and M. B. Thomas. 2002. The risks of biocontrol: transient impacts and minimum nontarget densities. *Ecol. Appl.* 12: 1872–1882.

McKelvey, K., B. R. Noon, and R. H. Lamberson. 1992. Conservation planning for species occupying fragmented landscapes: the case of the northern spotted owl. In: *Biotic Interactions and Global Change*. P. M. Kareiva, J. G. Kingsolver, and R. B. Huey (Eds.). Sinauer Assoc. Sunderland, Mass, pp. 424–450.

Nagylaki, T. 1992. *Introduction to Theoretical Population Genetics*. Springer-Verlag, Berlin, Heidelberg, New York.

Neubert, M. G. and M. Kot. 1992. The subcritical collapse of predator population in discrete-time predator-prey models. *Math. Biosci.* 110: 45–66.

Nicholson, A. J. and Bailey, V. A. 1935. The balance of animal populations. Part I. *Proc. Zool. Soc. Lond.* 3: 551–598.

Noble, B. 1969. *Applied Linear Algebra*. Prentice Hall, Englewood Cliffs, N.J.

Olesiuk, P. F., M. A. Bigg, and G. M. Ellis. 1990. Life history and population dynamics of resident killer whales (*Orcinus orca*) in the coastal waters of British Columbia and Washington State. *Report of the International Whaling Commission, Special Issue*. 12: 209–243.

Ortega, J. M. 1987. *Matrix Theory: A Second Course*. Plenum Press, New York.

Pakes, A. G. and R. A. Maller. 1990. *Mathematical Ecology of Plant Species Competition*. Cambridge Univ. Press.

Pielou, E. C. 1977. *Mathematical Ecology*. Wiley-Interscience, New York.

Prober, C. G. 1999. Evidence shows genetics, not MMR vaccine, determines autism. *AAP News* December 1999, p. 24.

Ricker, W. E. 1954. Stock and recruitment. *Journal of the Fisheries Research Board of Canada* 11: 559–623.

Sasaki, A. 2002. Coevolution in gene-for-gene systems. In: *Adaptive Dynamics of Infectious Diseases. In Pursuit of Virulence Management*. U. Dieckmann, J. A. J. Metz, M. W. Sabelis, and K. Sigmund (Eds.). Cambridge Univ. Press, pp. 233–247.

Silva, J. A. L. and T. G. Hallam. 1992. Compensation and stability in nonlinear matrix models. *Math. Biosci.* 110: 67–101.

Silva, J. A. L. and T. G. Hallam. 1993. Effects of delay, truncations and density dependence in reproduction schedules on stability of nonlinear Leslie matrix models. *J. Math. Biol.* 31: 367–395.

Sykes, Z. M. 1969. On discrete stable population theory. *Biometrics* 25: 285–293.

Thieme, H. R. 2003. *Mathematics in Population Biology*. Princeton Univ. Press.

Thomas, J. W., E. D. Forsman, J. B. Lint, E. C. Meslow, B. R. Noon, and J. Verner. 1990. A conservation strategy for the northern spotted owl. 1990-791-171/20026. U.S. Government Printing Office, Washington, D.C.

Thompson, J. N. and J. J. Burdon. 1992. Gene-for-gene coevolution between plants and parasites. *Nature* 360: 121–125.

United States Fish and Wildlife Service. 2004. Waterfowl population status, 2004. U.S. Department of the Interior, Washington, D.C.

United States Fish and Wildlife Service. 2005. Whooping Crane (*Grus americana*) Draft Revised International Recovery Plan, January 2005, www.npwrc.usgs.gov/resource/distr/birds/cranes/grusamer.htm

Vanderplank, J. E. 1984. *Disease Resistance in Plants*. Academic Press, London.

Várkonyi, G., I. Hanski, M. Rost, and J. Itãmies. 2002. Host-parasitoid dynamics in periodic boreal moths. *Oikos* 98: 421–430.

## 3.12 Appendix for Chapter 3

### 3.12.1 Maple Program: Nicholson-Bailey Model

Maple commands and the corresponding output for calculating and evaluating the Jacobian matrix for the Nicholson-Bailey model are given.

```
> with(linalg):
> F:=r*N*exp(-a*P): G:=e*N*(1-exp(-a*P)):
> A:=vector([F,G]);
```

$$A := [rN\exp(-aP), eN(1 - \exp(-aP))]$$

```
> J:=jacobian(A,[N,P]);
```

$$J := \begin{bmatrix} r\exp(-aP) & -rNa\exp(-aP) \\ e(1-\exp(-aP)) & eNa\exp(-aP) \end{bmatrix}$$

```
> J0:=simplify(subs({N=0,P=0},jacobian(A,[N,P])));
```

$$J0 := \begin{bmatrix} r & 0 \\ 0 & 0 \end{bmatrix}$$

```
> J1:=simplify(subs({P=ln(r)/a,  N=r*ln(r)/((r-1)*a*e)},jacobian(A,[N,P])));
```

$$J1 := \begin{bmatrix} 1 & -\dfrac{r\ln(r)}{(r-1)e} \\ \dfrac{(r-1)e}{r} & \dfrac{\ln(r)}{r-1} \end{bmatrix}$$

### 3.12.2 Whooping Crane Data

Data on the whooping crane are available from the United States Fish and Wildlife Service (2005) at the Internet site www.npwrc.usgs.gov/resource/distr/birds/cranes/grusamer.htm. Data for the whooping crane winter numbers in the Wood Buffalo/Aransas population are given in Table 3.3.

**Table 3.3** Whooping crane winter population sizes for the North American population, breeding in Wood Buffalo National Park and overwintering in Aransas National Wildlife Refuge (United States Fish and Wildlife Service, 2005).

| Year | Size | Year | Size | Year | Size |
|---|---|---|---|---|---|
| 1938 | 18 | 1960 | 36 | 1982 | 73 |
| 1939 | 22 | 1961 | 39 | 1983 | 75 |
| 1940 | 26 | 1962 | 32 | 1984 | 86 |
| 1941 | 16 | 1963 | 33 | 1985 | 97 |
| 1942 | 19 | 1964 | 42 | 1986 | 110 |
| 1943 | 21 | 1965 | 44 | 1987 | 134 |
| 1944 | 18 | 1966 | 43 | 1988 | 138 |
| 1945 | 22 | 1967 | 48 | 1989 | 146 |
| 1946 | 25 | 1968 | 50 | 1990 | 146 |
| 1947 | 31 | 1969 | 56 | 1991 | 132 |
| 1948 | 30 | 1970 | 57 | 1992 | 136 |
| 1949 | 34 | 1971 | 59 | 1993 | 143 |
| 1950 | 31 | 1972 | 51 | 1994 | 133 |
| 1951 | 25 | 1973 | 49 | 1995 | 158 |
| 1952 | 21 | 1974 | 49 | 1996 | 160 |
| 1953 | 24 | 1975 | 57 | 1997 | 180 |
| 1954 | 21 | 1976 | 69 | 1998 | 183 |
| 1955 | 28 | 1977 | 72 | 1999 | 188 |
| 1956 | 24 | 1978 | 75 | 2000 | 180 |
| 1957 | 26 | 1979 | 76 | 2001 | 176 |
| 1958 | 32 | 1980 | 78 | 2002 | 185 |
| 1959 | 33 | 1981 | 73 | 2003 | 194 |

### 3.12.3 Waterfowl Data

Waterfowl data are available from the United States Fish and Wildlife Service (2004). Data on the total duck breeding population in Minnesota (in thousands) are given in Table 3.4. Some of the duck species included in the survey are mallard (*Anas platyrhynchos*), gadwall (*A. strepera*), American wigeon (*A. americana*), green-winged teal (*A. crecca*), Northern shovelers (*A. clypeata*), and Northern pintail (*A. acuta*).

**Table 3.4** Breeding population estimates (in thousands) for total ducks in Minnesota from 1968 through 2004 (United States Fish and Wildlife Service, 2004).

| Year | Size | Year | Size | Year | Size |
|---|---|---|---|---|---|
| 1968 | 368.5 | 1981 | 515.2 | 1994 | 1320.1 |
| 1969 | 345.3 | 1982 | 558.4 | 1995 | 912.2 |
| 1970 | 343.8 | 1983 | 394.2 | 1996 | 1062.4 |
| 1971 | 286.9 | 1984 | 563.8 | 1997 | 953.0 |
| 1972 | 237.6 | 1985 | 580.3 | 1998 | 739.6 |
| 1973 | 415.6 | 1986 | 537.5 | 1999 | 716.5 |
| 1974 | 332.8 | 1987 | 614.9 | 2000 | 815.3 |
| 1975 | 503.3 | 1988 | 752.8 | 2001 | 761.3 |
| 1976 | 759.4 | 1989 | 1021.6 | 2002 | 1224.1 |
| 1977 | 536.6 | 1990 | 886.8 | 2003 | 748.9 |
| 1978 | 511.3 | 1991 | 868.2 | 2004 | 1099.3 |
| 1979 | 901.4 | 1992 | 1127.3 | | |
| 1980 | 740.7 | 1993 | 875.9 | | |

Chapter 4

# Linear Differential Equations: Theory and Examples

## 4.1 Introduction

When changes such as births and deaths occur continuously, then generations overlap and a continuous-time model (differential equations) is more appropriate than a discrete-time model (difference equations). In differential equations, the time interval is continuous and can be either finite or infinite in length $[t_0, T)$ for $t_0 < T < \infty$ or $[t_0, \infty)$, respectively, as opposed to difference equations, where time is a set of discrete values, $t = 0, 1, 2, \ldots$.

The most well-known population growth model and one of the simplest is due to Malthus (1798). The model for Malthusian growth is a differential equation. The Malthusian model assumes the rate of growth is proportional to the size of the population. Hence, if $x(t)$ is the population size, then

$$\frac{dx}{dt} = rx, \quad x(t_0) = x_0, \tag{4.1}$$

where $r > 0$ is referred to as the *per capita growth rate* or the *intrinsic growth rate.* The solution to this differential equation is found by separating variables,

$$\frac{dx}{x} = r\,dt, \quad \int_{x_0}^{x(t)} \frac{dy}{y} = \int_{t_0}^{t} r\,d\tau, \quad \ln[x(t)/x_0] = r(t - t_0).$$

Finally,

$$x(t) = x_0 \exp(r(t - t_0))$$

or $x(t) = x_0 e^{rt}$ when $t_0 = 0$. The population grows exponentially over time. Note also that the differential equation (4.1) is linear in $x$. The exponential growth exhibited by the solution of the differential equation (4.1) is comparable to the geometric growth exhibited by the solution to the linear difference equation, $x_{t+1} = ax_t$, where $x_t = x_0 a^t$. The constant $a = e^r$. If $a > 1$ (or $r > 0$), then there is exponential growth.

In this chapter, basic notation and definitions are given for first- and higher-order differential equations, as well as first-order systems. We concentrate on linear differential equations. Criteria are stated for solutions to approach the zero solution. These criteria are known as the Routh-Hurwitz criteria and are the analogue of the Jury conditions for difference equations. Techniques for analyzing a system of two first-order equations (in the plane) and the behavior exhibited by these types of systems are discussed. A biological example of a linear differential system, known as a pharmacokinetics model, is presented and analyzed. In the pharmacokinetics model, a drug is administered to an individual and the concentration of the drug in different compartments of the body is followed over time. The final section of this chapter gives a brief introduction to linear delay differential equations. In a delay model, the rate of change of state $dx(t)/dt$ depends on the state at a prior time, $t - \tau$, that is, it depends on $x(t - \tau)$. Thus, the dynamics of $x(t)$ are delayed by $\tau$ time units.

## 4.2 Basic Definitions and Notation

Differential equations are classified according to their order, whether they are linear or nonlinear, and whether they are autonomous or nonautonomous. These classifications schemes are similar to the ones defined for difference equations.

**Definition 4.1.** A *differential equation of order n* is an equation of the form

$$f(x, dx/dt, d^2x/dt^2, \ldots, d^nx/dt^n, t) = 0.$$

If this differential equation does not depend explicitly on $t$, then it is said to be *autonomous*; otherwise it is *nonautonomous*.

If an $n$th-order differential equation can be expressed as follows:

$$\frac{d^nx}{dt^n} + a_1(t)\frac{d^{n-1}x}{dt^{n-1}} + \cdots + a_{n-1}(t)\frac{dx}{dt} + a_n(t)x = g(t), \tag{4.2}$$

then it is referred to as linear.

**Definition 4.2.** The differential equation (4.2) is said to be *linear* if the coefficients $a_i$, $i = 1, \ldots, n$, and $g$ are either constant or functions of $t$ but not functions of $x$ or any of its derivatives. Otherwise, the differential equation (4.2) is said to be *nonlinear*. The linear differential equation (4.2) is said to be *homogeneous* if $g(t) \equiv 0$ and *nonhomogeneous* otherwise.

It will always be assumed that the functions $f$, $g$, and $a_i$ are real valued. Analogous definitions can be stated for systems. A first-order system of differential equations satisfies

$$\frac{dX}{dt} = F(X(t), t), \tag{4.3}$$

where the vector $X(t) = (x_1(t), x_2(t), \ldots, x_n(t))^T$, $F = (f_1, f_2, \ldots, f_n)^T$, and

$$f_i \equiv f_i(x_1(t), x_2(t), \ldots, x_n(t), t).$$

**Definition 4.3.** The system of differential equations (4.3) is said to be *autonomous* if the right-hand side of (4.3) does not depend explicitly on $t$; otherwise it is said to be *nonautonomous*.

**Definition 4.4.** The first-order system (4.3) is said to be *linear* if it can be expressed as

$$\frac{dx_i}{dt} = \sum_{j=1}^{n} a_{ij}(t)x_j + g_i(t), \tag{4.4}$$

$i = 1, \ldots, n$. If not, then it is *nonlinear*. If the system is linear and $g_i(t) \equiv 0$, then the system is said to be *homogeneous*; otherwise it is *nonhomogeneous*.

**Definition 4.5.** A *solution* of a differential equation or system is a scalar function $x(t)$ or vector function $X(t)$, respectively, which when substituted into the differential equation or system makes it an identity.

Suppose, in addition to the $n$th-order differential equation, the solution satisfies $n$ initial conditions. That is,

$$x(t_0) = x_0, \quad \frac{dx(t_0)}{dt} = x_1, \quad \cdots \quad \frac{d^{n-1}x(t_0)}{dt^{n-1}} = x_{n-1}. \tag{4.5}$$

Then the solution satisfying the differential equation and the initial conditions is known as the solution to an *initial value problem* (IVP). For a first-order system, an initial value problem has the form

$$\frac{dX}{dt} = F(X(t), t), \quad t > t_0, \quad X(t_0) = X_0. \tag{4.6}$$

The notation $dx(t_0)/dt$ means differentiation of $x$, then evaluation at $t_0$.

It is important to know conditions on the coefficients so that solutions to initial value problems exist and are unique. In the case of linear differential equations, the existence and uniqueness conditions are straightforward.

**Theorem 4.1**

*(i) Let*

$$\frac{d^n x}{dt^n} + a_1(t)\frac{d^{n-1}x}{dt^{n-1}} + \cdots + a_{n-1}(t)\frac{dx}{dt} + a_n(t)x = g(t), \tag{4.7}$$

$$x(t_0) = x_0, \quad \frac{dx(t_0)}{dt} = x_1, \quad \cdots \quad \frac{d^{n-1}x(t_0)}{dt^{n-1}} = x_{n-1}. \tag{4.8}$$

*If the coefficients $a_i$ and $g$, $i = 1, 2, \ldots, n-1$ are continuous on some interval containing $t_0$, $\alpha < t_0 < \beta$, then there exists a unique solution to the initial value problem (4.7) and (4.8) on this interval.*

*(ii) Let*

$$\frac{dx_i}{dt} = \sum_{j=1}^{n} a_{ij}(t)x_j + g_i(t), \quad x_i(t_0) = x_{i0}. \tag{4.9}$$

*for $i = 1, 2, \ldots, n$. If the coefficients $a_{ij}$ and $g_i$, $i, j = 1, 2, \ldots, n$ are continuous on some interval $\alpha < t_0 < \beta$, then there exists a unique solution to the initial value problem (4.9) on this interval.* □

**Example 4.1** Consider the initial value problem

$$\frac{d^2x}{dt^2} + \frac{3}{t}\frac{dx}{dt} + \frac{x}{t^2} = 0,$$

where $x(1) = 0$ and $dx(1)/dt = 1$. This differential equation is second order, linear, nonautonomous, and homogeneous. The coefficients are continuous on $(0, \infty)$. Hence, applying Theorem 4.1, there exists a unique solution to this initial value problem. Two linearly independent solutions (see Section 4.4) to the differential equation are $x_1(t) = t^{-1}$ and $x_2(t) = t^{-1}\ln|t|$. The general solution to the differential equation is $x(t) = c_1x_1(t) + c_2x_2(t)$. Applying the initial conditions leads to the unique solution to the initial value problem: $x(t) = t^{-1}\ln|t|$. The differential equation in this example is a special type of equation known as a Cauchy-Euler differential equation. (See Exercise 4.) ■

For additional information on the theory of differential equation, consult a textbook on ordinary differential equations listed in the references (Brauer and Nohel, 1969; Cushing, 2004; Sánchez, 1968; Waltman, 1986).

## 4.3 First-Order Linear Differential Equations

In this section, we review how to solve first-order linear differential equations. This method involves an integrating factor.

An initial value problem for a first-order linear differential equation has the following form:

$$\frac{dx}{dt} + a_1(t)x = g(t), \quad x(t_0) = x_0.$$

Assume that $a_1$ and $g$ are continuous for all $t \geq t_0$. The solution to this differential equation can be found with the use of an integrating factor. Let $I(t) = \exp(\int_{t_0}^{t} a_1(\tau)\,d\tau)$. The function $I(t)$ is known as an *integrating factor*. There are an infinite number of integrating factors, since any constant multiple of $I(t)$ is also an integrating factor. Multiplying both sides of the differential equation by the integrating factor $I(t)$ yields

$$I(t)\frac{dx(t)}{dt} + a_1(t)I(t)x(t) = I(t)g(t).$$

The left-hand side is an exact derivative,

$$\frac{d[x(t)I(t)]}{dt} = I(t)g(t).$$

Integrating both sides and solving for $x$ gives the unique solution to the initial value problem,

$$x(t) = e^{\left(-\int_{t_0}^{t} a_1(\tau)\,d\tau\right)}\left[x_0 + \int_{t_0}^{t} e^{\left(\int_{t_0}^{\tau} a_1(u)\,du\right)} g(\tau)\,d\tau\right]. \tag{4.10}$$

Note that if $t = t_0$, then the solution satisfies the initial condition $x(t_0) = x_0$.
If $a_1 =$ constant and $t_0 = 0$, then the solution simplifies to

$$x(t) = e^{-a_1 t}\left[x_0 + \int_0^t e^{a_1 \tau} g(\tau)\,d\tau\right].$$

In addition, if the equation is homogeneous, $g \equiv 0$, then the solution is given by

$$x(t) = x_0 e^{-a_1 t}.$$

If $a_1 < 0$, then the preceding solution represents Malthusian or exponential growth. If $a_1 > 0$, then $\lim_{t\to\infty} x(t) = 0$.

**Example 4.2** Let

$$\frac{dx}{dt} - \frac{x}{t} = te^{3t}, \quad x(1) = 2.$$

An integrating factor for this differential equation is $I(t) = e^{-\ln t} = t^{-1}$. Then

$$\frac{d(xt^{-1})}{dt} = e^{3t}.$$

Integrating both sides and solving for $x$ yields the general solution

$$x(t) = ct + t\frac{e^{3t}}{3}. \tag{4.11}$$

Applying the initial condition gives the unique solution to the initial value problem,

$$x(t) = \left(2 - \frac{e^3}{3}\right)t + t\frac{e^{3t}}{3}.$$ ∎

The solution (4.11) is the sum of two terms. The first term is the general solution to the homogeneous differential equation; that is, $x_h(t) = ct$ is the general solution to $dx/dt = x/t$. The second term is a particular solution to the nonhomogeneous differential equation, that is, $x_p(t) = te^{3t}/3$ is a solution to $dx/dt - x/t = te^{3t}$. The sum of these two solutions, $x(t) = x_h(t) + x_p(t)$, forms the *general solution* to the nonhomogeneous differential equation. These ideas form the basis of the solution method for higher-order linear differential equations.

## 4.4 Higher-Order Linear Differential Equations

The general solution to an $n$th-order, linear nonhomogeneous differential equation is the sum of two solutions, a general solution to the homogeneous differential equation and a particular solution to the nonhomogeneous differential equation,

$$x(t) = x_h(t) + x_p(t).$$

The general solution to an $n$th-order, homogeneous differential equation is the sum of $n$ linearly independent solutions, $\phi_1(t), \ldots, \phi_n(t)$, that is,

$$x_h(t) = \sum_{i=1}^{n} c_i \phi_i(t).$$

The linearly independent set $\{\phi_i(t)\}_{i=1}^n$ is called a *fundamental set of solutions* to (4.2). [Recall that solutions $\phi_1(t), \ldots, \phi_n(t)$ are *linearly independent* if $\sum_{i=1}^n \alpha_i \phi_i(t) = 0$ implies $\alpha_i = 0$, $i = 1, 2, \ldots, n$.] Therefore, the general solution to the nonhomogeneous differential equation (4.2) is $x(t) = \sum_{i=1}^n c_i \phi_i(t) + x_p(t)$. Various methods can be used to find the particular solution (e.g., variation of constants, method of undetermined coefficients).

### 4.4.1 Constant Coefficients

The special case of a linear homogeneous differential equation, where the coefficients are constant, is discussed in more detail. For this special case, there is a well-known method for solving homogeneous differential equations.

Suppose the coefficients of a linear homogeneous differential equation are constant. Then the differential equation (4.2) has the following form:

$$\frac{d^n x}{dt^n} + a_1 \frac{d^{n-1} x}{dt^n} + \cdots + a_{n-1} \frac{dx}{dt} + a_n x = 0. \tag{4.12}$$

There exist $n$ linearly independent solutions to this differential equation that exist for all time $(-\infty, \infty)$. To find these solutions, assume that $x(t) = e^{\lambda t}$. Note that $\lambda$ can be real or complex. Substituting $x(t) = e^{\lambda t}$ into the homogeneous differential equation (4.12) yields

$$e^{\lambda t}(\lambda^n + a_1 \lambda^{n-1} + \cdots + a_{n-1}\lambda + a_n) = 0.$$

The resulting polynomial

$$P(\lambda) = \lambda^n + a_1 \lambda^{n-1} + \cdots + a_{n-1}\lambda + a_n$$

is known as the *characteristic polynomial* and the equation $P(\lambda) = 0$ is known as the *characteristic equation* of the differential equation (4.12). The roots of $P(\lambda)$ are the *eigenvalues*. The solution form taken by $e^{\lambda t}$ depends on whether the eigenvalues are real or complex.

In the case of a second-order linear differential equation, $n = 2$, the forms of the solution are summarized.

1. The eigenvalues $\lambda_i$, $i = 1, 2$ are real and distinct. The general solution is

$$x(t) = c_1 e^{\lambda_1 t} + c_2 e^{\lambda_2 t}.$$

2. The eigenvalues $\lambda_1 = \lambda_2$ are real and equal. The general solution is

$$x(t) = c_1 e^{\lambda_1 t} + c_2 t e^{\lambda_1 t}.$$

3. The eigenvalues $\lambda_{1,2} = u \pm iv$ are complex conjugates. The general solution is

$$x(t) = e^{ut}[c_1 \cos(vt) + c_2 \sin(vt)].$$

In the case of complex conjugates, $u \pm iv$, the identity $e^{ivt} = \cos(vt) + i\sin(vt)$ is used to express the solutions in terms of sines and cosines.

Higher-order, linear differential equations with constant coefficients may have a characteristic polynomial with an eigenvalue $\lambda$ repeated $r$ times (a root of multiplicity $r$). There must be $r$ linearly independent solutions associated with that root. It can be shown that additional solutions are found by multiplying by powers of $t$. When the eigenvalue is real, then there are $r$ linearly independent solutions given by

$$e^{\lambda t}, te^{\lambda t}, \ldots, t^{r-1}e^{\lambda t}.$$

When the eigenvalue is complex, $\lambda = u + iv$, and is a root of multiplicity $r$, the complex conjugate $u - iv$ is also a root of multiplicity $r$. There must be $2r$ linearly independent solutions. Additional solutions independent from $e^{ut}\cos(vt)$ and $e^{ut}\sin(vt)$ are found by multiplying by powers of $t$:

$$te^{ut}\cos(vt), \quad te^{ut}\sin(vt), \quad \cdots, \quad t^{r-1}e^{ut}\cos(vt), \quad t^{r-1}e^{ut}\sin(vt).$$

The Wronskian can be used to verify that the $n$ solutions of an $n$th-order linear differential equation are independent.

**Definition 4.6.** If $x_1(t), \ldots, x_n(t)$ are $n$ functions with $n-1$ continuous derivatives, then the determinant

$$W(x_1, \ldots, x_n)(t) = \det\begin{pmatrix} x_1(t) & \cdots & x_n(t) \\ x_1'(t) & \cdots & x_n'(t) \\ \vdots & \cdots & \vdots \\ x_1^{(n-1)}(t) & \cdots & x_n^{(n-1)}(t) \end{pmatrix}$$

is called the *Wronskian* of $x_1, \ldots, x_n$. The primes (′) denote differentiation with respect to $t$.

The following theorem states that if the Wronskian is nonzero for some $t$ on the interval of existence, then the $n$ solutions are linearly independent.

**Theorem 4.2** *Suppose $\phi_1(t), \ldots, \phi_n(t)$ are $n$ solutions of the nth-order linear differential equation on the interval $I$,*

$$\frac{d^n x}{dt^n} + a_1(t)\frac{d^{n-1}x}{dt^{n-1}} + \cdots + a_{n-1}(t)\frac{dx}{dt} + a_n(t)x = 0.$$

*Then the $n$ solutions are linearly independent iff the Wronskian $W(\phi_1, \ldots, \phi_n)(t) \neq 0$ for some $t \in I$.* □

A proof of this result can be found in many ordinary differential equation texts.

**Example 4.3** Consider the differential equation $x'''(t) - 4x''(t) = 0$, where $x'' = d^2x/dt^2$, and so on. The characteristic equation is given by

$$\lambda^3 - 4\lambda^2 = 0.$$

Hence, the roots or eigenvalues are 0, 0, and 4 and the three linearly independent solutions are 1, $t$, and $e^{4t}$, respectively. The general solution is

$$x(t) = c_1 + c_2 t + c_3 e^{4t}.$$

To verify that these three solutions are linearly independent, we compute the Wronskian,

$$W(1, t, e^{4t}) = \det\begin{pmatrix} 1 & t & e^{4t} \\ 0 & 1 & 4e^{4t} \\ 0 & 0 & 16e^{4t} \end{pmatrix} = 16e^{4t} \neq 0.$$

■

**Example 4.4** Suppose the characteristic polynomial for a seventh-order, linear homogeneous differential equation is

$$P(\lambda) = (\lambda^2 - 3\lambda + 4)^2(\lambda - 2)^3 = 0.$$

Then the roots are $\lambda_{1,2,3,4} = 3/2 \pm i\sqrt{7}/2$ and $\lambda_{5,6,7} = 2$. The general solution of the differential equation is

$$x(t) = e^{3t/2}[c_1 \cos(\sqrt{7}t/2) + c_2 \sin(\sqrt{7}t/2) + c_3 t \cos(\sqrt{7}t/2) + c_4 t \sin(\sqrt{7}t/2)] + e^{2t}[c_5 + c_6 t + c_7 t^2].$$

The coefficients $c_i$ can be uniquely determined from the initial conditions. ■

**Example 4.5** Consider the fourth-order linear differential equation,

$$\frac{d^4x}{dt^4} + 6\frac{d^3x}{dt^3} + 10\frac{d^2x}{dt^2} + 6\frac{dx}{dt} + 9x = 0.$$

The characteristic polynomial satisfies

$$\lambda^4 + 6\lambda^3 + 10\lambda^2 + 6\lambda + 9 = (\lambda^2 + 1)(\lambda + 3)^2 = 0.$$

The eigenvalues are $\pm i, -3, -3$. Hence, the general solution satisfies

$$x(t) = c_1 \cos(t) + c_2 \sin(t) + c_3 e^{-3t} + c_4 t e^{-3t}.$$

If the initial conditions are $x(0) = 1$, $dx(0)/dt = 2$, $d^2x(0)/dt^2 = 1$, and $d^3x(0)/dt^3 = 0$, then the four constants $c_1, c_2, c_3$, and $c_4$ can be found by solving the following linear system:

$$\begin{aligned} c_1 + c_3 &= 1, \\ c_2 - 3c_3 + c_4 &= 2, \\ -c_1 + 9c_3 - 6c_4 &= 1, \\ -c_2 - 27c_3 + 27c_4 &= 0. \end{aligned}$$

The constants are $c_1 = 8/25$, $c_2 = 81/25$, $c_3 = 17/25$, and $c_4 = 4/5$. ■

It is important to note that an $n$th-order, linear homogeneous differential equation always has a solution equal to zero, $x(t) \equiv 0$. If all of the initial conditions are zero, then this is the unique solution to the initial value problem. In the case that all of the coefficients of the homogeneous, linear differential equation are constant, then we can determine whether the zero solution is "stable," that is, whether a solution to the initial value problem will tend to zero. Stability of the zero solution depends on the eigenvalues, the roots $\lambda_i$ of the characteristic equation $P(\lambda) = 0$. Because solutions have the form $e^{\lambda_i t}$, it follows that solutions to initial value problems approach zero if the $\lambda_i$ are negative real numbers or are complex numbers having negative real parts.

The distinction in behavior between linear difference and linear differential equations lies in the form of their solution. In difference equations, the solutions are linear combinations of $\lambda_i^t$ whereas in differential equations they are linear combinations of $e^{\lambda_i t}$. Solutions to a linear homogeneous *difference equation* with constant coefficients tend to zero if the eigenvalues $\lambda_i$ have magnitude less than one, $|\lambda_i| < 1$, whereas solutions to the linear homogeneous *differential equation* with constant coefficients tend to zero if the eigenvalues $\lambda_i$ are negative or have negative real parts, $\lambda_i < 0$ or $u \pm iv, u < 0$. The following theorem shows that solutions approach zero at an exponential rate if the eigenvalues lie in the left half of the complex plane.

**Theorem 4.3** *If all of the roots of the characteristic polynomial $P(\lambda)$ are negative or have negative real parts, then given any solution $x(t)$ of the homogeneous differential equation (4.2), there exist positive constants $M$ and $b$ such that*

$$|x(t)| \le Me^{-bt} \text{ for } t > 0$$

*and*

$$\lim_{t\to\infty} |x(t)| = 0.$$

**Proof** Let the roots of $P(\lambda)$ be denoted as $\lambda_k = u_k + iv_k$, where $u_k < 0$, $k = 1, \ldots, n$. There exists a positive constant $b$ such that $u_k < -b$ or $u_k + b < 0$ for all $k = 1, \ldots, n$. Then

$$|e^{\lambda_k t} e^{bt}| = e^{(u_k+b)t},$$

which approaches zero as $t \to \infty$. Also, $|t^{r_k} e^{\lambda_k t} e^{bt}| = |t^{r_k} e^{(u_k+b)t}|$, where $r_k$ is a nonnegative integer. This latter expression approaches zero as $t \to \infty$. Thus, there exists a constant $M_k > 0$ such that $|t^{r_k} e^{\lambda_k t} e^{bt}| \le M_k$ or $|t^{r_k} e^{\lambda_k t}| \le M_k e^{-bt}$ for $t > 0$. Any solution $x(t)$ is the the sum of terms of the form $\phi_k(t) = t^{r_k} e^{\lambda_k t}$,

$$x(t) = \sum_{k=1}^{n} c_k \phi_k(t),$$

where $\phi_k(t)$ are the fundamental set of solutions and $c_k$ are constants, $k = 1, \ldots, n$. If $M_0 = \max_k |c_k|$ and $M = M_0[\sum_{k=1}^n M_k]$, then for $t \ge 0$,

$$|x(t)| \le \sum_{k=1}^{n} |c_k| |\phi_k(t)| \le M_0 \sum_{k=1}^{n} |\phi_k(t)|$$

$$\le M_0 \left[ \sum_{k=1}^{n} M_k \right] e^{-bt} = Me^{-bt}.$$

It follows that $\lim_{t\to\infty} |x(t)| = 0$. □

Theorem 4.3 shows that the rate of convergence to zero is exponential and is determined by the root with the largest negative real part.

## 4.5 Routh-Hurwitz Criteria

Important criteria that give necessary and sufficient conditions for all of the roots of the characteristic polynomial (with real coefficients) to lie in the left half of the complex plane are known as the *Routh-Hurwitz criteria*. The name refers to E. J. Routh and A. Hurwitz, who contributed to the formulation of these criteria. In 1875, Routh, a British mathematician, developed an algorithm to determine the number of roots that lie in the right half of the complex plane (Gantmacher, 1964). In 1895, Hurwitz, a German mathematician, verified the determinant criteria for roots to lie in the left half of the complex plane. According to Theorem 4.3, if the roots of the characteristic polynomial lie in the left half of the complex plane, then any solution to the linear, homogeneous differential equation converges to zero. The Routh-Hurwitz criteria for differential equations are analogous to the Jury conditions for difference equations. The Routh-Hurwitz criteria are used in Chapters 5 and 6 to determine local asymptotic stability of an equilibrium for nonlinear systems of differential equations. The Routh-Hurwitz criteria are stated in the next theorem.

**Theorem 4.4** **(Routh-Hurwitz Criteria).** *Given the polynomial,*

$$P(\lambda) = \lambda^n + a_1\lambda^{n-1} + \cdots + a_{n-1}\lambda + a_n,$$

*where the coefficients $a_i$ are real constants, $i = 1, \ldots, n$, define the $n$ Hurwitz matrices using the coefficients $a_i$ of the characteristic polynomial:*

$$H_1 = (a_1), \quad H_2 = \begin{pmatrix} a_1 & 1 \\ a_3 & a_2 \end{pmatrix}, \quad H_3 = \begin{pmatrix} a_1 & 1 & 0 \\ a_3 & a_2 & a_1 \\ a_5 & a_4 & a_3 \end{pmatrix},$$

*and*

$$H_n = \begin{pmatrix} a_1 & 1 & 0 & 0 & \cdots & 0 \\ a_3 & a_2 & a_1 & 1 & \cdots & 0 \\ a_5 & a_4 & a_3 & a_2 & \cdots & 0 \\ \vdots & \vdots & \vdots & \vdots & \cdots & \vdots \\ 0 & 0 & 0 & 0 & \cdots & a_n \end{pmatrix},$$

*where $a_j = 0$ if $j > n$. All of the roots of the polynomial $P(\lambda)$ are negative or have negative real parts iff the determinants of all Hurwitz matrices are positive:*

$$\det H_j > 0, \quad j = 1, 2, \ldots, n.$$

When $n = 2$, the Routh-Hurwitz criteria simplify to $\det H_1 = a_1 > 0$ and

$$\det H_2 = \det \begin{pmatrix} a_1 & 1 \\ 0 & a_2 \end{pmatrix} = a_1 a_2 > 0$$

or $a_1 > 0$ and $a_2 > 0$. For polynomials of degree $n = 2, 3, 4$ and 5, the Routh-Hurwitz criteria are summarized.

*Routh-Hurwitz criteria for $n = 2, 3, 4,$ and 5.*

$n = 2$: $a_1 > 0$ and $a_2 > 0$.

$n = 3$: $a_1 > 0$, $a_3 > 0$, and $a_1a_2 > a_3$.

$n = 4$: $a_1 > 0$, $a_3 > 0$, $a_4 > 0$, and $a_1a_2a_3 > a_3^2 + a_1^2a_4$.

$n = 5$: $a_i > 0$ $i = 1, 2, 3, 4, 5$, $a_1a_2a_3 > a_3^2 + a_1^2a_4$, and
$(a_1a_4 - a_5)(a_1a_2a_3 - a_3^2 - a_1^2a_4) > a_5(a_1a_2 - a_3)^2 + a_1a_5^2$.

For a proof of the Routh-Hurwitz criteria, please see Gantmacher (1964). Theorem 4.4 is verified in the case $n = 2$.

**Proof of Theorem 4.4** For $n = 2$, the Routh-Hurwitz criteria are just $a_1 > 0$ and $a_2 > 0$. The characteristic polynomial in the case $n = 2$ is

$$P(\lambda) = \lambda^2 + a_1\lambda + a_2 = 0.$$

The eigenvalues satisfy

$$\lambda_{1,2} = \frac{-a_1 \pm \sqrt{a_1^2 - 4a_2}}{2}.$$

Suppose $a_1$ and $a_2$ are positive. It is easy to see that if the roots are real, they are both negative, and if they are complex conjugates, they have negative real parts.

Next, to prove the converse, suppose the roots are either negative or have negative real parts. Then it follows that $a_1 > 0$. If the roots are complex conjugates, $0 < a_1^2 < 4a_2$, which implies that $a_2$ is also positive. If the roots are real, then since both of the roots are negative it follows that $a_2 > 0$. □

Necessary but not sufficient conditions for the roots of the polynomial $P(\lambda)$ to lie in the left half of the complex plane are that the coefficients of $P(\lambda)$ be strictly positive. This result is stated in the next corollary.

**Corollary 4.1** *Suppose the coefficients of the characteristic polynomial are real. If all of the roots of the characteristic polynomial*

$$P(\lambda) = \lambda^n + a_1\lambda^{n-1} + a_2\lambda^{n-2} + \cdots + a_n$$

*are negative or have negative real parts, then the coefficients $a_i > 0$ for $i = 1, 2, \ldots, n$.*

**Proof** The corollary is a direct consequence of the Routh-Hurwitz criteria but can be verified separately. The characteristic equation can be factored into the form

$$(\lambda + r_1)\cdots(\lambda + r_{k_1})(\lambda^2 + 2c_1\lambda + c_1^2 + d_1^2)\cdots(\lambda^2 + 2c_{k_2}\lambda + c_{k_2}^2 + d_{k_2}^2) = 0,$$

where the real roots are $-r_i < 0$ for $i = 1, \ldots, k_1$ and the complex roots are $-c_j \pm d_j i$ for $j = 1, \ldots, k_2$ and $k_1 + 2k_2 = n$. If all of the roots are either negative or have negative real parts, then $r_i > 0$ and $c_j > 0$ for all $i$ and $j$. Thus, all the coefficients in the factored characteristic equation are positive, which implies that if the characteristic equation is expanded and simplified,

$$\lambda^n + a_1\lambda^{n-1} + a_2\lambda^{n-2} + \cdots + a_n = 0,$$

then all of the coefficients must satisfy $a_i > 0, i = 1, \ldots, n$. ◀

As a consequence of this corollary, it follows that if any coefficient is zero in the characteristic polynomial, then at least one eigenvalue is either zero, is purely imaginary, or lies in the right half of the complex plane. For example, if $a_n = 0$, then there is a zero eigenvalue.

**Example 4.6** Consider the differential equation,

$$\frac{d^3x}{dt^3} + a_2\frac{dx}{dt} + a_3x = 0, \quad a_2, a_3 > 0.$$

Because $a_1 = 0$, it follows from the corollary that at least one eigenvalue is zero, is purely imaginary, or lies in the right half of the complex plane. For example, when $a_2 = 2$ and $a_3 = 1$, the roots of the characteristic polynomial

$$\lambda^3 + 2\lambda + 1 = 0$$

are approximately $-0.453$ and $0.227 \pm 1.468i$. There are two complex roots with positive real parts. ■

**Example 4.7** Consider the linear differential equation

$$\frac{d^3x}{dt^3} + 4\frac{d^2x}{dt^2} + \frac{dx}{dt} + ax = 0.$$

The characteristic polynomial is

$$P(\lambda) = \lambda^3 + 4\lambda^2 + \lambda + a.$$

According to the Routh-Hurwitz criteria for the roots to have negative real parts and the solution to approach zero, the coefficients must satisfy, $a_1 > 0$, $a_3 > 0$, $a_1a_2 > a_3$. But $a_1 = 4$, $a_2 = 1$, and $a_3 = a$ so that $a$ must satisfy $4 > a > 0$. ■

## 4.6 Converting Higher-Order Equations to First-Order Systems

A linear differential equation of the form (4.2),

$$\frac{d^nx}{dt^n} + a_1\frac{d^{n-1}x}{dt^{n-1}} + \cdots + a_{n-1}\frac{dx}{dt} + a_nx = g(t), \tag{4.13}$$

can be expressed as an equivalent first-order system. Define $n$ new variables, $x_1, \ldots, x_n$, as follows:

$$x_1 = x,$$
$$x_2 = \frac{dx}{dt},$$
$$\vdots$$
$$x_n = \frac{d^{n-1}x}{dt^{n-1}}.$$

Then

$$\frac{dx_1}{dt} = \frac{dx}{dt} = x_2,$$
$$\frac{dx_2}{dt} = \frac{d^2x}{dt^2} = x_3,$$
$$\vdots$$
$$\frac{dx_n}{dt} = \frac{d^nx}{dt^n} = -a_1x_n - a_2x_{n-1} - \cdots - a_{n-1}x_2 - a_nx_1 + g(t),$$

where the last equation follows from the differential equation (4.13). Written in matrix form, the first-order linear system can be expressed as

$$\frac{dX}{dt} = AX + G(t),$$

where $X(t) = (x_1(t), x_2(t), \ldots, x_n(t))^T$ and

$$A = \begin{pmatrix} 0 & 1 & 0 & \cdots & 0 \\ 0 & 0 & 1 & \cdots & 0 \\ \vdots & \vdots & \vdots & \cdots & \vdots \\ 0 & 0 & 0 & \ddots & 1 \\ -a_n & -a_{n-1} & -a_{n-2} & \cdots & -a_1 \end{pmatrix}$$

and $G(t) = (0, 0, 0, \ldots, g(t))^T$. Matrix $A$ is the called the *companion matrix* associated with the differential equation (4.13).

**Example 4.8** The following second-order equation can be converted to an equivalent first-order system of the form $dX/dt = AX + G(t)$:

$$\frac{d^2x}{dt^2} + 4\frac{dx}{dt} + 3x = \sin(t).$$

The matrices $A$ and $G$ satisfy

$$A = \begin{pmatrix} 0 & 1 \\ -3 & -4 \end{pmatrix} \quad \text{and} \quad G(t) = \begin{pmatrix} 0 \\ \sin(t) \end{pmatrix}.$$

■

A first-order system can sometimes be converted to a higher-order equation. This conversion cannot be done for all systems because first-order systems are more general than higher-order equations. We shall see where some of the problems lie in the next example. (A first-order differential system is *not* equivalent to a higher-order differential equation.)

Consider the case of a first-order system with constant coefficients,

$$\frac{dx}{dt} = a_{11}x + a_{12}y,$$

$$\frac{dy}{dt} = a_{21}x + a_{22}y.$$

In matrix notation, $dX/dt = AX$, where $A = (a_{ij})$ and $X = (x, y)^T$. Differentiating $dy/dt$ with respect to $t$,

$$\frac{d^2y}{dt^2} = a_{21}\frac{dx}{dt} + a_{22}\frac{dy}{dt}.$$

If $a_{21} = 0$, then our technique fails and we try differentiating $dx/dt$ with respect to $t$ (which then requires $a_{12} \neq 0$). Suppose $a_{21} \neq 0$. Then substituting $dx/dt = a_{11}x + a_{12}y$ and $a_{21}x = dy/dt - a_{22}y$ leads to a second-order differential equation in $y$,

$$\frac{d^2y}{dt^2} - (a_{11} + a_{22})\frac{dy}{dt} + (a_{11}a_{22} - a_{12}a_{21})y = 0.$$

Note that the coefficient of $dy/dt$ is $-\mathrm{Tr}(A)$ and the coefficient of $y$ is $\det(A)$. The characteristic equation for the differential equation in $y$ is

$$\lambda^2 - (a_{11} + a_{22})\lambda + a_{11}a_{22} - a_{12}a_{22} = \lambda^2 - \mathrm{Tr}(A)\lambda + \det(A) = 0.$$

Since the coefficients $a_{ij}$ are constants, the solution to $y$ can be obtained by finding the roots of the characteristic equation. Once $y$ is known, the solution to $x$ can be obtained from one of the original differential equations.

## 4.7 First-Order Linear Systems

The nonhomogeneous linear system has the form

$$\frac{dX}{dt} = A(t)X(t) + G(t).$$

The elements of the coefficient matrix $A(t)$ and the elements of the vector $G(t)$ are continuous on some interval containing the initial point $t_0$ so that there exists a unique solution to an IVP (Theorem 4.1). It follows from the theory for linear differential equations that the general solution to the nonhomogeneous system is the sum of the general solution to the homogeneous system and a particular solution to the nonhomogeneous system,

$$X(t) = X_h(t) + X_p(t).$$

The general solution to the homogeneous system consists of $n$ linearly independent solutions, $\phi_i(t)$, $i = 1, \ldots, n$. A *fundamental matrix* of solutions is $\Phi(t) = (\phi_1(t), \ldots, \phi_n(t))$, where the columns of $\Phi(t)$ are the vectors $\phi_i(t)$. Because the solutions are linearly independent, $\det\Phi(t) \neq 0$ for all $t$ on the interval of existence. (Compare with the Wronskian.) Hence, the inverse $\Phi^{-1}(t)$ exists for all $t$ on the interval of existence. The unique solution to the IVP for the linear nonhomogeneous system can be expressed in the form

$$X(t) = \Phi(t)\Phi^{-1}(t_0)X_0 + \Phi(t)\int_{t_0}^{t} \Phi^{-1}(s)G(s)\,ds,$$

where $X(t_0) = X_0$. For a proof of this result see Brauer and Nohel (1969) or Waltman (1986). Compare the solution $X(t)$ to the unique solution of the first order equation $x(t)$ given in (4.10).

### 4.7.1 Constant Coefficients

There are many methods that can be applied to find solutions to first-order linear, homogeneous systems with constant coefficients. This type of system will be especially important in the study of nonlinear autonomous systems of differential equations in the next chapter. Let

$$\frac{dX}{dt} = AX, \tag{4.14}$$

where $A = (a_{ij})$ is a constant matrix with real elements $a_{ij}$. First note that the zero solution, $X = \mathbf{0}$, is a fixed point of the differential equation. The zero solution is also referred to as an *equilibrium point*, a *steady state*, or a *critical point*.

In the simplest case of (4.14) when the system reduces to a scalar equation, $A = (a)$ is a $1 \times 1$ matrix, then the differential equation is

$$\frac{dx}{dt} = ax. \tag{4.15}$$

The general solution to (4.15) is $x(t) = ce^{at}$. If the initial condition $x(0) = x_0$, then $x(t) = x_0 e^{at}$ (as already shown in the Introduction). If $a < 0$, then $\lim_{t\to\infty} x(t) = 0$, but if $a > 0$, then the limit is infinite ($\pm\infty$ depending on the sign of $x_0$).

When the dimension of $A$ is greater than one, the general solution of (4.14) can be expressed in terms of the exponential of a matrix $A$. The general solution is

$$X(t) = e^{At}C,$$

where $e^{At}$ is an $n \times n$ matrix and $C$ is an $n \times 1$ vector. Matrix $\Phi(t) = e^{At}$ is known as the fundamental matrix with the property that $\Phi(0) = I$, the $n \times n$ identity matrix (the columns of $\Phi$ are $n$ linearly independent solutions of the differential equation). The matrix exponential $e^{At}$ is defined as follows:

$$e^{At} = I + At + A^2\frac{t^2}{2!} + A^3\frac{t^3}{3!} + \cdots = \sum_{k=0}^{\infty} A^k\frac{t^k}{k!},$$

Where the series converges for all $t$. There are many methods for computing the matrix exponential (see, e.g., Leonard, 1996; Moler and Van Loan, 1978, 2003; Waltman, 1986). Some of these methods are discussed in the Appendix for Chapter 4. If the elements of $A$ are known real values, then one may use computer algebra systems or numerical methods to compute $e^{At}$. In the next examples, the matrix exponential is computed directly from the definition of $e^{At}$.

**Example 4.9** Suppose $A$ is diagonal,

$$A = \begin{pmatrix} a_{11} & 0 \\ 0 & a_{22} \end{pmatrix}.$$

Then $A^k = \begin{pmatrix} a_{11}^k & 0 \\ 0 & a_{22}^k \end{pmatrix}$ and

$$e^{At} = \begin{pmatrix} \sum_{k=0}^{\infty} \frac{(a_{11}t)^k}{k!} & 0 \\ 0 & \sum_{k=0}^{\infty} \frac{(a_{22}t)^k}{k!} \end{pmatrix} = \begin{pmatrix} e^{a_{11}t} & 0 \\ 0 & e^{a_{22}t} \end{pmatrix}.$$

The solution to the system $dX/dt = AX$ is

$$X(t) = e^{At}X_0 = x_0 e^{a_{11}t}\begin{pmatrix} 1 \\ 0 \end{pmatrix} + y_0 e^{a_{22}t}\begin{pmatrix} 0 \\ 1 \end{pmatrix} = \begin{pmatrix} x_0 e^{a_{11}t} \\ y_0 e^{a_{22}t} \end{pmatrix},$$

where $X_0 = (x_0, y_0)^T$. ■

**Example 4.10** Suppose $A = \begin{pmatrix} 0 & 1 \\ 0 & 2 \end{pmatrix}$. Then $A^k = \begin{pmatrix} 0 & 2^{k-1} \\ 0 & 2^k \end{pmatrix}$ and

$$e^{At} = \begin{pmatrix} 1 & \frac{1}{2}\sum_{k=1}^{\infty} \frac{(2t)^k}{k!} \\ 0 & \sum_{k=0}^{\infty} \frac{(2t)^k}{k!} \end{pmatrix} = \begin{pmatrix} 1 & \frac{1}{2}[e^{2t} - 1] \\ 0 & e^{2t} \end{pmatrix}.$$

Thus, the solution to the linear system $dX/dt = AX$ is

$$X(t) = e^{At}X_0 = \begin{pmatrix} x_0 + \frac{1}{2}y_0[e^{2t} - 1] \\ y_0 e^{2t} \end{pmatrix}.$$ ■

The Maple commands for computing the exponential of a matrix are available in the linear algebra package. They are given below for the matrix in Example 4.10.

```
> with(linalg):
> A:=matrix(2,2,[0,1,0,2]);
```

$$A := \begin{bmatrix} 0 & 1 \\ 0 & 2 \end{bmatrix}$$

```
> eA:=exponential(A,t);
```

$$eA := \begin{bmatrix} 1 & \frac{1}{2}e^{2t} - \frac{1}{2} \\ 0 & e^{2t} \end{bmatrix}$$

A straightforward method to compute the general solution to $dX/dt = AX$, instead of computing $e^{At}$, is the same method that was used for the higher-order, constant coefficient differential equations. We need to find $n$ linearly independent solutions, $\phi_1(t), \ldots, \phi_n(t)$, which make up the fundamental matrix, $\Phi(t)$. Let $X = e^{\lambda t}V$. Then it follows that $AV = \lambda V$, where $\lambda$ is an eigenvalue of $A$ and $V$ is an eigenvector corresponding to $\lambda$. We summarize the form taken by the general solution in the case of a $2 \times 2$ matrix $A$. The eigenvalues are the solutions to the characteristic polynomial, $\det(A - \lambda I) = 0$. If $A = (a_{ij})$, then the characteristic polynomial has the following form:

$$\lambda^2 - (a_{11} + a_{22})\lambda + a_{11}a_{22} - a_{12}a_{21},$$

where the coefficients of the polynomial are the negative of the trace and the determinant of the matrix $A$,

$$\lambda^2 - \mathrm{Tr}(A)\lambda + \det(A) = 0.$$

The form taken by the solutions is summarized in the following three cases.

1. Eigenvalues $\lambda_i$, $i = 1, 2$ of $A$ are real and there exist two linearly independent eigenvectors $V_i$, $i = 1, 2$ corresponding to $\lambda_i$, $i = 1, 2$. The general solution to $dX/dt = AX$ has the form

$$X(t) = c_1 V_1 e^{\lambda_1 t} + c_2 V_2 e^{\lambda_2 t}.$$

2. Eigenvalues $\lambda_i$, $i = 1, 2$ of $A$ are real and equal ($\lambda_1 = \lambda_2$) and there exists only one linearly independent eigenvector $V_1$. The general solution to $dX/dt = AX$ has the form

$$X(t) = c_1 V_1 e^{\lambda_1 t} + c_2[V_1 t e^{\lambda_1 t} + P e^{\lambda_1 t}],$$

where the equation $(A - \lambda_1 I)P = V_1$ can be solved for $P$ (vector $P$ is known as a *generalized eigenvector*).

3. Eigenvalues $\lambda_i$, $i = 1, 2$ of $A$ are complex conjugate pairs, $\lambda_{1,2} = a \pm ib$, $b \neq 0$. The general solution to $dX/dt = AX$ has the form

$$X(t) = \begin{pmatrix} \alpha_1 \\ \alpha_2 \end{pmatrix} e^{at} \sin(bt) + \begin{pmatrix} \alpha_3 \\ \alpha_4 \end{pmatrix} e^{at} \cos(bt).$$

Actually, the solution involves only two independent arbitrary constants, $c_1$ and $c_2$. The four constants $\alpha_i$, $i = 1, 2, 3, 4$, depend on the associated eigenvalues and eigenvectors and can be found by substitution into the differential equation.

For more information about solutions to first-order linear systems, please consult a textbook on ordinary differential equations.

## 4.8 Phase Plane Analysis

The solution behavior for two-dimensional linear systems is studied in the phase plane, that is, in the $x$-$y$ plane. Let $X = (x, y)^T$ and $A = (a_{ij})$ so that $dX/dt = AX$ can be expressed as follows:

$$\frac{dx}{dt} = a_{11}x + a_{12}y,$$

$$\frac{dy}{dt} = a_{21}x + a_{22}y. \tag{4.16}$$

The origin, $x = 0$ and $y = 0$, is an equilibrium solution of system (4.16). Assume that $\det(A) \neq 0$. Then the origin is the unique equilibrium solution, an *isolated equilibrium*.

Solutions to the linear system (4.16) are characterized by the eigenvalues of the matrix $A$, which in turn depend on the trace and determinant of $A$. The origin will be classified as a node, saddle, spiral, or center. The origin is further classified as stable or unstable. We begin by defining these latter terms for a linear system. We distinguish between stable and asymptotically stable.

The origin is *asymptotically stable* if the eigenvalues of $A$ are negative or have negative real parts. The origin is *stable* if the eigenvalues of $A$ are nonpositive or have nonpositive real parts. The origin is *unstable* if at least one eigenvalue of $A$ is positive or has positive real part. Solutions approach the origin if the origin is asymptotically stable, $\lim_{t\to\infty}(x(t), y(t)) = (0, 0)$. Based on these definitions, it is easy to determine the stability of the origin once the eigenvalues are known. In addition, even without calculating the eigenvalues, the stability can be determined by applying the Routh-Hurwitz criteria to the characteristic polynomial of $A$:

$$\lambda^2 - \mathrm{Tr}(A)\lambda + \det(A) = 0.$$

Asymptotic stability is determined only by the trace and determinant because these two quantities are the coefficients of the characteristic polynomial. According to the Routh-Hurwitz criteria, the eigenvalues lie in the left half of the complex plane iff the coefficients are positive.

**Corollary 4.2** *Suppose $dX/dt = AX$, where $A$ is a constant $2 \times 2$ matrix with $\det(A) \neq 0$. The origin is asymptotically stable iff*

$$\mathrm{Tr}(A) < 0 \quad \text{and} \quad \det(A) > 0.$$

*The origin is stable iff $\mathrm{Tr}(A) \le 0$ and $\det(A) > 0$. The origin is unstable iff $\mathrm{Tr}(A) > 0$ or $\det(A) < 0$.* ◀

Now, we give specific criteria for the origin of a general linear differential system to be classified into one of four types: node, saddle, spiral, or center. Then we apply the previous results to classify the origin as stable or unstable. This classification scheme is based on the fact that the origin is the only fixed point or equilibrium solution of the linear system, $\det(A) \neq 0$. Matrix $A$ has no zero eigenvalues. The classification scheme depends on whether the eigenvalues are real or complex, whether the real eigenvalues are positive or negative, and whether the complex eigenvalues have negative real parts. References for the qualitative theory of differential equations can be found in many textbooks (see, e.g., Brauer and Nohel, 1969; Cushing, 2004; Sánchez, 1968).

## Real Eigenvalues:

In the case of real eigenvalues, $\lambda_1$ and $\lambda_2$, the corresponding eigenvectors $V_1$ and $V_2$ are directions along which solutions travel toward or away from the origin. For example, if $\lambda_1$ is positive, solutions will travel along $V_1$, away from the origin. If $\lambda_2$ is negative, solutions will travel along $V_2$, toward the origin. In general, solutions travel in a direction which is a linear combination of $V_1$ and $V_2$. The origin is classified as either a node or a saddle.

1. ***Node:*** Both eigenvalues have the same sign and may be distinct or equal, $\lambda_1 \le \lambda_2 < 0$ or $0 < \lambda_1 \le \lambda_2$. The origin can be further classified as proper or improper (Brauer and Nohel, 1969; Sánchez, 1968). A node is called

*proper* when the eigenvalues are equal and there are two linearly independent eigenvectors; otherwise it is called *improper*. A proper node is also referred to as a *star point* or *star solution* (Cushing, 2004; Gulick, 1992). The reason for this latter name can be seen in Figure 4.1 (solutions approach the origin in all directions). The term *degenerate node* is also used to refer to a node when the two eigenvalues of matrix $A$ are equal (Gulick, 1992). In Figure 4.1, the improper node in the upper left corner has two distinct eigenvalues; it is not degenerate. But the other two nodes, to the right of the node in the upper left corner, are degenerate nodes (the eigenvalues are equal). If there is only one independent eigenvector, the dynamics are illustrated in the center figure and if there are two independent eigenvectors, the dynamics are illustrated in the upper right corner (star solution).

2. ***Saddle:*** Eigenvalues $\lambda_1$ and $\lambda_2$ have opposite signs, $\lambda_1\lambda_2 < 0$ (e.g., $\lambda_1 < 0 < \lambda_2$).

## Complex Eigenvalues:

In the case of complex eigenvalues, $\lambda_{1,2} = a \pm ib, b \neq 0$. Because solutions to the linear system $dX/dt = AX$ include factors with $\cos(bt)$ and $\sin(bt)$, solutions spiral around the equilibrium. If the real part $a < 0$, then the solutions with $e^{at}\cos(bt)$ or $e^{at}\sin(bt)$ spiral inward, toward the origin. But if the real part $a > 0$, then solutions spiral outward, away from the origin. Finally,

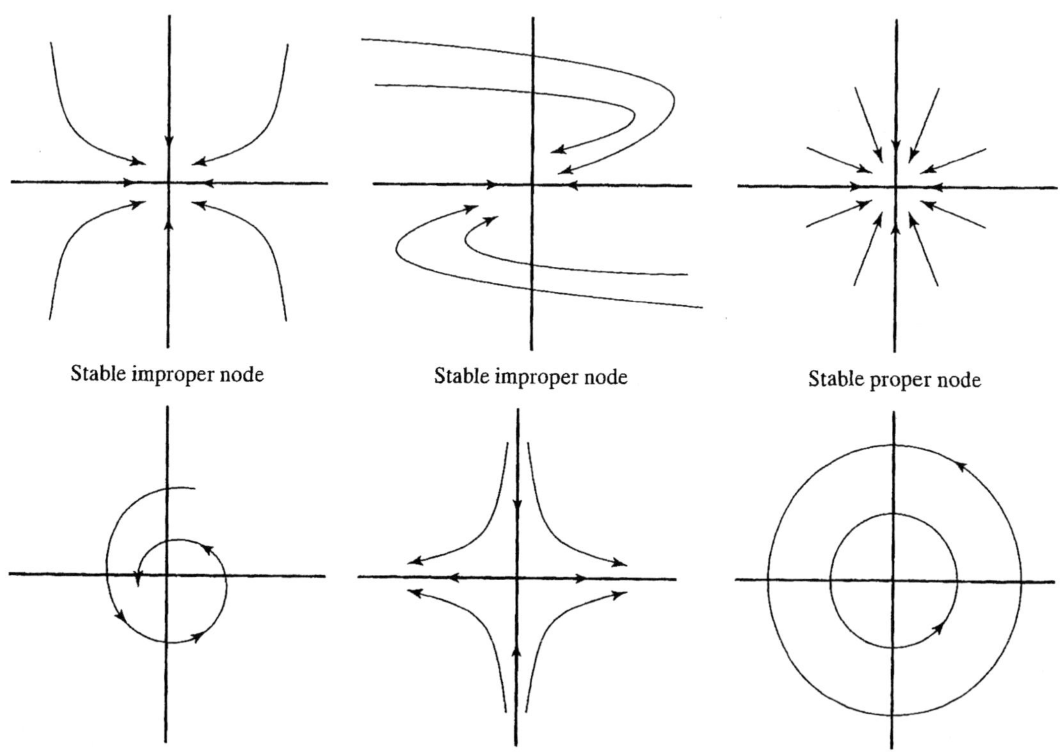


**Figure 4.1** Graphs of solutions for an improper and proper node, spiral, saddle, and center.

if the real part $a = 0$, then solutions are closed curves, encircling the origin. The origin is classified as a *spiral (or focus)* if $a \neq 0$ and a *center* if $a = 0$.

3. ***Spiral or Focus:*** Eigenvalues have nonzero real part ($a \neq 0$).
4. ***Center:*** Eigenvalues are purely imaginary ($a = 0$), $\lambda_{1,2} = \pm ib$.

A node or spiral can be classified as either asymptotically stable or unstable depending on whether the real part of the eigenvalue is negative or positive, respectively. A saddle point is always unstable and a center is neither asymptotically stable nor unstable. A center is sometimes called *neutrally stable* (it is stable, but not asymptotically stable).

**Example 4.11** Let $A = \begin{pmatrix} 0 & -1 \\ 1 & 0 \end{pmatrix}$. The eigenvalues of $A$ are $\pm i$, so that the origin is a center. The solution to $dX/dt = AX$ can be found directly by noting $dx/dt = -y$ and $dy/dt = x$ so that

$$\frac{dy}{dx} = \frac{dy/dt}{dx/dt} = -\frac{x}{y}.$$

Separating variables and integrating,

$$\frac{y^2}{2} + \frac{x^2}{2} = c.$$

This latter equation is a circle centered at the origin. Solutions travel in a counterclockwise direction on circles surrounding the origin. ■

**Example 4.12** Let $A = \begin{pmatrix} 0 & 1 \\ 1 & 0 \end{pmatrix}$. The eigenvalues of $A$ are $\pm 1$, so that the origin is a saddle. The system $dX/dt = AX$ can be written as $dy/dx = x/y$. Separating variables and integrating,

$$\frac{y^2}{2} - \frac{x^2}{2} = c.$$

This latter equation is a hyperbola with center at the origin. See Figure 4.2. ■

**Example 4.13** Let $A = \begin{pmatrix} 1 & 0 \\ 0 & 3 \end{pmatrix}$. The eigenvalues of $A$ are 1 and 3. The origin is an unstable node. Solutions can be found by integrating $dx/dt = x$ and $dy/dt = 3y$ so that $x(t) = x_0 e^t$ and $y(t) = y_0 e^{3t}$. Solutions in the phase plane have the form

$$X(t) = x_0 e^t \begin{pmatrix} 1 \\ 0 \end{pmatrix} + y_0 e^{3t} \begin{pmatrix} 0 \\ 1 \end{pmatrix}.$$

■

The classification schemes can be related to the signs of the trace and determinant of $A$ and the discriminant of the characteristic polynomial. Let

$$\tau = \mathrm{Tr}(A) \quad \text{and} \quad \delta = \det(A).$$

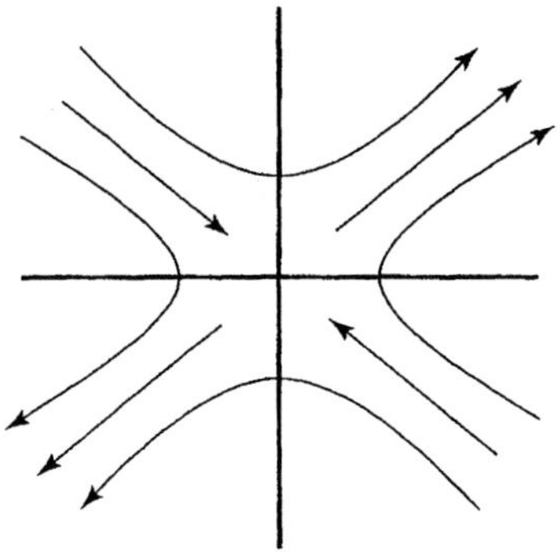

**Figure 4.2** Solutions to Example 4.12 are graphed in the phase plane. The origin is a saddle.

Recall that the eigenvalues, the roots of the characteristic polynomial $\lambda^2 - \text{Tr}(A)\lambda + \det(A) = \lambda^2 - \tau\lambda + \delta$, satisfy

$$\lambda_{1,2} = \frac{\tau \pm \sqrt{\tau^2 - 4\delta}}{2}.$$

The discriminant is denoted as $\gamma$ and defined as follows:

$$\gamma = \tau^2 - 4\delta.$$

The following classification scheme summarizes the dynamics according to the sign of the discriminant, $\gamma$, that is, according to whether the eigenvalues are real or complex conjugates. Improper and proper nodes are not distinguished.

## Eigenvalues are real ($\gamma \geq 0$):

*Unstable node* if $\tau > 0$ and $\delta > 0$ ($\lambda_{1,2} > 0$)
*Saddle point* if $\delta < 0$ ($\lambda_1 < 0 < \lambda_2$)
*Stable node* if $\tau < 0$ and $\delta > 0$ ($\lambda_{1,2} < 0$).

## Eigenvalues are complex conjugates $a \pm bi$ ($\gamma < 0$):

*Unstable spiral* if $\tau > 0$ ($a > 0$)
*Neutral center* if $\tau = 0$ ($a = 0$)
*Stable spiral or stable focus* if $\tau < 0$ ($a < 0$)

The classification scheme is illustrated in the $\tau$-$\delta$ plane in Figure 4.3. Note that asymptotic stability requires $\tau < 0$ and $\delta > 0$ (the trace is negative and the determinant is positive, Corollary 4.2).

**Example 4.14** Determine the conditions on $a$ so that the zero equilibrium of the following system is a stable spiral:

$$\frac{dx}{dt} = y, \quad \frac{dy}{dt} = -x + ay.$$

**Figure 4.3** Stability diagram in the $\tau$-$\delta$ plane.

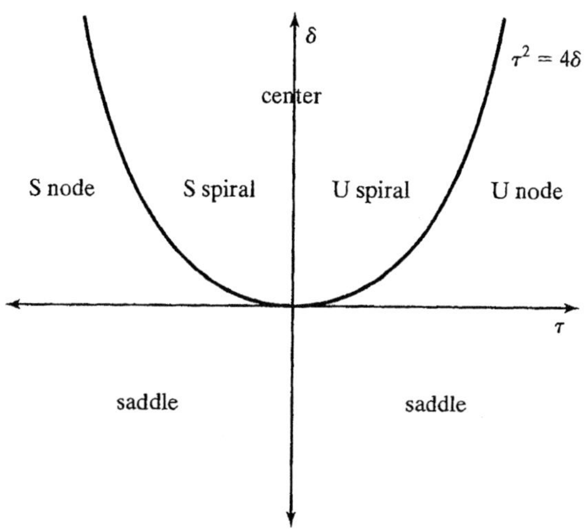

Matrix $A = \begin{pmatrix} 0 & 1 \\ -1 & a \end{pmatrix}$. For the zero equilibrium to be a stable spiral, the discriminant and trace $\tau$ must be negative. $\operatorname{Tr} A = \tau = a < 0$, $\det A = \delta = 1$, so that $\gamma = \tau^2 - 4\delta = a^2 - 4 < 0$. The two inequalities give the conditions for a stable spiral: $-2 < a < 0$. ∎

## 4.9 Gershgorin's Theorem

When the zero equilibrium is an isolated equilibrium for the system $dX/dt = AX$, then all solutions converge to the origin iff the eigenvalues of $A$ are negative or have negative real parts (lie in the left half of the complex plane). The Routh-Hurwitz criteria give necessary and sufficient conditions for the eigenvalues to lie in the left half of the complex plane. Another result which provides sufficient conditions for the eigenvalues to lie in the left half of the complex plane is known as Gershgorin's Theorem (see e.g., Noble, 1969; Ortega, 1987).

**Theorem 4.5** **(Gershgorin's Theorem).** *Let $A$ be an $n \times n$ matrix. Let $D_i$ be the disk in the complex plane with center at $a_{ii}$ and radius $r_i = \sum_{j=1, j\neq i}^{n} |a_{ij}|$. Then all eigenvalues of the matrix $A$ lie in the union of the disks $D_i$, $i = 1, 2, \ldots, n$, $\bigcup_{i=1}^{n} D_i$. In particular, if $\lambda$ is an eigenvalue of $A$, then for some $i = 1, 2, \ldots, n$,*

$$|\lambda - a_{ii}| \leq r_i.$$

A disk $D_i$ is graphed in Figure 4.4.

**Proof** Let $\lambda$ be any eigenvalue of $A$ and $V = (v_1, \ldots, v_n)^T$ an eigenvector corresponding to this eigenvalue. Then $AV = \lambda V$, which implies $\lambda v_i = \sum_{j=1}^{n} a_{ij} v_j$ or

$$(\lambda - a_{ii}) v_i = \sum_{j=1, j\neq i}^{n} a_{ij} v_j, \quad i = 1, \ldots, n.$$

Let $v_k$ denote the element of $V$ with greatest magnitude (i.e., $|v_k| \geq |v_j|, j \neq k$). Then $|v_j/v_k| \leq 1$ for all $j = 1, \ldots, n$, and

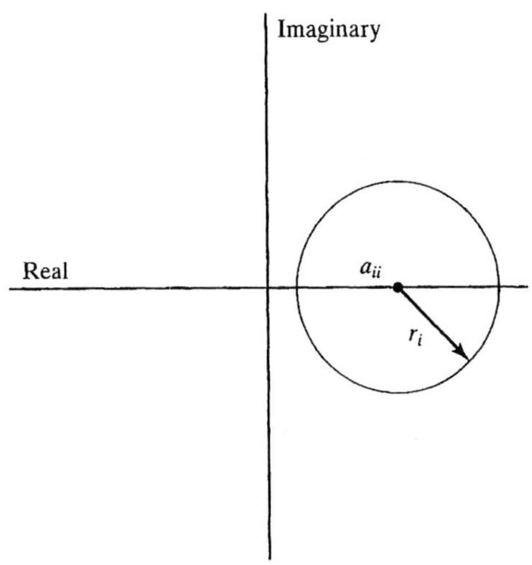


**Figure 4.4** Gershgorin's disk in the complex plane.

$$|\lambda - a_{kk}| \leq \sum_{j=1,j\neq i}^{n} |a_{kj}| \left|\frac{v_j}{v_k}\right| \leq \sum_{j=1,j\neq i}^{n} |a_{kj}|.$$

Hence, $\lambda$ lies in the disk $D_k$. The conclusion of the theorem follows. □

Gershgorin's Theorem applies to real and complex matrices $A$. When the entries of $A$ are real, it follows from Gershgorin's Theorem that the disks lie in the left half of the complex plane if

$$r_i + a_{ii} < 0$$

or $a_{ii} < -r_i$ for $i = 1, 2, \ldots, n$. The strict inequality guarantees the Gershgorin disk lies entirely in the left half of the complex plane. This result is stated in the next corollary.

**Corollary 4.3** *Let A be an $n \times n$ with real entries. If the diagonal elements of A satisfy*

$$a_{ii} < -r_i, \quad \textit{where } r_i = \sum_{j=1,j\neq i}^{n} |a_{ij}|$$

*for $i = 1, 2, \ldots, n$, then the eigenvalues of A are negative or have negative real parts.* ◀

**Example 4.15** Suppose matrix $A$ equals

$$\begin{pmatrix} a & -1 & 3 \\ 1/2 & b & -2 \\ 0 & -5 & c \end{pmatrix}.$$

The values of radii $r_i$ are given by $r_1 = 4$, $r_2 = 5/2$, and $r_3 = 5$. From Corollary 4.3, it follows that if $a < -4$, $b < -5/2$, and $c < -5$, then the eigenvalues are negative or have negative real parts. ■

The next example illustrates that the conditions in Corollary 4.3 are sufficient, but not necessary.

**Example 4.16** Suppose matrix $A$ equals

$$\begin{pmatrix} -2 & 1 & 0 \\ 1 & -2 & 1 \\ 0 & 1 & -2 \end{pmatrix}.$$

For this matrix, $r_1 = 1 = r_3$ and $r_2 = 2$ but $a_{22} = -r_2$. However, the eigenvalues of $A$ are negative, $\lambda_{1,2,3} = -2, -2 \pm \sqrt{2}$. ■

## 4.10 An Example: Pharmacokinetics Model

A model for the ingestion of a drug into the body is constructed (Yeargers et al., 1996). The drug is taken orally and is delivered to the gastrointestinal (GI) tract. The drug, then, moves into the blood stream, without delay, at a rate proportional

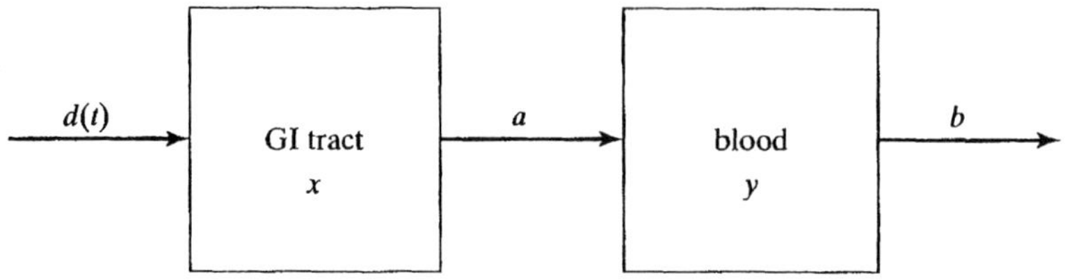


**Figure 4.5** Compartmental diagram of drug concentration in the GI tract and blood.

to its concentration in the GI tract and independent of its concentration in the blood. The drug is metabolized and cleared from the blood at a rate proportional to its concentration there. The model is based on the two compartments, GI tract and blood. Let $x(t)$ denote the concentration of the drug in the GI tract and $y(t)$ the concentration in the blood. In addition, let $d(t)$ denote the drug dosage. Figure 4.5 represents the two-compartmental model.

The two compartments can be modeled as a system of linear, nonhomogeneous differential equations:

$$\frac{dx}{dt} = -ax + d(t),$$

$$\frac{dy}{dt} = ax - by, \quad a,b > 0, a \neq b.$$

Because this system is linear, we know that the general solution to this model is

$$X(t) = e^{At}X_0 + e^{At}\int_0^t e^{-As}G(s)\,ds,$$

where $G(s) = (d(s), 0)^T$ and $A = \begin{pmatrix} -a & 0 \\ a & -b \end{pmatrix}$. Since $A$ has two negative eigenvalues, $-a$ and $-b$, $\lim_{t\to\infty} e^{At}X_0 = \mathbf{0}$. The homogeneous solution represents a transient solution. Therefore,

$$\lim_{t\to\infty} X(t) = \lim_{t\to\infty} e^{At}\int_0^t e^{-As}G(s)\,ds,$$

where $e^{At}$ equals

$$\begin{pmatrix} e^{-at} & 0 \\ a\dfrac{e^{-bt} - e^{-at}}{a-b} & e^{-bt} \end{pmatrix}.$$

Another method to solve this system is to solve for $x$ first, a first-order nonhomogeneous equation, then use $x$ to solve for $y$.

Suppose there is continuous release of a drug into the GI tract, that is, suppose $d(t)$ is constant. Let $d(t) = 1$ and initially, $x(0) = 0 = y(0)$. The solution to the system can be shown to satisfy

$$x(t) = \frac{1}{a}(1 - e^{-at}),$$

$$y(t) = \frac{1}{b} + \frac{e^{-at}}{a-b} - \frac{ae^{-bt}}{b(a-b)}.$$

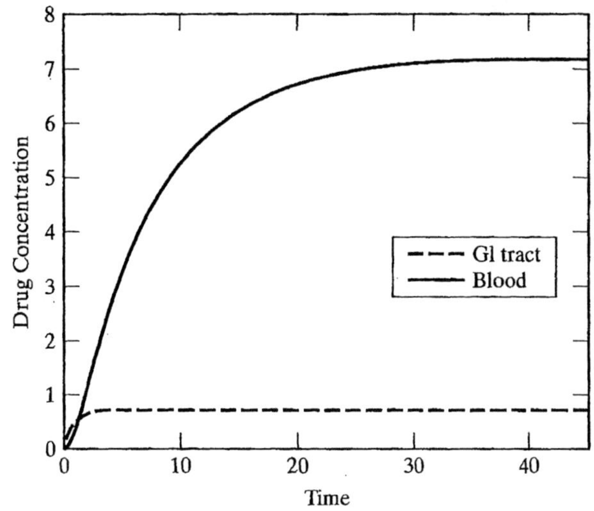


**Figure 4.6** Drug concentration in the GI tract and blood for the pharmacokinetics model.

Then

$$\lim_{t\to\infty} x(t) = \frac{1}{a} \quad \text{and} \quad \lim_{t\to\infty} y(t) = \frac{1}{b}.$$

The half-life of the drug in particular compartments can be estimated. For example, in the GI tract, where $dx/dt = -ax$, $x(t) = x_0e^{-at}$, the half-life is the value of $t$ where $x(t) = x_0/2$. Thus, the half-life is $t = \ln(2)/a$. Suppose time is measured in hours and the half-life of a particular drug in the GI tract is 1/2 hour and in the blood it is 5 hours, then

$$a = 2\ln(2) \quad \text{and} \quad b = \frac{\ln(2)}{5}.$$

The solution for the pharmacokinetics model is graphed in Figure 4.6.

Consider the case where drug injection is periodic. This is a reasonable situation because prescription drugs are often taken at specific intervals of time. Suppose a drug is prescribed every six hours. Then a reasonable assumption about the drug injection is as follows:

$$d(t) = \begin{cases} 2, & 0 \le t \le 1/2, \\ 0, & 1/2 < t < 6, \end{cases}$$

where $d(t + 6) = d(t)$ (Yeargers et al., 1996). The drug is taken orally every six hours and released into the GI tract over a half-hour period. Important and interesting questions arise with periodic drug injection. For example, if the dose is changed or the period is changed (e.g., every 12 hours), how does the maximum or minimum concentration of the drug change over time? For the drug to be effective, a certain minimum concentration needs to be maintained. In addition, to minimize side effects of the drug, a maximum concentration should not be exceeded. These questions are considered in the problem in Exercise 18.

## 4.11 Discrete and Continuous Time Delays

The response of a biological system to a particular input or stimulus is often not immediate but is delayed. For example, in the pharmacokinetics model, there may be a delay before the drug enters the blood stream. If there is a fixed time until the reaction occurs following a stimulus, the delay may be modeled as a fixed time or a *discrete delay*. If the reaction to the stimulus does not occur after a fixed period of time but occurs over a continuous range of times, it is called a *continuous delay*.

A first-order linear differential equation with a discrete delay $\tau > 0$ has the following form:

$$\frac{dx}{dt} = ax(t) + bx(t - \tau) + f(t). \tag{4.17}$$

A first-order linear differential equation with a continuous delay over an interval $[0, \tau]$ becomes an integrodifferential equation

$$\frac{dx}{dt} = ax(t) + b\int_0^{\tau} x(t - s)\, ds + f(t).$$

Making a change of variable in the integral $u = t - s$ leads to the integrodifferential equation

$$\frac{dx}{dt} = ax(t) + b\int_{t-\tau}^{t} x(u)\, du + f(t).$$

In this section, we show how to solve linear discrete delay differential equations, differential equations of the form (4.17).

Delay differential equations of the form (4.17) can be solved in a stepwise fashion. First, it is necessary that the initial condition be given on the interval $[-\tau, 0]$, an interval whose length equals the length of the delay. Suppose $x(t) = \phi_0(t)$ on $[-\tau, 0]$. The solution is found on successive intervals, $[0, \tau]$, then $[\tau, 2\tau]$, and so on. This method of solution is referred to as the *method of steps* (Jacquez, 1996; Kuang, 1993). On the interval $[0, \tau]$, the differential equation is a nonhomogeneous linear differential equation without a delay because $\phi_0(t - \tau)$ on $[0, \tau]$ is a known function,

$$\frac{dx}{dt} = ax(t) + b\phi_0(t - \tau) + f(t),$$

with initial condition $x(0) = \phi_0(0)$. This differential equation can be solved to obtain $x(t) = \phi_1(t)$ on $[0, \tau]$. Then, on the interval $[\tau, 2\tau]$, the differential equation satisfies

$$\frac{dx}{dt} = ax(t) + b\phi_1(t - \tau) + f(t),$$

with initial condition $x(\tau) = \phi_1(\tau)$. Continuing in this stepwise fashion, solutions $\phi_n(t)$, $n = 0, 1, 2, \ldots$ can be found on the intervals $[(n - 1)\tau, n\tau]$ applying techniques from ordinary differential equations.

**Example 4.17** Solve the following delay differential equation:

$$\frac{dx\ (t)}{dt} = x(t - 1) + 1$$

with initial condition $x(t) = 0 = \phi_0(t)$ for $t \in [-1, 0]$. First, we solve the following initial value problem on $[0, 1]$:

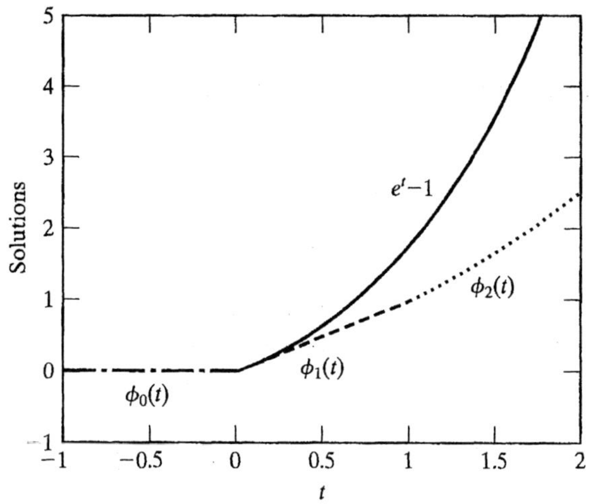


**Figure 4.7** The solution of the delay differential equation is compared to the solution of the nondelay differential equation, $x(t) = e^t - 1$.

$$\frac{dx(t)}{dt} = \phi_0(t-1) + 1 = 1, \quad x(0) = \phi_0(0) = 0.$$

The solution to this initial value problem on [0,1] is $x(t) = t = \phi_1(t)$. On the interval [1, 2], we solve

$$\frac{dx(t)}{dt} = \phi_1(t-1) + 1 = t, \quad x(1) = \phi_1(1) = 1.$$

The solution to this initial value problem is $x(t) = \phi_2(t) = t^2/2 + 1/2$. The solution to the delay differential equation on $[-1, 2]$ is

$$x(t) = \begin{cases} 0, & t \in [-1, 0], \\ t, & t \in [0, 1], \\ \dfrac{t^2}{2} + \dfrac{1}{2}, & t \in [1, 2]. \end{cases}$$

The solution to the IVP without delay, $dx/dt = x(t) + 1$, $x(0) = 0$, is $x(t) = e^t - 1$. Note how the delay affects the solution behavior. The delay causes the solution to lag behind the nondelay differential equation (see Figure 4.7). ■

**Example 4.18** Suppose the delay differential equation (4.17) has the form

$$\frac{dx(t)}{dt} = -x(t) + \frac{1}{2}x(t-1)$$

with initial condition $x(t) = 1 = \phi_0(t)$ for $t \in [-1, 0]$. For $t \in [0, 1]$, we solve the initial value problem,

$$\frac{dx(t)}{dt} = -x(t) + \frac{1}{2}\phi_0(t-1) = -x(t) + \frac{1}{2}, \quad x(0) = \phi_0(0) = 1.$$

The solution on [0, 1] is

$$x(t) = \phi_1(t) = \frac{1}{2}e^{-t} + \frac{1}{2}.$$

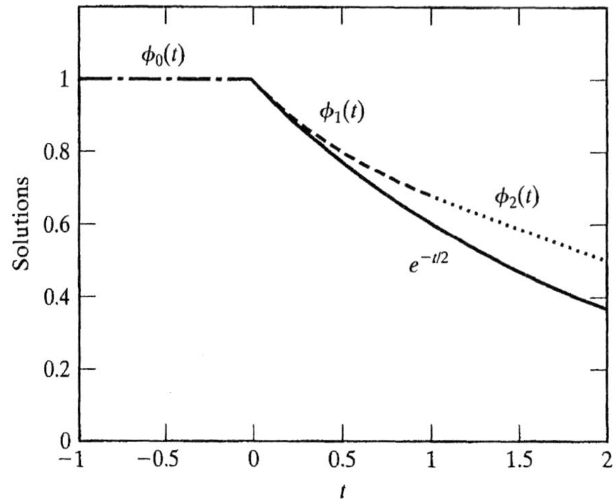


**Figure 4.8** The solution of the delay differential equation is compared to the solution of the nondelay differential equation, $x(t) = e^{-t/2}$.

Next, for $t \in [1, 2]$, we solve

$$\frac{dx(t)}{dt} = -x(t) + \frac{1}{2}\phi_1(t-1) = -x(t) + \frac{1}{4}e^{-(t-1)} + \frac{1}{4}$$

with initial condition

$$x(1) = \phi_1(1) = \frac{1}{2}e^{-1} + \frac{1}{2}.$$

It can be shown that the solution is

$$x(t) = \phi_2(t) = \frac{1}{4}[1 + te^{1-t}] + \frac{1}{2}e^{-t}.$$

The solution to the delay differential equation on $[-1, 2]$ is

$$x(t) = \begin{cases} 1, & t \in [-1,0], \\ \frac{1}{2}e^{-t} + \frac{1}{2}, & t \in [0,1], \\ \frac{1}{4}[1 + te^{1-t}] + \frac{1}{2}e^{-t}, & t \in [1,2]. \end{cases}$$

The solution to the nondelay differential equation $dx/dt = -x(t)/2$, $x(0) = 1$, is $x(t) = e^{-t/2}$. The delay and nondelay solutions are compared in Figure 4.8. It can be seen that the solution to the delayed equation lags behind the nondelay equation. ■

Inclusion of discrete and continuous delays often makes biological models more realistic. In practice there may be multiple delays. See Cushing (1977), Gopalsamy (1992), Jacquez (1996), or Kuang (1993) for an introduction to delay differential equations with applications to biology. In Section 5.9, the dynamics of a nonlinear delay differential equation are studied.

## 4.12 Exercises for Chapter 4

1. Use an integrating factor to find the unique solution to the following initial value problems.

   (a) $\dfrac{dx}{dt} + 3t^2x = 4te^{-t^3}, \quad x(0) = 1$

   (b) $\dfrac{dx}{dt} + \dfrac{2}{t}x = 2t + 5, \quad x(1) = 1$

2. Find the unique solution to the following initial value problems, $x' = dx/dt$, $x'' = d^2x/dt^2$, $x''' = d^3x/dt^3$, $x^{(4)} = d^4x/dt^4$, and $y' = dy/dt$.

   (a) $x^{(4)} - 4x = 0$. At $t = 0$, $x(0) = x'(0) = x''(0) = 0$, and $x'''(0) = 1$.

   (b) $x' = 4x$, $y' = -y$. At $t = 0$, $x(0) = 1 = y(0)$.

   (c) $x'' + 6x' + 9x = 0$. At $t = 0$, $x(0) = 5$, and $x'(0) = 0$.

3. Find the general solution to the following differential equations, $x' = dx/dt$, $x'' = d^2x/dt^2$, and $x''' = d^3x/dt^3$.

   (a) $x'' - x' - 6x = 0$

   (b) $x'' - 4x' + 5x = 0$

   (c) $x''' - 5x'' + 3x' + 9x = 0$

   (d) $x''' + 16x' = 0$

   (e) $x'' + 2ax' + (a^2 + b^2)x = 0, \; a, b \neq 0$

4. A second-order Cauchy-Euler differential equation has the form

$$at^2\frac{d^2x}{dt^2} + bt\frac{dx}{dt} + cx = 0.$$

   The method for finding solutions to the Cauchy-Euler differential equation is to assume $x = t^m$. Substituting this solution into the differential equation leads to the auxiliary equation,

$$am(m - 1) + bm + c = 0.$$

   The roots $m_1$ and $m_2$ of the auxiliary equation lead to the solution of the Cauchy-Euler equation. If the roots are real and distinct, then $t^{m_1}$ and $t^{m_2}$ are two linearly independent solutions. If the roots are real and equal, then $t^{m_1}$ and $t^{m_1}\ln|t|$ are two linearly independent solutions. Finally, if the roots are complex, $u \pm vi$, then the solutions are $t^u\cos(v\ln|t|)$ and $t^u\sin(v\ln|t|)$. Find the general solution to the following Cauchy-Euler equations.

   (a) $t^2\dfrac{d^2x}{dt^2} - 2x = 0$

   (b) $t^2\dfrac{d^2x}{dt^2} + 5t\dfrac{dx}{dt} + 4x = 0$

   (c) $t^2\dfrac{d^2x}{dt^2} - t\dfrac{dx}{dt} + 2x = 0$

5. First, find a constant coefficient, homogeneous differential equation whose characteristic polynomial has the following roots. Then write down the general solution to this differential equation.

   (a) $\lambda = a \pm bi$

   (b) $\lambda = 0, 0, \pm a, \pm a, \; a \neq 0$

**6.** A second-order linear, homogeneous differential equation with constant coefficients has two solutions $\{x_1, x_2\}$ which may take one of three forms:

$$\{e^{\lambda_1 t}, e^{\lambda_2 t}\}, \quad \{e^{\lambda_1 t}, te^{\lambda_1 t}\}, \quad \text{or} \quad \{e^{ut}\cos(vt), e^{ut}\sin(vt)\}.$$

In each case, compute the Wronskian and show that the solutions are linearly independent; the Wronskian does not equal zero.

**7.** For the linear differential systems $dX/dt = AX$, the matrices $A$ are given below. For each system, find the eigenvalues and eigenvectors of $A$. Then find the general solution to each differential system.

(a) $A = \begin{pmatrix} 6 & 2 \\ 2 & 3 \end{pmatrix}$

(b) $A = \begin{pmatrix} 1 & 2 \\ 4 & 3 \end{pmatrix}$

(c) $A = \begin{pmatrix} 1 & 1 & -1 \\ 0 & 2 & 0 \\ 0 & 1 & -1 \end{pmatrix}$

**8.** Consider the following initial value problem:

$$\frac{d^3x}{dt^3} + a\frac{d^2x}{dt^2} + b\frac{dx}{dt} + x = 0,$$

$$\left.\frac{d^2x}{dt^2}\right|_{t=0} = \alpha, \quad \left.\frac{dx}{dt}\right|_{t=0} = \beta, \quad \text{and} \quad x(0) = \gamma.$$

(a) For what values of $a$ and $b$ does the solution $x(t)$ approach zero for any initial conditions?

(b) Give an example, where $a$ and $b$ are specified constants, to show that $\lim_{t\to\infty} x(t)$ may not exist.

**9.** A chemical compound undergoes the transformation

$$A \xrightarrow{k_1} B \xrightarrow{k_2} C.$$

This process is described by the following kinetic equations:

$$\frac{dA}{dt} = -k_1 A,$$

$$\frac{dB}{dt} = k_1 A - k_2 B,$$

$$\frac{dC}{dt} = k_2 B,$$

where $k_1 \neq k_2$ and $k_1, k_2 > 0$. The variables $A$, $B$, and $C$ represent the concentration of each of the chemicals (Kaplan and Glass, 1995). The initial conditions are $A(0) = N$, $B(0) = 0$, and $C(0) = 0$.

(a) Find the solution for $A(t)$.

(b) Use (a) to find the solution for $B(t)$.

(c) Use (b) to find the solution for $C(t)$.

(d) If $k_1 = 2k_2$, at what time is $B$ a maximum?

(e) Find the following limits, if they exist, $\lim_{t\to\infty} A(t)$, $\lim_{t\to\infty} B(t)$, and $\lim_{t\to\infty} C(t)$.

**10.** Let $dX/dt = AX$, where $X_0 = (2,1)^T$ and

$$A = \begin{pmatrix} 1 & 1 \\ 0 & 1 \end{pmatrix} \quad \text{or} \quad A = \begin{pmatrix} 2 & 1 \\ 1 & 2 \end{pmatrix}.$$

For each matrix $A$, find the eigenvalues and eigenvectors of $A$. Then find the fundamental matrix $e^{At}$ and the solution to the initial value problem.

**11.** Use the methods discussed in the Appendix to find the exponential of the following matrix:

$$A = \begin{pmatrix} 1 & 1 & 0 \\ 0 & 3 & -2 \\ 0 & 3 & -4 \end{pmatrix}.$$

(a) Use the expression given by (4.18) to find $e^{At}$.

(b) Use the expression given by (4.19) to find $e^{At}$.

(c) Use a computer algebra system or another method to find $e^{At}$ and verify your answers in (a) and (b).

**12.** Let $dX/dt = AX$, where $A$ is given in Exercise 11.

(a) Use the exponential of matrix $A$ found in Exercise 11 to write down the general solution to the linear system.

(b) If $X_0 = (2, 1, 0)^T$, find the solution to the initial value problem.

(c) If $X_0 = (2, 1, -1)^T$, find the solution to the initial value problem.

**13.** Show that the zero equilibrium is always unstable if

$$\frac{dx}{dt} = ax + y, \quad a \neq -1,$$

$$\frac{dy}{dt} = -x + y.$$

When is the zero equilibrium a saddle point? Unstable spiral?

**14.** For the following linear differential system, determine conditions on the parameter $a$ such that the origin is a (a) saddle, (b) stable node, or (c) stable spiral.

$$\frac{dx}{dt} = (a - 1)x + y,$$

$$\frac{dy}{dt} = -ay$$

**15.** Determine whether the origin is stable or unstable, a node, spiral, saddle, or center for the system $dX/dt = AX$.

(a) $A = \begin{pmatrix} 6 & 2 \\ 2 & 3 \end{pmatrix}$

(b) $A = \begin{pmatrix} -2 & 4 \\ -1 & 1 \end{pmatrix}$

(c) $A = \begin{pmatrix} -1 & a \\ 0 & a \end{pmatrix}, a \neq 0$

**16.** Apply Gershgorin's Theorem to determine sufficient conditions on the parameters $a, b,$ and $c$ such that the eigenvalues of $A$ are negative or have negative real parts.

(a) $A = \begin{pmatrix} a & -1 & 0 \\ -1 & b & 1 \\ 0 & -2 & c \end{pmatrix}$

(b) $A = \begin{pmatrix} a & b \\ c & -2 \end{pmatrix}$

**17.** Show that the solution to the pharmacokinetics model is

$$x(t) = \frac{1}{a}(1 - e^{-at}),$$
$$y(t) = \frac{1}{b} + \frac{e^{-at}}{a - b} - \frac{ae^{-bt}}{b(a - b)}.$$

**18.** Suppose the drug in the pharmacokinetics model is administered periodically, $d(t) = 2$ for $t \in [6t, 6t + 0.5)$, $t = 0, 1, \ldots,$ and $d(t) = 0$ elsewhere. Let $a = 2/\ln(2)$ and $b = \ln(2)/5$. Let the initial conditions satisfy $x(0) = 0$ and $y(0) = 0$. The differential equation for $x(t)$ can be expressed in terms of the Heaviside function, $H(t) = 1$ for $t \geq 0$ and $H(t) = 0$ for $t < 0$. For example, on the interval $t \in [0, 48)$,

$$\frac{dx(t)}{dt} = -ax(t) + 2\sum_{k=0}^{7} H(t - 6k) - 2\sum_{k=0}^{7} H(t - 6k - 0.5).$$

(a) Use a differential equation solver to graph the solution during the first 48 hours, $t \in [0, 48)$. See the Maple program in the Appendix.

(b) How does the drug concentration change (maximum and minimum after 24 hours) if the dose is changed from every 6 hours to every 12 hours?

(c) How does the drug concentration change (maximum and minimum after 24 hours) if the dose is reduced to $d(t) = 1$?

**19.** Show that the solutions to the delay differential equation on $[0, 2]$ ($\tau = 1$) in Example 4.18 are

$$\phi_1(t) = \frac{1}{2}e^{-t} + \frac{1}{2} \quad \text{and} \quad \phi_2(t) = \frac{1}{4}[1 + te^{1-t}] + \frac{1}{2}e^{-t}.$$

**20.** Solve the following delay differential equations on the interval $[0, 2]$ by the method of steps.

(a) $\dfrac{dx(t)}{dt} = x(t - 1) + 2t$, with initial condition $x(t) = 0$ for $t \in [-1, 0]$.

(b) $\dfrac{dx(t)}{dt} = x(t - 1) + x(t)$ with initial condition $x(t) = 1$ for $t \in [-1, 0]$.

(c) $\dfrac{d^2x(t)}{dt^2} = 2x(t - 1) + 1$ with initial conditions $x(t) = 0$ for $t \in [-1, 0]$ and $dx/dt = 0$ for $t = 0$. This is a second-order delay differential equation, so the derivative at zero needs to be specified.

**21.** The whooping crane data in the Appendix for Chapter 3 appears to increase exponentially. Fit the whooping crane data to the exponential curve $x(t) = x_0e^{at}$ using a least squares approximation. Use *polyfit* in MATLAB with the curve $\ln(x(t)) = \ln(x_0) + at$, where $x(t)$ is the whooping crane winter population size. What are the estimates for $x_0$ and $a$? These estimates should agree with those for the discrete model $x(t) = x_0\lambda^t$, where $\lambda = e^a$. See Example 3.1 in Chapter 3.

## 4.13 References for Chapter 4

Brauer, F. and J. A. Nohel. 1969. *Qualitative Theory of Ordinary Differential Equations.* W. A. Benjamin, Inc., New York. Reprinted: Dover, 1989.

Cushing, J. M. 1977. *Integrodifferential Equations and Delay Models in Population Dynamics.* Lecture Notes in Biomathematics # 20. S. Levin (Ed.) Springer-Verlag, Berlin.

Cushing, J. M. 2004. *Differential Equations: An Applied Approach*. Prentice Hall, Upper Saddle River, N.J.

Gantmacher, F. R. 1964. *Matrix Theory, Vol. II*. Chelsea Pub. Co., New York.

Gopalsamy, K. 1992. *Stability and Oscillations in Delay Differential Equations of Population Dynamics*. Kluwer Academic Pub., The Netherlands.

Gulick, D. 1992. *Encounters with Chaos*. McGraw-Hill, Inc., N. Y.

Jacquez, J. 1996. *Compartmental Analysis in Biology and Medicine*. 3rd ed. BioMedware, Ann Arbor, Mich.

Kaplan, D. and L. Glass. 1995. *Understanding Nonlinear Dynamics*. Springer-Verlag, New York.

Kuang, Y. 1993. *Delay Differential Equations with Applications in Population Dynamics*. Academic Press, Inc., San Diego.

Leonard, I. E. 1996. The matrix exponential. *SIAM Review* 39: 507–512.

Malthus, T. 1798. An essay on the principle of population, as it affects the future improvement of society, with remarks on the speculations of Mr. Godwin, M. Condorcet and other writers. J. Johnson, London.

Moler, C. and C. Van Loan. 1978. Nineteen dubious ways to compute the exponential of a matrix. *SIAM Review* 20: 801–836.

Moler, C. and C. Van Loan. 2003. Nineteen dubious ways to compute the exponential of a matrix, twenty-five years later. *SIAM Review* 45: 3–49.

Noble, B. 1969. *Applied Linear Algebra*. Prentice Hall, Inc., Englewood Cliffs, N.J.

Ortega, J. M. 1987. *Matrix Theory: A Second Course*. Plenum Press, New York.

Sánchez, D. 1968. *Ordinary Differential Equations and Stability Theory: An Introduction*. W. H. Freeman and Co., San Francisco.

Waltman, P. 1986. *A Second Course in Elementary Differential Equations*. Academic Press, New York.

Yeargers, E. K., R. W. Shonkwiler, and J. V. Herod. 1996. *An Introduction to the Mathematics of Biology*. Birkhäuser, Boston.

## 4.14 Appendix for Chapter 4

### 4.14.1 Exponential of a Matrix

The matrix exponential $e^{At}$ can be computed in several ways. We discuss two methods here.

If matrix $A$ is an $n \times n$ diagonalizable matrix, then, in theory, it is straightforward to compute the exponential of matrix $A$. Recall that a matrix $A$ is diagonalizable iff $A$ has $n$ linearly independent eigenvectors (Ortega, 1987). In addition, if all of the eigenvalues of $A$ are distinct, then the eigenvectors are linearly independent. Suppose matrix $A$ is diagonalizable and the eigenvalues of $A$ are $\lambda_i, i = 1, 2, \ldots, n$. Then $A^k$ can be expressed in terms of the eigenvectors of $A$,

$$A^k = H\Lambda^k H^{-1},$$

where $\Lambda = \text{diag}(\lambda_1, \lambda_2, \ldots, \lambda_n)$ and the columns of $H$ are the right eigenvectors of $A$, ordered corresponding to their associated eigenvalues. Then $e^{At}$ simplifies to

$$e^{At} = H\sum_{k=0}^{\infty} \Lambda^k \frac{t^k}{k!} H^{-1} = H\,\text{diag}(e^{\lambda_1 t}, e^{\lambda_2 t}, \ldots, e^{\lambda_n t})H^{-1}. \tag{4.18}$$

Another method for computing $e^{At}$ is due to Leonard (1996). This method does not require $A$ to be diagonalizable. However, it requires solving an $n$th-order differential equation. This is sometimes easier than solving the linear system $dX/dt = AX$ consisting of $n$ equations. Suppose $A$ is an $n \times n$ matrix with characteristic equation

$$\det(\lambda I - A) = \lambda^n + a_1\lambda^{n-1} + \cdots + a_n = 0.$$

This polynomial equation is also a characteristic equation of an $n$th-order scalar differential equation of the form

$$x^{(n)}(t) + a_1 x^{(n-1)}(t) + \cdots a_n x(t) = 0.$$

To find a formula for $e^{At}$ it is necessary to find $n$ linearly independent solutions to this $n$th-order scalar differential equation, $x_1(t), x_2(t), \ldots, x_n(t)$, with initial conditions

$$\left.\begin{matrix} x_1(0) = 1 \\ x_1'(0) = 0 \\ \vdots \\ x_1^{(n-1)}(0) = 0 \end{matrix}\right\}, \quad \left.\begin{matrix} x_2(0) = 0 \\ x_2'(0) = 1 \\ \vdots \\ x_2^{(n-1)}(0) = 0 \end{matrix}\right\}, \quad \ldots, \quad \left.\begin{matrix} x_n(0) = 0 \\ x_n'(0) = 0 \\ \vdots \\ x_n^{(n-1)}(0) = 1 \end{matrix}\right\}.$$

Then

$$e^{At} = x_1(t)I + x_2(t)A + \cdots + x_n(t)A^{n-1}, \quad -\infty < t < \infty. \tag{4.19}$$

Verification of equation (4.19) can be found in Leonard (1996). These latter two methods are illustrated in the following example.

**Example 4.19.** Suppose matrix $A$ is given by

$$A = \begin{pmatrix} 1 & -2 \\ -2 & 1 \end{pmatrix}.$$

The eigenvalues of $A$ are $\lambda_1 = -1$ and $\lambda_2 = 3$ with corresponding eigenvectors $(1,1)^T$ and $(-1,1)^T$, respectively. Matrix $A$ is diagonalizable. Both of the methods discussed previously can be applied. Matrices

$$H = \begin{pmatrix} 1 & -1 \\ 1 & 1 \end{pmatrix} \quad \text{and} \quad H^{-1} = \begin{pmatrix} \frac{1}{2} & \frac{1}{2} \\ -\frac{1}{2} & \frac{1}{2} \end{pmatrix}.$$

Then

$$e^{At} = H\begin{pmatrix} e^{-t} & 0 \\ 0 & e^{3t} \end{pmatrix} H^{-1}$$

$$= \begin{pmatrix} \frac{1}{2}[e^{-t} + e^{3t}] & \frac{1}{2}[e^{-t} - e^{3t}] \\ \frac{1}{2}[e^{-t} - e^{3t}] & \frac{1}{2}[e^{-t} + e^{3t}] \end{pmatrix} \tag{4.20}$$

To apply the method of Leonard given by (4.19), we find the characteristic polynomial of $A$: $\lambda^2 - 2\lambda - 3 = 0$. Then the corresponding second-order differential equation, $x''(t) - 2x'(t) - 3x(t) = 0$, has a general solution $x(t) = c_1 e^{-t} + c_2 e^{3t}$. Applying the initial conditions to find the constants $c_1$ and $c_2$, the solutions $x_1(t)$ and $x_2(t)$ are

$$x_1(t) = \frac{3}{4}e^{-t} + \frac{1}{4}e^{3t} \quad \text{and} \quad x_2(t) = -\frac{1}{4}e^{-t} + \frac{1}{4}e^{3t},$$

respectively. Then applying the identity (4.19) gives the solution

$$e^{At} = x_1(t)I + x_2(t)A$$

$$= \begin{pmatrix} \frac{1}{2}[e^{-t} + e^{3t}] & \frac{1}{2}[e^{-t} - e^{3t}] \\ \frac{1}{2}[e^{-t} - e^{3t}] & \frac{1}{2}[e^{-t} + e^{3t}] \end{pmatrix}$$

which agrees with (4.20).

### 4.14.2 Maple Program: Pharmacokinetics Model

The following Maple commands use the DEtools package to numerically approximate the solution to the pharmacokinetics model in Exercise 18.

```
> with(DEtools):
> a:=2*ln(2); b:=ln(2)/5;
> xeq:=diff(x(t),t)=-a*x(t)+2*sum(Heaviside(t-6*k),k=0..7)
      -2*sum(Heaviside(t-6*k-0.5),k=0..7);
> yeq:=diff(y(t),t)=a*x(t)-b*y(t);
> ic:=x(0)=0, y(0)=0;
> DEplot([xeq,yeq],[x(t),y(t)],t=0..48,[[ic]],
      stepsize=0.025,scene=[t,x]);
> DEplot([xeq,yeq],[x(t),y(t)],t=0..48,[[ic]],
      stepsize=0.025,scene=[t,y]);
```

Chapter 5

# NONLINEAR ORDINARY DIFFERENTIAL EQUATIONS: THEORY AND EXAMPLES

## 5.1 Introduction

Some theory and techniques useful in the analysis of nonlinear ordinary differential equations are introduced in this chapter. We concentrate on autonomous differential equations and systems. In particular, an equilibrium solution and local stability of an equilibrium solution for an autonomous system are defined. For a scalar differential equation, local stability is studied via a phase line diagram, and for a system of two differential equations, local stability is studied via a phase plane analysis. The classification scheme in Chapter 4 for analyzing the stability and behavior of solutions near the origin in two-dimensional linear systems, $dX/dt = AX$, is useful for nonlinear two-dimensional autonomous systems as well. An important theorem from differential equations is stated for two-dimensional autonomous systems the Poincaré-Bendixson Theorem. Under the conditions stated in the Poincaré-Bendixson Theorem, the asymptotic behavior of solutions can be predicted. In addition, Bendixson's and Dulac's criteria are stated, criteria that if satisfied by the autonomous system guarantees that the system will not have any periodic solutions.

We give a brief introduction to bifurcation theory for differential equations in Section 5.8. We show that three types of bifurcations are possible: saddle node, transcritical, and pitchfork bifurcations. This is the same as the classification scheme for scalar difference equations. However, for scalar differential equations, period-doubling bifurcations do not occur. For a system of two or more differential equations, there is a fourth type of bifurcation known as a Hopf bifurcation. When a Hopf bifurcation occurs, there exist periodic solutions.

Local stability in a nonlinear delay differential equation model is introduced in Section 5.9. We use the delay logistic model as an example. Some other techniques and theory useful in the study of differential equations are introduced in the remaining sections: qualitative matrix stability, Liapunov stability, and persistence theory. Qualitative matrix stability is another method that can be used to show local stability of a positive equilibrium using the properties of a matrix. Liapunov stability is a technique that can be used to show global stability of an equilibrium. Persistence theory is more general than stability theory and is particularly important for biological systems. A system is persistent if solutions stay away from zero (e.g., the population or system does not become extinct).

## 5.2 Basic Definitions and Notation

We discuss only first-order differential equations and systems because a higher-order differential equation of the form $d^n x/dt^n = g(d^{n-1}x/dt^{n-1}, \ldots, x)$ can be expressed as a first-order system. Recall that an *autonomous* system of differential equations has the form

$$\frac{dX}{dt} = F(X), \tag{5.1}$$

where $X = (x_1, \ldots, x_n)^T$, $F(X) = (f_1(x_1, \ldots, x_n), \ldots, f_n(x_1, \ldots, x_n))^T$, and $F$ does not depend explicitly on $t$. An initial value problem satisfies (5.1) with a given initial condition, $X(t_0) = X_0$. We will assume, unless stated otherwise, that a unique solution exists to initial value problems and that the interval of existence is $[t_0, \infty)$. For example, the following theorem gives sufficient conditions on the vector function $F$ for existence and uniqueness of solutions to initial value problems (see, e.g., Brauer and Nohel, 1969; Coddington and Levinson, 1955).

**Theorem 5.1** *Suppose $F$ and $\partial F/\partial x_i$ for $i = 1, \ldots, n$ are continuous functions of $(x_1, \ldots, x_n)$ on $\mathbf{R}^n$. Then a unique solution exists to the initial value problem*

$$\frac{dX}{dt} = F(X), \quad X(t_0) = X_0$$

*for any initial value $X_0 \in \mathbf{R}^n$.* □

Although a unique solution exists, the maximal interval of existence $[t_0, t)$ may be finite, $T < \infty$, unless the solution is bounded.

**Example 5.1** Let $dx/dt = x^2$ and $x(0) = x_0$. According to Theorem 5.1, there exists a unique solution for any initial value $x_0$. For this differential equation it is easy to find the solution by separation of variables:

$$x(t) = \frac{x_0}{1 - tx_0}.$$

However, the solution blows up at a finite time, $t = 1/x_0$. Thus, the interval of existence for the solution is $[0, 1/x_0)$ for $x_0 > 0$. ■

The solution $X(t) = (x_1(t), x_2(t), \ldots, x_n(t))$ describes parametrically a curve lying in $\mathbf{R}^n$. This curve is called a *trajectory* (*orbit* or *path*) of the system. The motion of the solution in $\mathbf{R}^n$ is described by the velocity $dX/dt$. The region $\mathbf{R}^n$, where the solution is graphed, is called *phase space* when $n = 3$, *phase plane* when $n = 2$, and *phase line* when $n = 1$.

An important fact about unique solutions to initial value problems is that any two distinct solution trajectories or orbits *cannot* intersect. We show that if solutions intersect they must be translations of each other. Therefore, any two solutions must either follow the same path in $\mathbf{R}^n$ or follow distinct paths that do not intersect. The following result follows from the existence and uniqueness theory.

**Corollary 5.1** *Suppose $F$ and $\partial F/\partial x_i$ for $i = 1, \ldots, n$ are continuous functions of $(x_1, \ldots, x_n)$ on $\mathbf{R}^n$. In addition, suppose $X_1(t)$ and $X_2(t)$ are two solutions satisfying the differential system $dX/dt = F(X)$ with the initial conditions*

$$X_1(0) = X_0 \quad \text{and} \quad X_2(t_0) = X_0.$$

*Then*

$$X_2(t) = X_1(t - t_0).$$

**Proof** Because of the assumptions on $F$, solutions to an initial value problem $dX/dt = F(X)$ for any initial conditions are unique. Let $Y(t) = X_1(t - t_0)$ and $u = t - t_0$. Then

$$\begin{aligned}\frac{dY(t)}{dt} &= \frac{dX_1(u)}{du}\frac{du}{dt} = \frac{dX_1(u)}{du} \\ &= F(X_1(u)) = F(X_1(t - t_0)) \\ &= F(Y(t)).\end{aligned}$$

In addition, $Y(t_0) = X_1(0) = X_0$. Because $Y$ satisfies the same IVP as $X_2$, by uniqueness, it follows that $Y(t) = X_2(t)$. The conclusion of the corollary holds. ◀

Next, an important type of solution is defined, a constant solution known as an equilibrium solution.

**Definition 5.1.** An *equilibrium solution* (*steady-state solution*, *fixed point*, or *critical point*) of the differential system (5.1) is a constant solution $\bar{X}$ satisfying

$$F(\bar{X}) = 0.$$

**Example 5.2** Equilibrium solutions of the logistic differential equation,

$$\frac{dx}{dt} = rx\left(1 - \frac{x}{K}\right), \quad r, K > 0,$$

are $x = 0$ and $x = K$. ■

Denote the Euclidean distance between two points $X_1 = (x_{11}, \ldots, x_{1n})^T$ and $X_2 = (x_{21}, \ldots, x_{2n})^T$ in $\mathbf{R}^n$ as $\|X_1 - X_2\|_2 = \sqrt{\sum_{i=1}^n (x_{1i} - x_{2i})^2}$. We state the definition for local stability of an equilibrium solution to the system (5.1).

**Definition 5.2.** An equilibrium solution $\bar{X}$ of (5.1) is said to be *locally stable* if for each $\epsilon > 0$ there exists a $\delta > 0$ with the property that every solution $X(t)$ of (5.1) with initial condition $X(t_0) = X_0$,

$$\|X_0 - \bar{X}\|_2 < \delta,$$

satisfies the condition that

$$\|X(t) - \bar{X}\|_2 < \epsilon$$

for all $t \geq t_0$. If the equilibrium solution is not locally stable it is said to be *unstable*.

**Definition 5.3.** An equilibrium solution $\bar{X}$ is said to be *locally asymptotically stable* if it is locally stable and if there exists $\gamma > 0$ such that $\|X_0 - \bar{X}\|_2 < \gamma$ implies

$$\lim_{t\to\infty} \|X(t) - \bar{X}\|_2 = 0.$$

The following example illustrates a locally stable and locally asymptotically stable equilibrium solution to a linear system of differential equations.

**Example 5.3** Consider the initial value problem, $dX/dt = AX$, $X(0) = X_0$, where

$$A = \begin{pmatrix} a_{11} & 0 \\ 0 & a_{22} \end{pmatrix}.$$

The equilibrium solution to this system is the origin, $X = \mathbf{0}$, and the solution $X(t) = (x(t), y(t))^T$ satisfies

$$X(t) = x_0 e^{a_{11}t} \begin{pmatrix} 1 \\ 0 \end{pmatrix} + y_0 e^{a_{22}t} \begin{pmatrix} 0 \\ 1 \end{pmatrix},$$

where $X_0 = (x_0, y_0)^T$. If $a_{ii} \leq 0$, $i = 1, 2$, then $\|X_0\|_2 < \epsilon$ implies $\|X(t)\|_2 < \epsilon$, and it follows that the zero equilibrium is locally stable. If $a_{ii} < 0$, $i = 1, 2$, then the zero equilibrium is locally asymptotically stable. However, the zero equilibrium is more than just locally stable, it is globally asymptotically stable; that is, solutions converge to zero for any initial condition. If $a_{ii} > 0$ for some $i$, then the zero equilibrium is unstable (a saddle point or unstable node). For example, if $a_{11} > 0$ and $x_0 > 0$, then $\lim_{t\to\infty} x(t) = \infty$. There exists $\epsilon > 0$ such that for any $\delta > 0$ and initial condition $\|X_0\|_2 < \delta$ with $x_0 > 0$, $\|X(t)\|_2 > \epsilon$ for some $t$. ■

Another important type of solution is a periodic solution. In general, a periodic solution is bounded and continuous, so the interval of existence is $(-\infty, \infty)$.

**Definition 5.4.** A *periodic solution* of the differential system (5.1) is a nonconstant solution $X(t)$ satisfying $X(t + T) = X(t)$ for all $t$ on the interval of existence for some $T > 0$. The minimum value of $T > 0$ is called the *period* of the solution.

**Example 5.4** The following initial value problem

$$\frac{dx}{dt} = y, \quad \frac{dy}{dt} = -x, \quad x(0) = 0, \quad \text{and} \quad y(0) = 1$$

has the unique solution $x(t) = \sin(t)$ and $y(t) = \cos(t)$. The solution is periodic with period $2\pi$. ■

An interesting result about autonomous scalar equations is that they cannot have periodic solutions. Periodic solutions require two or more differential equations. This behavior is distinct from scalar difference equations.

**Theorem 5.2** *Assume $f(x)$ is continuous for $x \in (-\infty, \infty)$. Then the autonomous differential equation $dx/dt = f(x)$ has no periodic solutions.*

**Proof** Suppose $x(t)$ is a periodic solution with period $T > 0$ ($dx/dt \not\equiv 0$ and $dx/dt$ is continuous because $f$ is continuous). Multiplying both sides of the differential equation by $dx/dt$ and integrating from $t$ to $t + T$ yields

$$\int_t^{t+T} \left(\frac{dx(s)}{ds}\right)^2 ds = \int_t^{t+T} f(x(s))\frac{dx(s)}{ds}\, ds = \int_{x(t)}^{x(t+T)} f(u)\, du,$$

where a change of variables is made in the integral $u = x(s)$. Because $x(t + T) = x(t)$, the last integral on the right side must be zero. However, the integral on the left side is strictly positive because $(dx/dt)^2$ is not identically zero. Thus, we have a contradiction. □

Periodic solutions can be asymptotically stable, stable, or unstable. In the phase plane, periodic solutions are represented by closed curves. Therefore, if the initial value of a solution is in the interior of the closed curve, the solution may or may not approach the curve. Likewise, if the initial value of a solution is exterior to the closed curve, the solution may or may not approach the curve. Therefore, a periodic solution can be asymptotically stable, stable, or unstable from the interior or from the exterior. The solution in Example 5.4 is stable. All solutions to the differential equation in this example are periodic solutions with the exception of the equilibrium point at the origin. Therefore, the equilibrium point is stable and the periodic solutions are stable (often referred to as neutrally stable).

## 5.3 Local Stability in First-Order Equations

We give a simple criterion for determining the local asymptotic stability of an equilibrium solution to a first order autonomous scalar differential equation. Suppose the autonomous differential equation

$$\frac{dx}{dt} = f(x)$$

has an equilibrium at $\bar{x}$. We use Taylor's formula to expand about the equilibrium $\bar{x}$. Let $u(t) = x(t) - \bar{x}$ and assume $f$ has two continuous derivatives in an interval containing $\bar{x}$. Then $du/dt = dx/dt$. Expanding $f$ about $\bar{x}$ using Taylor's formula with remainder yields

$$\frac{du}{dt} = f(\bar{x}) + f'(\bar{x})(x - \bar{x}) + f''(\xi)\frac{(x - \bar{x})^2}{2},$$

where $\xi$ is some number between $x$ and $\bar{x}$. Since $\bar{x}$ is an equilibrium, $f(\bar{x}) = 0$. The *linearization* of the above equation is then defined to be

$$\frac{du}{dt} = f'(\bar{x})u. \tag{5.2}$$

Equation (5.2) approximates the solution dynamics of the original differential equation for $x$ near $\bar{x}$ since the term with $(x - \bar{x})^2$ can be made sufficiently small. The linear equation (5.2) determines the local stability of $\bar{x}$. When $f'(\bar{x}) < 0$, the linearized solution approaches zero. Hence, when $x(t_0)$ is near $\bar{x}$, $x(t)$ approaches $\bar{x}$. Also, when $f'(\bar{x}) > 0$, there exist initial conditions where $x(t)$ does not approach $\bar{x}$.

**Theorem 5.3** *Suppose $f'$ is continuous on an open interval $I$ containing $\bar{x}$, where $\bar{x}$ is an equilibrium of $dx/dt = f(x)$. Then $\bar{x}$ is locally asymptotically stable if*

$$f'(\bar{x}) < 0$$

*and unstable if $f'(\bar{x}) > 0$.* □

For a rigorous proof of this result, please consult Hale and Koçak (1991). Stability is indeterminate in the case $f'(\bar{x}) = 0$. As was shown for difference equations, higher-order terms in the Taylor series expansion need to be considered when $f'(\bar{x}) = 0$. For example, if $f'(\bar{x}) = 0$ and $f''(\bar{x}) \neq 0$, the linearization yields

$$\frac{du}{dt} = \frac{f''(\bar{x})}{2}u^2.$$

The equilibrium $u = 0$ is unstable (see Exercise 1). The equilibrium dynamics (stability and instability) as a function of a parameter are studied in Section 5.8.

**Definition 5.5.** The equilibrium $\bar{x}$ of $dx/dt = f(x)$ is called *hyperbolic* if $f'(\bar{x}) \neq 0$. Otherwise, it is called *nonhyperbolic.*

The value of $f'(\bar{x})$ is known as the *eigenvalue* of the linearized equation. Nonhyperbolic equilibria have a zero eigenvalue, and hence their local stability is indeterminate.

### 5.3.1 Application to Population Growth Models

Some well-known equations for population growth are used to illustrate the stability of the equilibria. Malthusian or exponential growth is described first, then Gompertz and logistic growth. The last example describes what has been referred to as an Allee effect, a threshold population level below which the population growth is limited.

**Example 5.5** **(Malthusian Growth).** In *Malthusian growth*, the rate of growth is proportional to the population size. The name "Malthusian" honors the contributions of the English economist Thomas Malthus (1755–1845) to the theory of population growth. Let $a = b - d$ the birth rate minus the death rate. Then

$$\frac{dx}{dt} = ax = f(x).$$

This differential equation is linear and $f'(x) = a$. The equilibrium is at zero and the solution is

$$x(t) = x(0)e^{at}.$$

There is exponential growth if $a > 0$ and exponential decay if $a < 0$. The zero equilibrium is asymptotically stable if $a < 0$ and unstable if $a > 0$. The *doubling time* ($a > 0$) is the time $t$ such that $x(t) = 2x(0)$ or $t = \ln 2/a$. The *half-life* ($a < 0$) is the time $t$ such that $x(t) = 0.5x(0)$ or $t = -\ln 2/a$. Note that the doubling time and half-life are inversely proportional to the reproductive constant $a$. In general, the reproductive rate may depend on time or population size, $a(t)$ or $a(x)$, so that the rate of growth is nonlinear. ■

**Example 5.6** **(Gompertz Growth).** The *Gompertz* growth model has been applied to the growth of solid tumors and is named for the English actuary Benjamin Gompertz, 1779–1865 (see Applebaum, 2001). Solid tumors do not grow exponentially with time. As the tumor becomes larger, the doubling time of the total tumor volume continuously increases. Let $x(t)$ = volume of dividing cells at time $t$. Then $x(t)$ satisfies the differential equation

$$\frac{dx}{dt} = k\exp(-\alpha t)x \quad \text{or} \quad \frac{dx}{dt} = a(t)x, \quad \text{where} \quad \frac{da}{dt} = -\alpha a$$

and $a(t) = k\exp(-\alpha t)$, $\alpha > 0$. This equation is linear and nonautonomous. Theorem 5.3 cannot be applied. There exists one equilibrium, the zero equilibrium. The solution to the preceding differential equation can be found by separation of variables,

$$x(t) = x(0)\exp\left(\frac{k}{\alpha}[1 - \exp(-\alpha t)]\right).$$

After solving for $\exp(-\alpha t)$, the reproductive rate $a$ can be written as a function of tumor size $x$: $a(x(t)) = k - \alpha\ln(x(t)/x(0))$, making the differential equation nonlinear. As $t \to \infty$, $a(t) \to 0$, $x(t) \to x(0)\exp(k/\alpha)$, and the doubling time, $\ln 2/a \to \infty$. The zero equilibrium is unstable. Some applications of the Gompertz equation to tumor or cancer cell growth include Aroesty et al. (1973), Gyllenberg and Webb (1989), Kozusko and Bajzer (2003), Laird (1964) and references therein. In Figure 5.1, the logarithm of $x(t)/x(0)$ is graphed, $\ln[x(t)/x(0)] = (k/\alpha)[1 - \exp(-\alpha t)]$. ■

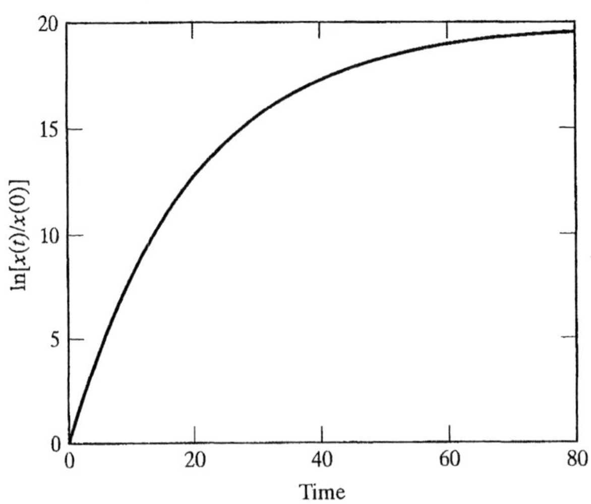


**Figure 5.1** Solution $\ln[x(t)/x(0)]$ to the Gompertz differential equation.

**Example 5.7** **(Logistic Growth).** The differential equation for logistic growth is one of the most well-known models for population growth. This equation was first introduced in 1837 by the Belgian mathematician Pierre Verhulst (1804–1849). The differential equation is often referred to as *Verhulst's equation*. In this model, it is assumed that the rate of growth $a$ is a decreasing function of the population size, $a(x) = r(1 - x/K)$. For population sizes greater than $K$, the rate of growth is negative and if $x < K$ it is positive. The logistic differential equation is

$$\frac{dx}{dt} = r\left(1 - \frac{x}{K}\right)x,$$

where $r > 0$ is the intrinsic growth rate and $K > 0$ is the carrying capacity. As shown in Example 5.2, there are two equilibria, $x = 0$ and $x = K$. The differential equation can be solved by separation of variables

$$x(t) = \frac{x(0)K}{x(0) + (K - x(0))e^{-rt}}. \tag{5.3}$$

If $x(0) > 0$, the population size approaches the carrying capacity, $\lim_{t\to\infty} x(t) = K$. Thus, for positive initial conditions, the equilibrium $x = K$ is globally asymptotically stable. The zero equilibrium is unstable. Since $f'(0) = r > 0$ and $f'(K) = -r$, where $f'(x) = r(1 - 2x/K)$, Theorem 5.3 shows that the zero equilibrium is unstable and the positive equilibrium $K$ is locally asymptotically stable. However, direct calculation of the solution shows that the behavior is more than just local since all solutions with positive initial conditions approach $K$. See Figure 5.2. ■

**Example 5.8** **(Allee Effect).** The Allee effect is named for Warder C. Allee, a U.S. zoologist known for his research on social behavior and aggregation of animals (Allee, 1931). The Allee effect refers to reduced fitness or decline in population growth at low population sizes or densities. At low population densities the population is so widely dispersed that reproductive contacts are restricted and infrequent (Hallam and Levin, 1986). In population models, the Allee effect is often modeled as a threshold, below which there is population extinction.

Suppose the per capita reproductive rate is a quadratic function of the population size or density $x$: $a(x) = a_1 + a_2x + a_3x^2$ (as opposed to linear in

**Figure 5.2** Solutions to the logistic differential equation with $K = 100$ and various initial conditions.

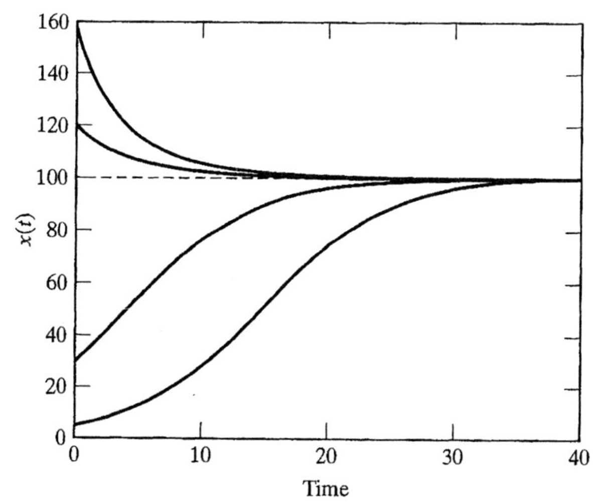

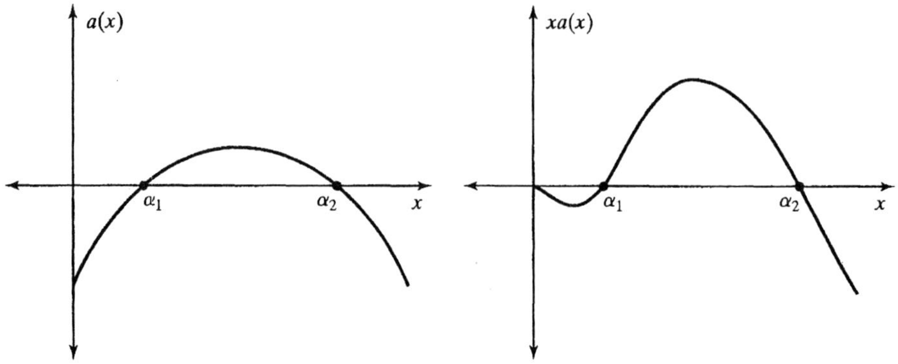


**Figure 5.3** Schematics of $a(x)$ (left) and $xa(x)$ (right) for the Allee differential equation.

the logistic model). Then a population growth model with an Allee effect may have the form

$$\frac{dx}{dt} = xa(x) = x(a_1 + a_2 x + a_3 x^2),$$

where $a_1 < 0, a_2 > 0$ and $a_3 < 0$. There is a positive maximum intrinsic growth rate at intermediate densities so that $a(x) = a_3(x - \alpha_1)(x - \alpha_2), 0 < \alpha_1 < \alpha_2$. This model has three equilibria, $x = 0, x = \alpha_1$, and $x = \alpha_2$. We show that $x = 0$ and $x = \alpha_2$ are locally asymptotically stable and $x = \alpha_1$ is unstable. An implicit solution for the differential equation can be found via separation of variables.

Differentiating $f'(x) = a_3[(x - \alpha_2)x + (x - \alpha_1)x + (x - \alpha_1)(x - \alpha_2)]$. Evaluating the derivative at $\bar{x} = 0$ gives $f'(0) = a_3(-\alpha_1)(-\alpha_2) < 0$ because $a_3 < 0$, $\alpha_1 > 0$, and $\alpha_2 > 0$. Thus $\bar{x} = 0$ is locally asymptotically stable. Evaluating the derivative at $\bar{x} = \alpha_1$ gives $f'(\alpha_1) = a_3(\alpha_1 - \alpha_2)\alpha_1 > 0$, which implies $\bar{x} = \alpha_1$ is unstable. Evaluating the derivative at $\bar{x} = \alpha_2$ gives $f'(\alpha_2) = a_3(\alpha_2 - \alpha_1)\alpha_2 < 0$, which implies $\bar{x} = \alpha_2$ is locally asymptotically stable. Graphs of $a(x)$ and $xa(x) = dx/dt$ are given in Figure 5.3. ■

## 5.4 Phase Line Diagrams

There is an alternate geometric approach that is useful for checking local stability in the case of first-order autonomous differential equations. It involves graphing the direction of flow on a phase line diagram.

The general form for a first-order nonautonomous differential equation is

$$\frac{dx}{dt} = f(x, t).$$

Note that $f(x, t)$ is the slope of the line tangent to the solution $x(t)$ at the point $(t, x)$. We can use this information to construct a *direction field* in the $t$-$x$ plane, the set of all directions or slopes given by $f(x, t)$ at the point $(t, x)$. Figure 5.4 shows the graph of the direction field for the ordinary differential equation $dx/dt = x^2 - t = f(x, t)$. Graphs of a direction field are generally too tedious to draw by hand. The direction fields in Figures 5.4 and 5.5 were generated using the MATLAB program *pplane6* written by John C. Polking at Rice University.

For *autonomous* differential equations, $dx/dt = f(x)$, the direction of flow does not change with $t$. For autonomous equations, it is only necessary to determine the direction of flow on the $x$-axis. In particular, if the sign of $dx/dt$ is positive, the direction of flow is in the positive direction, and if it is negative, the flow is in the negative direction.

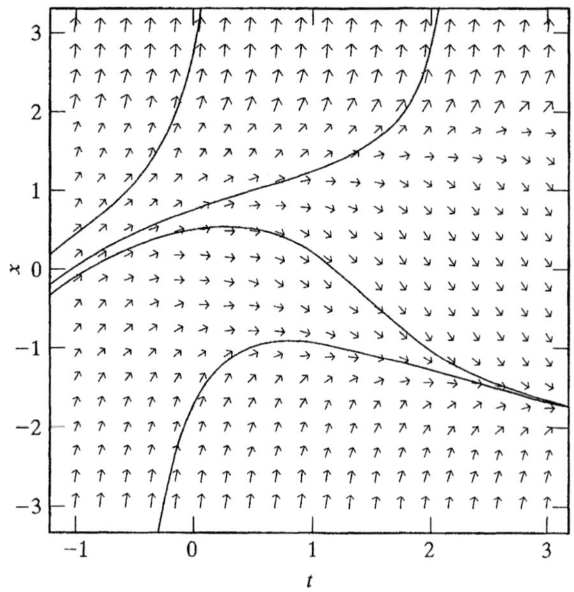


**Figure 5.4** Direction field and solution trajectories for $dx/dt = x^2 - t = f(x, t)$.

**Example 5.9** Consider the following autonomous differential equation:

$$\frac{dx}{dt} = x(1 - x)(2 - x) = f(x).$$

The direction field is graphed in Figure 5.5. By projecting the direction of flow onto the $x$-axis, a *one-dimensional phase portrait* or a *phase line diagram* is obtained. ■

For autonomous equations, is not necessary to draw the direction field in the $t$-$x$ plane before drawing the phase line diagram. One may just graph $dx/dt = f(x)$. When $f(x) > 0$, then $x$ is increasing and when $f(x) < 0$, then $x$ is decreasing. In the case $f(x) = 0$, $x$ is an equilibrium; there is no change in the dynamics. In Example 5.9, the equilibrium $\bar{x} = 1$ is locally asymptotically stable

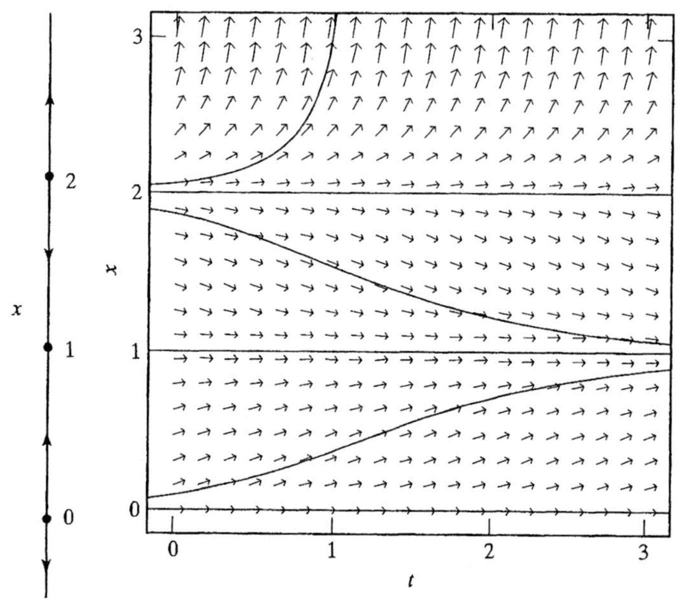


**Figure 5.5** Direction field, solution trajectories, and phase line diagram for $dx/dt = x(1 - x)(2 - x)$.

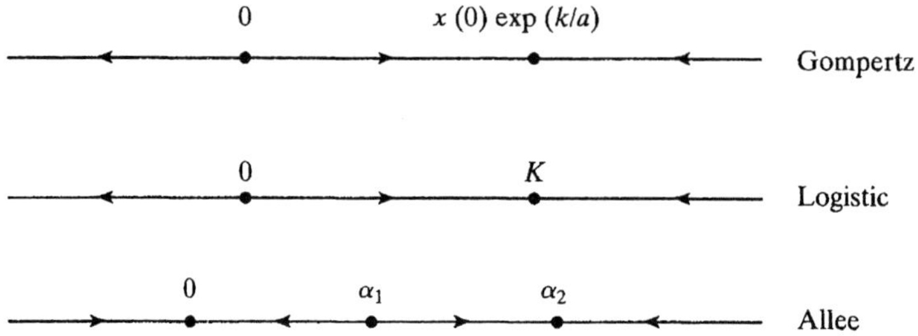


**Figure 5.6** Phase line diagrams for the Gompertz, logistic, and Allee differential equations.

for initial conditions in the interval $(0, 2)$ and the equilibrium $\bar{x} = 2$ is unstable since initial conditions in the interval $(2, \infty)$, solutions tend to infinity.

Phase line diagrams corresponding to Examples 5.6, 5.7, and 5.8, that is, Gompertz growth, logistic growth, and Allee effect, respectively, are illustrated in the following example.

**Example 5.10** Gompertz growth satisfies $dx/dt = x(k - \alpha \ln(x/x(0)))$. The equilibria are $\bar{x} = 0$ and $\bar{x} = x(0)\exp(k/\alpha)$. If initial conditions $x(0)$ are in the interval $(0, \infty)$, then $\lim_{t\to\infty} x(t) = x(0)\exp(k/\alpha)$.

The equilibria for the logistic differential equation, $dx/dt = r(1 - x/K)x$, are $x = 0$ and $x = K$. If initial conditions lie in the interval $(0, \infty)$, then solutions satisfy $\lim_{t\to\infty} x(t) = K$.

The equilibria for the model with an Allee effect, $dx/dt = a_3 x(x - \alpha_1)(x - \alpha_2)$, are $0$, $\alpha_1$, and $\alpha_2$. If initial conditions lie in the interval $(-\infty, \alpha_1)$, then solutions satisfy $\lim_{t\to\infty} x(t) = 0$. However, if initial conditions lie in the interval $(\alpha_1, \infty)$, then $\lim_{t\to\infty} x(t) = \alpha_2$. The equilibria $0$ and $\alpha_2$ are locally asymptotically stable and $\alpha_1$ is unstable. Phase line diagrams for the Gompertz, logistic, and Allee differential equations are graphed in Figure 5.6. ■

## 5.5 Local Stability in First-Order Systems

Consider the first-order autonomous differential system with two variables, $x$ and $y$,

$$\frac{dx}{dt} = f(x, y), \tag{5.4}$$

$$\frac{dy}{dt} = g(x, y).$$

Recall that the *equilibria* (steady states, fixed points, or critical points) of this system are solutions $(\bar{x}, \bar{y})$ satisfying $f(\bar{x}, \bar{y}) = 0$ and $g(\bar{x}, \bar{y}) = 0$.

The local stability of an equilibrium is determined by the eigenvalues of the Jacobian matrix. The functions $f$ and $g$ are expanded using Taylor formula about the equilibrium, $(\bar{x}, \bar{y})$, where $u = x - \bar{x}$ and $v = y - \bar{y}$. Assume that $f$ and $g$ have continuous second-order partial derivatives in an open set containing the point $(\bar{x}, \bar{y})$. Then

$$\begin{aligned}\frac{du}{dt} &= f(\bar{x}, \bar{y}) + f_x(\bar{x}, \bar{y})u + f_y(\bar{x}, \bar{y})v \\ &\quad + f_{xx}(\bar{x}, \bar{y})\frac{u^2}{2} + f_{xy}(\bar{x}, \bar{y})uv + f_{yy}(\bar{x}, \bar{y})\frac{v^2}{2} + \dots,\end{aligned}$$

$$\frac{dv}{dt} = g(\bar{x}, \bar{y}) + g_x(\bar{x}, \bar{y})u + g_y(\bar{x}, \bar{y})v$$
$$+ g_{xx}(\bar{x}, \bar{y})\frac{u^2}{2} + g_{xy}(\bar{x}, \bar{y})uv + g_{yy}(\bar{x}, \bar{y})\frac{v^2}{2} + \ldots,$$

where, for example, $f_x(\bar{x}, \bar{y})$ means $\partial f(x, y)/\partial x|_{x=\bar{x}, y=\bar{y}}$. We used the fact that $f(\bar{x}, \bar{y}) = 0$ and $g(\bar{x}, \bar{y}) = 0$. The *system linearized about the equilibrium* $(\bar{x}, \bar{y})$ is

$$\frac{dZ}{dt} = JZ,$$

where $Z = (u, v)^T$ and $J$ is the Jacobian matrix evaluated at the equilibrium,

$$J = \begin{pmatrix} f_x(x, y) & f_y(x, y) \\ g_x(x, y) & g_y(x, y) \end{pmatrix}\Bigg|_{x=\bar{x},\, y=\bar{y}}.$$

From Chapter 4, we know that solutions to the linear system $dZ/dt = JZ$ approach zero iff the eigenvalues have negative real parts. Also, from the Routh-Hurwitz criteria, the eigenvalues have negative real parts iff the coefficients of the characteristic polynomial are both positive. Since the characteristic polynomial of matrix $J$ is

$$\lambda^2 - \mathrm{Tr}(J)\lambda + \det(J),$$

then the eigenvalues have negative real parts iff $\mathrm{Tr}(J) < 0$ and $\det(J) > 0$. The stability results are summarized in the following theorem. Stability only requires that the first-order partial derivatives of $f$ and $g$ be continuous.

**Theorem 5.4** *Assume the first-order partial derivatives of f and g are continuous in some open set containing the equilibrium* $(\bar{x}, \bar{y})$ *of system (5.4). Then the equilibrium is locally asymptotically stable if*

$$\mathrm{Tr}(J) < 0 \quad \text{and} \quad \det(J) > 0,$$

*where J is the Jacobian matrix evaluated at the equilibrium. In addition, the equilibrium is unstable if either* $\mathrm{Tr}(J) > 0$ *or* $\det(J) < 0$. □

We use the classification scheme developed in Chapter 4 for linear systems to identify the types of equilibria (node, saddle, spiral) for nonlinear systems. Because the linearization is only an approximation to the nonlinear system, the nonlinear system may behave differently from the linear system in three cases.

(i) $\det(J) = 0$. There is at least one zero eigenvalue. In the linear system, the equilibria are not isolated. In the nonlinear system, this may be the case also. If there is an isolated equilibrium, it may be a node, spiral, or saddle.

(ii) $\mathrm{Tr}(J) = 0$ and $\det(J) > 0$. The eigenvalues are purely imaginary. The equilibrium is a center in the linear system but may be a center or spiral in the nonlinear system.

(iii) $\mathrm{Tr}(J)^2 = 4\det(J)$. This represents the borderline between complex and real eigenvalues. Therefore, in the nonlinear system the equilibrium could be a node or a spiral.

The reason for ambiguity in these cases is clear from the classification scheme in $\tau$-$\delta$ parameter space (see Figure 5.7). The curves $\det(J) = 0$ $(\delta = 0)$,

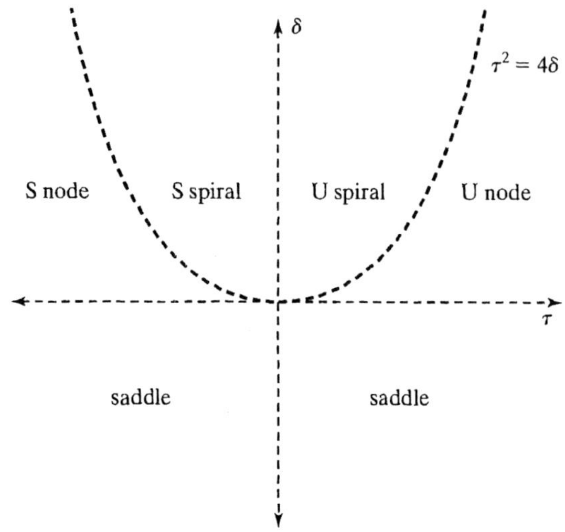


**Figure 5.7** Stability diagram in the $\tau$-$\delta$ plane, $\tau = \text{Tr}(J)$, and $\delta = \det(J)$. The dashed curves represent the three cases, (i)–(iii).

$\text{Tr}(J) = 0$ and $\det(J) > 0$ ($\tau = 0$ and $\delta > 0$), and $\text{Tr}(J)^2 = 4\det(J)$ ($\tau^2 = 4\delta$) are transition regions between two types of behavior.

> **Definition 5.6.** Let $\bar{X}$ denote an equilibrium of the system $dX/dt = F(X)$ and $J$ denote the Jacobian matrix of $F(X)$ evaluated at $\bar{X}$. Then the equilibrium $\bar{X}$ is said to be *hyperbolic* if the eigenvalues of the Jacobian matrix $J$ have nonzero real parts. Otherwise, it is said to be *nonhyperbolic*.

In the case of nonhyperbolic equilibria, the local stability criteria are indeterminate. When the $\text{Tr}(J) = 0$ and the $\det(J) > 0$ or when the $\det(J) = 0$, matrix $J$ either has complex conjugate eigenvalues with zero real parts or has a zero eigenvalue. In either case, the equilibrium is nonhyperbolic.

**Example 5.11** Consider a predator-prey model,

$$\frac{dx}{dt} = x\left(r - r\frac{x}{K} - ay\right), \quad r, K, a > 0,$$

$$\frac{dy}{dt} = y(-b + cx), \quad b, c > 0.$$

The variable $x$ represents the density of the prey species and the variable $y$ the predator. It is assumed, in the absence of the predator, that the prey grows logistically. Also, in the absence of the prey, the predator dies out (exponentially). The term $ay$ represents the per capita loss of prey to the predator and $cx$ represents the per capita gain to the predator.

First, it should be noted that if $x(0) = 0$, then $x(t) = 0$ for all time and if $y(0) = 0$, then $y(t) = 0$ for all time. Since solutions to this predator-prey model are unique, no solution beginning with $x(0) > 0$ and $y(0) > 0$ will cross the $x$- or $y$-axes. Thus, we say that the positive quadrant, where $x \geq 0$ and $y \geq 0$, is *positively invariant*. That is, if $(x(0), y(0)) \in \mathbf{R}_+^2 = \{(x, y) | x \geq 0, y \geq 0\}$, then $(x(t), y(t)) \in \mathbf{R}_+^2$ for $t \geq 0$. It is important that solutions cannot become negative because the variables represent population densities. It should be verified that solutions remain nonnegative in all models of biological systems.

The predator-prey model has three equilibria:

$$(\bar{x}, \bar{y}) = (0, 0), \quad (\bar{x}, \bar{y}) = (K, 0),$$

and

$$(\bar{x}, \bar{y}) = \left(\frac{b}{c}, \frac{r(cK - b)}{acK}\right).$$

This latter equilibrium is positive only if $K > b/c$. Therefore, a positive equilibrium exists (both prey and predator survive) iff the carrying capacity of the prey $K$ is sufficiently large ($> b/c$). Otherwise the prey population is too small to support the predator population. If the carrying capacity $K \leq b/c$, then only the prey survives. The Jacobian matrix for this system has the form

$$J = \begin{pmatrix} r - 2\frac{r}{K}x - ay & -ax \\ cy & -b + cx \end{pmatrix}.$$

At the origin, the Jacobian is a diagonal matrix with eigenvalues $\lambda_{1,2} = r, -b$. Because one eigenvalue is positive and the other negative, the origin is a saddle point which is unstable.

At the equilibrium $(K, 0)$, the Jacobian matrix is upper triangular. Therefore, the eigenvalues are along the diagonal: $-r$ and $-b + cK$. If $K < b/c$, then both eigenvalues are negative and the equilibrium is a stable node. However, if $K > b/c$, then one eigenvalue is positive and one is negative so the equilibrium is a saddle point.

At the equilibrium $\bar{x} = b/c$ and $\bar{y} = r(cK - b)/(acK)$, where $K > b/c$, the Jacobian matrix is given by

$$J(\bar{x}, \bar{y}) = \begin{pmatrix} -\frac{rb}{Kc} & -\frac{ab}{c} \\ \frac{r(cK - b)}{aK} & 0 \end{pmatrix}.$$

It is straightforward to check that $\text{Tr}(J) < 0$ and $\det(J) > 0$. Thus, the positive equilibrium is locally asymptotically stable. The type of equilibrium, node or spiral, depends on the sign of the discriminant,

$$\gamma = \text{Tr}(J)^2 - 4\det(J) = \left[-\frac{rb}{cK}\right]^2 - 4\frac{rb(cK - b)}{cK}.$$

The discriminant may be positive, negative, or zero. If it is negative, then the positive equilibrium is a stable spiral, but if it is positive, then it is a stable node. In any case, if $K > b/c$, the positive equilibrium is locally asymptotically stable.

Summarizing the results, if $K < b/c$, then the equilibrium $(K, 0)$ is locally asymptotically stable (only the prey survives), and if $K > b/c$, then the equilibrium $(b/c, r(cK - b)/(acK))$ is locally asymptotically stable (both prey and predator survive). ■

For a system consisting of more than two differential equations, local asymptotic stability depends on the Routh-Hurwitz criteria described in Section 4.5. The stability criteria depend on the eigenvalues of the Jacobian matrix evaluated at $\bar{X}, J(\bar{X})$. If all of the eigenvalues have negative real parts, then the equilibrium is locally asymptotically stable. The eigenvalues are determined by finding the roots of the characteristic equation, then the Routh-Hurwitz criteria can be applied to show local asymptotic stability.

**Theorem 5.5** *Suppose $dX/dt = F(X)$ is a nonlinear first-order autonomous system with an equilibrium $\bar{X}$. Denote the Jacobian matrix of $F$ evaluated at $\bar{X}$ as $J(\bar{X})$. If the characteristic equation of the Jacobian matrix $J(\bar{X})$,*

$$\lambda^n + a_1\lambda^{n-1} + a_2\lambda^{n-2} + \cdots + a_{n-1}\lambda + a_n = 0,$$

*satisfies the conditions of the Routh-Hurwitz criteria in Theorem 4.4, that is, the determinants of all of the Hurwitz matrices are positive, $det(H_j) > 0$, $j = 1, \ldots, n$, then the equilibrium $\bar{X}$ is locally asymptotically stable. If $det(H_j) < 0$ for some $j = 1, \ldots, n$, then the equilibrium $\bar{X}$ is unstable.* □

**Example 5.12** Consider the three-dimensional competitive system,

$$\frac{dx_1}{dt} = x_1(a_{10} - a_{11}x_1 - a_{12}x_2 - a_{13}x_3),$$

$$\frac{dx_2}{dt} = x_2(a_{20} - a_{21}x_1 - a_{22}x_2 - a_{23}x_3),$$

$$\frac{dx_3}{dt} = x_3(a_{30} - a_{31}x_1 - a_{32}x_2 - a_{33}x_3),$$

Assume $a_{ij} > 0$, for $i = 1, 2, 3$ and $j = 0, 1, 2, 3$. Each species $x_i$ in the absence of the other two species grows logistically to a carrying capacity given by $a_{i0}/a_{ii}$. The presence of all three species decreases the rate of growth of each of the species; the species compete for common resources. Note that if $x_i(0) \geq 0$ for all $i$, then the solutions cannot become negative. The positive octant, $\mathbf{R}^3_+ = \{(x_1, x_2, x_3) : x_i \geq 0, i = 1, 2, 3\}$, is positively invariant. See Exercise 17.

This model has several equilibria: a zero equilibrium, three one-species equilibria, three two-species equilibria, and a three-species equilibrium. Note that the three-species equilibrium is found by solving $A\bar{X} = B$, where $A = (a_{ij})$ is a $3 \times 3$ matrix and $B = (a_{i0})$ is a $3 \times 1$ matrix. The solution $\bar{X}$ is unique if $A$ is nonsingular. (However, existence of a solution does not guarantee that the solution is nonnegative.)

We analyze the stability of the zero equilibrium: $\bar{x}_i = 0, i = 1, 2, 3$, and the one-species equilibria: $\bar{x}_i = a_{i0}/a_{ii}, \bar{x}_j = 0, j \neq i$. The Jacobian matrix evaluated at the zero equilibrium satisfies

$$J(0,0,0) = \begin{pmatrix} a_{10} & 0 & 0 \\ 0 & a_{20} & 0 \\ 0 & 0 & a_{30} \end{pmatrix}.$$

Since $a_{i0} > 0, i = 1, 2, 3$, are the eigenvalues of the matrix, it follows that the zero equilibrium is unstable. The Jacobian matrix evaluated at $\bar{x}_1 = a_{10}/a_{11}$ and $\bar{x}_2 = 0 = \bar{x}_3$ is

$$J\left(\frac{a_{10}}{a_{11}}, 0, 0\right) = \begin{pmatrix} -a_{11}\bar{x}_1 & -a_{12}\bar{x}_1 & -a_{13}\bar{x}_1 \\ 0 & a_{20} - a_{21}\bar{x}_1 & 0 \\ 0 & 0 & a_{30} - a_{31}\bar{x}_1 \end{pmatrix}.$$

Since the matrix is upper triangular, the eigenvalues are along the diagonal. If $\bar{x}_1 > \max\{a_{20}/a_{21}, a_{30}/a_{31}\}$, then the eigenvalues are all negative and the equilibrium $(a_{10}/a_{11}, 0, 0)$ is locally asymptotically stable. However, if the inequality is reversed, the equilibrium is unstable.

Due to the symmetry of the equations, the other two equilibria $\bar{x}_j = a_{j0}/a_{jj}$, $j = 2, 3$ and $\bar{x}_i = 0, i \neq j$ have a similar analysis. The equilibrium $\bar{x}_j = a_{j0}/a_{jj}$, $\bar{x}_i = 0,\ i \neq j$ is locally asymptotically stable if $\bar{x}_j > \max\{a_{k0}/a_{kj}, a_{l0}/a_{lj}\}$, $k, l \neq j$. ∎

For a system of two autonomous differential equations, there exists a powerful theory based on phase plane methods that allows us to analyze further the behavior of solutions. For some autonomous systems, this theory allows us to conclude for all positive initial conditions that solutions converge to an equilibrium.

## 5.6 Phase Plane Analysis

In this section and the next, the analysis is restricted to two-dimensional autonomous systems of the form

$$\frac{dx}{dt} = f(x, y) \quad \text{and} \quad \frac{dy}{dt} = g(x, y). \tag{5.5}$$

It is assumed that $f$ and $g$ have continuous first-order partial derivatives. Solutions to initial value problems exist and are unique. There is only one solution curve or trajectory passing through a given point $(x_0, y_0)$ in the $x$-$y$ plane. The solution curve $(t, x(t), y(t))$ in three dimensions doesn't need to be considered because the direction of flow is independent of $t$. Thus, we need to consider only the solution curve or trajectory $(x(t), y(t))$ in the $x$-$y$ plane or in the *phase plane*. We have already seen that direction fields in the $t$-$x$ plane for an autonomous differential equation, $dx/dt = f(x)$, can be simplified to a phase line diagram. For two autonomous differential equations, the dynamics can be simplified to a phase plane. Solution curves or trajectories $(x(t), y(t))$ are parametric equations with $t$ acting as the parameter.

For any point $(x, y)$,

$$\frac{dy}{dx} = \frac{g(x, y)}{f(x, y)}$$

gives the slope of the trajectory in the $x$-$y$ plane and the tangent vector $(f(x, y), g(x, y))^T$ gives the direction of the trajectory. With the exception of equilibria (also called *singular points*), $(\bar{x}, \bar{y})$, where $f(\bar{x}, \bar{y}) = 0 = g(\bar{x}, \bar{y})$, there is a unique direction associated with each point $(x, y)$, given by the vector $(f(x, y), g(x, y))^T$. At equilibria or singular points, there is no direction because the flow stops at these points; they are fixed points. The collection of vectors defines a *direction field*. The direction field can be used as a visual aid in sketching a family of solution curves called a *phase plane portrait* or *phase plane diagram*. The task of creating a direction field for two autonomous equations can be very tedious unless it is computer generated. A more efficient method to determine direction of flow than by creating the entire direction field is to analyze the direction of flow along the $x$- and $y$-zero isoclines or nullclines.

> **Definition 5.7.** The *x-zero isocline* or *nullcline* for system (5.5) is the set of all points in the $x$-$y$ plane satisfying $f(x, y) = 0$. The *y-zero isocline* or *nullcline* is the set of all points satisfying $g(x, y) = 0$.

In general, an $x$ isocline is a curve in the $x$-$y$ plane satisfying $f(x, y) = c_1 =$ constant and a $y$ isocline is a curve satisfying $g(x, y) = c_2 =$ constant. The nullclines are the particular curves where the constants $c_1$ and $c_2$ are zero.

On an $x$-nullcline, the tangent vector $(0, g)$ is parallel to the $y$-axis. On the $y$-nullcline, the tangent vector $(f, 0)$ is parallel to the $x$-axis. Thus, we need to check only if the direction of flow is up or down ($\uparrow, \downarrow$) on the $x$ nullcline or left or right ($\leftarrow, \rightarrow$) on the $y$-nullcline. When the $x$- and $y$-nullclines intersect, there is an equilibrium which indicates no direction change in the tangent vector.

**Example 5.13** Consider the following autonomous system:

$$\frac{dx}{dt} = xy - y = y(x - 1) = f(x, y),$$

$$\frac{dy}{dt} = 2x - xy = x(2 - y) = g(x, y).$$

The $x$-nullcline is $y(x - 1) = 0$ so that $y = 0$ or $x = 1$. The $y$-nullcline is $x(2 - y) = 0$ so that $x = 0$ or $y = 2$. The equilibria are the points where $x$- and $y$-nullclines intersect [i.e., $(0, 0)$ and $(1, 2)$]. Next we determine the direction of flow along the nullclines. On the $x$-nullcline the direction of flow is either up or down depending on whether $g$ is positive or negative, respectively. On the $y$-nullcline the direction of flow is either right or left depending on whether $f$ is positive or negative, respectively. See Table 5.1.

The direction along the $x$-nullcline is graphed as short up or down arrows and the direction along the $y$-nullcline is graphed as short right or left arrows. See Figure 5.8. ■

It is important to note the following facts about the direction of flow. Direction vectors vary continuously from one point to the next except at equilibria. A change in orientation (up or down, right or left) can take place only at equilibria. When a nullcline passes through an equilibrium, the orientation of arrows is reversed if the Jacobian matrix $J$ satisfies $\det(J) \neq 0$. Local changes in behavior can take place

**Table 5.1** Direction of flow in the $x$-$y$ plane for the autonomous system in Example 5.13.

| $x$-nullcline | $g(x, y)$ | up $(+)$ | down $(-)$ |
|---|---|---|---|
| $x = 1$ | $2 - y$ | $y < 2$ | $y > 2$ |
| $y = 0$ | $2x$ | $x > 0$ | $x < 0$ |
| **$y$-nullcline** | **$f(x, y)$** | **right $(+)$** | **left $(-)$** |
| $x = 0$ | $-y$ | $y < 0$ | $y > 0$ |
| $y = 2$ | $2(x - 1)$ | $x > 1$ | $x < 1$ |

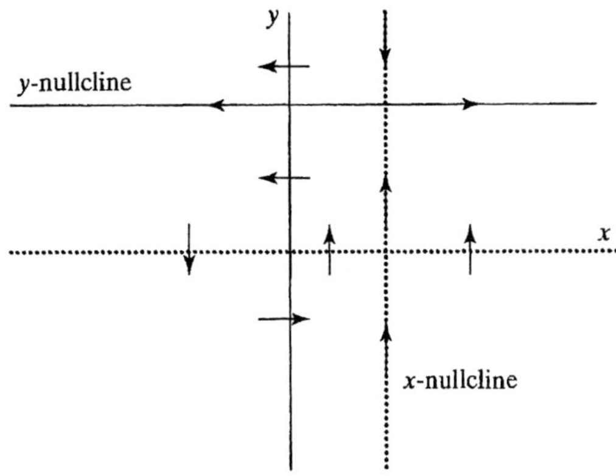


**Figure 5.8** Direction of flow along the $x$- and $y$-nullclines for the autonomous system in Example 5.13.

close to equilibria. Thus, it is important to know the local behavior of the system near each equilibrium (i.e., the behavior of the linearized system $\dot{Z} = JZ$, where $J$ is the Jacobian matrix evaluated at the equilibrium of interest). In this model, the Jacobian matrix is

$$J = \begin{pmatrix} y & x-1 \\ 2-y & -x \end{pmatrix}.$$

At the zero equilibrium,

$$J(0,0) = \begin{pmatrix} 0 & -1 \\ 2 & 0 \end{pmatrix}.$$

Since $\text{Tr}(J) = 0$ and $\det(J) > 0$, the local stability criterion tells us nothing; this is because the eigenvalues are purely imaginary, $\lambda_{1,2} = \pm i\sqrt{2}$. If this were a linear system, the origin would be a center, but for a nonlinear system, we need some additional information before determining stability. We will return to this example later when we examine Lotka-Volterra predator-prey systems and it will be shown using another method that the origin does behave as a center.

At the equilibrium $(1,2)$,

$$J(1,2) = \begin{pmatrix} 2 & 0 \\ 0 & -1 \end{pmatrix},$$

$\lambda_{1,2} = 2, -1$. Thus, the point $(1,2)$ is unstable; it is a saddle point. Also, note that $\text{Tr}(J) = 1 > 0$ and $\det(J) = -2 < 0$. Local asymptotic stability requires $\text{Tr}(J) < 0$ and $\det(J) > 0$. The eigenvector corresponding to $\lambda_1 = 2$ is $V_1 = (1,0)^T$ and the eigenvector corresponding to $\lambda_2 = -1$ is $V_2 = (0,1)^T$. The solution to the linearized system is

$$c_1 e^{2t}\begin{pmatrix} 1 \\ 0 \end{pmatrix} + c_2 e^{-t}\begin{pmatrix} 0 \\ 1 \end{pmatrix}.$$

It can be seen that there is a solution trajectory moving away from the equilibrium $(1, 2)$ in the horizontal direction [direction $(1,0)^T$, an *unstable manifold*] and that there is a solution trajectory moving toward $(1, 2)$ in the vertical direction [direction $(0,1)^T$, a *stable manifold*] because the equilibrium $(1,2)$ is a saddle point.

The behavior near an equilibrium in a nonlinear system looks similar to the behavior of the linearized system with the three exceptions (three cases listed after Theorem 5.4). The direction field and solution trajectories for this example are graphed in the phase plane in Figure 5.9.

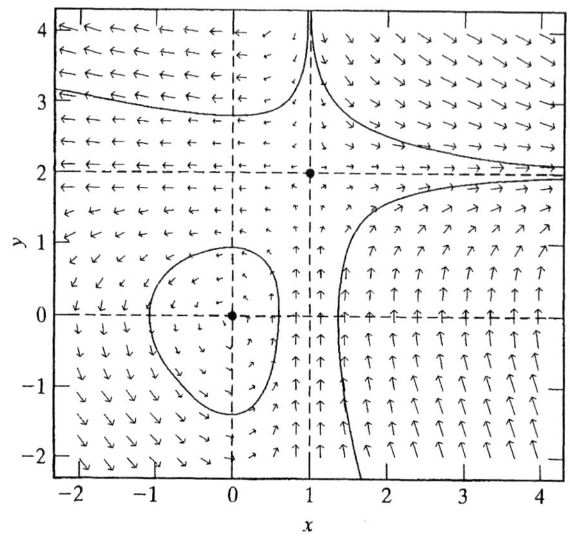

**Figure 5.9** Direction field and solution trajectories for the autonomous system in Example 5.13. The nullclines are the dashed curves and the solutions are the solid curves.

## 5.7 Periodic Solutions

Analytical results concerning periodic solutions for two-dimensional autonomous systems $dx/dt = f(x, y)$ and $dy/dt = g(x, y)$ are stated in this section. The first result is known as the Poincaré-Bendixson Theorem. This theorem states that bounded solutions whose limiting set does not contain any equilibria must approach a periodic solution. There are two other important theorems known as Bendixson's and Dulac's criteria. Each of these theorems give a criterion such that if it is satisfied, then the system will not have any periodic solutions.

### 5.7.1 Poincaré-Bendixson Theorem

Some terminology and notation are introduced in regard to a phase plane analysis. The term "trajectory" is used synonymously with *orbit* in the phase plane. The notation $\Gamma(X_0, t)$ is used to denote a solution trajectory as a function of time $t$ beginning at the initial point $X_0 = (x(t_0), y(t_0)) = (x_0, y_0)$. In addition, $\Gamma^+(X_0, t)$ denotes that part of the solution trajectory where $t \geq t_0$, a *positive orbit*, and $\Gamma^-(X_0, t)$ denotes that part of the solution where $t \leq t_0$, a *negative orbit*. If solutions are bounded, then their negative and positive orbits approach limiting sets as $t \to -\infty$ or as $t \to +\infty$. The *α-limit set*, denoted $\alpha(X_0)$, refers to the set of points in the plane that are approached by the negative orbit, $\Gamma^-(X_0, t)$, as $t \to -\infty$ [i.e., $(x_l, y_l) \in \alpha(X_0)$ iff there exists a sequence of decreasing times $\{t_i\}_{i=1}^{\infty}$, $t_i \to -\infty$ as $i \to \infty$, such that $\lim_{i\to\infty}(x(t_i), y(t_i)) = (x_l, y_l)$]. The *ω-limit set*, denoted $\omega(X_0)$, refers to the set of points in the plane that are approached by the positive orbit, $\Gamma^+(X_0, t)$, as $t \to \infty$.

A very important result in the theory of two-dimensional autonomous systems is known as the Poincaré-Bendixson Theorem. This theorem states conditions for existence of periodic solutions to the system (5.5). The names Poincaré and Bendixson refer to the contributions made by the well-known French mathematician Jules Henri Poincaré (1854–1912) and the Swedish mathematician Ivar O. Bendixson (1861–1935).

**Theorem 5.6** **(Poincaré-Bendixson Theorem).** *Let $\Gamma^+(X_0, t)$ be a positive orbit of (5.5) that remains in a closed and bounded region of the plane. Suppose the $\omega$-limit set does not contain any equilibria. Then either*

*(i) $\Gamma^+(X_0, t)$ is a periodic orbit ($\Gamma^+(X_0, t) = \omega(X_0)$) or*

*(ii) the $\omega$-limit set, $\omega(X_0)$, is a periodic orbit.* □

For a proof of this result, consult Coddington and Levinson (1955). An important consequence of this theorem is known as the Poincaré-Bendixson trichotomy, which states that bounded solutions containing only a finite number of equilibria can behave in one of only three ways (Coddington and Levinson, 1955; Smith and Waltman, 1995).

**Theorem 5.7** **(Poincaré-Bendixson Trichotomy).** *Let $\Gamma^+(X_0, t)$ be a positive orbit of (5.5) that remains in a closed and bounded region $B$ of the plane. Suppose $B$ contains only a finite number of equilibria. Then the $\omega$-limit set takes one of the following three forms:*

*(i) $\omega(X_0)$ is an equilibrium.*

*(ii) $\omega(X_0)$ is a periodic orbit.*

*(iii) $\omega(X_0)$ contains a finite number of equilibria and a set of trajectories $\Gamma_i$ whose $\alpha$- and $\omega$-limit sets consist of one of these equilibria for each trajectory $\Gamma_i$.* □

An important assumption in both of these theorems is that solutions are bounded. In Case (ii), if $\Gamma^+(X_0, t) \neq \omega(X_0)$ but approaches the periodic orbit, then the periodic orbit may be a *limit cycle.* In Case (iii), the limiting set is referred to as a *cycle graph.* The cycle graph may consist of either an equilibrium and a *homoclinic orbit* (connecting an equilibrium to itself) or several equilibria and *heteroclinic orbits* (connecting two different equilibria). Examples of $\omega$-limit sets are graphed in Figure 5.10.

An important fact is that a periodic orbit must enclose at least one equilibrium point. A periodic orbit changes direction as it follows a closed curve in the $x$-$y$ plane and a change in direction can only occur at an equilibrium point. Another important fact concerning periodic orbits is that if there exists exactly one equilibrium point inside a periodic orbit, it cannot be a saddle point. But it can be a node or a spiral. The direction of flow around a saddle point does not allow periodic orbits to encircle it. The direction of flow around any closed curve in the plane can be classified according to the *index* of a closed curve. [See, for example, Coddington and Levinson (1955) or Strogatz (2000).]

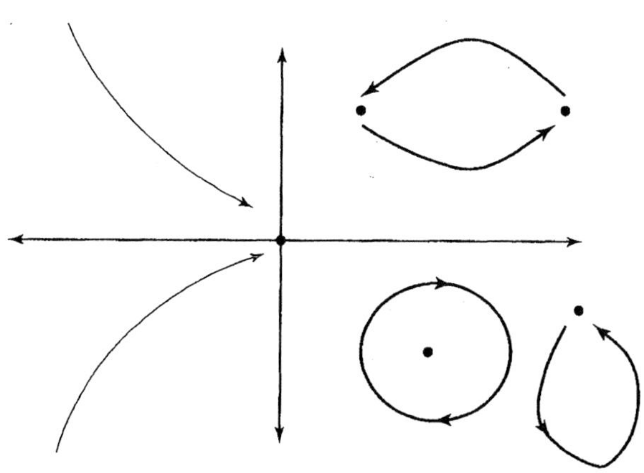

**Figure 5.10** Examples of $\omega$-limit sets in the phase plane.

**Example 5.14** Consider the following nonlinear system:

$$\frac{dx}{dt} = 8x - y^2, \quad \frac{dy}{dt} = -y + x^2. \tag{5.6}$$

The $x$-nullcline is $y^2 = 8x$ and the $y$-nullcline is $y = x^2$. The nullclines intersect at two equilibria $(0, 0)$ and $(2, 4)$. On the $x$-nullcline, $dy/dt$ satisfies

$$\left.\frac{dy}{dt}\right|_{y^2=8x} = -y + \frac{y^4}{64} = y\left(-1 + \frac{y^3}{64}\right).$$

When $y < 0$ or $y > 4$, then $dy/dt > 0$ and when $0 < y < 4$, $dy/dt < 0$. On the $y$-nullcline, $dx/dt$ satisfies

$$\left.\frac{dx}{dt}\right|_{y=x^2} = 8x - x^4 = x(8 - x^3).$$

When $x < 0$ or $x > 2$, then $dx/dt < 0$, and when $0 < x < 2$, then $dx/dt > 0$. The direction of flow along the nullclines is sketched in Figure 5.11.

Next, we determine the behavior near the equilibria. The Jacobian matrix

$$J = \begin{pmatrix} 8 & -2y \\ 2x & -1 \end{pmatrix}.$$

At the origin,

$$J(0, 0) = \begin{pmatrix} 8 & 0 \\ 0 & -1 \end{pmatrix}.$$

Because one eigenvalue is positive ($\lambda_1 = 8$) and one is negative ($\lambda_2 = -1$), the origin is a saddle point. Solutions move away from the origin along the $x$-axis (*unstable manifold*) and move toward the origin along the $y$-axis (*stable manifold*). This behavior can be seen if we solve the linear system $dZ/dt = J(0, 0)Z$,

$$Z = c_1 e^{8t}\begin{pmatrix} 1 \\ 0 \end{pmatrix} + c_2 e^{-t}\begin{pmatrix} 0 \\ 1 \end{pmatrix}.$$

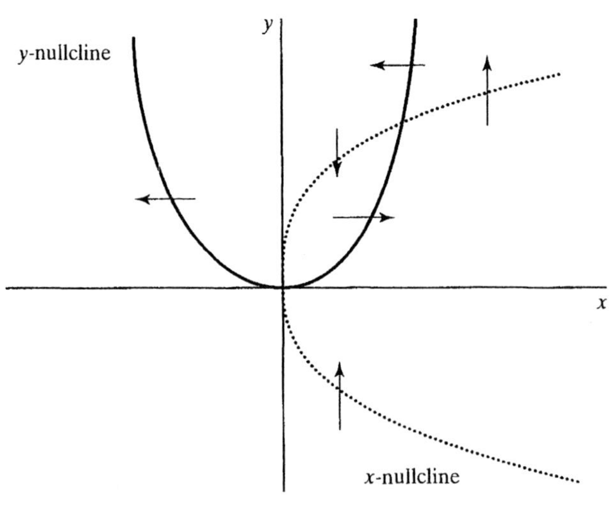


**Figure 5.11** Direction of flow along the nullclines for system (5.6).

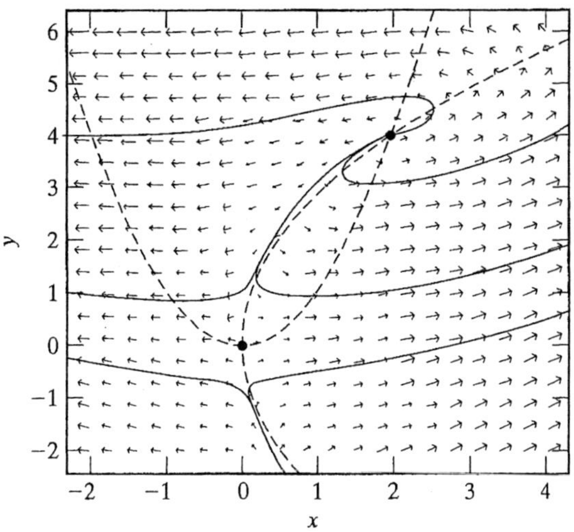

**Figure 5.12** Direction field and some solution trajectories for the system (5.6). The nullclines are the dashed curves and the solutions are the solid curves.

At the equilibrium (2, 4),

$$J(2,4) = \begin{pmatrix} 8 & -8 \\ 4 & -1 \end{pmatrix}.$$

The characteristic equation is $\lambda^2 - 7\lambda + 24 = 0$ or $\lambda_{1,2} = 7/2 \pm i\sqrt{47}/2$. The equilibrium (2, 4) is an unstable spiral point. The direction field and some solution trajectories are graphed in Figure 5.12 for the system (5.6). ■

The Poincaré-Bendixson Theorem can be applied to the nonlinear system in the last example if solutions are bounded. But it is easy to show that this is not the case for system (5.6). For $x(0) < 0$, $dx/dt < 0$ so that $x(t)$ is decreasing. In addition, $dx/dt < 8x$. Thus, $x(t) < x(0)e^{8t} \to -\infty$ as $t \to \infty$. Because solutions are not bounded, the Poincaré-Bendixson Theorem cannot be applied to check for periodic solutions. However, the next two results can be applied.

### 5.7.2 Bendixson's and Dulac's Criteria

Two important mathematical results give sufficient conditions that rule out the possibility of periodic solutions. They are Bendixson's criterion and Dulac's criterion. First, we define a simply connected set. A *simply connected set* $D \subset \mathbf{R}^2$ is a connected set having the property that every simple closed curve in $D$ can be continuously shrunk (within $D$) to a point (Rudin, 1974). For example, the entire plane, $\mathbf{R}^2$, is a simply connected set. Geometrically, a simply connected set is one without any holes.

**Theorem 5.8** **(Bendixson's Criterion).** *Suppose $D$ is a simply connected open subset of $\mathbf{R}^2$. If the expression $div(f, g) \equiv \partial f/\partial x + \partial g/\partial y$ is not identically zero and does not change sign in D, then there are no periodic orbits of the autonomous system (5.5) in D.*

**Proof** Assume that there is a periodic solution $C$ (a simple closed curve) in the simply connected region $D$. Let $S$ denote the interior of $C$. When $C$ is transversed counterclockwise, Green's Theorem in the plane (integration by parts in two dimensions) gives the following identity:

$$\int_C f(x,y)\,dy - g(x,y)\,dx = \iint_S \left(\frac{\partial f}{\partial x} + \frac{\partial g}{\partial y}\right) dx\,dy. \tag{5.7}$$

Note that the right-hand side does not equal zero by hypothesis. The autonomous system satisfies

$$\frac{dx}{dy} = \frac{f(x,y)}{g(x,y)} \quad \text{or} \quad g(x,y)\,dx = f(x,y)\,dy.$$

Thus, the integral on the left side of (5.7) must be zero, which leads to a contradiction. □

**Example 5.15** Bendixson's criterion can be applied to the system in Example 5.14. In this example, $f(x,y) = 8x - y^2$ and $g(x,y) = -y + x^2$. Let $D$ be any open region in $\mathbf{R}^2$. Then $\operatorname{div}(f,g) = 8 - 1 = 7 \neq 0$. Bendixson's criterion implies there are no periodic solutions in $D$. ■

A simple but important generalization of Bendixson's criterion is known as Dulac's criterion.

**Theorem 5.9** **(Dulac's Criterion).** *Suppose $D$ is a simply connected open subset of $\mathbf{R}^2$ and $B(x,y)$ is a real-valued $C^1$ function in $D$. If the expression*

$$\operatorname{div}(Bf, Bg) = \frac{\partial(Bf)}{\partial x} + \frac{\partial(Bg)}{\partial y}$$

*is not identically zero and does not change sign in $D$, then there are no periodic solutions of the autonomous system (5.5) in $D$.* □

The function $B$ is called a *Dulac function*. Dulac's criterion simplifies to Bendixson's criterion in the special case $B(x,y) \equiv 1$. There is no general method for determining an appropriate Dulac function for a given system. The difficulty in finding a Dulac function is similar to the difficulty in finding an appropriate "integrating factor" when solving differential equations (Hale and Koçak, 1991). Note that Dulac's and Bendixson's criteria give sufficient but not necessary conditions for the nonexistence of periodic solutions. If neither of these criteria are satisfied, there may or may not be periodic solutions.

**Example 5.16** Suppose $f$ and $g$ are linear functions:

$$\frac{dx}{dt} = ax + by,$$

$$\frac{dy}{dt} = cx + dy.$$

Applying Bendixson's criterion, $\partial f/\partial x + \partial g/\partial y = a + d$. If $a + d \neq 0$, then there are no periodic solutions in the entire plane. But if $a + d = 0$, then Bendixson's criterion does not apply. The following linear system

$$\frac{dx}{dt} = y,$$

$$\frac{dy}{dt} = -x,$$

has $a = 1$ and $d = -1$, so that Bendixson's criterion does not apply. By separating variables and solving for $x$ and $y$, this system satisfies $x^2 + y^2 = C =$ constant; the origin is a center. Every solution, not beginning at the origin, is a periodic solution. The following linear system

$$\frac{dx}{dt} = y,$$

$$\frac{dy}{dt} = x,$$

has $a = 0 = d$, so again Bendixson's criterion does not apply. However, for this system, there do not exist any periodic solutions; the origin is a saddle point ($x^2 - y^2 = C =$ constant). ∎

**Example 5.17** Consider the predator-prey model, where prey and predator grow logistically in the absence of the other species,

$$\frac{dx}{dt} = x(1 - ax - by),$$

$$\frac{dy}{dt} = y(1 + cx - dy),$$

where $a, b, c, d > 0$. Let $B(x, y) = 1/(xy)$. Note that $B$ is continuously differentiable in the positive quadrant, $D = \{(x, y)|x > 0, y > 0\}$. Thus, $\text{div}(Bx(1 - ax - by), By(1 + cx - dy)) = -a/y - d/x < 0$ in $D$. Dulac's criterion implies there does not exist any periodic solutions in $D$. ∎

The Poincarè-Bendixson Theorem and Dulac's and Bendixon's criteria apply only in two dimensions. However, there is a generalization of Dulac's criteria to three dimensions in some special cases (Busenberg and van den Driessche, 1990). This generalization involves finding a vector function $g$ such that along solutions, the dot product of the curl of $g$ and the unit normal vector, on the surface of a region in $\mathbf{R}^3_+$, is negative.

## 5.8 Bifurcations

If a parameter is allowed to vary, the dynamics of the differential system may change. An equilibrium can become unstable and a periodic solution may appear or a new stable equilibrium may appear making the previous equilibrium unstable. The value of the parameter at which these changes occur is known as a *bifurcation value* and the parameter that is varied is known as the *bifurcation parameter*. We discuss several types of bifurcations: saddle node, transcritical, pitchfork, and Hopf bifurcations. The first three types of bifurcations occur in scalar and in systems of differential equations. The fourth type, Hopf, does not occur in scalar differential equations because this type of bifurcation involves a change to a periodic solution. Scalar autonomous differential equations cannot have periodic solutions. Excellent introductions to the theory of nonlinear dynamical systems and bifurcation theory in differential equations include the books by Hale and Koçak (1991) and Strogatz (2000).

## 5.8.1 First-Order Equations

First, we discuss bifurcations in the case of scalar differential equations. Consider the scalar differential equation

$$\frac{dx}{dt} = f(x, r), \tag{5.8}$$

where $r$ is the bifurcation parameter and $\bar{x}(r)$ is an equilibrium solution which depends on $r$. There are three different types of bifurcations:

I. saddle node
II. pitchfork
III. transcritical

These three types of bifurcations occur in scalar difference equations also. However, in scalar difference equations, there is additional type of bifurcation known as a period-doubling bifurcation.

At the bifurcation value $\bar{r}$, it is the case that the equilibrium changes stability. In particular, for $r = \bar{r}$ and $x = \bar{x}(\bar{r})$,

$$\left.\frac{df(x,r)}{dx}\right|_{(x,r)=(\bar{x}(\bar{r}),\bar{r})} = 0. \tag{5.9}$$

We discuss briefly the dynamics for each of three types of bifurcations for scalar differential equations. The bifurcation dynamics are similar to the dynamics in the case of difference equations, discussed in Chapter 2. In a saddle node bifurcation, as the bifurcation parameter passes through the bifurcation point, two equilibria disappear, so that there are no equilibria afterward. One of the two equilibria is stable and the other one is unstable, before they disappear. This type of bifurcation is sometimes referred to as a *blue sky bifurcation* (Strogatz, 2000) because equilibria appear as "out of the clear blue sky." In a pitchfork bifurcation, there are two stable equilibria separated by an unstable equilibrium. A system where there are two different stable equilibria is said to have the property of *bistability*. When the bifurcation point is passed, there is only one stable equilibrium. This type of bifurcation is referred to as a *supercritical pitchfork bifurcation*. There is also a *subcritical pitchfork bifurcation*. In a subcritical pitchfork bifurcation, the stability is the reverse of the supercritical bifurcation, that is, there are two unstable equilibria separated by a stable equilibrium, until the bifurcation point is passed. Then there is only one unstable equilibrium. The diagram looks like a "pitchfork." The diagram in Figure 2.10 II is a supercritical pitchfork bifurcation. In a transcritical bifurcation, there are two equilibria, one stable and one unstable. When the bifurcation point is passed, there is an exchange of stability; the unstable equilibrium becomes stable and the stable one becomes unstable.

The following three examples are canonical examples of these three types of bifurcations.

I. $\dfrac{dx}{dt} = r + x^2$

II. $\dfrac{dx}{dt} = rx - x^3$

III. $\dfrac{dx}{dt} = rx + x^2$

In each case, the bifurcation value is at $r = 0$. At $r = 0$, there is a change in the stability of the equilibrium. The criterion in (5.9) is satisfied for $\bar{r} = 0$ and $\bar{x}(\bar{r}) = 0$. A bifurcation diagram illustrating bifurcations of type I, II, and III is the same as the one for difference equations. See Chapter 2, Figure 2.10.

**Example 5.18** Consider the canonical differential equation of type I: $dx/dt = r + x^2 = f(x, r)$. The equilibria satisfy $x^2 = -r$ or $\bar{x}(r) = \pm\sqrt{-r}$. When $r > 0$, there are no equilibria. When $r = 0$, there is one equilibrium. Finally, when $r < 0$, there are two equilibria. Also, $df(x, r)/dx = 2x$ evaluated at $\bar{x}$(r) equals $\pm 2\sqrt{-r}$. The positive equilibrium is unstable and the negative one is stable. There is a saddle node bifurcation at $\bar{r} = 0$. ■

These types of bifurcations also occur in higher-dimensional systems of differential equations. For example, in the two-dimensional system with equations

$$\frac{dx}{dt} = r + x^2,$$

$$\frac{dy}{dt} = -y,$$

there is a saddle node bifurcation at $r = 0$.

### 5.8.2 Hopf Bifurcation Theorem

A fourth type of bifurcation occurs in systems of differential equations consisting of two or more equations. This fourth type is known as a *Hopf bifurcation*. It is also referred to as a *Poincaré-Andronov-Hopf bifurcation* (Hale and Koçak, 1991) to acknowledge the contributions to the theory by French mathematician Jules Henri Poincaré (1854–1912), Russian mathematician Alexander A. Andronov (1901–1952), and German mathematician Heinz Hopf (1894–1971). We have seen in Chapter 3 that a similar type of bifurcation occurred in a predator-prey system modeled by a system of difference equations (known as a Neimark-Sacker bifurcation).

The Hopf Bifurcation Theorem stated here is for a system of two differential equations. There is a Hopf Bifurcation Theorem for higher dimensions also (see Marsden and McCracken, 1976). This Hopf Bifurcation Theorem states sufficient conditions for the existence of periodic solutions. As one parameter is varied, the dynamics of the system change from a stable spiral to a center to an unstable spiral. The eigenvalues of the linearized system change from having negative real parts to zero real parts to positive real parts. Under certain conditions, there exist periodic solutions.

Consider a system of autonomous differential equations given by

$$\frac{dx}{dt} = f(x, y, r) \quad \text{and} \quad \frac{dy}{dt} = g(x, y, r), \tag{5.10}$$

where the functions $f$ and $g$ depend on the bifurcation parameter $r$. Suppose there exists an equilibrium $(\bar{x}(r), \bar{y}(r))$ of system (5.10) and the Jacobian matrix evaluated at this equilibrium has eigenvalues $\alpha(r) \pm i\beta(r)$. In addition, suppose a change in stability occurs at the value of $r = r^*$, where $\alpha(r^*) = 0$. If $\alpha(r) < 0$ for values of $r$ close to $r^*$ but for $r < r^*$ and if $\alpha(r) > 0$ for values of $r$ close to $r^*$ but for $r > r^*$ (also $\beta(r^*) \neq 0$), then the equilibrium changes from a stable spiral to an unstable spiral as $r$ passes through $r^*$. The Hopf Bifurcation Theorem states that there exists a periodic orbit near $r = r^*$ for any neighborhood of the equilibrium in $\mathbf{R}^2$. The parameter $r$ is the *bifurcation parameter* and $r^*$ is the *bifurcation value*. The theorem is valid only when the bifurcation parameter has values close to the bifurcation value.

Before we state the theorem, a simple example is presented which exhibits a Hopf bifurcation.

**Example 5.19** Consider the linear system

$$\frac{dx}{dt} = rx - y,$$

$$\frac{dy}{dt} = x + ry.$$

The origin is an equilibrium. The trace and determinant of the Jacobian matrix evaluated at the origin are $2r$ and $r^2 + 1$, respectively. Since the discriminant of the Jacobian matrix is negative, $(2r)^2 - 4r^2 - 4 = -4$, the eigenvalues are $r \pm i$. If $r < 0$, the origin is a stable spiral. If $r = 0$, the origin is a center, and if $r > 0$, it is an unstable spiral. The bifurcation value is at $r = r^* = 0$. Recall the stability diagram, where stability is graphed as a function of the trace $\tau$ and determinant $\delta$. The bifurcation in this example occurred because $r$ crossed the $\delta$-axis where $\delta > 0$. A Hopf bifurcation occurs. As the bifurcation parameter $r$ increases through the bifurcation value $r^* = 0$, the equilibrium $(0, 0)$ changes from a stable spiral to a neutral center to an unstable spiral. There are infinitely many periodic solutions at the bifurcation value $r^* = 0$. Solutions to $dx/dt = -y$ and $dy/dt = x$ are of the form $x^2(t) + y^2(t) = c$, where $c$ is a constant that depends on initial conditions. ■

The linear example illustrates the change in stability as the bifurcation parameter $r$ is varied. In general, at a Hopf bifurcation, as $r$ passes through the bifurcation value $r^*$, there are three possible dynamics that may occur.

(i) At the bifurcation value $r^*$ infinitely many neutrally stable concentric closed orbits encircle the equilibrium.

(ii) A stable spiral changes to a stable limit cycle for values of the parameter close to $r^*$ (*supercritical bifurcation*).

(iii) A stable spiral and unstable limit cycle change to an unstable spiral for values of the parameter close to $r^*$ (*subcritical bifurcation*).

Example 5.19 illustrates a change of stability of type (i). Figure 5.13 illustrates a supercritical and a subcritical bifurcation in $x$-$y$-$r$ space. Stable solutions are identified by solid curves and unstable solutions by dashed curves.

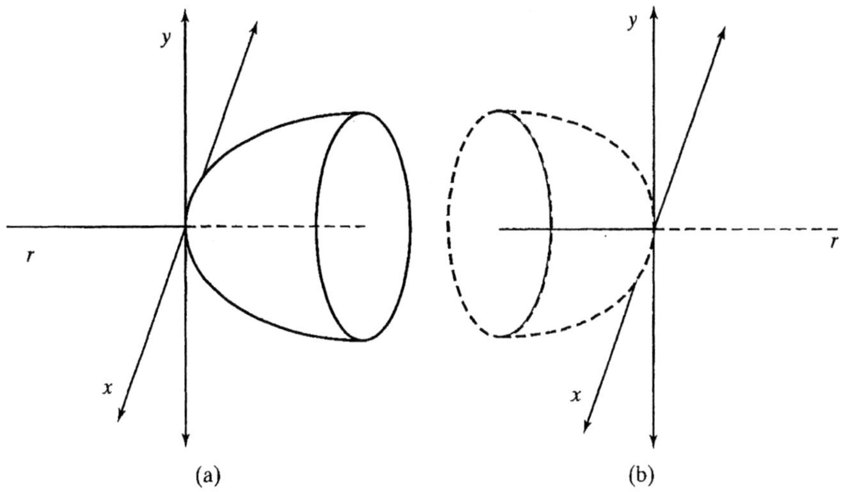


**Figure 5.13** (a) Supercritical and (b) subcritical bifurcations in $x$-$y$-$r$ space. Solid curves circling or on the $r$-axis are stable. Dashed curves are unstable.

The Hopf Bifurcation Theorem is stated as given by Hale and Koçak (1991). For a proof of this theorem see Hale and Koçak (1991) or Marsden and McCracken (1976). First the system is transformed so that the equilibrium is at the origin and the parameter $r$ at $r^* = 0$ gives purely imaginary eigenvalues. System (5.10) is rewritten as follows:

$$\frac{dx}{dt} = a_{11}(r)x + a_{12}(r)y + f_1(x, y, r)$$

$$\frac{dy}{dt} = a_{21}(r)x + a_{22}(r)y + g_1(x, y, r). \tag{5.11}$$

The linearization of system (5.11) about the origin is given by $dZ/dt = J(r)Z$, where $Z = (x, y)^T$ and

$$J(r) = \begin{pmatrix} a_{11}(r) & a_{12}(r) \\ a_{21}(r) & a_{22}(r) \end{pmatrix} \tag{5.12}$$

is the Jacobian matrix evaluated at the origin.

**Theorem 5.10** **(Hopf Bifurcation Theorem).** *Let $f_1$ and $g_1$ in system (5.11) have continuous third-order derivatives in x and y. Assume that the origin (0, 0) is an equilibrium of (5.11) and that the Jacobian matrix $J(r)$, defined in (5.12), is valid for all sufficiently small $|r|$. In addition, assume that the eigenvalues of matrix $J(r)$ are $\alpha(r) \pm i\beta(r)$ with $\alpha(0) = 0$ and $\beta(0) \neq 0$ such that the eigenvalues cross the imaginary axis with nonzero speed (transversal),*

$$\left.\frac{d\alpha}{dr}\right|_{r=0} \neq 0.$$

*Then, in any open set U containing the origin in $\mathbf{R}^2$ and for any $r_0 > 0$, there exists a value $\bar{r}$, $|\bar{r}| < r_0$ such that the system of differential equations (5.11) has a periodic solution for $r = \bar{r}$ in U (with approximate period $T = 2\pi/\beta(0)$).* □

**Example 5.20** Consider the linear system in Example 5.19 with bifurcation parameter $r$. We show that the conditions of the Hopf Bifurcation Theorem hold.

$$\frac{dx}{dt} = rx - y,$$

$$\frac{dy}{dt} = x + ry.$$

In this case, $f_1 = 0 = g_1$. The Jacobian matrix is

$$J(r) = \begin{pmatrix} r & -1 \\ 1 & r \end{pmatrix}.$$

with eigenvalues equal to $r \pm i$. Since $\alpha(r) = r$ and $\beta(r) = 1$, it follows that $\alpha(0) = 0$, $\beta(0) \neq 0$, and $d\alpha/dr = 1 \neq 0$. The conditions of the Hopf Bifurcation Theorem hold. In fact, we know that there exists a periodic solution for $r = 0$ in every neighborhood of the origin. ■

A computational method can be applied to determine whether a supercritical or subcritical bifurcation occurs. This method is given in the Appendix to this chapter.

**Example 5.21** Consider the system

$$\frac{dx}{dt} = rx + y,$$

$$\frac{dy}{dt} = -x + ry - y^3. \tag{5.13}$$

There is an equilibrium at $(0, 0)$. The Jacobian matrix is

$$J(r) = \begin{pmatrix} r & 1 \\ -1 & r - 3y^2 \end{pmatrix}\Bigg|_{(0,0)} = \begin{pmatrix} r & 1 \\ -1 & r \end{pmatrix}.$$

The eigenvalues of $J(r)$ are $r \pm i$. The Hopf Bifurcation Theorem can be applied. In addition, a test for a supercritical or a subcritical bifurcation can be applied (Appendix) to show that the bifurcation at $r = 0$ is supercritical (Exercise 32). If the parameter $r$ is sufficiently small and positive, then the system of differential equations has a stable periodic solution in a neighborhood of the origin. Figure 5.14 illustrates the dynamics of this system for $r = -1/2$ and $r = 1/2$. Note for $r = 1/2$ that there are three equilibria. ■

## 5.9 Delay Logistic Equation

The continuous logistic equation $dx/dt = rx(1 - x/K)$ has a discrete approximation which is given by the following difference equation: $x_{t+1} = x_t + rx_t(1 - x_t/K)$. This latter equation represents a delay in the growth, since the change in the population size does not occur until one unit later, $t$ to $t + 1$. In the continuous logistic equation, the change in growth is instantaneous. As we have seen in Chapter 2, the equilibrium $\bar{x} = K$ can be destabilized in the difference equation model as $r$ increases. The equilibrium $\bar{x} = K$ in the continuous logistic

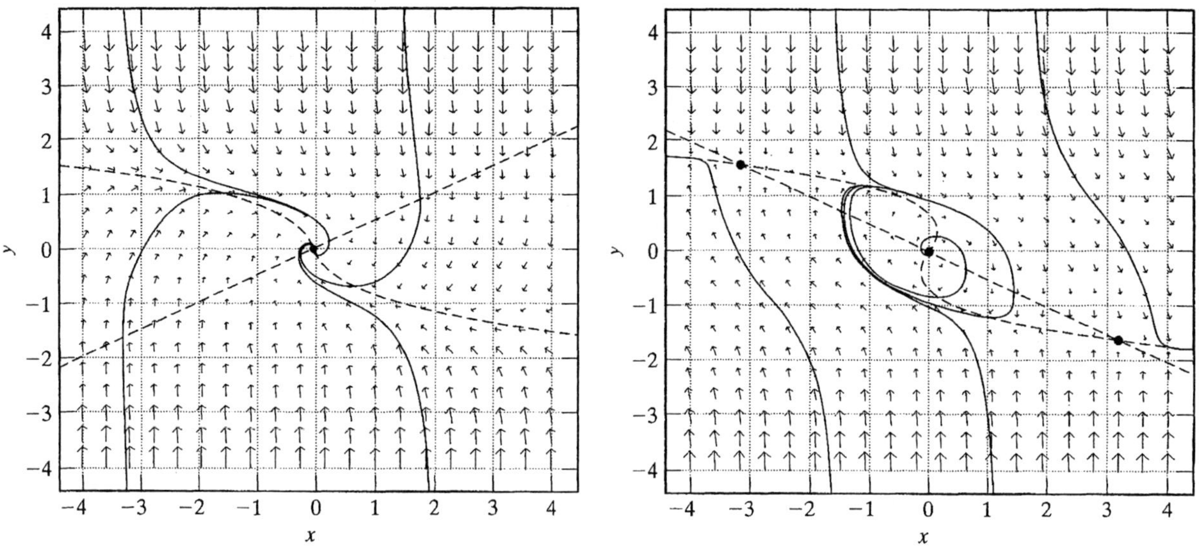


**Figure 5.14** Dynamics of system (5.13) in the phase plane when $r = -1/2$ (left figure) and $r = 1/2$ (right figure). The equilibrium $(0,0)$ is stable in the figure on the left, but in the figure on the right solutions near the origin converge to a stable periodic solution.

equation is asymptotically stable if $r, K > 0$ and $x(0) > 0$. However, in the discrete case, the equilibrium $\bar{x} = K$ is only locally asymptotically stable if $0 < r < 2$. For values of $r$ satisfying $r > 2$, solutions become periodic (period-doubling) and chaotic. Delays often change the stability of an equilibrium. We will show that putting a delay of length $T$ in the density-dependent term in the continuous logistic equation changes the range of values $r$ for which the equilibrium is stable.

Consider the logistic equation with a delay of $T$ in the density-dependent factor,

$$\frac{dx\,(t)}{dt} = rx(t)\left(1 - \frac{x(t-T)}{K}\right). \tag{5.14}$$

The parameters $r, K$, and $T$ are positive. Parameter $r$ is the intrinsic growth rate, $K$ is the carrying capacity, and $T$ is the delay parameter. Note that equation (5.14) still has two constant solutions or equilibria: $\bar{x} = 0$ and $\bar{x} = K$. The density-dependent factor, $1 - x(t-T)/K$, which regulates the rate of growth, is not instantaneous but depends on the population at an earlier time $t - T$. For example, the population size which affects food resources may not be immediately felt by the population but only after a period of time $T$. Equation (5.14) is sometimes referred to as the *Hutchinson-Wright equation* because it was first studied by the ecologist Hutchinson (1948) and the mathematician Wright (1946) (see also Kot, 2001). Note that to compute the solution to this discrete-delay equation for $t > 0$ it is necessary to know the value of $x(t)$ on the interval $[-T, 0]$. The method of steps can be applied to this delay model. To find the solution on the interval $[0, T]$, it is necessary to solve the following equation:

$$\frac{dx\,(t)}{dt} = rx(t)\left(1 - \frac{\phi_0(t-T)}{K}\right),$$

where $\phi_0(t) = x(t)$ on the interval $[-T, 0]$. We do not solve this differential equation by the method of steps, but determine the region of stability which depend on $r$ and $T$, where the equilibrium $K$ is locally asymptotically stable.

The discrete-delay differential equation (5.14) is a simplification of a more general model where the delay is over all past populations. A more general model is an integrodifferential equation, where the delay is continuous (discussed at the end of Chapter 4). A model with a continuous delay in the density-dependent factor with a logistic growth takes the following form:

$$\frac{dx(t)}{dt} = rx(t)\left(1 - K^{-1}\int_{-\infty}^{t} k(t-s)x(s)\,ds\right), \tag{5.15}$$

where $k(t)$ is a weighting factor which says how much weight should be given to past populations, also referred to as the *kernel* of the integral. For example, Cushing (1977) refers to a kernel of the form

$$k_1(t) = \frac{1}{T}\exp(-t/T)$$

as a "weak" delay kernel. The maximum of $k_1(t)$ is at $t = 0$; there is little dependence on past populations. However, the kernel

$$k_2(t) = \frac{t}{T^2}\exp(-t/T)$$

is a referred to as a "strong" delay kernel (Cushing, 1977). The strong delay kernel has a maximum at $t = T$. See Figure 5.15. Both of these kernels satisfy

$$\int_0^{\infty} k_i(t)\,dt = 1, \quad i = 1, 2.$$

When $k(t)$ is the Dirac delta function, $k(t) = \delta(t - T)$, equation (5.15) reduces to the logistic model with a discrete delay at $t = T$. The Dirac delta function satisfies

$$\delta(t-T) = 0 \text{ for } t \neq T, \quad \int_{-\infty}^{\infty} \delta(t-T)\,dt = 1,$$

and

$$\int_{-\infty}^{t} \delta(t - T - s)x(s)\,ds = x(t - T).$$

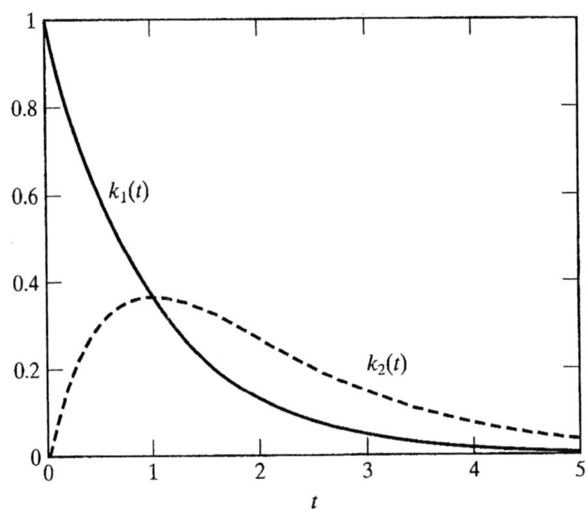


**Figure 5.15** Graphs of the delay kernels $k_1(t)$ and $k_2(t)$ when $T = 1$.

We concentrate on the logistic model with a discrete delay at $t = T$, model (5.14), and show that the equilibrium $K$ is locally asymptotically stable for small values of the delay $T$. Long delays are destabilizing. It will be shown that there exists a value $T_c$ such that for $0 \leq T < T_c$, the equilibrium $K$ of the delayed logistic equation is locally asymptotically stable. The value of $T_c$ depends on the parameter $r$.

The solution of a linear delay differential equation with an equilibrium at the origin approaches zero if the roots of the characteristic equation have negative real parts (Bellman and Cooke, 1963; Gopalsamy, 1992). The characteristic equation is obtained by assuming solutions have the form $e^{\lambda t}$. Thus, to determine local stability of the equilibrium $K$ for (5.14), the equation needs to be linearized. Before linearization, a change of variables is made to simplify the equation.

Let $\tau = rt, \hat{T} = rT$, and

$$y(\tau) = \frac{x(t)}{K}.$$

Note that the units of $x$ are the same as $K$ and the units of $t$ are the same as $1/r$; thus, $\tau$ and $y$ are dimensionless variables. Differentiating $y$ with respect to $\tau$,

$$\frac{dy}{d\tau} = \frac{1}{K}\frac{dx}{dt}\frac{dt}{d\tau} = \frac{1}{rK}\frac{dx}{dt}.$$

Since $x(t) = Ky(rt)$, it follows that $x(t - T) = Ky(r(t - T)) = Ky(rt - rT) = Ky(\tau - \hat{T})$. The discrete-delay differential equation (5.14) expressed in terms of $y$ satisfies

$$\frac{dy(\tau)}{d\tau} = y(\tau)(1 - y(\tau - \hat{T})). \tag{5.16}$$

The equilibrium $\bar{x} = K$ in the differential equation (5.14) corresponds to the equilibrium $\bar{y} = 1$ in the differential equation (5.16).

Next, equation (5.16) is linearized about the equilibrium $\bar{y} = 1$. Let $u(\tau) = x(\tau) - 1$. Then

$$\frac{du(\tau)}{d\tau} = \frac{dy(\tau)}{d\tau} = (u(\tau) + 1)[-u(\tau - \hat{T})] \approx -u(\tau - \hat{T}).$$

The linearization has the form

$$\frac{du(\tau)}{d\tau} = -u(\tau - \hat{T}). \tag{5.17}$$

The local asymptotic stability result is stated for the original nonlinear delay equation (5.14) but the linearized system (5.17) is used to verify this result (Cushing, 1977; Gopalsamy, 1992). We show that the equilibrium $\bar{y} = 1$ is locally asymptotically stable if $\hat{T} < \pi/2$. Expressed in terms of the original variables, the equilibrium $\bar{x} = K$ is locally asymptotically stable if $T < \pi/(2r)$ (or $r < \pi/(2T)$).

**Theorem 5.11** *The equilibrium $K$ of the delayed logistic differential equation (5.14) is locally asymptotically stable if $0 \leq T < \pi/(2r)$ and unstable if $T > \pi/(2r)$.*

**Proof** Assume solutions to (5.17) are of the form $u(\tau) = e^{\lambda\tau}$. Substituting $e^{\lambda\tau}$ into the differential equation (5.17) yields

$$\lambda e^{\lambda\tau} = -e^{\lambda(\tau - T)}.$$

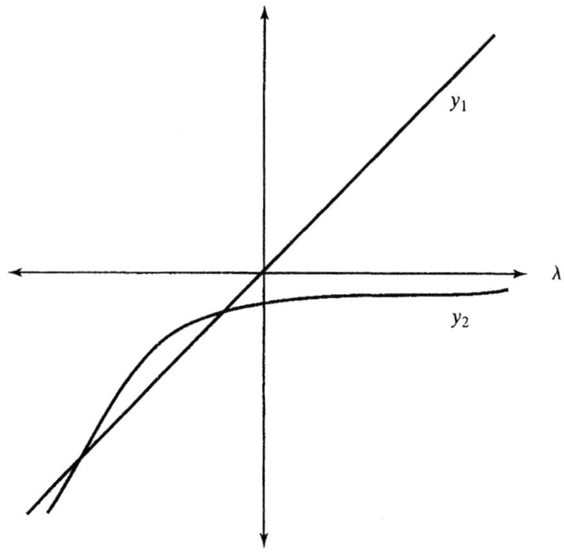


**Figure 5.16** Graphs of $y_1 = \lambda$ and $y_2 = -e^{-\lambda T}$. There are either none, one, or two points of intersection of these two curves. The points of intersection occur only at negative values of $\lambda$.

For simplicity, we let $\hat{T} = T$. Simplifying, the characteristic equation is

$$\lambda = -e^{-\lambda T}.$$

The characteristic equation is a transcendental equation in $\lambda$. The solutions $\lambda$ are the eigenvalues. If the eigenvalues are negative or have negative real parts, then the equilibrium $u = 0$ of the linearized equation is asymptotically stable. If $u = 0$ is asymptotically stable, then in terms of the original variables, equation (5.14), the equilibrium $x = K$ is locally asymptotically stable. Two cases are considered.

Case 1 Assume the eigenvalues $\lambda$ are real. Then it is easy to see that all solutions $\lambda$ of $\lambda = -e^{-\lambda T}$ are negative for $T \geq 0$. See Figure 5.16.

Case 2 Assume the eigenvalues $\lambda$ are complex, that is, $\lambda = a \pm ib$, $b \neq 0$. Applying Euler's formula to the transcendental equation,

$$a + ib = -e^{-(a+ib)T} = -e^{-aT}[\cos(bT) - i\sin(bT)]$$

($b$ may be positive or negative). Equating real and imaginary parts,

$$a = -e^{-aT}\cos(bT) \quad \text{and} \quad b = e^{-aT}\sin(bT). \tag{5.18}$$

It remains to be shown for $0 \leq T < \pi/2$ that the solutions to the equations in (5.18) satisfy $a < 0$.

If a solution exists for $b > 0$ to the equations in (5.18), then the same solution holds for $-b$ and conversely. Therefore, without loss of generality, assume $b > 0$. Simplifying the equations in (5.18) yields

$$a = -b\cot(bT) \quad \text{and} \quad b = \sin(bT)\, e^{bT\cot(bT)}. \tag{5.19}$$

Let $s = bT$. Then the latter equation for $b$ can be expressed as

$$\frac{s}{T} = \sin(s)e^{s\cot(s)}.$$

The points of intersection of the curves $y_1 = s/T$ and $y_2 = \sin(s)\, e^{s\cot(s)}$ in the $s$-$y$ plane can be seen in Figure 5.17. When $0 < T \leq e^{-1}$, there is no point of intersection for $s \in (0, \pi)$ but there are points of intersection for $s \in (2\pi, 5\pi/2)$, $(4\pi, 9\pi/2), \ldots$, and so on. (When $T = 1/e$ the graphs of $y_1$ and $y_2$ are tangent at the origin.) Note for each of the points of intersection

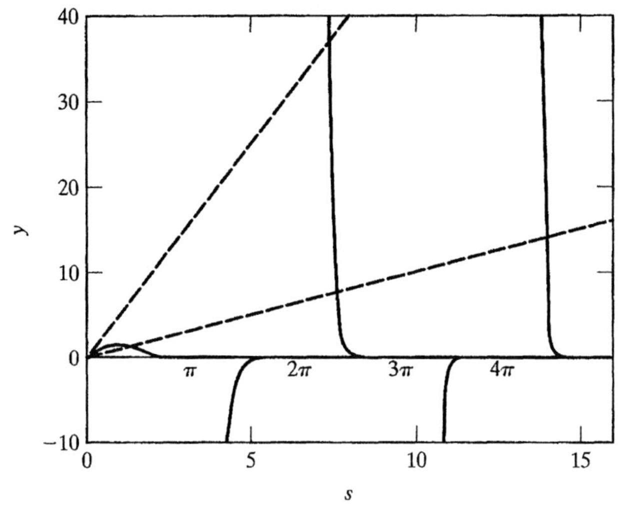


**Figure 5.17** Graphs of $y_1 = s/T$ (dashed curves), where $0 < T < 1/e$ and $1/e < T < \pi/2$ and $y_2 = \exp(s\cot(s))\sin(s)$ (solid curves) in the $s$-$y$ plane. The horizontal axis is $s$ and the vertical axis is $y$.

$bT = s \in (2\pi, 5\pi/2), (4\pi, 9\pi/2), \ldots$, that $a < 0$. For $e^{-1} < T < \pi/2$, there is an additional point of intersection for $s \in (0, \pi)$. For $s = \pi/2$, $a = 0$. Thus, if $e^{-1} < T < \pi/2$, the intersection of the curves $y_1$ and $y_2$ for $0 < s < \pi/2$ implies $a < 0$ [follows from the first equation in (5.19)]. Consequently, in all cases, if $0 \le T < \pi/2$, then $a < 0$; the eigenvalues have negative real parts. Summarizing the results, the equilibrium $u = 0$ of the linearized equation is locally asymptotically stable if $0 \le T < \pi/2$.

In the case $T > \pi/2$ there is an intersection of $y_1$ and $y_2$, where $\pi/2 < s < \pi$. Consequently $a > 0$; there is an eigenvalue with positive real part. Hence, if $T > \pi/2$, the equilibrium $u = 0$ is unstable.

Expressing the inequalities in terms of the original variables, the delay $T$ is replaced by $rT$ and $T_c = \pi/(2r)$. In terms of the original differential equation (5.14), equilibrium $\bar{x} = K$ is locally asymptotically stable if

$$0 \le rT < \frac{\pi}{2}$$

and unstable if $rT > \pi/2$. □

The global asymptotic stability of $K$ (for $x(t) \ge 0$, $x(t) \ne 0$ on $[-T, 0]$) was proved by Wright in 1955 for the differential delay equation (5.14) when $rT < 3/2$. A proof of this result is given by Kuang (1993). Kuang states that the global asymptotic stability result can be extended to $rT < 37/24$, and probably to $rT < 1.567$ (note that $\pi/2 = 1.571\ldots$). Wright's conjecture that the equilibrium $K$ is globally asymptotically stable for $rT < \pi/2$ is still an open problem.

An intuitive argument shows that periodic solutions are possible and that the period is approximately $4T$ near the bifurcation value. Suppose at some time $t = t_1$ that $x(t_1) = K$ and for $t - T \le t < t_1$, $x(t) < K$. Then the delay differential equation satisfies $dx(t)/dt > 0$ at $t = t_1$; the solution increases above $K$. When time reaches $t = t_1 + T$, then $x(t - T) = x(t_1) = K$ and $dx(t)/dt = 0$. For $t$ slightly greater than $t_1 + T$, $dx(t)/dt < 0$, the solution starts decreasing. There exists a time $t_2$, such that $x(t_2) = K$, $x(t - T) > K$, and $dx(t)/dt < 0$, for $t_1 + T < t < t_2 + T$. Thus, $x(t)$ is increasing on the interval $(t_1 - T, t_1 + T)$ and decreasing on the interval $(t_1 + T, t_2 + T)$. At $t_2 + T$, $dx(t)/dt = 0$ and $x(t)$ changes direction again. Thus, the period of the solution is approximately $4T$.

Data collected by Nicholson (1957) on sheep blowflies show cyclic behavior (see Figure 5.18). The Australian sheep blowfly (*Lucilia cuprina*) is a major

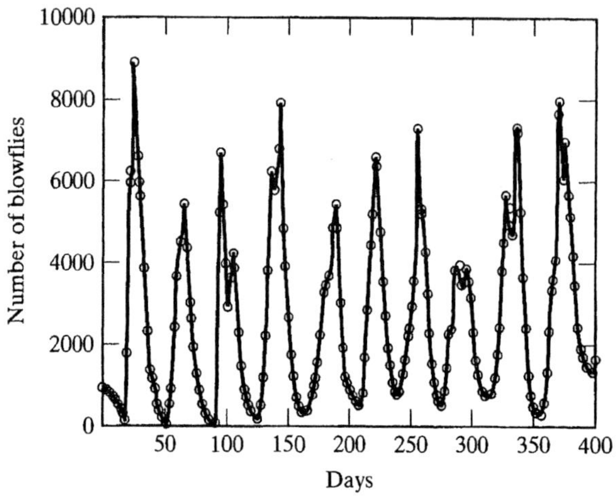


**Figure 5.18** Nicholson's data on blowflies are graphed in the figure on the left. Data were collected bidaily for 722 days (shown are the first 400 days). An approximate solution to the logistic delay differential equation is graphed in the figure on the right, $rT = 2.1$ and $K = 2500$. Data are available from Brillinger et al. (1980).

insect pest of sheep in Australia. From egg to adult stage, there is a developmental delay, so the delay differential equation is a reasonable model to apply to these data. May (1975) fit the data to the logistic delay differential equation (5.14), where $rT = 2.1 > \pi/2$. In this case, there is a periodic solution whose period is approximately $4.54T$ (Murray, 1993, 2002). The observed period is about 40 days; thus, the delay $T \approx 40/4.54 \approx 9$ days. An approximate solution to the delay differential equation (5.14) with $rT = 2.1$ and $K = 2500$ is graphed in Figure 5.18.

As a final example in this section, we study the stability of the equilibrium $K$ for the delay integrodifferential equation (5.15) when there is a weak delay kernel, $k(t) = T^{-1}\exp(-t/T)$. It will be shown for the weak delay that the equilibrium $K$ is always locally asymptotically stable.

**Example 5.22** Let

$$\frac{dx(t)}{dt} = rx(t)\left(1 - \frac{1}{K}\int_{-\infty}^{t} k(t-s)x(s)\,ds\right).$$

Making the change of variable $\tau = t - s$ in the integral leads to

$$\frac{dx(t)}{dt} = rx(t)\left(1 - \frac{1}{K}\int_{0}^{\infty} k(\tau)x(t-\tau)\,d\tau\right).$$

Linearizing the equation about $K$, $u(t) = x(t) - K$, yields

$$\begin{aligned}\frac{du(t)}{dt} &= r[u(t) + K]\left(1 - \frac{1}{K}\int_{0}^{\infty} k(\tau)[u(t-\tau) + K]\right)d\tau.\\ &= r[u(t) + K]\left(-\frac{1}{K}\int_{0}^{\infty} k(\tau)u(t-\tau)\,d\tau\right)\\ &\approx -r\int_{0}^{\infty} k(\tau)u(t-\tau)\,d\tau.\end{aligned}$$

Here we have used the fact that $\int_0^\infty k(\tau)\,d\tau = 1$. Assume that $u(t) = e^{\lambda t}$ and $k(t) = T^{-1}\exp(-t/T)$, a weak delay kernel. Substituting these functions into the linearized equation and simplifying, the following identity is obtained:

$$\lambda = -\frac{r}{T}\int_0^\infty e^{-\lambda\tau}e^{-\tau/T}\,d\tau. \tag{5.20}$$

Two cases are considered.

Case 1 Suppose $\lambda$ is real and $\lambda > 0$. Then integration of (5.20) leads to

$$\lambda = -\frac{r}{\lambda T + 1}.$$

The characteristic equation can be expressed as

$$T\lambda^2 + \lambda + r = 0.$$

Because all of the coefficients are positive, the solution $\lambda$ must be negative (a contradiction).

Case 2 Suppose $\lambda = a + ib$ and $a > 0$. Then $e^{-\lambda\tau} = e^{-a\tau}(\cos(b\tau) - i\sin(b\tau))$. Separating real and imaginary parts in (5.20), the real part of the equation satisfies

$$a = -\frac{r}{T}\int_0^\infty e^{-\tau(a+1/T)}\cos(b\tau)\,d\tau.$$

Integrating this latter equation leads to

$$a = -\frac{r(aT + 1)}{(aT + 1)^2 + b^2T^2}.$$

This equation can be expressed as a third-degree polynomial in $a$,

$$T^2a^3 + 2Ta^2 + (1 + b^2T^2 + rT)a + r = 0.$$

Because all of the coefficients are positive, the solution $a$ must be negative (a contradiction).

These two cases show that the delay logistic equation with a weak delay kernel has a locally asymptotically stable equilibrium at $K$ with no restrictions imposed on the delay $T$. ■

Stability in the case of the strong delay kernel, $k(t) = tT^{-2}\exp(-t/T)$, is left as an exercise (Exercise 25). In this case, the equilibrium $K$ is locally asymptotically stable if $T < 2/r$.

## 5.10 Stability Using Qualitative Matrix Stability

Qualitative stability is a matrix property. Qualitative stability is determined only from the signs of the Jacobian matrix (assuming all entries of the matrix are real numbers). For example, any matrix of the form

$$Q = \text{sign}(J) = \begin{pmatrix} - & + \\ 0 & - \end{pmatrix}$$

with negative values along the diagonal and zero below the diagonal will have negative eigenvalues. Therefore, in this example, if $J$ is the Jacobian matrix evaluated at an equilibrium, the equilibrium for the corresponding differential equations will be locally asymptotically stable. The term "qualitative" implies qualitative information (signs) is used to determine stability rather than quantitative information. A definition of qualitative stability is given below.

**Definition 5.8.** A square matrix $J$ whose sign pattern is $Q = \text{sign}(J)$ is said to be *qualitatively stable* if all matrices with the same sign pattern have negative eigenvalues or eigenvalues with negative real parts.

Qualitative stability of a matrix $J$, a Jacobian matrix evaluated at an equilibrium, implies local asymptotic stability of the equilibrium. However, the converse is certainly *not* true. A Jacobian matrix can have negative eigenvalues, but not all matrices with that sign pattern will necessarily have negative eigenvalues or eigenvalues with negative real parts. Therefore, an equilibrium can be locally asymptotically stable, but the associated Jacobian matrix may not be qualitatively stable. Qualitative stability of a Jacobian matrix is determined by checking whether the signs of the matrix satisfy certain properties.

One advantage of a test for qualitative matrix stability over local asymptotic stability is that the criteria for qualitative matrix stability are often easier to apply. The Routh-Hurwitz criteria for local asymptotic stability requires calculation of the characteristic equation and the determinants of the $n$ Hurwitz matrices. For large $n$ the Routh-Hurwitz criteria can lead to many tedious calculations. Another advantage of qualitative matrix stability over local asymptotic stability is that if the equilibrium is one where all of the states are *positive*, then the signs of the Jacobian matrix can often be determined from a digraph (directed graph) rather than from calculating partial derivatives. A disadvantage of the test for qualitative matrix stability versus local asymptotic stability is that when the test for qualitative matrix stability fails, we don't know whether the equilibrium is locally asymptotically stable. The equilibrium may or may not be locally asymptotically stable.

A system of differential equations having the following form:

$$\frac{dx_i}{dt} = x_i\left(a_{i0} + \sum_{j=1}^{n} a_{ij}x_j\right), \quad i = 1, \ldots, n, \tag{5.21}$$

is often referred to as a *Lotka-Volterra* system. The name refers to Alfred Lotka (1880–1949) and Vito Volterra (1860–1940) because of their contributions to the study of these types of equations. Note that the system $(dx_i/dt)/x_i$ is affine. An equilibrium solution to this Lotka-Volterra system is a solution $\bar{X}$ to $A\bar{X} = b$, where $A = (a_{ij})$ is the $n \times n$ matrix of coefficients in (5.21) and $b = -(a_{i0})$ is the $n$-vector of the negative of the intrinsic growth rates. Suppose $\det(A) \neq 0$, so that there is a unique equilibrium solution. If the equilibrium solution has all coordinates positive, $\bar{X} = (\bar{x}_1, \ldots, \bar{x}_n)^T > 0$, it can be shown that the Jacobian matrix has a special form. The Jacobian matrix of a Lotka-Volterra system evaluated at a positive equilibrium satisfies

$$J(\bar{X}) = \text{diag}(\bar{x}_i)A = \begin{pmatrix} \bar{x}_1 a_{11} & \bar{x}_1 a_{12} & \cdots & \bar{x}_1 a_{1n} \\ \vdots & \vdots & \cdots & \vdots \\ \bar{x}_n a_{n1} & \bar{x}_n a_{n2} & \cdots & \bar{x}_n a_{nn} \end{pmatrix}.$$

The matrix $A$ is sometimes referred to as the *interaction matrix*. The sign pattern of $J$ is determined by the sign pattern of $A$, that is,

$$Q = \text{sign}(J) = \text{sign}(A).$$

Thus, to find $Q$ we just need to find the sign pattern of $A$.

**Example 5.23** Consider the following one-predator, two-prey Lotka-Volterra system (Edelstein-Keshet, 1988):

$$\frac{dx_1}{dt} = x_1(-a_{10} + a_{12}x_2 + a_{13}x_3)$$

$$\frac{dx_2}{dt} = x_2(a_{20} - a_{21}x_1)$$

$$\frac{dx_3}{dt} = x_3(a_{30} - a_{31}x_1 - x_3),$$

where all of the coefficients given in the preceding system are positive, $a_{ij} > 0$. First, a positive equilibrium must satisfy $AX = b$ or

$$\bar{x}_1 = \frac{a_{20}}{a_{21}}, \quad \bar{x}_3 = a_{30} - a_{31}\bar{x}_1, \quad \bar{x}_2 = \frac{a_{10} - a_{13}\bar{x}_3}{a_{12}}.$$

The equilibrium will be positive if $\bar{x}_2 > 0$ and $\bar{x}_3 > 0$. We assume the equilibrium is positive. Since $Q = \text{sign}(J) = \text{sign}(A)$, the signed matrix of the Jacobian matrix $J$ satisfies

$$Q = \text{sign}(J) = \begin{pmatrix} 0 & + & + \\ - & 0 & 0 \\ - & 0 & - \end{pmatrix}.$$

The signed directed graph (signed digraph) for matrix $Q$ provides this information in a graphical format (Figure 5.19). In the signed digraph, each state is represented by a circle and referred to as a *node*. An arrow from state $x_i$ to state $x_j$ has a positive sign if the element $q_{ji} = +$ in matrix $Q$ and negative if $q_{ji} = -$. For example, in Figure 5.19, the arrow from $x_2$ to $x_1$ is marked with a positive sign because $q_{12} = +$ and the arrow from $x_1$ to $x_3$ is marked with a negative sign because $q_{31} = -$. There are no arrows connecting $x_2$ and $x_3$ because $q_{32} = 0 = q_{23}$. A *feedback loop* is an arrow from $x_i$ to $x_i$. The feedback loop is positive if

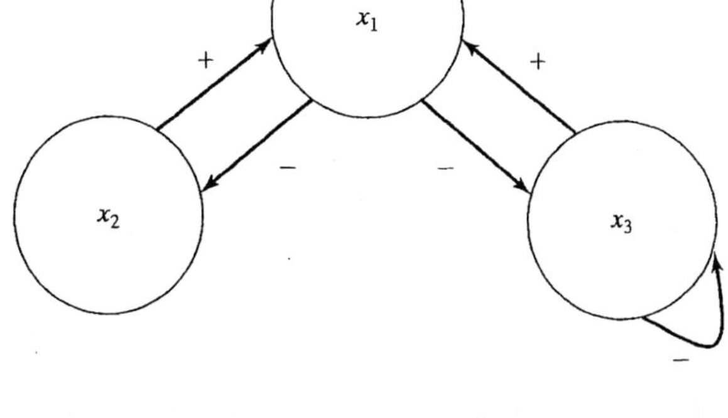


**Figure 5.19** A signed digraph for the one-predator, two-prey model.

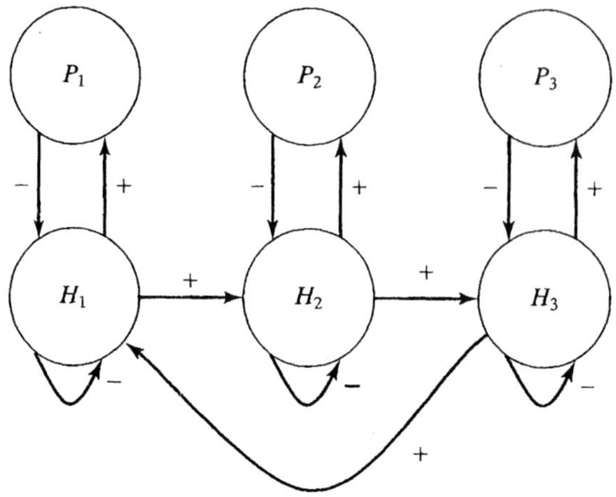


**Figure 5.20** A signed digraph for the three-parasitoids, three-hosts model.

$q_{ii} = +$ and negative if $q_{ii} = -$. Note that in Figure 5.19 there are two arrows between $x_1$ and $x_2$; one is marked with $+$ and the other $-$. We refer to these pair of arrows as having *opposite effects*. ■

---

**Example 5.24** Suppose there are three parasitoids ($P_1, P_2, P_3$) that attack three different stages in the life cycle of a host ($H_1, H_2, H_3$) (Edelstein-Keshet, 1988). For example, in an insect population, the three stages may represent larvae, pupae, and adults. In addition, suppose the system of differential equations modeling these six states is a Lotka-Volterra system. The signed digraph in Figure 5.20 indicates the relationships among the six states and the signs of the interaction matrix where the state vector is $(P_1, P_2, P_3, H_1, H_2, H_3)$.

The signed matrix $Q$ can be easily found from the signed digraph,

$$Q = \text{sign}(J) = \begin{pmatrix} 0 & 0 & 0 & + & 0 & 0 \\ 0 & 0 & 0 & 0 & + & 0 \\ 0 & 0 & 0 & 0 & 0 & + \\ - & 0 & 0 & - & 0 & + \\ 0 & - & 0 & + & - & 0 \\ 0 & 0 & - & 0 & + & - \end{pmatrix}.$$

■

We state necessary and sufficient conditions for a matrix to be qualitatively stable. First, we state five necessary conditions. These five conditions must be satisfied if the matrix is to be qualitatively stable. Second, one more condition is stated which together with the five conditions gives sufficient conditions for qualitative stability of a matrix. See Jeffries et al. (1977) for necessary and sufficient conditions and a more thorough treatment of qualitative stability. Edelstein-Keshet (1988) and Pielou (1977) provide good introductions to qualitative stability and some additional examples. Let $q_{ij}$ denote the signs of the elements in the matrix $Q = \text{sign}(J)$, either 0, $+$, or $-$. In the case of a Lotka-Volterra system, we assume that there exists a unique positive equilibrium.

---

**Theorem 5.12** **(Necessary Conditions for Qualitative Stability).** *If an $n \times n$ matrix $J$ is qualitatively stable and $Q = \text{sign}(J) = (q_{ij})$, then*

1. *$q_{ii}$ is either negative or zero for all i.*
2. *$q_{ii}$ is negative for some i.*

3. *$q_{ij}$ and $q_{ji}$ must have opposite signs for all $i \neq j$, that is, one must be negative and the other positive, unless they are both zero.*
4. *Given any sequence, $q_{ij}, q_{jk}, \ldots, q_{lr}, q_{ri}$ containing three or more distinct indices $i, j, k, \ldots, l, r$, at least one of the elements in the sequence must be zero.*
5. $\det(J) \neq 0$. □

Condition 5 implies that the underlying Jacobian matrix is nonsingular. If the equilibrium is hyperbolic, condition 5 is satisfied. In some of our examples, we are given only the signed matrix $Q$. Therefore, we will assume that condition 5 is satisfied in our examples.

The first four conditions can be interpreted in terms of the signed digraph. The first and second conditions state that the signed digraph has no positive feedback loop and there is at least one negative feedback loop (arrow from $x_i$ to $x_i$ with a negative sign). The third condition states that any pair of interacting nodes $x_i$ and $x_j$, $i \neq j$, must have opposite sign patterns $(+, -)$ or $(-, +)$, that is, in the signed digraph there are two arrows connecting $x_i$ and $x_j$, one with a positive sign and one with a negative sign. The fourth condition helps eliminate cyclic behavior; there can be no closed loop among three or more distinct nodes. The last condition helps guarantee the equilibrium is isolated and is hyperbolic (indeterminate cases).

These five conditions are due to Quirk and Ruppert (1965). Example 5.23 satisfies the five conditions. Example 5.24 does not satisfy condition 4 because there is a closed loop between the $H_i$. Thus, the matrix in Example 5.24 is not qualitatively stable.

At this point, it is not possible to say whether the matrix in Example 5.23 is qualitatively stable (the five conditions do not distinguish between neutral and asymptotic stability). One more condition is needed to obtain sufficient conditions for qualitative stability. The signed matrix $Q$ must *fail the color test.* The color test is performed on the signed digraph. This additional condition is due to Jeffries (1974). Before we state the color test, we need to define two terms: predation link and predation community.

**Definition 5.9.** A *predation link* in a signed digraph is a pair of nodes connected by one edge with a $+$ sign and another edge with a $-$ sign. A *predation community* is a subgraph consisting of all connected predation links.

For example, in Figure 5.19, there are two predation links between $x_1$ and $x_2$ and between $x_1$ and $x_3$. Since these two links are connected by node $x_1$, the entire signed digraph forms a predation commmunity.

## Color Test:

Color each node with a negative feedback loop gray and all other nodes white. For each predation community the following *color test* is applied.

(i) There is at least one white node.

(ii) Each white node is connected by a predation link to one other white node.

(iii) Each gray node connected by a predation link to one white node is also connected by a predation link to another white node.

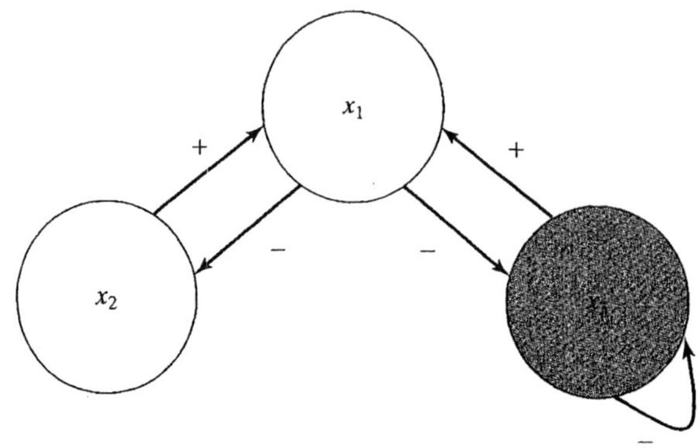


**Figure 5.21** Signed digraph with the color test applied for Example 5.25.

As an example, consider Figure 5.19. Node $x_3$ is gray and the other two nodes are white. There is at least one white node and each white node is connected by a predation link to one other white node. However, condition (iii) is not satisfied because the gray node is only connected to one white node. Now we state the criteria for qualitative stability.

**Theorem 5.13** *Let $Q = sign\,(J) = (q_{ij})$. If matrix $J$ satisfies the five necessary conditions for qualitative stability and if, in addition, each predation community in the digraph associated with matrix $Q$ fails at least one of the three conditions (i)–(iii) in the color test, then matrix $J$ is qualitatively stable.* □

The conditions for qualitative stability can be applied to Example 5.23.

**Example 5.25** **(Example 5.23 Revisited).** The predation community in this example is given in Figure 5.21. We have already shown that the five criteria are satisfied for the matrix $Q$ and that conditions (i) and (ii) are satisfied in the color test. Since condition (iii) is not satisfied, the signed digraph fails the color test. Hence, the underlying Jacobian matrix $J$ corresponding to the predator-prey system, where $Q = sign\,(J)$, is qualitatively stable. Thus, if the predator-prey system has a positive equilibrium with Jacobian matrix $J$, then the equilibrium is locally asymptotically stable. ■

## 5.11 Global Stability and Liapunov Functions

An important technique in stability theory for differential equations is known as the *direct method of Liapunov*. A function with particular properties known as a *Liapunov function* is constructed to prove stability or asymptotic stability of an equilibrium in a given region. The construction of Liapunov functions is often difficult for particular systems, but for Lotka-Volterra systems, there has been some success.

A procedure referred to as the direct method of Liapunov for studying the stability of a equilibrium is discussed in this section (Hale and Koçak, 1991; LaSalle and Lefschetz, 1961). This method has practical importance because estimates for the *basin of attraction* of the equilibrium can be obtained. (A basin of attraction is a subset $U$ in $\mathbf{R}^n$ containing the equilibrium with the property that solutions beginning in $U$ approach the equilibrium.) The method can be applied to autonomous and nonautonomous systems consisting of $n$ differential equations. The method is demonstrated for the following two-dimensional autonomous system:

$$\frac{dx}{dt} = f(x, y) \quad \text{and} \quad \frac{dy}{dt} = g(x, y). \tag{5.22}$$

The objective is to find a particular function, a Liapunov function, having certain properties in relation to system (5.22). In the following discussion, it is assumed that the equilibrium of interest is at the origin. If the equilibrium is not at the origin, a change of variable translates the equilibrium to the origin, $u = x - \bar{x}$ and $v = y - \bar{y}$.

**Definition 5.10.** Let $U$ be an open subset of $\mathbf{R}^2$ containing the origin. A real-valued $C^1(U)$ function $V$, $V: U \rightarrow \mathbf{R}$, $[(x, y) \in U, V(x, y) \in \mathbf{R})]$ is said to be *positive definite* on the set $U$ if the following two conditions hold:

(i) $V(0, 0) = 0$.

(ii) $V(x, y) > 0$ for all $(x, y) \in U$ with $(x, y) \neq (0, 0)$.

The function $V$ is said to be *negative definite* if $-V$ is positive definite.

**Example 5.26** The function $V(x, y) = x^2 + y^2$ is positive definite on all of $\mathbf{R}^2$, while the function $V(x, y) = x^2 + y^2 - y^3$ is positive definite only near the $x$-axis. On the other hand, the functions $V(x, y) = x + y^2$, $V(x, y) = (x + y)^2$, and $V(x, y) = x^2$ are not positive definite in any open neighborhood of the origin. Figure 5.22 shows the set of points in the $x$-$y$ plane at which $V(x, y) = 0$ for these latter three examples. ■

If $V$ is a positive definite function on the set $U$, then $V$ has a minimum at the origin. This extreme point of $V$ is isolated so that the surface $z = V(x, y)$, near the origin, has the general shape of a paraboloid with vertex at the origin. The intersections of this graph with the horizontal plane $z = k$, that is, the *level sets* of $V$,

$$V^{-1}(k) \equiv \{(x, y) \in \mathbf{R}^2: V(x, y) = k\},$$

are closed curves for small $k > 0$. The projection of these level sets onto the $x$-$y$ plane results in concentric ovals encircling the origin. See Figure 5.23.

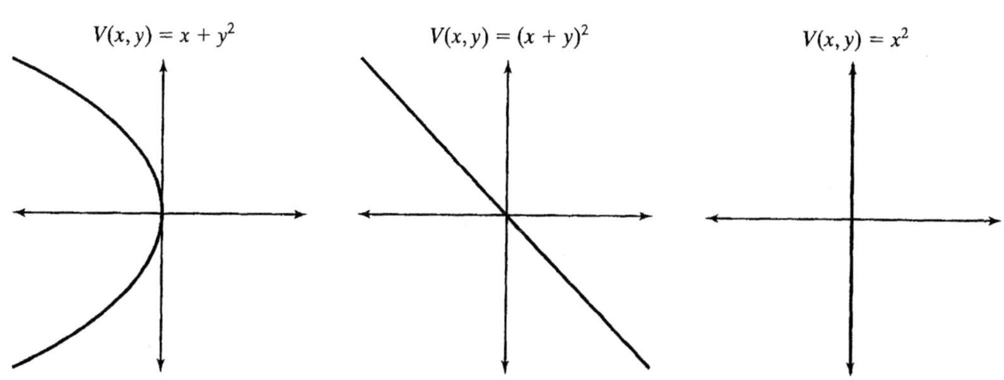


**Figure 5.22** The set in the $x$-$y$ plane (bold curves) such that $V(x, y) = 0$.

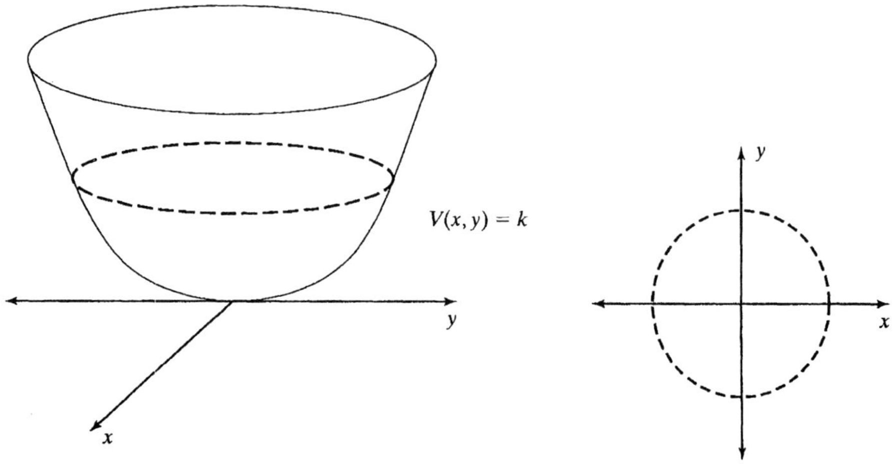


**Figure 5.23** The level sets of the function $V$: $V(x, y) =$ constant.

It is important to know how solutions of the two-dimensional autonomous system cross the level sets of a positive definite function $V$. Let $(x(t), y(t))$ be a solution of the differential system. Then

$$\frac{dV(x(t), y(t))}{dt} = \frac{\partial V}{\partial x}\frac{dx}{dt} + \frac{\partial V}{\partial y}\frac{dy}{dt}.$$

The preceding expression is the inner product of the vector $(f, g)$ with the gradient vector,

$$\frac{dV(x, y)}{dt} = (f(x, y), g(x, y)) \cdot \nabla V(x, y) = \|(f(x, y), g(x,y))\| \, \|\nabla V(x, y)\| \cos\theta,$$

where $\theta$ is the angle between $(f, g)$ and the gradient of $V$, and $\|(a, b)\|$ means the Euclidean norm, $\|(a, b)\| = \sqrt{a^2 + b^2}$. The gradient vector is the outward normal vector to the level curve of $V$ at $(x, y)$. Thus, if $dV(x, y)/dt < 0$, then the angle between $(f, g)$ and $\nabla V$ at $(x, y)$ is obtuse, which implies that the orbit through $(x, y)$ is crossing the level curve from the outside to the inside. Similarly, if $dV(x, y)/dt = 0$, then the orbit is tangent to the level curve; if $dV(x, y)/dt > 0$, the orbit is crossing the level curve from the inside to the outside. See Figure 5.24.

The definition of a Liapunov function is given next. Then Liapunov's stability theorem is stated and verified. Although the definitions and theorem are stated for the equilibrium at the origin, remember that the results can be applied to any equilibrium.

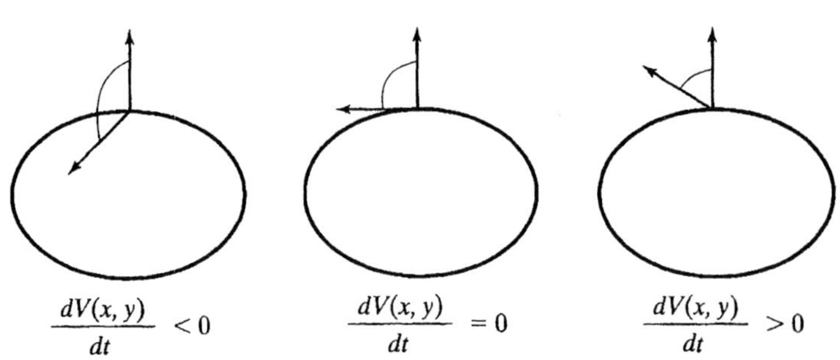


**Figure 5.24** The gradient $\nabla V(x, y)$ is normal to the curve of $V(x, y) = k$. The sign of $dV(x, y)/dt$ depends on the angle between $\nabla V(x, y)$ and the orbit $(x(t), y(t))$ as the orbit crosses the curve $V(x, y) = k$.

**Definition 5.11.** A positive definite function $V$ in an open neighborhood of the origin is said to be a *Liapunov function* for the autonomous differential system, $dx/dt = f(x, y), dy/dt = g(x, y)$, if $dV(x, y)/dt \leq 0$ for all $(x, y) \in U - \{(0, 0)\}$. If $dV(x, y)/dt < 0$ for all $(x, y) \in U - \{(0, 0)\}$, the function $V$ is called a *strict Liapunov function.*

**Theorem 5.14** **(Liapunov's Stability Theorem).** *Let $(0, 0)$ be an equilibrium of the autonomous system (5.22) and let $V$ be a positive definite $C^1$ function in a neighborhood $U$ of the origin.*

(i) *If $dV(x, y)/dt \leq 0$ for $(x, y) \in U - \{(0, 0)\}$, then (0, 0) is stable.*

(ii) *If $dV(x, y)/dt < 0$ for $(x, y) \in U - \{(0, 0)\}$, then (0, 0) is asymptotically stable.*

(iii) *If $dV(x, y)/dt > 0$ for $(x, y) \in U - \{(0, 0)\}$, then (0, 0) is unstable.*

In Case (i) the function $V$ is a *Liapunov function* and in Case (ii) $V$ is a *strict Liapunov function.*

**Proof** Cases (i) and (ii) are verified.

Case (i) Let $\epsilon > 0$ be sufficiently small so that the neighborhood of the origin consisting of the points $\|(x, y)\| \leq \epsilon$ is contained in $U$ ($\|\cdot\|$ denotes the Euclidean norm). Let $m$ be the minimum value of $V$ on the boundary of the neighborhood, $\|(x, y)\| = \epsilon$. Since $V$ is positive definite and the set $\|(x, y)\| = \epsilon$ is closed and bounded, it follows that $m > 0$. Now, choose a $\delta > 0$ with $0 < \delta \leq \epsilon$ such that $V(x, y) < m$ for $\|(x, y)\| \leq \delta$. Such a $\delta$ always exists because $V$ is continuous with $V(0, 0) = 0$. If $\|(x_0, y_0)\| \leq \delta$, then the solution with initial conditions $(x_0, y_0)$ satisfies $\|(x, y)\| \leq \epsilon$ for $t \geq 0$ since $dV/dt \leq 0$ implies that $V(x(t), y(t)) \leq V(x_0, y_0) < m$ for $t \geq 0$. The origin is stable.

Case (ii) The function $V(x(t), y(t))$ decreases along solutions that lie in $U$. Thus, as $t \to \infty$, $V(x(t), y(t))$ approaches a limit. Suppose $V \to l > 0$. Then it follows from the uniform continuity of $dV(x(t), y(t))/dt$ (solutions are bounded and $f$ and $g$ are $C^1$) that $dV(x(t), y(t))/dt \to 0$ in an annular region excluding the origin. This is impossible, since $-dV/dt$ is positive definite, $dV/dt = 0$ only at the origin, and $(x(t), y(t))$ does not tend to the origin when $V \to l$. It follows that $V(x(t), y(t))$ approaches 0, which implies $(x(t), y(t))$ approaches $(0, 0)$. The origin is asymptotically stable. □

The difficulty in verifying stability using Liapunov's direct method is finding a suitable Liapunov function $V$.

**Example 5.27** Consider the logistic differential equation,

$$\frac{dx}{dt} = rx\left(1 - \frac{x}{K}\right),$$

where $r, K > 0$. There are two equilibria $\bar{x} = 0, K$. We know from previous analyses that $K$ is globally asymptotically stable for positive initial conditions. Let $U = (0, \infty) = \mathbf{R}_+$, the positive $x$ axis. A strict Liapunov function is given by

$$V(x) = (x - K)^2.$$

Since

$$\frac{dV(x)}{dt} = 2(x - K)\frac{dx}{dt} = 2(x - K)rx\left(1 - \frac{x}{K}\right) = -2rx\frac{(x - K)^2}{K}.$$

The function $V$ is a $C^1(U)$ function that is positive except at $x = K, V(K) = 0$. Also, $-dV/dt$ is positive in $U$ except at $x = K, dV(K)/dt = 0$. Thus, according to part (ii) of Liapunov's Stability Theorem, the equilibrium $x = K$ is asymptotically stable for all initial values in $U$. ■

**Example 5.28** Consider the Lotka-Volterra predator-prey system,

$$\frac{dx}{dt} = x(a - y)$$

$$\frac{dy}{dt} = y(-b + x), \quad a, b > 0$$

The positive equilibrium is $(\bar{x}, \bar{y}) = (b, a)$. The equilibrium is stable but not asymptotically stable. A Liapunov function for this system has the form

$$V(x, y) = \left(x - b - b\ln\left(\frac{x}{b}\right)\right) + \left(y - a - a\ln\left(\frac{y}{a}\right)\right).$$

First note that $V(x, y)$ is continuous and differentiable for $x, y > 0$ and $V(b, a) = 0$. Next, we show that $V(x, y) > 0$ for $x, y > 0$ and $x \neq b, y \neq a$. Note that the partial derivatives are $V_x = 1 - b/x$, $V_y = 1 - a/y$, $V_{xx} = b/x^2$, $V_{yy} = a/y^2$, and $V_{xy} = 0$. Thus, a critical point of $V$ occurs when $V_x = 0 = V_y$ at $x = b$ and $y = a$. This critical point is a local minimum because we can apply the following test: $V_{xx}V_{yy} - V_{xy}^2 > 0$ and $V_{xx} > 0$. At the local minimum $V(b, a) = 0$ and $(b, a)$ is the only local minimum for $x, y > 0$ so it is a global minimum for $x, y > 0$. Thus, $V(x, y) > 0$ for $x, y > 0$ and $x \neq b, y \neq a$ (positive definite in $\mathbf{R}_+^2$). Next, we calculate the derivative of $V$ along solutions:

$$\begin{aligned}\frac{dV}{dt} &= V_x\frac{dx}{dt} + V_y\frac{dy}{dt} \\ &= \left(1 - \frac{b}{x}\right)x(a - y) + \left(1 - \frac{a}{y}\right)y(-b + x) \\ &= (x - b)(a - y) + (y - a)(x - b) = 0.\end{aligned}$$

Thus, according to part (i) of Liapunov's stability theorem, the equilibrium $(b, a)$ is stable in the region $\mathbf{R}_+^2 = \{(x, y)|x > 0, y > 0\}$; that is, globally stable in $\mathbf{R}_+^2$. ■

**Example 5.29** Goh (1977) has shown that in many Lotka-Volterra systems with a unique positive equilibrium given by $(\bar{x}_1, \ldots, \bar{x}_n)$, there exists a Liapunov function in $\mathbf{R}_+^n$ which has a form similar to the one given in the previous example. The Liapunov function has the form

$$V(x_1, \ldots, x_n) = \sum_{i=1}^{n} c_i\left[x_i - \bar{x}_i - \bar{x}_i\ln\left(\frac{x_i}{\bar{x}_i}\right)\right],$$

where the constants $c_i$ are positive and are chosen dependent on the parameters of the particular system. ■

## 5.12 Persistence and Extinction Theory

We end this chapter by discussing the concept of persistence and extinction, an important concept for biological systems. Basically, persistence of a system means no state of the system approaches zero, that is, there can be no extinction of any of the populations that make up the biological system.

**Definition 5.12.** Given a system of differential equations, $dX/dt = F(X, t)$, $X(0) = X_0$, where $X(t) = (x_1(t), x_2(t), \ldots, x_n(t))^T$, the system is said to be *persistent* if for any positive initial conditions, $X_0 > 0$, the solution $X(t)$, satisfies

$$\liminf_{t\to\infty} x_i(t) > 0$$

for $i = 1, 2, \ldots, n$.

There are other definitions of persistence that either weaken or strengthen the previous definition. For example, the system is said to be *weakly persistent* if

$$\limsup_{t\to\infty} x_i(t) > 0$$

for $i = 1, 2, \ldots, n$; *uniformly persistent* if there exists $\delta > 0$ such that

$$\liminf_{t\to\infty} x_i(t) > \delta$$

for $i = 1, 2, \ldots, n$; *permanent* if there exists a time $T > 0$ and a compact set $K$ in the interior of the positive cone, $\mathbf{R}^n_+ = \{(x_1, x_2, \ldots, x_n) \in \mathbf{R}^n | x_i > 0, i = 1, 2, \ldots, n\}$ such that $X(t) \in K$ for $t > T$. (Solutions enter the compact set $K$ and remain in $K$.) Persistence or extinction of a subset of the set $\{x_i\}_{i=1}^n$ can also be defined (e.g., some populations survive and some do not).

Weak persistence and persistence are generally not very good indications of population survival because solutions may be initial condition dependent. For example, in the case of persistence, there could be a set of initial conditions $\{X_0^k\}_{k=1}^\infty$ such that the corresponding solution $X^k(t) = (x_i^k(t))$ satisfies

$$\epsilon_k > \liminf_{t\to\infty} x_i^k(t) > 0,$$

where $\epsilon_k \to 0$ as $k \to \infty$ for some $i$. Even uniform persistence and permanence may not be very good measures of survival since solutions may approach very close to the extinction boundaries if $\delta$ is small or the compact set $K$ is close to the extinction boundaries. Another more reasonable type of persistence criterion is referred to as *practical persistence*. Practical persistence requires that the bounds on the solutions be specified a priori (dependent on population data). Given $L_i > 0$ and $M_i > 0$, solutions $x_i(t)$ exhibit *practical persistence* if $0 < L_i < \liminf_{t\to\infty} x_i(t) \le \limsup_{t\to\infty} x_i(t) \le M_i$, $i = 1, 2, \ldots, n$ (Cantrell and Cosner, 1996; Cao and Gard, 1997).

In general, practical persistence implies permanence. Persistence implies weak persistence. If solutions are uniformly bounded, $\limsup_{t\to\infty} x_i(t) < M$, $i = 1, \ldots, n$, then uniform persistence and permanence are equivalent. If a system has a globally stable equilibrium in $\mathbf{R}^n_+$, then it is permanent. The converse of this statement is not true. If a system is permanent, it may not have a globally stable equilibrium. Further discussion and examples of systems that are permanent or persistent may be found in Hofbauer and Sigmund (1988, 1998) or Freedman and Moson (1990).

**Example 5.30** A simple Lotka-Volterra food chain has the following form:

$$\frac{dx_1}{dt} = x_1(a_{10} - a_{11}x_1 - a_{12}x_2)$$

$$\frac{dx_2}{dt} = x_2(-a_{20} + a_{21}x_1 - a_{23}x_3) \tag{5.23}$$

$$\vdots$$

$$\frac{dx_{n-1}}{dt} = x_{n-1}(-a_{n-1,0} + a_{n-1,n-2}x_{n-2} - a_{n-1,n}x_n)$$

$$\frac{dx_n}{dt} = x_n(-a_{n0} + a_{n,n-1}x_{n-1}),$$

where all of the coefficients are strictly positive except possibly $a_{11} \geq 0$. Assume $x_i(0) > 0$ for $i = 1, 2, \ldots, n$. In this model, $x_1$ is at the bottom of the food chain, $x_2$ feeds on $x_1$, $x_3$ feeds on $x_2$, and so on. The top predator is $x_n$. Also note that the per capita growth rate for all populations is negative, except for the bottom of the food chain $x_1$. In the absence of $x_1$, the food chain collapses [i.e., all other populations $x_i(t), i = 2, \ldots, n$ approach zero]. Gard and Hallam (1979) give conditions to show that this system is weakly persistent [i.e., $\limsup_{t\to\infty} x_i(t) > 0,\ i = 1, 2 \ldots, n$]. ■

We verify weak persistence in the case of a three-trophic-level food chain model ($n = 3$) when $a_{11} > 0$. For example, at the bottom of the food chain $x_1$ is a plant, $x_2$ is a herbivore that eats the plant, and $x_3$ is a predator that eats the herbivore (e.g., $x_1$ = grass, $x_2$ = rabbits, and $x_3$ = coyotes). A *persistence function* is used to verify weak persistence but it requires that a particular constant be positive,

$$\mu = a_{10} - \frac{a_{11}}{a_{21}}a_{20} - \frac{a_{12}}{a_{32}}a_{30}. \tag{5.24}$$

It can be shown that for $\mu > 0$ there are exactly four nonnegative equilibria: $E_0 = (0, 0, 0)$, $E_1 = (x_1^*, 0, 0)$, $E_2 = (\bar{x}_1, \bar{x}_2, 0)$, and $E_3 = (\bar{x}_1, \bar{x}_2, \bar{x}_3)$. In addition, for $\mu > 0$, $E_0$, $E_1$, and $E_2$ are unstable and $E_3$ is locally asymptotically stable. We verify the following persistence result. See Gard and Hallam (1979) or Hallam and Levin (1986) for more details of the proof in the general case.

**Theorem 5.15** *Consider the food chain (5.23) with $n = 3$ and $a_{11} > 0$. Let $\mu$ be defined as in (5.24). If $\mu > 0$, then the food chain (5.23) is weakly persistent, and if $\mu < 0$, then the food chain is not persistent.*

**Proof** First, note that all solutions remain in the positive cone $\mathbf{R}^3_+ = \{(x_1, x_2, x_3) | x_i > 0, i = 1, 2, 3\}$, if initial conditions are positive. Second, the solution $x_1(t)$ is bounded. Boundedness of $x_1$ follows from

$$\frac{dx_1}{dt} \leq x_1(a_{10} - a_{11}x_1).$$

By comparison with the logistic differential equation (Corduneanu, 1977), it can be shown that

$$\limsup_{t\to\infty} x_1(t) \leq \frac{a_{10}}{a_{11}}.$$

Third, it can be shown that every solution $x_i(t)$ is bounded, $i = 1, 2, 3$. Define

$$u = a_{21}a_{32}x_1 + a_{12}a_{32}x_2 + a_{12}a_{23}x_3.$$

Then

$$\begin{aligned}\frac{du}{dt} &= a_{21}a_{32}x_1(a_{10} - a_{11}x_1) + a_{12}a_{32}x_2(-a_{20}) + a_{12}a_{23}x_3(-a_{30}) \\ &= a_{21}a_{32}x_1(-a_{10}) + a_{12}a_{32}x_2(-a_{20}) + a_{12}a_{23}x_3(-a_{30}) \\ &\quad + a_{21}a_{32}x_1(2a_{10} - a_{11}x_1) \\ &\leq -mu + b,\end{aligned}$$

where $m = \min\{a_{10}, a_{20}, a_{30}\}$ and $b = \max_{x_1}\{a_{21}a_{32}x_1(2a_{10} - a_{11}x_1)\}$. By comparison, the solution $u(t)$ is bounded. But $u(t)$ is bounded iff $x_i(t)$ is bounded for $i = 1, 2, 3$.

Next, we define a persistence function

$$\rho(t) = x_1^{r_1}x_2^{r_2}x_3^{r_3},$$

where $r_1 = 1$, $r_2 = a_{11}/a_{21}$, $r_3 = a_{12}/a_{32}$. Since solutions are bounded, $\rho(t) \to 0$ if $x_i(t) \to 0$ for some $i = 1, 2, 3$.

Suppose $\mu > 0$. Assume for some $i$, $i = 1, 2, 3$, that $\limsup_{t\to\infty} x_i(t) = 0$. Then $\lim_{t\to\infty} x_i(t) = 0$ since solutions are positive for all time. Thus, $\rho(t) \to 0$. We obtain a contradiction to this assumption when $\mu > 0$. If $i < 3$, then it can be seen for some time $T$, $t > T$,

$$\frac{dx_{i+1}}{dt} \leq -\frac{a_{i+1,0}}{2}x_{i+1},$$

since $x_i$ can be made sufficiently small: $x_i(t) \leq a_{i+1,0}/(2a_{i+1,i})$ for $t > T$. Then $x_{i+1}$ is an exponentially declining population, so that $\lim_{t\to\infty} x_{i+1}(t) = 0$. Thus, if $i = 2$, then the top predator $x_3(t)$ dies out. If $i = 1$, then $x_2(t)$ dies out, but by a similar argument it can be seen that $x_3(t) \to 0$. In any case, the top predator dies out. Now, we take the derivative of $\rho$,

$$\begin{aligned}\frac{d\rho}{dt} &= r_1x_1^{r_1-1}\dot{x}_1\, x_2^{r_2}x_3^{r_3} + r_2x_2^{r_2-1}\dot{x}_2x_1^{r_1}x_3^{r_2} + r_3x_3^{r_3-1}\dot{x}_3x_1^{r_1}x_2^{r_2} \\ &= \rho(\mu - r_2a_{23}x_3),\end{aligned}$$

where $\dot{x}_i$ means $dx_i/dt$. Since $x_3(t) \to 0$, it follows for sufficiently large $t$, $\dot{\rho} \geq \rho(\mu/2)$. Thus, $\rho(t) \to \infty$, a contradiction. Hence, $\limsup_{t\to\infty} x_i(t) > 0$ for $i = 1, 2, 3$, the system is weakly persistent.

Suppose $\mu < 0$. Then as shown above,

$$\frac{d\rho}{dt} \leq \rho(\mu - r_2a_{23}x_3) \leq \mu\rho.$$

Hence, $\rho(t) \to 0$. Suppose the system is persistent, so that $\liminf_{t\to\infty} x_i(t) > 0$ for $i = 1, 2, 3$. This contradicts the fact that $\rho(t)$ approaches zero. Hence, the system is not persistent. □

In the general food chain model (5.23) with $a_{11} > 0$ it can be shown that there exists a positive equilibrium $\bar{X} = (\bar{x}_1, \bar{x}_2, \ldots, \bar{x}_n)$. A strict Liapunov function can be constructed to show global asymptotic stability of the positive equilibrium (Harrison, 1979). An equilibrium of (5.23) satisfies

$$\begin{aligned}-a_{i0} &= -a_{i,i-1}\bar{x}_{i-1} + a_{i,i+1}\bar{x}_{i+1}, \\ a_{10} &= a_{11}\bar{x}_1 + a_{12}\bar{x}_2\end{aligned}$$

for $i = 2, \ldots, n$. Notice that if $\bar{x}_1 = 0$, then $\bar{x}_i = 0$ for $i = 2, \ldots, n$. Rewrite the differential equations as

$$\frac{dx_i}{dt} = x_i(a_{i,i-1}[x_{i-1} - \bar{x}_{i-1}] - a_{ii}[x_i - \bar{x}_i] - a_{i,i+1}[x_{i+1} - \bar{x}_{i+1}]).$$

Define the Liapunov function,

$$V(x_1, x_2, \ldots, x_n) = \sum_{i=1}^{n} C_i\left[x_i - \bar{x}_i - \bar{x}_i \ln\left(\frac{x_i}{\bar{x}_i}\right)\right],$$

where the $C_i$ are chosen such that $C_i a_{i,i+1} = C_{i+1} a_{i+1,i}$ (Hallam and Levin, 1986; Harrison, 1979). Then it can be shown that $dV/dt$ satisfies

$$\frac{dV}{dt} = -\sum_{i=1}^{n} C_i a_{ii}[x_i - \bar{x}_i]^2 \le 0.$$

Since $a_{11} > 0$, $dV/dt = 0$ iff $x_1 = \bar{x}_1$. It can be shown that the only invariant set of the differential equations is the equilibrium; then $\lim_{t\to\infty} x_i(t) = \bar{x}_i$ (Harrison, 1979). For another variation on a simple food chain model, see Jang and Baglama (2000).

## 5.13 Exercises for Chapter 5

**1.** Suppose $f$ can be expanded using Taylor's formula with a remainder:

$$f(x) = f(\bar{x}) + f'(\bar{x})(x - \bar{x}) + \frac{f''(\bar{x})}{2!}(x - \bar{x})^2 + \frac{f'''(\bar{x})}{3!}(x - \bar{x})^3 + \frac{f^{(4)}(\xi)}{4!}(x - \bar{x})^4,$$

where $\xi$ is between $x$ and $\bar{x}$. Suppose $dx/dt = f(x)$ and $\bar{x}$ is an equilibrium solution. Verify the following.

(a) If $f'(\bar{x}) = 0$ and $f''(\bar{x}) \ne 0$, then $\bar{x}$ is unstable.

(b) Suppose $f'(\bar{x}) = 0 = f''(\bar{x})$. If $f'''(\bar{x}) < 0$, then $\bar{x}$ is locally asymptotically stable, and if $f'''(\bar{x}) > 0$, then $\bar{x}$ is unstable.

**2.** (a) Make the following change of variables in the logistic differential equation, $u(t) = (K - x(t))/x(t)$, to show that

$$\frac{du}{dt} = -ru, \quad u(0) = \frac{K - x(0)}{x(0)}.$$

(b) Solve the differential equation for $u(t)$; then find $x(t)$ and show that $x(t)$ satisfies equation (5.3).

**3.** A mathematical model for the growth of a population is

$$\frac{dx}{dt} = \frac{2x^2}{1 + x^4} - x = f(x), \quad x(0) \ge 0,$$

where $x$ is the population density. Find the equilibria and determine their stability. Sketch $f(x)$.

**4.** For the following differential equations, find the equilibria; then graph the phase line diagrams. Use the phase line diagrams to determine the stability of the equilibria.

(a) $dx/dt = \sin(x)\cos(x)$

(b) $dx/dt = x(a - x)(x - b)^2, \quad 0 < a < b$

(c) $dx/dt = x(e^x - x - 2)$

**5.** A spatially implicit model for the proportion of islands occupied by a species was modeled by Levins (1969, 1970) and Hanski (1999). The model takes the following form:

$$\frac{dp}{dt} = (m + cp)(1 - p) - ep = f(p),$$

where $p(t)$ is the fraction of occupied islands at time $t$, $m \geq 0, c > 0$, and $e > 0$. The constants $m$, $c$, and $e$ are the rates of immigration from the mainland, colonization, and extinction, respectively.

(a) Suppose there is no immigration from the mainland, $m = 0$, and the colonization rate is greater than the extinction rate, $c > e$. In Levins original model, $m = 0$. For this model, find the nonnegative equilibria and determine their stability. What happens if the colonization rate is less than the extinction rate, $c < e$?

(b) Suppose there is immigration from the mainland, $m > 0$. For this model, show that there exists a unique positive equilibrium which is asymptotically stable for all initial values $0 \leq p(0) \leq 1$. Note that $f(p)$ is quadratic in $p$. Use a phase line diagram.

**6.** Growth of a population is modeled by the following differential equation:

$$\frac{dN}{dt} = \frac{\alpha n_2 + \beta n_1}{N} - (\alpha + \beta),$$

where $\alpha, \beta, n_1$, and $n_2$ are positive constants.

(a) Find the equilibrium solution for this model; then draw a phase line diagram.

(b) If $N(0) > 0$, find $\lim_{t\to\infty} N(t)$.

**7.** A model with an Allee effect is derived by Thieme (2003). The model is based on the fact that unmated females must find a mate and during the time they are searching, before reproduction, they experience an increased mortality due to predation. The model for the population size $N(t)$ at time $t$ satisfies the following differential equation:

$$\frac{dN}{dt} = rN\left(1 - \frac{N}{K} - \frac{a}{1 + bN}\right), \quad N(0) > 0,$$

where $r$, $a$, $b$, and $K$ are positive parameters. We use $a$ to refer to the Allee effect. If $a = 0$, the model is the logistic growth equation.

(a) Make the following change of variables: $\tau = rt$ and $x(\tau) = bN(t)$. Then show that the differential equation can be expressed in the following form:

$$\dot{x} = x\left(1 - \frac{x}{M} - \frac{a}{1 + x}\right), \quad x(0) > 0, \tag{5.25}$$

where $M = bK$ and $\dot{x} = dx/d\tau$. Equation (5.25) is in dimensionless form. Assume $M > 1$.

(b) For $0 < a < 1$, show that model (5.25) has a unique stable positive equilibrium, $\bar{x} < M$.

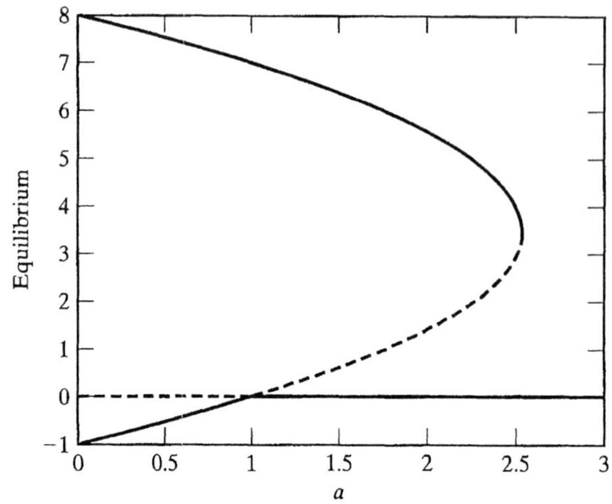


**Figure 5.25** A bifurcation diagram for model (5.25). The dashed curves represent unstable equilibria and the solid curves represent stable equilibria.

(c) For $1 < a < (M + 1)^2/(4M)$, show that model (5.25) has two positive equilibria, $0 < \bar{x}_1 < \bar{x}_2$. Show that $\bar{x}_1$ is unstable and $\bar{x}_2$ is locally stable. Thieme (2003) refers to $\bar{x}_1$ as a *watershed equilibrium* or as a *breakpoint density*. This equilibrium divides the solution behavior into two types. When initial conditions are below this equilibrium value, solutions approach zero, and above this equilibrium value, solutions approach the stable equilibrium $\bar{x}_2$.

(d) For $a > (M + 1)^2/(4M)$, find $\lim_{\tau\to\infty} x(\tau)$.

**8.** In Exercise 7, it can be seen that the parameter $a$ plays the role of a bifurcation parameter in model (5.25), that is, the solution behavior changes as $a$ varies. Assume $M = 8$.

(a) Show that the bifurcation diagram based on the parameter $a$ in model (5.25) is given by the graph in Figure 5.25.

(b) Show that there is a transcritical bifurcation when $a = 1$ near the equilibrium $\bar{x} = 0$.

(c) Show there is a saddle node bifurcation when $a = (M + 1)^2/(4M) = 81/32$ near the equilibrium $\bar{x} = (M - 1)/2 = 3.5$.

**9.** The Jacobian matrices evaluated at a positive equilibrium $(\bar{x}, \bar{y}, \bar{z})$ for two different systems of differential equations are computed below. Give conditions for stability of the equilibrium $(\bar{x}, \bar{y}, \bar{z})$.

(a) $J = \begin{pmatrix} a - \bar{x}/2 & -3 & a \\ 0 & b - \bar{y}/3 & 1 \\ 0 & 0 & c - \bar{z} \end{pmatrix}, \quad a, b, c > 0$

(b) $J = \begin{pmatrix} -b\bar{x} & -3 & a \\ 0 & -\bar{y} & b\bar{y} \\ 0 & \bar{z} & -\bar{z} \end{pmatrix}, \quad b > 0$

**10.** A competition model satisfies

$$\frac{dx}{dt} = r_1 x(1 - x - a_1 y),$$

$$\frac{dy}{dt} = r_2 y(1 - y - a_2 x),$$

where the intrinsic growth rates $r_i > 0, i = 1, 2,$ and the competition coefficients satisfy $0 < a_i < 1, i = 1, 2$ (weak competition).

(a) Find all of the equilibria.

(b) Determine the conditions for local asymptotic stability for each of the equilibria.

**11.** Make the following change of variables from rectangular coordinates to polar coordinates:

$$x = r\cos\theta,\quad y = r\sin\theta,\quad r^2 = x^2 + y^2,\quad \theta = \arctan(y/x).$$

Then show that

$$r\frac{dr}{dt} = x\frac{dx}{dt} + y\frac{dy}{dt} \quad \text{and} \quad r^2\frac{d\theta}{dt} = x\frac{dy}{dt} - y\frac{dx}{dt}.$$

**12.** (a) Show that the origin is a nonhyperbolic equilibrium for the following system:

$$\frac{dx}{dt} = y + x(x^2 + y^2)$$

$$\frac{dy}{dt} = -x + y(x^2 + y^2)$$

(b) Make the change of variables given in Exercise 11 to express the system in part (a) in terms of polar coordinates,

$$\frac{dr}{dt} = r^3 \quad \text{and} \quad \frac{d\theta}{dt} = -1.$$

Graph the phase line diagram for $r$. Then find the solution in terms of $r$ and $\theta$ when $x(0) = 1$ and $y(0) = 0$. Describe the solution behavior.

**13.** (a) For the following system determine the local asymptotic stability of the origin.

$$\frac{dx}{dt} = y + x(x^2 + y^2 - 4)$$

$$\frac{dy}{dt} = -x + y(x^2 + y^2 - 4)$$

(b) Make the change of variables given in Exercise 11 to express the system in part (a) in terms of polar coordinates,

$$\frac{dr}{dt} = r(r^2 - 4),\quad \frac{d\theta}{dt} = -1.$$

(c) Graph the phase line diagram for $r$. Describe the solution behavior when $0 < r < 2$. When $r > 2$.

**14.** Suppose after a change of variables, the differential equations in terms of polar coordinates satisfy

$$\frac{dr}{dt} = (r - 1)(r - 2) \quad \text{and} \quad \frac{d\theta}{dt} = 2.$$

Graph the solution behavior on a phase line diagram for $r$. What is the solution behavior if $0 < r < 1$? If $1 < r < 2$? If $r > 2$?

**15.** Lotka-Volterra competition equations satisfy

$$\frac{dx_1}{dt} = x_1(K_1 - x_1 - ax_2),$$

$$\frac{dx_2}{dt} = x_2(K_2 - bx_1 - x_2),$$

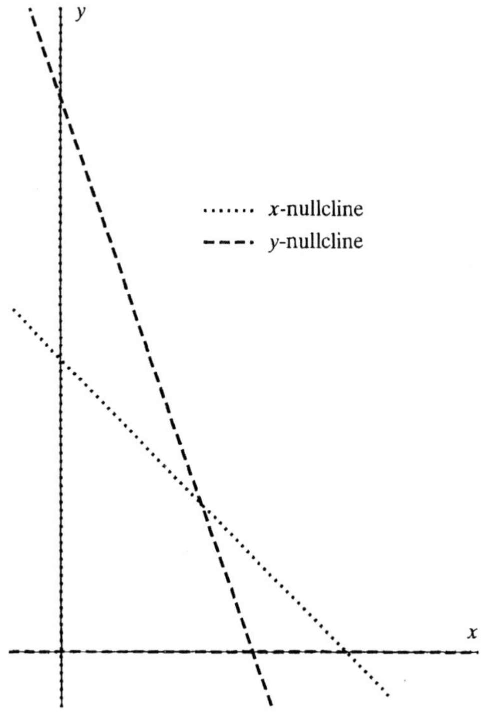


**Figure 5.26** Nullclines for $x$ and $y$.

where the parameters are all positive: $K_1, K_2, a, b > 0$. Apply Dulac's criterion to show there are no limit cycles for $x_1, x_2 > 0$. [*Hint:* Use $B(x_1, x_2) = 1/(x_1x_2)$.]

**16.** A Lotka-Volterra competitive system,

$$\frac{dx}{dt} = x(4 - x - y),$$
$$\frac{dy}{dt} = y(8 - 3x - y),$$

represents the change in densities of two competing species $x$ and $y$. The $x$- and $y$-nullclines are graphed in Figure 5.26.

(a) Graph the direction of flow along each of the nullclines.

(b) Find the coordinates $(\bar{x}, \bar{y})$ of the four equilibria.

(c) Determine the stability of each equilibrium and state whether the equilibrium is a node, spiral, and so on.

**17.** Consider the $n$-species competition model given by

$$\frac{dx_i}{dt} = x_i\left(a_{i0} - \sum_{j=1}^{n} a_{ij}x_j\right), \quad i = 1, 2, \ldots, n,$$

where $a_{i0} > 0$ and $a_{ij} > 0, i, j = 1, 2, \ldots, n$. Show that the region $\mathbf{R}^n_+ = \{(x_1, x_2, \ldots, x_n) | x_i \geq 0, i = 1, 2, \ldots, n\}$ is positively invariant. [*Hint:* Determine the direction of flow on each of the $n$ faces: $\Omega_i = \{(x_1, x_2, \ldots, x_n) \in \mathbf{R}^n_+ | x_i = 0, x_j \geq 0, j \neq i\}, i = 1, 2, \ldots, n$.]

**18.** A predator-prey model satisfies the differential equations,

$$\frac{dH}{dt} = rH\left(1 - \frac{H}{K}\right) - \alpha\frac{PH}{H + \beta}$$
$$\frac{dP}{dt} = \gamma P\left(-1 + \delta\frac{H}{H + \beta}\right),$$

where $H$ is the prey density and $P$ is the predator density. The parameters are $r, K, \alpha, \beta, \gamma, \delta > 0$; $H(0)$, and $P(0) > 0$.

(a) Make the following change of variables to put the model in dimensionless form:

$$x = \frac{H}{K}, \quad y = \frac{\alpha}{rK}P, \quad \text{and} \quad \tau = rt.$$

Then

$$\frac{dx}{d\tau} = x\left(1 - x - \frac{y}{x + b}\right),$$
$$\frac{dy}{d\tau} = cy\left(-1 + a\frac{x}{x + b}\right).$$

(b) How are the new parameters, $a$, $b$, and $c$, defined in terms of the old parameters?

(c) Show that if $a < 1$, the only nonnegative equilibria are $(0, 0)$, $(1, 0)$.

(d) In the case $a < 1$, there is no limit cycle since there is no positive equilibrium. Use the Poincaré Bendixson trichotomy to show that $\lim_{t\to\infty} y(t) = 0$ and $\lim_{t\to\infty} x(t) = 1$. (*Hint:* First show solutions are bounded.)

(e) When $a > 1 + b$ graph the nullclines, identify the equilibria by circles and label the direction of flow along the nullclines.

**19.** Consider an epidemic model, where $S + I + R = N =$ constant, so that $R = N - S - I$. The original model with three differential equations can be simplified to one with just two differential equations,

$$\frac{dS}{dt} = -\frac{\beta}{N}SI + \nu(N - S - I),$$
$$\frac{dI}{dt} = I\left(\frac{\beta}{N}S - \gamma\right).$$

The parameters are all positive: $\beta, N, \gamma, \nu > 0$. The parameter $\beta$ represents the contact rate, $N$ the total population size, $\gamma$ the recovery rate, and $\nu$ the rate of loss of immunity. Define the basic reproduction number

$$\mathcal{R}_0 = \frac{\beta}{\gamma},$$

the number of secondary infections caused by one infectious individual.

(a) If $\mathcal{R}_0 < 1$, find all of the nonnegative equilibria and determine the local asymptotic stability for each of the equilibria. Do a phase plane analysis for initial conditions satisfying

$$S(0) > 0, \quad I(0) > 0, \quad \text{and} \quad S(0) + I(0) \le N.$$

(b) Do the same as in part (a) but assume $\mathcal{R}_0 > 1$.

**20.** For the following nonlinear differential equations, find the stable equilibria as a function of the bifurcation parameter $r$. Then draw the bifurcation diagram. Show that the bifurcation diagrams have a form corresponding to Figure 2.10 II and III.

(a) $dx/dt = rx - x^3$

(b) $dx/dt = rx + x^2$

**21.** For the following nonlinear differential equations, find the stable equilibria as a function of the bifurcation parameter $r$. Then draw the bifurcation diagram.

(a) $dx/dt = rx + x^3$ (pitchfork)

(b) $dx/dt = r - x^2$ (saddle node)

(c) $dx/dt = r + 2x + x^2$ (saddle node)

(d) $dx/dt = rx - x^2$ (transcritical)

(e) $dx/dt = x(r - e^x)$ (transcritical)

**22.** Show that the following system has a saddle node bifurcation at $r = 0$:

$$\frac{dx}{dt} = r + x^2,$$

$$\frac{dy}{dt} = -y.$$

**23.** The following system represents the concentrations of two chemicals, $x$ and $y$ (Hale and Koçak, 1991, pages 360–361):

$$\frac{dx}{dt} = 1 - (r + 1)x + x^2y,$$

$$\frac{dy}{dt} = x(r - xy),$$

where $r > 0$.

(a) Show that there exists a unique positive equilibrium, $(\bar{x}, \bar{y}) = (1, r)$.

(b) Determine the local asymptotic stability of the positive equilibrium.

(c) Show that there is a Hopf bifurcation at $r = 2$.

**24.** Consider the logistic delay differential equation, where the delay is not in the density-dependent term but in the linear growth term,

$$\frac{dx\,(t)}{dt} = rx(t - T)\left(1 - \frac{x(t)}{K}\right), \quad r, K, T > 0.$$

The following problems will show that the local stability of this delay differential equation is the same as the nondelay equation.

(a) The linearization of the differential equation about $\bar{x} = 0$ is $du/dt = ru(t - T)$. Show that the characteristic equation for the linear equation is $\lambda = re^{-\lambda T}$. Verify that the zero equilibrium is unstable.

(b) Let $u(t) = x(t) - K$. Linearize the differential equation about $\bar{x} = K$. Then find the characteristic equation and verify that the equilibrium $\bar{x} = K$ is locally asymptotically stable.

**25.** Suppose the continuous delay logistic equation (5.15) has a strong delay kernel, $k(t) = \frac{t}{T^2}\exp(-t/T)$. Note that $\int_0^\infty k(t)\,dt = 1$. For the equilibrium $K$, use the method of Example 5.22 to verify that there are no positive real eigenvalues $\lambda$ for the linearization of this system near $K$. (In the Appendix to this chapter it is shown that there do not exist any complex eigenvalues with positive real parts if $T < 2/r$. Hence, $K$ is locally asymptotically stable if $T < 2/r$.)

**26.** Suppose a Lotka-Volterra system

$$\frac{dx_i}{dt} = x_i\left(a_{i0} + \sum_{j=1}^{n} a_{ij}x_j\right)$$

has a positive equilibrium $\bar{X} = (\bar{x}_1, \ldots, \bar{x}_n)$ which satisfies $A\bar{X} = b$, where $A = (a_{ij})$ and $b = (-a_{i0})$. Show that the Jacobian matrix evaluated at $\bar{X}$ satisfies $J(\bar{X}) = \text{diag}(\bar{x}_i)A$.

**27.** (a) Draw a signed digraph for each of the following signed matrices $Q_i, i = 1, 2, 3$.
(b) Determine whether the matrix $J_i$ with signed matrix $Q_i$ is qualitatively stable, $i = 1, 2, 3$. Assume $\det(J_i) \neq 0$.

$$Q_1 = \begin{pmatrix} 0 & + & 0 \\ - & - & - \\ 0 & + & 0 \end{pmatrix}, \quad Q_2 = \begin{pmatrix} 0 & + & 0 & 0 \\ - & + & 0 & - \\ 0 & 0 & 0 & + \\ 0 & + & - & 0 \end{pmatrix},$$

$$Q_3 = \begin{pmatrix} - & - & 0 & 0 & + \\ + & - & - & 0 & 0 \\ 0 & + & - & - & 0 \\ 0 & 0 & + & - & - \\ - & 0 & 0 & + & 0 \end{pmatrix}$$

**28.** Suppose there are five species, where species 2 preys on 1, species 3 on 2, species 4 on 3, and species 5 on 4. Species 3 is also self-regulating (Edelstein-Keshet, 1988). The model is a Lotka-Volterra system of differential equations. The signed digraph is graphed in Figure 5.27. Construct the signed matrix $Q$ for this signed digraph; then determine whether matrix $Q$ is qualitatively stable.

**29.** For the predator-prey system,

$$\frac{dx}{dt} = x(1 - ax - y), \quad 0 < a < 1,$$

$$\frac{dy}{dt} = y(-1 + x),$$

show that the following function is a strict Liapunov function and that $(1, 1 - a)$ is locally asymptotically stable for $x, y > 0$.

$$V(x, y) = x - 1 - \ln(x) + y - [1 - a] - [1 - a]\ln\left(\frac{y}{1 - a}\right).$$

**30.** For the food chain Example 5.30, suppose $a_{11} = 0$. Show that the food chain (5.23) with $n = 3$ is weakly persistent if

$$\mu_0 = a_{10} - a_{30}\frac{a_{12}}{a_{32}} > 0.$$

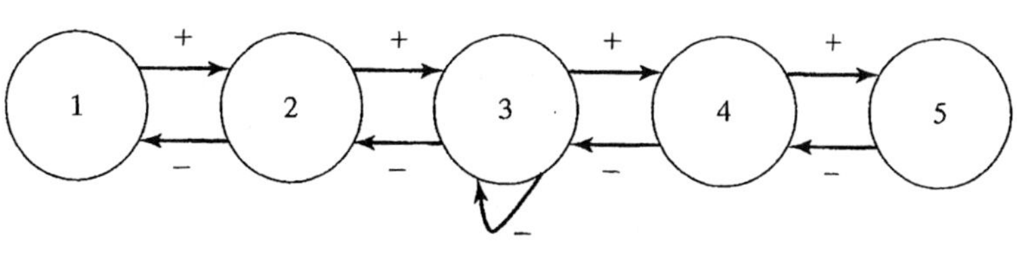


**Figure 5.27** A signed digraph for the five-species model.

In this case, solutions may be unbounded. However, assume when one component goes to zero, $x_i(t) \to 0$, then all solutions $x_j(t)$ are bounded, $x_j(t) < K$ for $j = 1, 2, \ldots, n$. [For a proof see Gard and Hallam (1979).]

**31.** Consider the system of differential equations,

$$\frac{dx}{dt} = y,$$
$$\frac{dy}{dt} = -x + 2\mu y - x^2 y,$$

where $\mu < 1$.

(a) Find the Jacobian matrix evaluated at (0, 0) and show that the eigenvalues satisfy $\mu \pm i\sqrt{1 - \mu^2}$.

(b) Show that this system satisfies the Hopf Bifurcation Theorem. This system is related to van der Pol's equation, discussed in Chapter 6.

**32.** For Example 5.21, apply the criteria in the Appendix to show that the bifurcation at $r = 0$ is a supercritical bifurcation.

**33.** The following system has been transformed so that the bifurcation value is at $r = 0$ and the equilibrium is at the origin,

$$\frac{dx}{dt} = rx + y + x^3$$
$$\frac{dy}{dt} = -x + ry + yx^2.$$

Show that the Hopf Bifurcation Theorem holds. Then apply the criteria in the Appendix to show that the bifurcation is subcritical.

## 5.14 References for Chapter 5

Allee, W. C. 1931. *Animal Aggregations, a Study in General Sociology*. Univ. Chicago Press, Chicago.

Applebaum, E. B. 2001. Models for growth. *The College Mathematics Journal* 32: 258–259.

Aroesty, J., T. Lincoln, N. Shapiro, and G. Boccia, 1973. Tumor growth and chemotherapy: Mathematical methods, computer simulations, and experimental foundations. *Math. Biosci.* 17: 243–300.

Bellman, R. and K. L. Cooke. 1963. *Differential Difference Equations.* Academic Press, New York.

Brauer, F. and J. A. Nohel. 1969. *Qualitative Theory of Ordinary Differential Equations.* W. A. Benjamin, Inc., New York. Reprinted: Dover, 1989.

Brillinger, D. R., J. Guckenheimer, P. Guttorp, and G. Oster. 1980. Empirical modelling of population time series data: The case of age and density dependent vital rates. *Lecture Notes in the Life Sciences*, Vol. 13. AMS, Providence, R. I., pp. 65–90.

Busenberg, S. and P. van den Driessche. 1990. Analysis of a disease transmission model in a population with varying size. *J. Math. Biol.* 28: 257–290.

Cantrell, R. S., and C. Cosner. 1996. Practical persistence in ecological models via comparison methods. *Proc. Roy. Soc. Edinburgh* 126A: 247–272.

Cao, Y. and T. C. Gard. 1997. Practical persistence for differential delay models of population interactions. In: J. Graef, R. Shivaji, B. Soni, and J. Shu (Eds.), Differential Equations and Computational Simulations III, *Electronic Journal of Differential Equations*, Conference 01, pp. 41–53.

Coddington, E. A. and N. Levinson. 1955. *Theory of Ordinary Differential Equations*. McGraw-Hill, New York.

Corduneanu, C. 1977. *Principles of Differential and Integral Equations.* 2nd ed. Chelsea Pub. Co., The Bronx, N.Y.

Cushing, J. M. 1977. *Integrodifferential Equations and Delay Models in Population Dynamics.* Springer-Verlag, Berlin.

Edelstein-Keshet, L. 1988. *Mathematical Models in Biology*. The Random House/Birkhäuser Mathematics Series, New York.

Freedman, H. I. and P. Moson. 1990. Persistence definitions and their connections. *Proc. Amer. Math. Soc.* 109: 1025–1033.

Gard, T. C. and T. G. Hallam. 1979. Persistence in food webs: I Lotka-Volterra food chains. *Bull. Math. Biol.* 41: 877–891.

Goh, B. S. 1977. Global stability in many-species systems. *Amer. Natur.* 111: 135–143.

Gopalsamy, K. 1992. *Stability and Oscillations in Delay Differential Equations of Population Dynamics.* Kluwer Academic Pub., The Netherlands.

Gyllenberg, M. and G. F. Webb. 1989. Quiescence as a explanation of Gompertzian tumor growth. *Growth, Dev. Aging.* 53: 25–33.

Hale, J. and H. Koçak. 1991. *Dynamics and Bifurcations.* Springer-Verlag: New York.

Hallam, T. G. and S. A. Levin (Eds.) 1986. *Mathematical Ecology: An Introduction.* Biomathematics Vol. 17. Springer-Verlag, Berlin, Heidelberg.

Hanski, I. 1999. *Metapopulation Ecology*. Oxford Univ. Press Inc., New York.

Harrison, G. W. 1979. Global stability of predator-prey interactions. *J. Math. Biol.* 8: 159–171.

Hofbauer, J. and K. Sigmund. 1988. *The Theory of Evolution and Dynamical Systems.* Cambridge University Press, Cambridge, U. K.

Hofbauer, J. and K. Sigmund. 1998. *Evolutionary Games and Population Dynamics.* Cambridge University Press, Cambridge, U. K.

Hutchinson, G. E. 1948. Circular causal systems in ecology. *Ann. N.Y. Acad. Sci.* 50: 221–246.

Jang, S. R.-J. and J. Baglama. 2000. Qualitative behavior of a variable-yield simple food chain with an inhibiting nutrient. *Math. Biosci.* 164: 65–80.

Jeffries, C. 1974. Qualitative stability and digraphs in model ecosystems. *Ecology.* 55: 1415–1419.

Jeffries, C., V. Klee, and P. van den Driessche. 1977. When is a matrix sign stable? *Can. J. Math.* 29: 315–326.

Kot, M. 2001. *Elements of Mathematical Ecology*. Cambridge Univ. Press, Cambridge, U. K.

Kozusko, F. and Bajzer, Ž. 2003. Combining Gompertzian growth and cell population dynamics. *Math. Biosci.* 185: 153–167.

Kuang, Y. 1993. *Delay Differential Equations with Applications in Population Dynamics.* Academic Press, San Diego.

Laird, A. K. 1964. Dynamics of tumor growth. *Brit. J. Cancer.* 18: 490–502.

LaSalle, J. and S. Lefschetz. 1961. *Stability by Liapunov's Direct Method.* Academic Press, New York.

Levins, R. 1969. Some demographic and genetic consequences of environmental heterogeneity for biological control. *Bull. Entomol. Soc. Am.* 15: 227–240.

Levins, R. 1970. Extinction. *Lecture Notes Math.* 2: 75–107.

May, R. M. 1975. *Stability and Complexity in Model Ecosystems.* Princeton University Press, Princeton, N.J.

Marsden, J. E. and M. McCracken. 1976. *The Hopf Bifurcation and Its Applications.* Applied Math. Sciences, Vol. 19, Springer-Verlag, New York.

Miller, R. E. 1971. *Nonlinear Volterra Integral Equations.* Benjamin Press, Menlo Park, Calif.

Murray, J. D. 1993. *Mathematical Biology.* 2nd ed. Springer-Verlag, New York.

Murray, J. D. 2002. *Mathematical Biology: I An Introduction.* 3rd ed. Springer-Verlag, New York.

Nicholson, A. J. 1957. The self adjustment of population to change. *Cold Spring Harb. Symp. Quant. Biol.* 22: 153–173.

Pielou, E. C. 1977. *Mathematical Ecology.* 2nd ed. John Wiley & Sons, New York.

Quirk, J. and R. Ruppert. 1965. Qualitative economics and the stability of equilibrium. *Rev. Econ. Stud.* 32: 311–326.

Rudin, W. 1974. *Real and Complex Analysis.* 2nd ed. McGraw-Hill, Inc., N.Y.

Smith, H. and P. Waltman. 1995. *The Theory of the Chemostat.* Cambridge Univ. Press, Cambridge, U.K.

Strogatz, S. H. 2000. *Nonlinear Dynamics and Chaos with Applications to Physics, Biology, Chemistry and Engineering.* Perseus Pub., Cambridge, Mass.

Thieme, H. R. 2003. *Mathematics in Population Biology.* Princeton Univ. Press, Princeton N. J. and Oxford.

Wright, E. M. 1946. The non-linear difference-differential equation. *Quarterly J. Math.* 17: 245–252.

Wright, E. M. 1955. A non-linear difference-differential equation. *J. Reine Angew. Math.* 194: 66–87.

## 5.15 Appendix for Chapter 5

### 5.15.1 Subcritical and Supercritical Hopf Bifurcations

A computational method can be applied to test if the Hopf bifurcation is supercritical or subcritical. First, a transformation is made to put the system in canonical form. The canonical form for the differential system is as follows:

$$\frac{dx}{dt} = \alpha(r)x + \beta(r)y + f_1(x, y, r) = f(x, y, r),$$

$$\frac{dy}{dt} = -\beta(r)x + \alpha(r)y + g_1(x, y, r) = g(x, y, r).$$

The Jacobian matrix evaluated at the origin is

$$J(r) = \begin{pmatrix} \alpha(r) & \beta(r) \\ -\beta(r) & \alpha(r) \end{pmatrix},$$

where the eigenvalues are $\alpha(r) \pm i\beta(r)$, $\alpha(0) = 0$, and $\beta(0) > 0$. Thus at $r = 0$ (bifurcation value) the Jacobian matrix $J(0)$ has two purely imaginary eigenvalues $\pm\beta(0)i$.

To test whether the bifurcation is supercritical or subcritical, the signs of two quantities must be checked, $d\alpha(0)/dr$ and a quantity denoted here as $C$. The partial derivatives of $f$ and $g$ are calculated, then evaluated at the bifurcation value $r = 0$ and at the equilibrium $x = 0$ and $y = 0$. Then the value of the following expression $C$ is computed (Hale and Koçak, 1991):

$$C = (f_{xxx} + f_{xyy} + g_{xxy} + g_{yyy}) + \frac{1}{\beta(0)}[-f_{xy}(f_{xx} + f_{yy}) + g_{xy}(g_{xx} + g_{yy}) + f_{xx}g_{xx} - f_{yy}g_{yy}].$$

In the particular case $d\alpha(0)/dr > 0$, the following additional conditions result in either a subcritical or a supercritical bifurcation (see Figure 5.13):

1. If $C < 0$, then for $r < 0$ the origin is a stable spiral but for $r > 0$ there exists a stable periodic solution and the origin is unstable—a supercritical bifurcation.
2. If $C > 0$, then for $r < 0$ there exists an unstable periodic solution and the origin is stable but for $r > 0$ the origin is unstable—a subcritical bifurcation.
3. If $C = 0$, the test is inconclusive.

### 5.15.2 Strong Delay Kernel

In the continuous-delay logistic model with a strong delay kernel, we show that there do not exist any complex eigenvalues with positive real parts when $T < 2/r$. Let

$$\frac{dx(t)}{dt} = rx(t)\left(1 - \frac{1}{K}\int_{-\infty}^{t} k(t - s)x(s)\,ds\right),$$

where $k(t) = tT^{-2}\exp(-t/T)$. The integral expression is known as a *Volterra integral*. A change of variable, linearization about $K$, and substitution of $x = e^{\lambda t}$ lead to the characteristic equation

$$\lambda = -\frac{r}{T^2}\int_0^{\infty} e^{-\lambda\tau}\tau e^{-\tau/T}\,d\tau.$$

Suppose $\lambda = a + ib$, where $a > 0$ and $b \neq 0$. Then, applying $e^{-\lambda\tau} = e^{-a\tau}(\cos(b\tau) - i\sin(b\tau))$, the two equations satisfied by the real and imaginary parts are

$$a = -\frac{r[(aT + 1)^2 - b^2T^2]}{[(aT + 1)^2 + b^2T^2]^2}$$

and

$$b = \frac{2rbT(aT + 1)}{[(aT + 1)^2 + b^2T^2]^2}.$$

Simplifying these two equations leads to a fifth-degree polynomial in $a$ and a fourth-degree polynomial in $b$,

$$p(a) = T^4a^5 + 4T^3a^4 + \cdots + r(1 - b^2T^2) = 0$$

and

$$[(aT + 1)^2 + T^2b^2]^2 - 2rT(aT + 1) = 0. \tag{5.26}$$

All of the coefficients of $p(a)$ are positive except possibly the coefficient of $a^0$: $r(1 - b^2T^2)$. Therefore, if this latter coefficient is positive, $b^2T^2 < 1$, then there are no real solutions $a$ such that $a > 0$. But if $b^2T^2 > 1$, then there is one positive real root (by Descartes's Rule of Signs). Now we show that $b^2T^2 < 1$ if $rT < 2$. From equation (5.26) we see that

$$b^2T^2 = -(aT + 1)^2 \pm \sqrt{2rT(aT + 1)}.$$

Suppose $rT < 2$. We need to show that $b^2T^2 < 1$ or, equivalently, that

$$2rT(aT + 1) < [1 + (aT + 1)^2]^2.$$

Applying $T < 2/r$ to the left side of the last inequality leads to

$$2rT(aT + 1) < 4(aT + 1).$$

Now, it is easy to show that $4(aT + 1) < [1 + (aT + 1)^2]^2$.

See, for example, Cushing (1977) and Miller (1971) for a discussion of more general types of continuous-delay models and Volterra integral equations.

Chapter 6

# Biological Applications of Differential Equations

## 6.1 Introduction

We introduce a variety of biological models in this chapter that can be expressed in terms of differential equations. We present some well-known biological models of population and community dynamics such as predator-prey and competition models. In addition, we present some more recent models that have been developed for specific systems such as the spruce budworm model and the chemostat model. The techniques developed in the previous chapters will be useful for the analysis of these models.

We begin with a simple population model with harvesting in Section 6.2. This model has applications to animal and fish populations that are harvested or culled to control their population levels. In the sections following the harvest models, we discuss models for two or more populations, predator-prey and competition models. These types of models are very general but are useful for demonstrating some of the properties of predator-prey and competitive systems. For example, periodic cycles are seen in the predator-prey models and extinction of one of the species in the competition models (competitive exclusion). These features are typical of natural predator-prey and competitive systems. A model developed specifically for an insect pest of balsam fir trees is known as the spruce budworm model. This model and generalizations of it are studied in Section 6.5. After the spruce budworm model, we study models for population movement between spatial regions or patches. These models are known as metapopulation models. Several different types of metapopulation models are described in Section 6.6. Another model developed specifically for studying a particular biological application is the chemostat model. A chemostat is a laboratory device for growing microorganisms in a controlled environment. The chemostat has many applications ranging from fermentation to wastewater treatment. We study a simple model for a continuous culture chemostat in Section 6.7. Then, in Section 6.8, a variety of epidemic models are described that have different epidemiological features such as immunity versus no immunity. We compare the dynamics of these different models. In addition, we consider the impact of vaccination on the model dynamics. Finally, in the last section, we discuss briefly some well-known models for electrical impulses in the nervous system. These models are known as the Hodgkin-Huxley and FitzHugh-Nagumo models.

The biological examples studied in this chapter represent only a minute proportion of the rapidly growing number of mathematical models that have been and are being applied to biological systems. Additional applications can also be found in the exercises at the end of this chapter. The references, at the end of this chapter and other chapters, are a source for many other biological applications of differential equations.

## 6.2 Harvesting a Single Population

It is important to develop an ecologically acceptable strategy for harvesting renewable resources. Some important renewable resources are animals, fish, and plants. One of the goals is to maximize yield while maintaining the population from year to year—*maximum sustainable yield*. We use the logistic equation and include either a harvest rate proportional to the population size (constant effort) or a constant harvest rate (constant yield) to study maximum sustainable yield. In these simple models, it is shown that the constant effort strategy is better than the constant yield strategy.

First, we consider the constant effort strategy. Let $N \equiv N(t)$ represent the population size at time $t$ and

$$\frac{dN}{dt} = rN\left(1 - \frac{N}{K}\right) - EN = rN\left(1 - \frac{E}{r} - \frac{N}{K}\right),$$

where $EN$ is the harvest yield per time and $E$ is a positive parameter which is a measure of the effort expended. The equilibrium $K$ changes when there is harvesting,

$$\bar{N} = N_h(E) = K\left(1 - \frac{E}{r}\right), \quad \text{if } E < r,$$

and in this case the yield at equilibrium is

$$Y(E) = E\bar{N} = EK\left(1 - \frac{E}{r}\right).$$

If $E \geq r$, then $\bar{N} = 0$ is the only feasible equilibrium. The population is harvested to extinction. The phase line diagrams are graphed in Figure 6.1.

The maximum yield as a function of effort ($Y'(E) = 0$) occurs at $E = r/2$. Thus, the maximum sustainable yield and new harvesting equilibrium are given by

$$Y_M = Y(E)|_{E=r/2} = \frac{rK}{4}, \quad \bar{N}|_{E=r/2} = N_h(r/2) = \frac{K}{2}.$$

Let's consider the time to recovery after harvesting. The *recovery time* ($T_R$) is defined as the time required so that $\tilde{N}(T_R)/\tilde{N}(0) \approx e^{-1}$, where $\tilde{N}$ is the linear approximation about $\bar{N}$, that is, $\tilde{N} = N - \bar{N}$. In general,

$$\tilde{N}(t)/\tilde{N}(0) \approx \exp(-at),$$

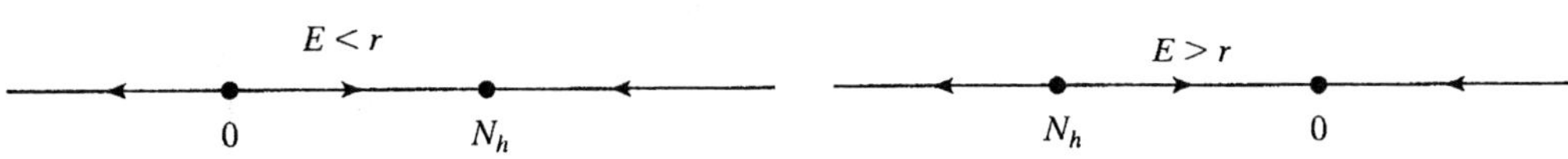


**Figure 6.1** Phase line diagrams for the constant effort model when $E < r$ and when $E > r$.

where $-a$ is the eigenvalue of the linearization, $a > 0$, $d\tilde{N}/dt \approx -a\tilde{N}$. The parameter $a$ is the intrinsic rate of decrease in the population at the equilibrium $\tilde{N}$ ($a$/time). Thus, $1/a$ is the time it takes for a decrease of $e^{-1}$ and is referred to as the recovery time,

$$T_R \approx \frac{1}{a}.$$

We will determine a formula for the recovery time $T_R$ in terms of the effort $E, T_R(E)$.

In the case that no effort is expended, $E = 0$, nothing is harvested. A small perturbation away from $K$ satisfies

$$\frac{d\tilde{N}}{dt} \approx -r\tilde{N} \quad \text{or} \quad \tilde{N}(t) \approx \tilde{N}(0)\exp(-rt).$$

Thus, the recovery time when there is no effort is

$$T_R(0) \approx \frac{1}{r}.$$

However, if $0 < E < r$, then the recovery time in a harvesting situation is

$$T_R(E) \approx \frac{1}{r - E}$$

and the ratio of the recovery time with harvesting to one without harvesting is

$$\frac{T_R(E)}{T_R(0)} = \frac{1}{1 - \dfrac{E}{r}}.$$

For a fixed $r$, the larger the value of $E$, $E < r$, the greater the recovery time. Therefore, there is a price to pay with harvesting in that the population takes longer to return to its previous population size if perturbed. When $E = r/2$, the value of $E$ which gives the maximum sustainable yield $Y_M$, $T_R(E) \approx 2T_R(0)$. It takes twice as long to recover when harvesting at maximum sustainable yield than when there is no harvesting.

A second type of harvesting strategy is constant yield. The model in this case has the following form:

$$\frac{dN}{dt} = rN\left(1 - \frac{N}{K}\right) - Y_0,$$

where $Y_0 > 0$ is the constant yield per unit time. There are either two, one, or no positive equilibria for this model. The maximum value of $f(N) = rN(1 - N/K)$ is $rK/4$. If $0 < Y_0 < rK/4$, there are two positive equilibria. Denote these two equilibria as $N_1$ and $N_2$, $N_1 < N_2$ (see Figure 6.2). There is only one positive equilibrium if $Y_0 = rK/4$, namely $N_1$, and no positive equilibrium if $Y_0 > rK/4$. In this latter case, the population is harvested to extinction. However, note that solutions become negative rather than reaching zero. For example, when $Y_0 > rK/4$, $dN/dt \le rK/4 - Y_0 = -c$ so that $N(t) \le N(0) - ct$ and $\lim_{t\to\infty} N(t) = -\infty$. The model is not biologically reasonable in cases where the solution is negative. This is one reason why the model for constant effort strategy is better than the model for constant yield strategy.

**Figure 6.2** There are two positive equilibria, $N_1$ and $N_2$, for the constant yield harvesting model when $0 < Y_0 < rK/4$.

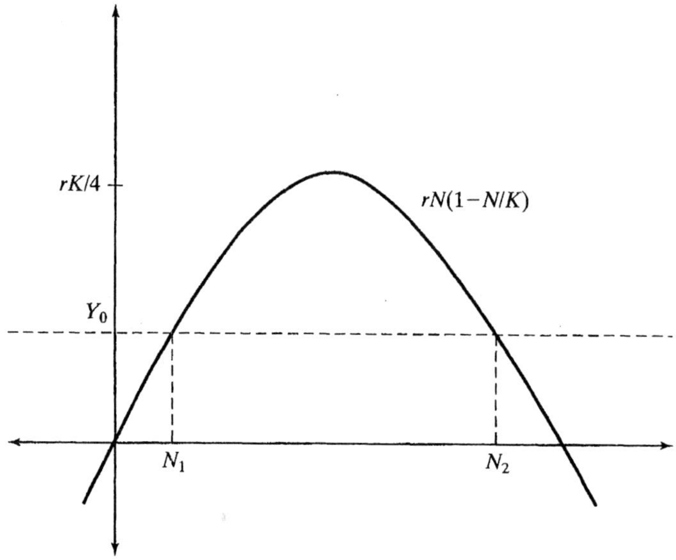


In the case $0 < Y_0 < rK/4$, the equilibrium $N_2$ is locally asymptotically stable. Equilibrium $N_2$ can be found by setting the right-hand side of the differential equation to zero,

$$N_2(Y_0) = \frac{K}{2}\left(1 + \left[1 - \frac{4Y_0}{Kr}\right]^{1/2}\right), \quad 0 < Y_0 < \frac{rK}{4}.$$

We will determine the time to recovery for a yield $Y_0$ when $N = N_2$. Let $\tilde{N} = N - N_2(Y_0)$. Then

$$\frac{d\tilde{N}}{dt} \approx \tilde{N}\frac{\partial f}{\partial N}\bigg|_{N=N_2(Y_0)} = -\tilde{N}r\left[1 - \frac{4Y_0}{rK}\right]^{1/2},$$

$$T_R(Y_0) \approx \frac{1}{r\sqrt{1 - \dfrac{Y_0}{Y_M}}}$$

where $Y_M = rK/4$ and

$$\frac{T_R(Y_0)}{T_R(0)} \approx \frac{1}{\sqrt{1 - \dfrac{Y_0}{Y_M}}}.$$

For this model, the time to recovery goes to infinity as $Y_0 \to Y_M^-$. This is another reason why the constant yield harvesting strategy is not as biologically reasonable as the constant effort strategy. For harvesting strategies that incorporate economic costs and an optimal harvesting strategy, please consult Clark (1976) and Kot (2001).

## 6.3 Predator-Prey Models

One of the most well-known predator-prey models is named after the two scientists, Alfred Lotka (1880–1949) and Vito Volterra (1860–1940), who were among the first to study and apply this model. Lotka was born in Poland to

American missionaries and moved to the United States in 1902. Many of his contributions can be found in the book *Elements of Physical Biology* (1925). Volterra, an Italian mathematician, published work in partial differential equations, integral equations, and functional analysis.

The Lotka-Volterra predator-prey model consists of two differential equations, one equation for the prey $x$ and the second equation for the predator $y$. The two differential equations are

$$\frac{dx}{dt} = ax - bxy = x(a - by),$$

$$\frac{dy}{dt} = -cy + exy = y(-c + ex), \tag{6.1}$$

where $a, b, c, d > 0$. Parameter $a$ is the intrinsic growth rate of the prey $x$, whereas $c$ is the death rate of the predator. In the absence of the prey, the predator population $y$ approaches zero because $dy/dt = -cy$. The parameter $b$ is the per capita reduction in prey per predator and $d$ is the per capita increase in predator per prey. In models where $x$ and $y$ represent biomass of prey and predator, the constant $d$ is often assumed to be constant multiple of $c$, $d = c/\gamma$, where $\gamma$ is related to the conversion of prey biomass into predator biomass. The terms $bxy$ and $dxy$ are referred to as *functional* and *numerical responses*, respectively (Hassell, 1978).

An interesting historical discussion about the Lotka-Volterra predator-prey model is given by Braun (1975). We summarize some of this discussion. In the 1920s, the Italian biologist Umberto D'Ancona was studying variations in fish populations. He came across data that were recorded on the percentages of total catch for several species of fish caught at Mediterranean ports during the period of World War I (WWI, 1914–1918). The data, recorded in Table 6.1 (Braun, 1975), give the percentage of total catch of selachians, (sharks, skates, rays, etc.), which, at that time, were not used for food consumption. Based on the data, D'Ancona did not understand why there should be such a large increase in the percentage of selachians during WWI. He thought the reason for this increase was the decreased amount of fishing during this period, but he could not understand how this would affect the fish populations. Therefore, he consulted Vita Volterra to help him interpret the data.

The model Volterra formulated is given by (6.1). The selachians are the predators $y$ and the food fish, or edible fish, are the prey $x$. We analyze the Lotka-Volterra predator-prey model and then consider how fishing might change the model dynamics.

Before we begin the analysis, note that this model can be put in a form equivalent to the model discussed in Example 5.13,

$$\frac{dx}{dt} = xy - y = y(x - 1),$$

$$\frac{dy}{dt} = 2x - xy = x(2 - y).$$

**Table 6.1** Percentage of selachians caught as part of the total fish caught (Braun, 1975).

| 1914 | 1915 | 1916 | 1917 | 1918 | 1919 | 1920 | 1921 | 1922 | 1923 |
|---|---|---|---|---|---|---|---|---|---|
| 11.9% | 21.4% | 22.1% | 21.2% | 36.4% | 27.3% | 16.0% | 15.9% | 14.8% | 10.7% |

Making the following change of variables in the preceding equations, $\tilde{x} = 1 - x$ and $\tilde{y} = 2 - y$, then

$$\frac{d\tilde{x}}{dt} = \tilde{x}(2 - \tilde{y}),$$

$$\frac{d\tilde{y}}{dt} = \tilde{y}(\tilde{x} - 1). \tag{6.2}$$

Model (6.2) agrees with the Lotka-Volterra predator-prey model (6.1), where $a = 2$, $b = 1$, $c = 1$, and $e = 1$. In Example 5.13, it was shown that the equilibrium $(1, 2)$ is a saddle but the origin is indeterminate (the eigenvalues are purely imaginary). Here, we show that the origin behaves like a center or, equivalently, that the positive equilibrium $(c/e, a/b)$ for (6.1) has periodic solutions encircling it.

To simplify the analysis of (6.1), the model is put in a dimensionless form. Let

$$u(\tau) = \frac{ex(t)}{c}, \quad v(\tau) = \frac{by(t)}{a}, \quad \text{and} \quad \tau = at.$$

Then

$$\frac{du}{d\tau} = u(1 - v),$$

$$\frac{dv}{d\tau} = \alpha v(u - 1), \tag{6.3}$$

where $\alpha = c/a$. The dimensionless equations have two equilibria: $(0, 0)$ and $(1, 1)$. The Jacobian matrix of the dimensionless system is

$$J(u, v) = \begin{pmatrix} 1 - v & -u \\ \alpha v & \alpha(u - 1) \end{pmatrix}.$$

Evaluated at the two equilibria, the Jacobian matrices are

$$J(0, 0) = \begin{pmatrix} 1 & 0 \\ 0 & -\alpha \end{pmatrix} \quad \text{and} \quad J(1, 1) = \begin{pmatrix} 0 & -1 \\ \alpha & 0 \end{pmatrix}.$$

Matrix $J(0, 0)$ has eigenvalues $\lambda_{1,2} = 1, -\alpha$ with corresponding eigenvectors $(1, 0)^T$ and $(0, 1)^T$. The zero equilibrium is a saddle point. Matrix $J(1, 1)$ has complex eigenvalues $\lambda_{1,2} = \pm i\sqrt{\alpha}$. In the linear case, the equilibrium $(1, 1)$ would be a neutral center. However, this is the ambiguous case; it could also be a spiral point (stable or unstable). It will be shown that every solution encircling the positive equilibrium is periodic. There is a unique periodic solution corresponding to each initial condition.

**Theorem 6.1** *Every solution of the Lotka-Volterra predator-prey system (6.3) with positive initial conditions is periodic.*

**Proof** First, it is important to note that if the equilibrium $(1, 1)$ is a spiral, then a solution $u(\tau)$ and $v(\tau)$ will intersect the line $u = 1$ an infinite number of times.

Let $u(\tau)$ and $v(\tau)$ represent a solution of (6.3). An implicit solution can be obtained from the differential equations by separation of variables:

$$\frac{dv}{du} = \frac{\alpha v(u - 1)}{u(1 - v)} \quad \text{or} \quad (1 - v)\frac{dv}{v} = \alpha(u - 1)\frac{du}{u}.$$

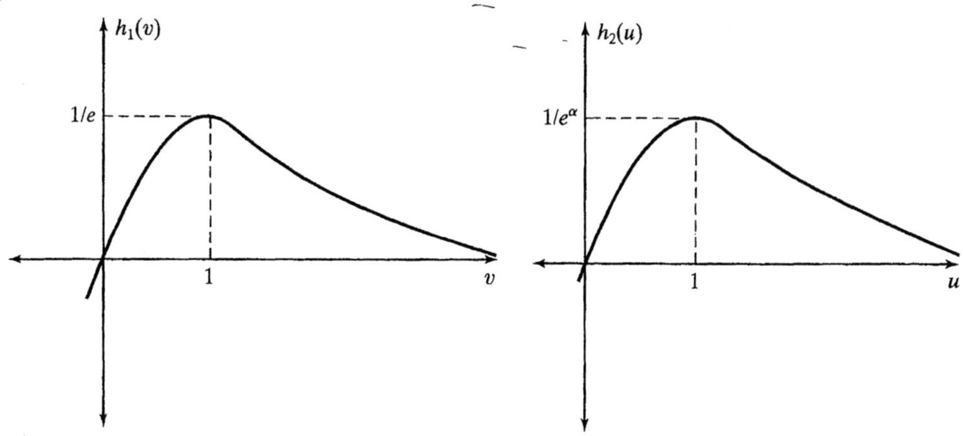


**Figure 6.3** Graphs of the functions $h_1$ and $h_2$.

Integration of the last expression yields

$$\ln(v) - v = \alpha u - \alpha \ln(u) + K \text{ or } ve^{-v} = K_1 u^{-\alpha} e^{\alpha u}.$$

The last expression can be written as follows:

$$\frac{v}{e^v}\frac{u^\alpha}{e^{\alpha u}} = h_1(v)h_2(u) = K_1, \tag{6.4}$$

where $h_1(v) = v/e^v$ and $h_2(u) = u^\alpha/e^{\alpha u}$. Both functions $h_i$ satisfy $h_i(0) = 0$, $h_i(\infty) = 0$, and they have a single maximum on $(0, \infty)$ when $h_i' = 0$. The maximum of $h_1$ occurs at $v = 1$ and the maximum of $h_2$ at $u = 1$ (the equilibrium values). The maximum values of these functions are $M_v = 1/e$ and $M_u = 1/e^\alpha$ (see Figure 6.3).

From equation (6.4) it follows that there is no solution if $K_1 > M_v M_u$ and only one solution if $K_1 = M_v M_u$, namely, $u = 1 = v$. Also, for $0 < K_1 = \tilde{c}M_u < M_v M_u$, where the solution for $h_2(u) = M_u$ is $u = 1$, has exactly two solutions for $v$. There is one solution, $v_m < 1$, and another solution, $v_M > 1$ ($\tilde{c} = v/e^v$). Also, $h_2(u) = \tilde{c}M_u/h_1(v)$ has no solution when $v \notin [v_m, v_M]$, a unique solution $u = 1$ when $v = v_m$ and $v = v_M$, and two solutions $u_1(v)$ and $u_2(v)$ when $v \in (v_m, v_M)$. The preceding argument shows that $u_1(v) < 1 < u_2(v)$, $u_1(v_m) = u_2(v_m)$, and $u_1(v_M) = u_2(v_M)$. See Figure 6.4. Thus, the solutions in the $u$-$v$ phase plane must be closed curves and cannot contain any equilibrium [unless $u(0) = 1 = v(0)$]. All solutions with $u(0), v(0) > 0$ must be periodic. □

Solutions to the predator-prey system (6.3) over time and in the phase plane are graphed in Figure 6.5.

The data of D'Ancona on the percentage of selachians caught are an average over one year, as shown in Table 6.1 (Braun, 1975). In order to compare D'Ancona's data with the solution values for the predator $y$ of Volterra's model, the *average values* of the model must be computed for any solution $x(t)$ and $y(t)$. Define the *average value* of $x(t)$ and $y(t)$ as follows:

$$\hat{x} = \frac{1}{T}\int_0^T x(t)\,dt \quad\text{and}\quad \hat{y} = \frac{1}{T}\int_0^T y(t)\,dt.$$

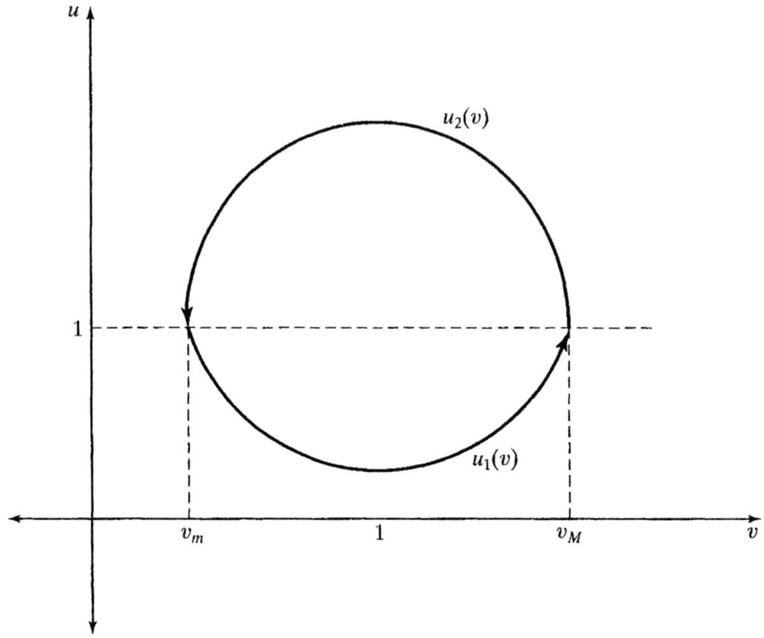


**Figure 6.4** Solutions to the dimensionless equation graphed in the $u$-$v$ plane.

**Theorem 6.2** *Let $x(t)$ and $y(t)$ be a periodic solution of the predator-prey system (6.1) with period $T > 0$. Then $\hat{x} = c/e$ and $\hat{y} = a/b$ (i.e., the average values are equal to the equilibrium values).*

**Proof** Rewrite the differential equations for $x$ and $y$ so that $dx/x = (a - by)\,dt$ and $dy/y = (-c + ex)\,dt$. Integrating from 0 to $T$ and equating the right-hand side to zero yields

$$\int_0^T \frac{dx}{x} = \ln x(T) - \ln x(0) = 0 \quad \text{and} \quad \int_0^T \frac{dy}{y} = \ln y(T) - \ln y(0) = 0.$$

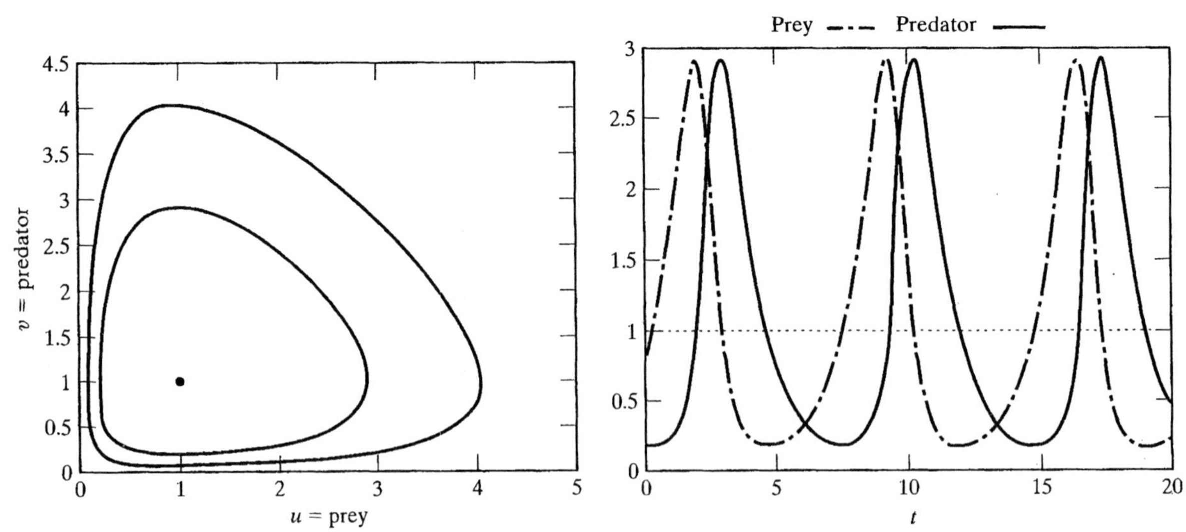


**Figure 6.5** Solutions to the Lotka-Volterra predator-prey model (6.3), $\alpha = 1$, in the phase plane and as functions of time.

Substituting $dx/x = a - by(t)$ in the first integral leads to

$$\int_0^T (a - by(t))\,dt = aT - b\int_0^T y(t)\,dt = 0 \quad \text{or} \quad \hat{y} = \frac{1}{T}\int_0^T y(t)\,dt = \frac{a}{b}.$$

A similar calculation for $dy/y$ leads to $\hat{x} = c/e$. □

When the effects of fishing are included in the Lotka-Volterra predator-prey model, a principle known as *Volterra's principle* is demonstrated. Assume that fishing has the same effect on the prey (food fish) as the predators (sharks) and the catch is proportional to population size. Thus, the rates of removal of prey and predator are $\epsilon x$ and $\epsilon y$, where $\epsilon$ reflects the intensity of fishing. The new model that includes fishing has the following form:

$$\frac{dx}{dt} = ax - bxy - \epsilon x = x(a - \epsilon - by),$$

$$\frac{dy}{dt} = -cy + exy - \epsilon y = y(-c - \epsilon + ex).$$

The new system has the form of (6.1) when $a$ is replaced by $a - \epsilon > 0$ ($\epsilon$ is small) and $c$ is replaced by $c + \epsilon$. There is a new equilibrium at

$$\bar{x} = \frac{c + \epsilon}{e}, \quad \bar{y} = \frac{a - \epsilon}{b}. \tag{6.5}$$

Volterra's principle states that a moderate amount of fishing ($a > \epsilon$) increases the number of prey on the average and decreases the number of predators. Conversely, a reduced level of fishing increases the number of predators (on the average) and decreases the number of prey. Volterra's principle has applications to insecticide or pesticide treatments which destroy both predators and their prey. Application of insecticides ($\epsilon > 0$) increases the population of the prey that were formerly kept under control by the predators.

Returning to D'Ancona's data, let's assume in the years 1915–1919 fishing was curtailed due to WWI but in the years 1914 and 1920–1923 fishing was not affected. We will apply the results from the Lotka-Volterra predator-prey model and assume that the fishing data are cyclic with a period of one year. The data in Table 6.1 are the annual percentage of predators in total fish caught [i.e., $P = \bar{y}/(\bar{x} + \bar{y}) \times 100\%$]. Then, in the war years (1915–1919), $P_0 \approx 25.7\%$ and in the other years (1914, 1920–1923), $P_\epsilon \approx 13.9\%$, almost a twofold increase in the predator during WWI. The actual model for the phenomenon observed in Table 6.1 is more complex than the Lotka-Volterra predator-prey model. However, the Lotka-Volterra model illustrates the oscillatory behavior observed in real-life predator-prey systems and the impact of a nonspecific predator on this relationship.

Probably the most well-known application of the Lotka-Volterra predator-prey equations is the lynx-hare cycles. The classical paper by Elton and Nicholson (1942) published data on the 10-year cycles of the Canadian lynx (*Lynx canadensis*) and snowshoe hare (*Lepus americanus*). Their lynx data are graphed in Figure 6.6 (Campbell and Walker, 1977). Data were based on Hudson's Bay Company pelt collections for snowshoe hares and lynx from the 1800s through early 1900. Canadian lynx feed almost exclusively on snowshoe hares when they are plentiful (Krebs et al., 2001). The hare cycles are driven primarily by food supplies and predation. Snowshoe hare populations are cyclic

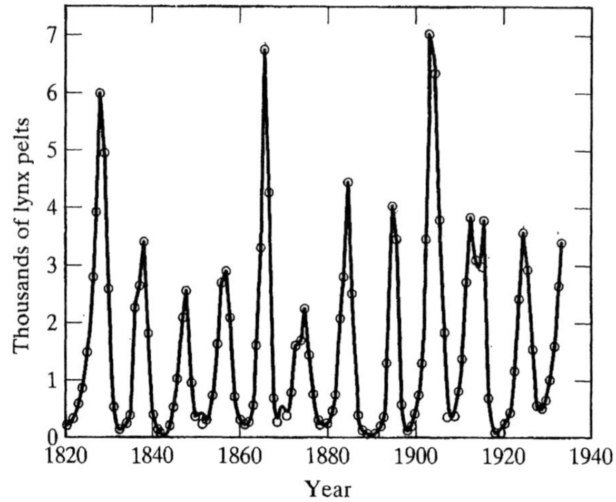


**Figure 6.6** Elton and Nicholson's lynx trapping data from 1821 through 1934 (Campbell and Walker, 1977).

with an approximate period of 10 years. Predators of the hare include hawks, owls, coyotes, foxes, and wolves, in addition to lynx (Krebs et al., 2001; Stenseth et al., 1997). Stenseth et al. (1997) suggest that three trophic levels govern the dynamics of the snowshoe hare population (vegetation-hare-predators) but that two trophic levels govern the lynx population (hare-lynx).

Additional data on lynx (*C. canadensis*) and red fox (*Vulpes fulva*), predators of the showshoe hare, are reported by Keith (1963) and graphed in Figure 6.7. The data on lynx and red fox fur pelts from 1919 through 1957 also show 10-year cycles. Data for Figure 6.7 are provided in the Appendix. See Exercise 4.

One criticism of the original Lotka-Volterra predator-prey equations is the exponential growth rate of the prey in the absence of the predator. A simple model that addresses this criticism is to assume logistic growth of the prey in the absence of the predator. In this case, the model has the following form (see Example 5.11):

$$\frac{dx}{dt} = ax - fx^2 - bxy = x(a - fx - by),$$

$$\frac{dy}{dt} = -cy + exy = y(ex - c),$$

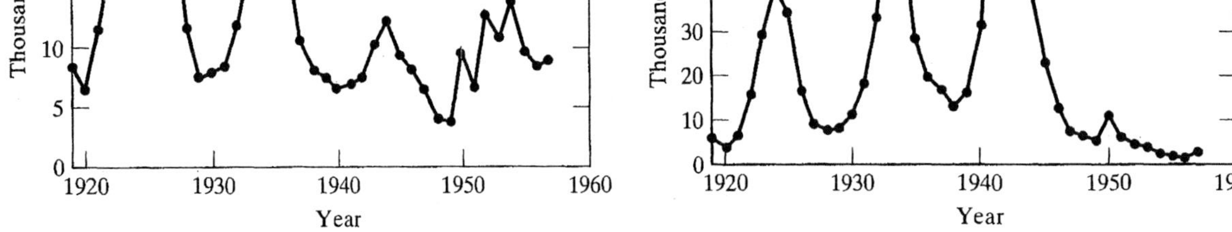


**Figure 6.7** Data on fur returns for lynx (*L. canadensis*) and red fox (*V. fulva*) for eight and seven Canadian provinces, respectively, from 1919 through 1957 (Keith, 1963).

where the term $fx^2$ represents intraspecific competition among the prey and $a, b, c, e, f > 0$.

In predator-prey differential equations there are two responses, referred to as a functional response and a numerical response. In the Lotka-Volterra prey differential equation, $dx/dt$, the term $bxy = F(x, y)$ is called the *functional response*. The expression $exy = G(x, y)$ in the predator differential equation $dy/dt$ is called the *numerical response*. Various functional and numerical responses have been considered for both predator-prey and host-parasite systems (Edelstein-Keshet, 1988; Hassell, 1978; May, 1976). Some numerical and functional responses that have been discussed in the literature include

$$\alpha y(1 - \exp(-\beta x)), \quad \alpha, \beta > 0, \quad \text{Ivlev},$$

$$\frac{\alpha xy}{x + \beta}, \quad \alpha, \beta > 0, \quad \text{Holling Type II},$$

$$\frac{\alpha yx^2}{x^2 + \beta^2}, \quad \alpha, \beta > 0, \quad \text{Holling Type III}.$$

The form $F(x, y) = bxy$ is called Holling Type I. The above forms incorporate the fact that the attack capacity is limited when the number of prey $x$ is very large. Also, Hassell (1978) states that Types I and II are more typical of invertebrate predators, whereas Type III is more typical of vertebrates.

A general form for a predator-prey system is one analyzed by Harrison (1979). Also, consult Hallam and Levin (1986). The model takes the form

$$\frac{dx}{dt} = a(x) - f(x)\,b(y),$$

$$\frac{dy}{dt} = n(x)g(y) + c(y).$$

The functions $f$ and $g$ are positive for $x \in (0, \infty)$; $a(x)$ is the growth rate of the prey in the absence of predation; $c(y)$ is the growth rate of the predator in the absence of the prey. The product $f(x)\,b(y)$ is the functional response and $n(x)g(y)$ is the numerical response. Assume the functions $n(x)$ and $b(y)$ are nondecreasing. Let $(\bar{x}, \bar{y})$ be a positive equilibrium of the preceding system. Then given some additional assumptions on the functions, $a, f, b, n, g$, and $c$, Harrison (1979) applies a Liapunov function of the form

$$V(x, y) = \int_{\bar{x}}^{x} \frac{n(s) - n(\bar{x})}{f(s)}\,ds + \int_{\bar{y}}^{y} \frac{b(s) - b(\bar{y})}{g(s)}\,ds,$$

to prove global asymptotic stability of the positive equilibrium.

---

**Example 6.1** Consider the following Lotka-Volterra model with one predator ($x_1$) and two prey ($x_2$ and $x_3$):

$$\frac{dx_1}{dt} = x_1(-a_{10} + a_{12}x_2 + a_{13}x_3),$$

$$\frac{dx_2}{dt} = x_2(a_{20} - a_{21}x_1),$$

$$\frac{dx_3}{dt} = x_3(a_{30} - a_{31}x_1 - x_3),$$

where $a_{ij} > 0$. The first prey $x_2$ grows exponentially in the absence of predator $x_1$. The second prey $x_3$ grows logistically to its carrying capacity $a_{30}$ in the absence of predator $x_1$. It is easy to see that the origin is unstable (eigenvalues are $-a_{10}$, $a_{20}$, $a_{30}$). There are other equilibria: $(a_{20}/a_{21}, a_{10}/a_{12}, 0)$, $(0, 0, a_{30})$, $(a_{30}/a_{31} - a_{10}/[a_{13}a_{31}], 0, a_{10}/a_{13})$, and

$$\bar{x}_1 = \frac{a_{20}}{a_{21}}, \quad \bar{x}_3 = a_{30} - a_{31}\bar{x}_1, \quad \bar{x}_2 = \frac{a_{10} - a_{13}\bar{x}_3}{a_{12}}$$

which is positive if $\bar{x}_2 > 0$ and $\bar{x}_3 > 0$.

The Jacobian matrix evaluated at the three-species equilibrium $(\bar{x}_1, \bar{x}_2, \bar{x}_3)$ is

$$J(\bar{x}_1, \bar{x}_2, \bar{x}_3) = \begin{pmatrix} 0 & a_{12}\bar{x}_1 & a_{13}\bar{x}_1 \\ -a_{21}\bar{x}_2 & 0 & 0 \\ -a_{31}\bar{x}_3 & 0 & -\bar{x}_3 \end{pmatrix}.$$

The characteristic equation of the Jacobian matrix is

$$\lambda^3 + \bar{x}_3\lambda^2 + (a_{12}a_{21}\bar{x}_1\bar{x}_2 + a_{13}a_{31}\bar{x}_1\bar{x}_3)\lambda + a_{12}a_{21}\bar{x}_1\bar{x}_2\bar{x}_3 = 0.$$

Necessary and sufficient conditions for asymptotic stability follow from the Routh-Hurwitz criteria: If the characteristic equation has the form $\lambda^3 + a_1\lambda^2 + a_2\lambda + a_3 = 0$, where $a_i > 0$ for $i = 1, 2, 3$ and $a_1a_2 > a_3$, then the equilibrium is locally asymptotically stable. The characteristic equation satisfies both of these conditions provided the equilibrium is positive. Thus, if the equilibrium is positive, it is locally asymptotically stable. Recall that local asymptotic stability can be determined using the criteria for qualitative stability. The following signed matrix, $Q$, shows that the positive equilibrium is qualitatively stable (see Example 5.25):

$$Q = \text{sign}(J) = \begin{pmatrix} 0 & + & + \\ - & 0 & 0 \\ - & 0 & - \end{pmatrix}.$$

■

For other functional forms for density dependence in the prey and predator populations and more general predator-prey models, consult the references (e.g., Ackleh et al., 2000; Edelstein-Keshet, 1988; Hallam and Levin, 1986; Hassell, 1978; Kuang and Beretta, 1998).

## 6.4 Competition Models

One of the most well-known competition models is Lotka-Volterra competition. The behavior of the two-species model is completely understood. The two-species model is discussed in the next section. The three-species competition model can exhibit some unexpected and interesting behavior. Some special cases of the three-species competition model are presented.

### 6.4.1 Two Species

A simple Lotka-Volterra model for two competing species illustrates the *principle of competitive exclusion*. When two or more species compete for the same basic resources, the "strongest" survives; the weaker species is driven to extinction.

The Lotka-Volterra model for competition between two species assumes that each species in the absence of the other grows logistically to a carrying capacity $K_i$. Hence, the model for two competing species has the following form:

$$\frac{dx_1}{dt} = r_1 x_1 \frac{K_1 - x_1 - \beta_{12}x_2}{K_1},$$

$$\frac{dx_2}{dt} = r_2 x_2 \frac{K_2 - x_2 - \beta_{21}x_1}{K_2},$$

where $r_i$, $K_i$, and $\beta_{ij}$ are positive parameters. We assume $x_1(0) > 0$ and $x_2(0) > 0$. The parameter $r_i$ represents the intrinsic growth rate of species $i$, $K_i$ is the carrying capacity of species $i$, and $\beta_{ij}/K_i$ is competition coefficient for species $i$. Note that the per capita growth rate $\dfrac{1}{x_i}\dfrac{dx_i}{dt} = f_i(x_1, x_2)$ is affine and $\partial f_i/\partial x_j < 0$ for $i \neq j$. The model has at most four nonnegative equilibria:

$$(0,0), \quad (K_1, 0), \quad (0, K_2), \quad \left(\frac{K_1 - \beta_{12}K_2}{1 - \beta_{12}\beta_{21}}, \frac{K_2 - \beta_{21}K_1}{1 - \beta_{12}\beta_{21}}\right),$$

if $\beta_{12}\beta_{21} < 1$. The $x_1$-nullclines are the lines $x_1 = 0$ and $K_1 = x_1 + \beta_{12}x_2$. The $x_2$-nullclines are the lines $x_2 = 0$ and $K_2 = x_2 + \beta_{21}x_1$. There are four cases to consider that are determined by how the nullclines intersect in the first quadrant.

It is time consuming but straightforward to complete the mathematical analysis for this model (linearize, determine the stability properties for all of the nonnegative equilibria, and complete the phase plane diagrams in each case). The asymptotic dynamics in each case are summarized in Figure 6.8.

**Case 1:** Solutions approach the equilibrium $(K_1, 0)$.

**Case 2:** Solutions approach the equilibrium $(0, K_2)$.

**Case 3:** Solutions approach either $(K_1, 0)$ or $(0, K_2)$. There are only two solution trajectories (separatrices) along which solutions approach the positive equilibrium

$$E = \left(\frac{K_1 - \beta_{12}K_2}{1 - \beta_{12}\beta_{21}}, \frac{K_2 - \beta_{21}K_1}{1 - \beta_{12}\beta_{21}}\right).$$

**Case 4:** Solutions approach the equilibrium

$$E = \left(\frac{K_1 - \beta_{12}K_2}{1 - \beta_{12}\beta_{21}}, \frac{K_2 - \beta_{21}K_1}{1 - \beta_{12}\beta_{21}}\right).$$

●

Cases 1 and 2 illustrate the principle of competitive exclusion. In these two cases, one species always dominates. In Case 3, the outcome depends on the initial conditions, referred to as a *founder effect*. The species first to establish itself (the founder) has an advantage and will be the superior competitor. There are some well-known laboratory experiments performed by the biologist G. F. Gause (1932) illustrating various competitive outcomes when the competing species are yeasts (e.g., *Saccharomyces cervisiae* and *Schizosaccharomyces kephir*).

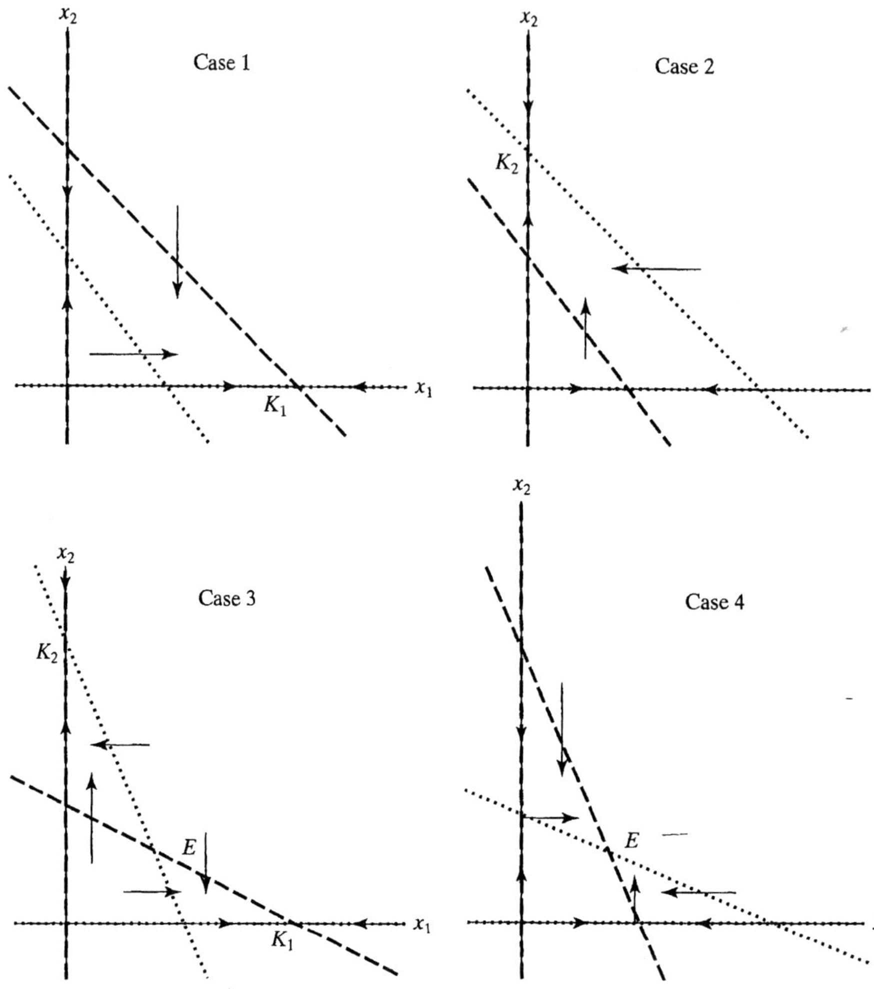


**Figure 6.8** The four cases of Lotka-Volterra competition for two competing species $x_1$ and $x_2$.

## 6.4.2 Three Species

Three-species Lotka-Volterra competition exhibits more interesting and diverse behavior than two-species competition. A three-species Lotka-Volterra competition model has the form

$$\frac{dx_i}{dt} = x_i\left(a_{i0} - \sum_{j=1}^{3} a_{ij}x_j\right),$$

where $a_{i0} > 0$ and $a_{ij} > 0$, $i, j = 1, 2, 3$. In the three-species competition model, it is possible for one- or two-species extinction, or global stability of a positive three-species equilibrium. This type of behavior is analogous to the two-species competition model. However, it can also be shown that solutions to the three-species model can be periodic (Hofbauer and So, 1994; van den Driessche and Zeeman, 1998; Zeeman, 1993) or have a stable heteroclinic orbit (Gilpin, 1975; May and Leonard, 1975). More complicated competition models with five species have also shown chaotic behavior (Huisman and Weissing, 2001). The competitive outcome often depends on the relationship between pairwise interactions. Here, we shall consider the case of a stable heteroclinic orbit for three species, a well-known example, first analyzed by May and Leonard in 1975.

In pairwise competition between three species, 1, 2, and 3, assume 2 eliminates 1, 1 eliminates 3, but 3 eliminates 2. This represents a nontransitive relationship.

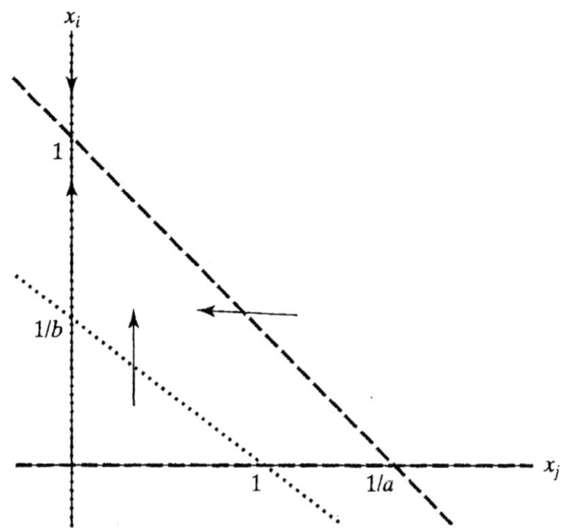


**Figure 6.9** The $x_i$-$x_j$ phase plane where species $x_i$ eliminates $x_j$. Solutions approach the equilibrium where $x_i = 1$ and $x_j = 0$.

An example of three species in which this type of relationship may occur is one where species 2 eliminates 1 in pairwise competition, and species 1 eliminates 3 in pairwise competition, but species 3 excretes a substance that is poisonous to species 2 but not to species 1. Therefore, species 3 eliminates 2. The Lotka-Volterra competition model with this property has the following form:

$$\begin{aligned}\frac{dx_1}{dt} &= x_1(1 - x_1 - bx_2 - ax_3),\\ \frac{dx_2}{dt} &= x_2(1 - ax_1 - x_2 - bx_3),\\ \frac{dx_3}{dt} &= x_3(1 - bx_1 - ax_2 - x_3),\end{aligned} \tag{6.6}$$

where $0 < a < 1$, $1 < b$, and $x_i(0) > 0$, $i = 1, 2, 3$. Model (6.6) is often referred to as the *May-Leonard competition model.* The system has five nonnegative equilibria: $(0, 0, 0)$, $(1, 0, 0)$, $(0, 1, 0)$, $(0, 0, 1)$, and $(1, 1, 1)/(1 + a + b)$. The $x_i$-$x_j$ phase planes for $i \neq j$ show that there are no two-species positive equilibria (see Figure 6.9).

The Jacobian matrix for the full three-dimensional system is

$$J = \begin{pmatrix} 1 - 2x_1 - bx_2 - ax_3 & -bx_1 & -ax_1 \\ -ax_1 & 1 - ax_1 - 2x_2 - bx_3 & -bx_2 \\ -bx_3 & -ax_3 & 1 - bx_1 - ax_2 - 2x_3 \end{pmatrix}.$$

It is easy to see that the eigenvalues of the Jacobian matrix at the origin are positive, $\lambda_i = 1, i = 1, 2, 3$. The origin is unstable.

For the single-species equilibria, the eigenvalues of the Jacobian matrix are $\lambda_1 = -1$, $\lambda_2 = 1 - a$ and $\lambda_3 = 1 - b$. Since $a < 1$, the single-species equilibria are unstable. Finally, the Jacobian matrix evaluated at the positive equilibrium $(1, 1, 1)/(1 + a + b)$ is

$$J_5 = -(1 + a + b)^{-1} \begin{pmatrix} 1 & b & a \\ a & 1 & b \\ b & a & 1 \end{pmatrix} = -\gamma A,$$

where $\gamma = (1 + a + b)^{-1}$. Matrices $A$ and $J_5$ are *circulant* matrices. There are explicit formulas for the eigenvalues of circulant matrices (Ortega, 1987).

The eigenvalues of $J_5$ are the eigenvalues of $A$ multiplied by $-\gamma$. The eigenvalues of $J_5$ are given by

$$\lambda_1 = -1, \quad \lambda_{2,3} = \frac{\gamma}{2}\left[a + b - 2 \pm (a - b)\sqrt{3}i\right].$$

Therefore, the three-species positive equilibrium is locally asymptotically stable if $a + b < 2$.

---

**Example 6.2** The following example illustrates the dynamics of the May-Leonard competition model when $a + b < 2$. See Figure 6.10.

$$\frac{dx_1}{dt} = x_1(1 - x_1 - 1.25x_2 - 0.5x_3),$$

$$\frac{dx_2}{dt} = x_2(1 - 0.5x_1 - x_2 - 1.25x_3),$$

$$\frac{dx_3}{dt} = x_3(1 - 1.25x_1 - 0.5x_2 - x_3),$$

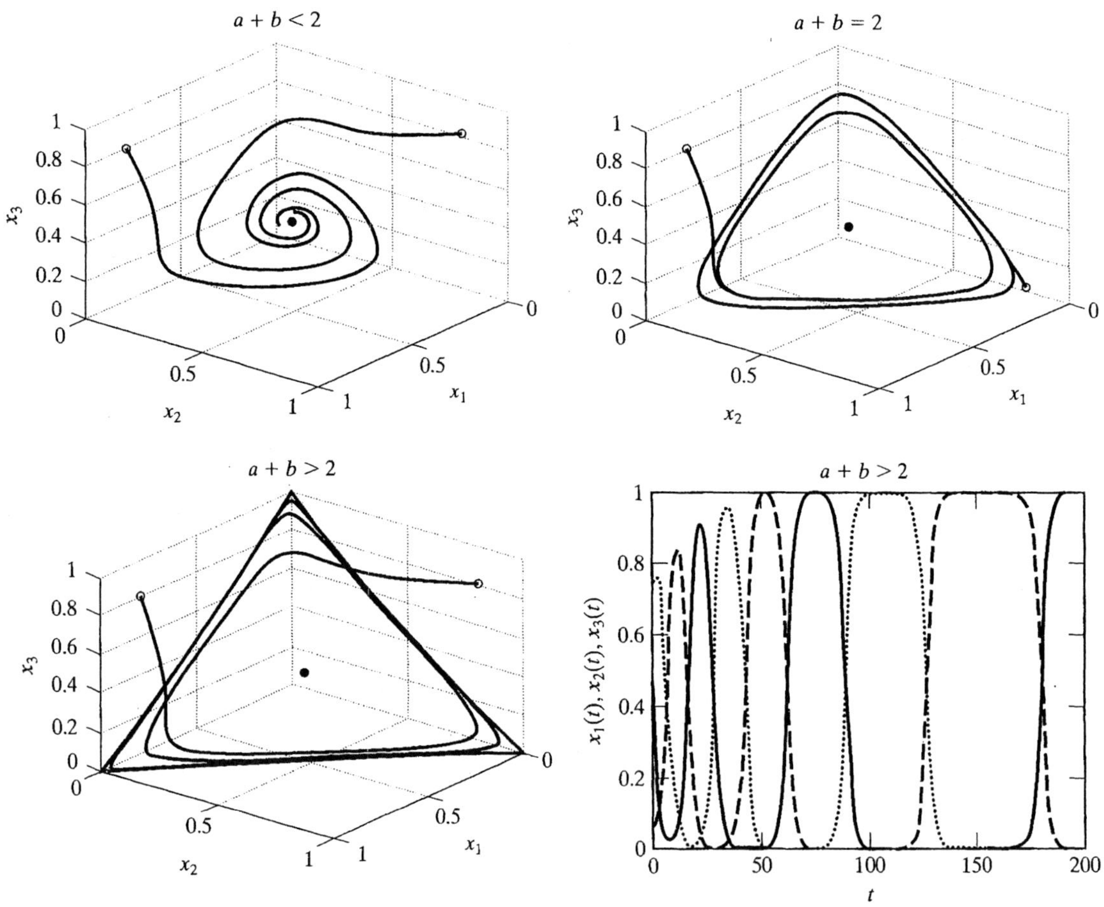


**Figure 6.10** The top two figures are graphs of two solutions to the May-Leonard competition model in the case $a + b = 1.75 < 2$ and $a + b = 2$. The bottom two figures are graphs of solutions to the May-Leonard competition model in the case $a + b = 2.5 > 2$. The bottom left figure graphs two solutions in phase space while the bottom right figure graphs one solution $(x_1(t), x_2(t), x_3(t))$ as a function of time.

where $a = 0.5$ and $b = 1.25$. The positive equilibrium $(4/11, 4/11, 4/11)^T$ is locally asymptotically stable. ■

We consider the case $a + b = 2$ briefly. Let $N(t) = x_1(t) + x_2(t) + x_3(t)$ and $P(t) = x_1(t)x_2(t)x_3(t)$. Sum the three differential equations in $x_1$, $x_2$, and $x_3$ to obtain

$$\frac{dN}{dt} = N(1 - N).$$

The variable $N$ represents the total population size of all three species; $N$ satisfies the logistic equation. We know that $\lim_{t\to\infty} N(t) = 1$. Thus, the asymptotic solution to the three-species competitive system lies on the plane $x_1 + x_2 + x_3 = 1$. In addition, it can be seen that

$$\frac{d[\ln P(t)]}{dt} = 3 - 3N(t) = 3\frac{d[\ln N(t)]}{dt}$$

for the case $a + b = 2$. Integrating,

$$\frac{P(t)}{P(0)} = \left(\frac{N(t)}{N(0)}\right)^3.$$

Thus, $\lim_{t\to\infty} P(t) = P(0)/[N(0)]^3 = C =$ constant. The asymptotic solution to the three-species competitive system lies on the hyperboloid $x_1x_2x_3 = C$, where the constant $C$ depends on the initial conditions.

Combining these two results, the asymptotic solution lies on the intersection of the surface $x_1x_2x_3 = C$ and the plane $x_1 + x_2 + x_3 = 1$. It can be shown that there are neutrally stable periodic solutions (see Chi et al., 1998; Schuster et al., 1979).

**Example 6.3** Solutions to the May-Leonard competition model when $a + b = 2$ are illustrated in Figure 6.10. Let

$$\frac{dx_1}{dt} = x_1(1 - x_1 - 1.5x_2 - 0.5x_3),$$

$$\frac{dx_2}{dt} = x_2(1 - 0.5x_1 - x_2 - 1.5x_3),$$

$$\frac{dx_3}{dt} = x_3(1 - 1.5x_1 - 0.5x_2 - x_3),$$

where $a = 0.5$ and $b = 1.5$. The positive equilibrium $(1/3, 1/3, 1/3)$ is neutrally stable. ■

In the case where $a + b > 2$, it can be shown that the asymptotic solution approaches the boundary of the plane $x_1 + x_2 + x_3 = 1$ in the first octant. Solutions cycle around the boundary. For example, the solution is close to $(1, 0, 0)$, then $(0, 1, 0)$, then $(0, 0, 1)$, and finally returns to the first point $(1, 0, 0)$ and continues cycling. The cycles increase in length over time. Thus, they are not periodic. The solutions approach heteroclinic orbits in each of the phase planes (see Chi et al., 1998; Schuster et al., 1979).

**Example 6.4** Solutions to the May-Leonard competition model when $a + b > 2$ are illustrated in Figure 6.10. Let

$$\frac{dx_1}{dt} = x_1(1 - x_1 - 2x_2 - 0.5x_3),$$

$$\frac{dx_2}{dt} = x_2(1 - 0.5x_1 - x_2 - 2x_3),$$

$$\frac{dx_3}{dt} = x_3(1 - 2x_1 - 0.5x_2 - x_3),$$

where $a = 0.5$, and $b = 2$. The positive equilibrium $(2/7, 2/7, 2/7)$ is unstable. ■

There are many other Lotka-Volterra-type models that are not discussed here (e.g., systems with mutualistic interactions and systems combining predation, competition, and mutualism). As the dimension of the system increases or the number of species included in the system increases, the analysis of the model becomes much more difficult and cannot be predicted easily. The May and Leonard competition model illustrates just some of the interesting behavior in three-species competition.

## 6.5 Spruce Budworm Model

The next model is applied to an insect pest, the spruce budworm, which feeds on the needles of balsam fir trees. Removal of the needles ultimately kills the trees. We describe a model for the spruce budworm developed by Ludwig et al. (1978). Tuchinsky (1981) provides a nice summary of the dynamics of Ludwig-Jones-Holling model and a historical introduction to the problems associated with outbreaks of the spruce budworm. In most years, spruce budworm densities are very low in the evergreen forests of Canada. For example, in one study, a thousand samples yielded only ten budworm larvae (Morris, 1963). However, in an outbreak year, spruce budworms devour all of the new needles on balsam fir and may kill up to 80% of the mature trees in a forest (Tuchinsky, 1981). Such outbreaks have destroyed mature forests of balsam fir in eastern Canada and the northeastern United States every 30 to 70 years since the early 1700s (Tuchinsky, 1981).

The spruce budworm model was developed by a mathematician, Ludwig, and two biologists, Jones and Holling (1978). The entire model consists of a system of three nonlinear differential equations. The three variables are defined as follows:

$B$ = budworm density (large budworm larvae/acre of land).

$S$ = number of ten square feet units (tsf) of branch surface area/acre (habitat space for the larvae).

$E$ = a measure on a scale from 0 to 1 of food energy reserves or energy reserve available to the budworm.

The variables $S$ and $E$ change slowly with respect to time, whereas the budworm density $B$ has much greater variation in time. An outbreak of budworm larvae in which balsam fir are denuded of foliage is about four years. The fir trees die and birch trees take over. The time scale for fir reforestation is on the order of

50 to 100 years. Thus, as a first approximation to the model of the spruce budworm–forest ecosystem, only the budworm density is modeled. In the first model, it is assumed that the variables $S$ and $E$ are constant. Later, we consider the dynamics of the variables $S$ and $E$.

The differential equation for budworm density is

$$\frac{dB}{dt} = r_B B\left(1 - \frac{B}{K_B}\right) - \beta\frac{B^2}{\alpha^2 + B^2}. \tag{6.7}$$

The first expression on the right side of (6.7) is logistic growth, $r_B$ is the intrinsic growth rate, and $K_B$ is the carrying capacity. The growth of the budworm density is limited by its food supply (the carrying capacity $K_B$ depends on $S$, the branch surface area/acre). The second expression on the right-hand side of (6.7) represents predation by birds and parasitoids. It is a *Holling Type III* functional response for predation (Hassell, 1978). At low budworm density, predation is low; birds find more abundant sources of prey. As budworm density increases, predation increases, until a maximum rate of predation $\beta$ is reached.

To simplify the analysis of the budworm differential equation, which has four different parameters, the equation is put in dimensionless form. This reduces the number of parameters to two. Let the dimensionless variables be denoted by $B^*$ and $t^*$ and assume

$$B = B^* b \quad \text{and} \quad t = t^*\tau,$$

where $b$ and $\tau$ are suitably chosen constants. Substitution of the new variables into the differential equation yields

$$\frac{dB^*}{dt^*} = r_B\frac{\tau}{b}B\left(1 - \frac{B}{K_B}\right) - \beta\frac{\tau B^2}{b(\alpha^2 + B^2)}.$$

If we choose $b = \alpha$ and $\tau = \alpha/\beta$, then

$$\frac{dB^*}{dt^*} = r_B\frac{\alpha}{\beta}B^*\left(1 - \frac{B^*\alpha}{K_B}\right) - \frac{B^{*2}}{1 + B^{*2}}.$$

Note that the variables $B^*$ and $t^*$ are dimensionless:

$$B^* = \frac{B}{\alpha} = \frac{\text{density}}{\text{density}} \quad \text{and} \quad t^* = \frac{t\beta}{\alpha} = \frac{\text{(time)(density/time)}}{\text{density}}.$$

We simplify the notation. Let $u = B^*$, $R = r_B\alpha/\beta$, $Q = K_B/\alpha$, and replace $t^*$ by $t$. The dimensionless budworm equation with only two parameters is given by

$$\frac{du}{dt} = Ru\left(1 - \frac{u}{Q}\right) - \frac{u^2}{1 + u^2} = u\left[R\left(1 - \frac{u}{Q}\right) - \frac{u}{1 + u^2}\right] = f(u). \tag{6.8}$$

The dimensionless differential equation (6.8) is now analyzed in terms of the parameters $R$ and $Q$. The equilibria are found by solving $f(u) = 0$ or by finding the points of intersections of the following two curves:

$$y_1 = R\left(1 - \frac{u}{Q}\right) \quad \text{and} \quad y_2 = \frac{u}{1 + u^2}.$$

Note that $u$ has been factored out; $u = 0$ is always one of the equilibria, but it is unstable. The analysis of the model depends on how the preceding equations cross in the $u$-$y$ plane.

For different values of $R$ and $Q$, there are one, two, or three positive equilibria in addition to the zero equilibrium $u = 0$. When $R$ is large the growth rate of the budworm is large and there is little predation, and when $Q$ is

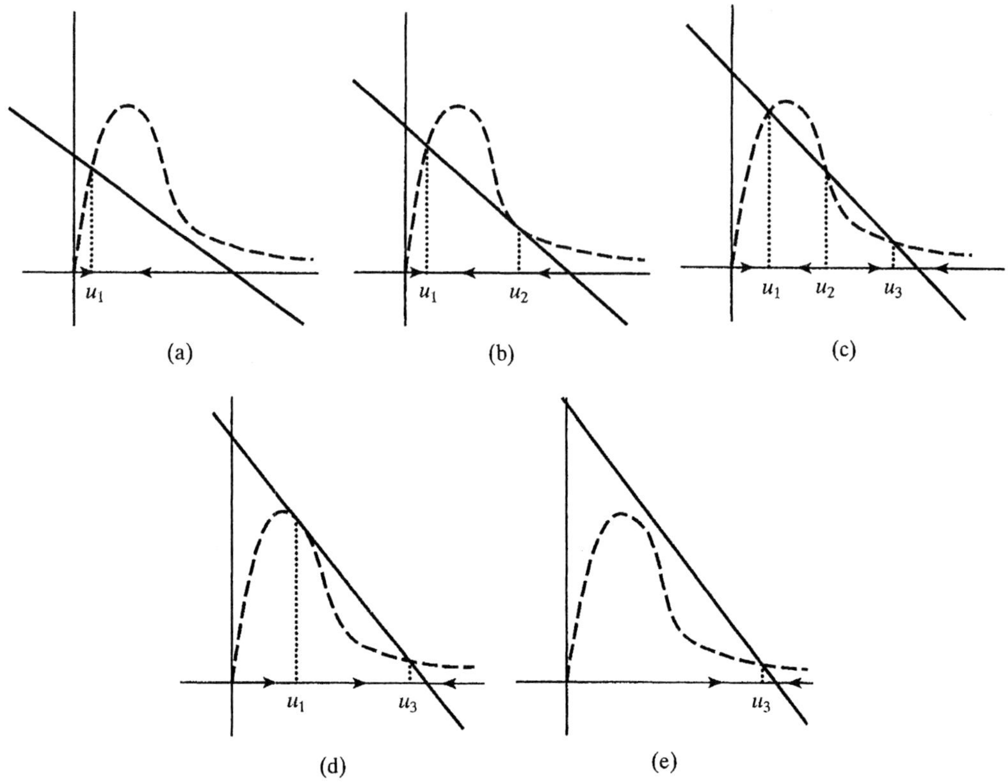


**Figure 6.11** Graphs of the curves $y_1 = R(1 - u/Q)$ (solid line) and $y_2 = u/(1 + u^2)$ (dashed curve). The points of intersection of these two curves are the positive equilibria, $u_i$, $i = 1, 2, 3$. The dynamics are illustrated in a series of figures (a)–(e), where the value of $R$ is increased. The phase line diagram is graphed on the horizontal axis. In (a) and (b) there is one stable equilibrium, $u_1$. In (c), there are two stable equilibria, $u_1$ and $u_3$, and in (d) and (e), there is one stable equilibrium, $u_3$. Equilibrium $u_1$ is the refuge equilibrium and $u_3$ is the outbreak equilibrium.

large the carrying capacity of the budworm is large. When either one of these parameters is large there is an outbreak; there are many spruce budworms attacking the trees. The dynamics are illustrated on a phase line diagram, graphed below the intersection of the two curves $y_1$ and $y_2$ in Figure 6.11. The population increases if the curve $y_1$ lies above $y_2$ and decreases if $y_1$ lies below $y_2$.

In Figure 6.11, the value of $R$ is steadily increased. At low levels of $R$, there is a single stable equilibrium at $u_1$, but as $R$ increases there are two, then three equilibria. When there are three equilibria, there are two stable ones. The equilibrium $u_1$ is called the *refuge equilibrium* and the equilibrium $u_3$ is called the *outbreak equilibrium*. The equilibrium $u_2$ is a threshold between these two equilibria. It is interesting to note that when the line $y_1$ is tangent to the curve $y_2$, there is a *saddle node* bifurcation. An equilibrium $u_2$ suddenly appears, and then there are two equilibria $u_2$ and $u_3$; $u_2$ is unstable and $u_3$ is stable.

It is possible to determine the values of $R$ and $Q$ that yield either one, two, or three equilibria. To do this, it is necessary to find where the curves $y_1$ and $y_2$ intersect and are tangent,

$$y_1 = y_2 \quad \text{and} \quad \frac{dy_1}{du} = \frac{dy_2}{du}. \tag{6.9}$$

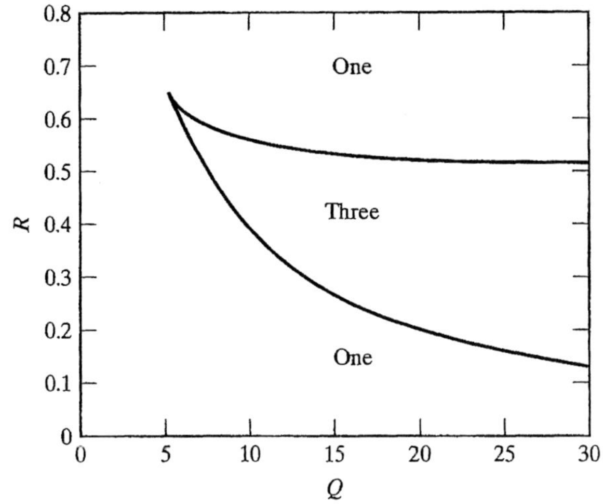


**Figure 6.12** Graphs of the parametric equations (6.10) identify the regions in $Q$-$R$ parameter space where there exist one, two, or three equilibria. Along the curves there exist exactly two equilibria. The lower curve is for the parameter range $u \in (\sqrt{3}, \infty)$ and the upper curve is for $u \in (1, \sqrt{3})$.

At values of $R$ and $Q$ satisfying (6.9) there are exactly two equilibria. The two equations in (6.9) can be expressed as

$$\frac{R}{Q}(Q - u) = \frac{u}{1 + u^2} \quad \text{and} \quad -\frac{R}{Q} = \frac{1 - u^2}{(1 + u^2)^2}.$$

These two equations can be solved for $R$ and $Q$ and expressed as functions of $u$,

$$Q = \frac{2u^3}{u^2 - 1} \quad \text{and} \quad R = \frac{2u^3}{(1 + u^2)^2}. \tag{6.10}$$

The two equations for $R$ and $Q$ can be thought of as parametric equations, where $u$ is the parameter and $u > 1$ (so that $Q > 0$). Graphing these parametric equations in $Q$-$R$ parameter space, we need to be careful, because they are not smooth functions of the parameter $u$. The derivative $dR/dQ$ is undefined when $u = \sqrt{3}$, that is,

$$\frac{dR}{dQ} = \frac{dR/du}{dQ/du} = \frac{u^2(6 - 2u^2)/(1 + u^2)^3}{2u^2(u^2 - 3)/(u^2 - 1)^2}.$$

Graphs of the parametric equations (6.10) for $u \in (1, \sqrt{3})$ and for $u \in (\sqrt{3}, \infty)$ are given in Figure 6.12.

The values of $R$ and $Q$ at which $y_1$ and $y_2$ are tangent form a cusp-shaped curve. Inside the cusp-shaped region, there are three equilibria, but outside this region, there is just one equilibrium. Suppose the value of $Q$ is fixed and $R$ increases from a value below the cusp to one in the cusp region the stable equilibrium switches from the low value $u_1$ to the high value $u_3$. The reverse behavior is observed if $R$ decreases from a value above the cusp to one in the cusp region: The stable equilibrium switches from the high value of $u_3$ to the low value $u_1$. This type of behavior has been called a *hysteresis effect* and is studied in catastrophe theory. The budworm density follows a cycle that traces one path as $u$ increases but a different path as $u$ decreases (Murray, 1993; Strogatz, 2000; Tuchinsky, 1981).

The results may be interpreted in terms of the outbreak behavior of the spruce budworm. The smaller equilibrium $u_1$ is the refuge equilibrium while $u_3$ is the outbreak equilibrium. How should the population be controlled to keep it in the refuge state and not allow it to reach the outbreak state? Much time, effort, and money is being spent on application and evaluation of various methods for controlling this pest. Some possible methods of control include (see Tuchinsky, 1981)

(1) Application of insecticides to kill the budworms (a temporary and expensive control).
(2) Biological controls that decrease reproduction such as by introducing sterile males or a new predator. In the model, this would reduce $r_B$ or increase $\beta$, which in either case reduces $R$.
(3) Modification of the forest (e.g., create empty spaces) to prevent spatial spread.

Next, we consider the slow variables, $S$ and $E$. Recall that these two variables represent the number of 10 square feet of branch surface area/acre or the habitat space for the larvae ($S$) and a measure of food energy reserves available to the budworm ($E$). The variables $S$ and $E$ change slowly with respect to time, whereas the budworm density $B$ has much greater variation in time. As described earlier, an outbreak of budworm larvae in which balsam fir are denuded of foliage is about four years. The time scale for fir reforestation is on the order of 50–100 years. We model the slow variables, $S$ and $E$, and assume $B$ is near an equilibrium. A logistic form for surface area $S$ is chosen,

$$\frac{dS}{dt} = r_S S\left(1 - \frac{S}{K_S \frac{E}{K_E}}\right).$$

The carrying capacity $K_S E/K_E$ is maximal at $K_S$ when $E = K_E$. The additional factor $E/K_E$ is included because $S$ does not increase (and may decrease) under conditions of stress when energy reserves $E$ are low. The branch surface area decreases because of death of foliage or death of a tree.

Energy reserve is self-limited and is also affected by stress. Energy is modeled by the equation

$$\frac{dE}{dt} = r_E E\left(1 - \frac{E}{K_E}\right) - p\frac{B}{S}.$$

In the equation for energy reserve, it is assumed that stress is proportional to $B/S$. The ratio $B/S$ measures the average number of budworm per average branch size.

The system of differential equations for the slow variables $S$ and $E$ is analyzed. The equilibria for this system may be either at an outbreak state or a state prior to outbreak where the density is low. The $S$-nullclines are given by

$$S = 0 \quad \text{and} \quad S = \frac{K_S E}{K_E},$$

and the $E$-nullcline is given by

$$S = \frac{pK_E}{r_E}\frac{B}{E(K_E - E)}.$$

As budworm density increases, the $U$-shaped nullcline $dE/dt = 0$ moves upward. At outbreak level, there are no equilibria and $S$ and $E$ decrease (in fact, $E$ can become negative). See Figure 6.13.

The dynamics of the entire spruce budworm system are represented by three differential equations for $B$, $S$, and $E$. We formulate a model for this system but do not analyze it (see Ludwig et al., 1978 or Tuchinsky, 1981). For the

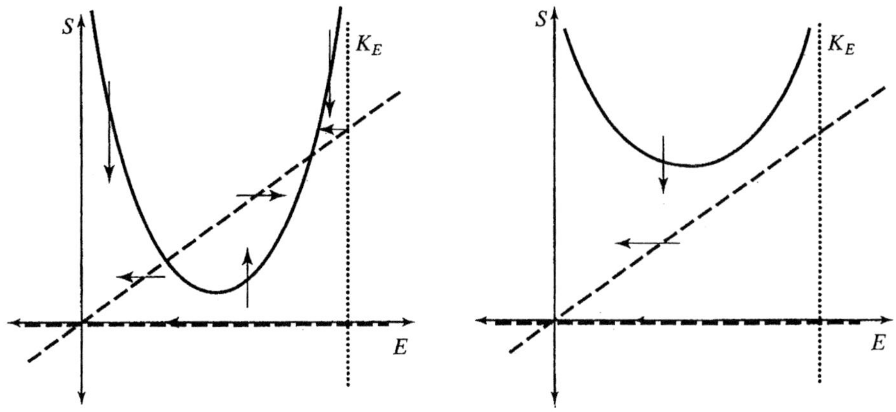


**Figure 6.13** Phase plane analysis for the slow variables $S$ and $E$ in the spruce budworm system.

three-dimensional system it is assumed that $K_B = KS$ (budworm carrying capacity is proportional to branch surface area) and $\alpha = \alpha_1 S$ (as the surface area increases, predators must search more foliage for prey; a higher level of $B$ can be supported at the same level of predation $\beta$). With these assumptions the system of three differential equations becomes

$$\frac{dB}{dt} = r_B B\left(1 - \frac{B}{KS}\right) - \beta\frac{B^2}{\alpha_1^2 S^2 + B^2},$$

$$\frac{dS}{dt} = r_S S\left(1 - \frac{S}{K_S \frac{E}{K_E}}\right),$$

$$\frac{dE}{dt} = r_E E\left(1 - \frac{E}{K_E}\right) - p\frac{B}{S}.$$

There are nine positive parameters, $r_B, K, \beta, \alpha_1, r_S, K_S, K_E, r_E$, and $p$. Some of these parameters have been estimated (Tuchinsky, 1981). There is a defect in the model in that there is only one generation of forest. If the forest dies, it does not recover ($S \to 0, E \to$ negative). To correct this defect, the model could be restarted after forest death. However, an improved model is to allow for recovery of the forest by reducing the carrying capacity of $B$, replacing $K_B = KS$ by

$$KS\frac{E^2}{T_E^2 + E^2}.$$

As $E$ decreases so does the budworm density. This factor should also be included in the stress factor, that is, replace $pB/S$ by

$$p\frac{B}{S}\frac{E^2}{T_E^2 + E^2}.$$

These two changes allow the forest to recover at low levels of $E$. The spruce budworm–forest model has provided some insight into the dynamics of this important ecological problem and has suggested some methods of control that can be examined in a more complex model.

Another well-studied insect pest–forest ecosystem involves the mountain pine beetle (MPB). Much of the MPB life cycle is spent as larvae feeding on the inner bark of its host—pine trees. This feeding activity eventually girdles the tree and kills it. Most western pine trees in the United States serve as host for this insect, but currently the primary hosts are ponderosa pine and lodgepole pine (Logan and Powell, 2001). Mathematical models based on timing, temperature, and spatial dynamics have helped bring greater understanding to some of of these complex interactions between the MPB and its host (see e.g., Logan et al., 1998; Powell et al., 2000).

## 6.6 Metapopulation and Patch Models

Models discrete in space but continuous in time can be modeled by a system of ordinary differential equations and are often referred to as *patch* models or *metapopulation* models (Hanski, 1999; Hanski and Gilpin, 1997 and references therein). A patch refers to a region in space that is suitable habitat for a particular species. Richard Levins in 1970 first used the term *metapopulation* to describe a population consisting of many local populations (Hanski, 1999). In Levins's (1969, 1970) first model, he describes the proportion of occupied patches $p(t)$ from a very large number of patches. Then $1 - p(t)$ is the proportion of patches that are not occupied. Levins's model is an ordinary differential equation. The model has the form

$$\frac{dp}{dt} = cp(1 - p) - ep, \tag{6.11}$$

where $c$ and $e$ are colonization and extinction parameters for the population, respectively. This model can be analyzed easily. There exist two equilibria, $\bar{p} = 0$ and $\bar{p} = 1 - e/c$; either all patches are empty or a proportion is occupied. All patches are empty if the colonization rate is less than the extinction rate, $c \leq e$, but a proportion of the patches is occupied if $c > e$. In this latter case, solutions to (6.11) converge to $\bar{p} = 1 - e/c$. (A phase line diagram can be used to verify this result.)

Keymer et al. (2000) extended model (6.11) to one with a dynamic landscape, where the patches themselves are not static but can change over time. Patches can change from being habitable to uninhabitable. For example, environmental or anthropogenic changes (e.g., climate change or clearcutting of forests for agricultural purposes) can result in landscape patterns where some patches are not suitable for certain species.

Let $p_0$ denote the proportion of uninhabitable patches and $1 - p_0$ denote those that are habitable. The habitable patches can be further subdivided into the proportion of patches that are not occupied, $p_1$, and those that are occupied, $p_2$. The system of differential equations formulated by Keymer et al. (2000) has the following form:

$$\begin{aligned}
\frac{dp_0}{dt} &= e(p_1 + p_2) - \lambda p_0, \\
\frac{dp_1}{dt} &= \lambda p_0 - \beta p_1 p_2 + \delta p_2 - e p_1, \\
\frac{dp_2}{dt} &= \beta p_1 p_2 - (\delta + e) p_2,
\end{aligned} \tag{6.12}$$

where $\lambda$ is the rate of patch creation, $e$ is the rate of patch destruction, $\delta$ is the rate of population extinction, and $\beta$ is the rate of propagule reproduction. Because $p_0 + p_1 + p_2 = 1$, the system (6.12) can be reduced to two equations. Let $p_0 = 1 - p_1 - p_2$. Then the dynamics of the habitable patches satisfy the following system of equations:

$$\frac{dp_1}{dt} = \lambda(1 - p_1 - p_2) - \beta p_1 p_2 + \delta p_2 - e p_1,$$

$$\frac{dp_2}{dt} = p_2(\beta p_1 - \delta - e). \tag{6.13}$$

It is straightforward to show that the system (6.13) has two equilibria, $(\bar{p}_1, \bar{p}_2)$, given by

$$\left(\frac{\lambda}{\lambda + e}, 0\right) \quad \text{and} \quad \left(\frac{\delta + e}{\beta}, 1 - \frac{e}{\lambda + e} - \frac{\delta + e}{\beta}\right).$$

A threshold for persistence can be defined for model (6.13):

$$\mathcal{R}_0 = \frac{\beta\lambda}{(\lambda + e)(\delta + e)}.$$

This threshold is the number of propagules needed during the species and the patch lifetime for the species to persist. This parameter has a similar interpretation in epidemiology, where $p_0$ is the proportion of immune hosts, $p_1$ is the proportion of susceptible hosts, and $p_2$ is the proportion of infected hosts. In this case, $\mathcal{R}_0$ is the average number of successful contacts resulting in infection of a susceptible (Keymer et al., 2000). If $\mathcal{R}_0 \leq 1$, then the first equilibrium with all patches empty is globally asymptotically stable. However, if $\mathcal{R}_0 > 1$, then the second equilibrium with a proportion of habitable patches occupied is globally asymptotically stable. See Exercise 12.

In another metapopulation model formulated by Hanski (1985, 1999), the dynamics are more complicated. Occupied patches are divided into patches with small and large population size. In addition, there is a term representing migration from small to large population sizes. Higher immigration with increasing population size in turn reduces the chance of extinction. This phenomenon has been referred to as the *rescue effect*. Suppose there are empty patches, patches with small population size, and patches with large population size. Let $E$, $S$, and $L$, denote the fraction of empty patches, patches with small populations, and patches with large populations, respectively. The proportion of occupied patches is $P = S + L$ and the proportion of empty patches is $E = 1 - P$. The model is described by the following equations:

$$\frac{dE}{dt} = e_S S - cLE,$$

$$\frac{dS}{dt} = cLE + e_L L - e_S S - rS - mLS, \tag{6.14}$$

$$\frac{dL}{dt} = rS + mLS - e_L L.$$

The extinction rate for a small population is $e_S S$. The colonization rate of an empty patch by a small population is $cLE$ (occurs only from large populations). A small population can grow to a large population and a large population can become small at rates $rS$ and $e_L L$, respectively. A large population does not become extinct without first becoming a small population. In addition, migration occurs between large and small populations at a rate $mLS$. We assume all parameters are positive. Hanski showed that with migration, $m > 0$, there is a range of parameter values where there are three equilibria. One equilibrium is

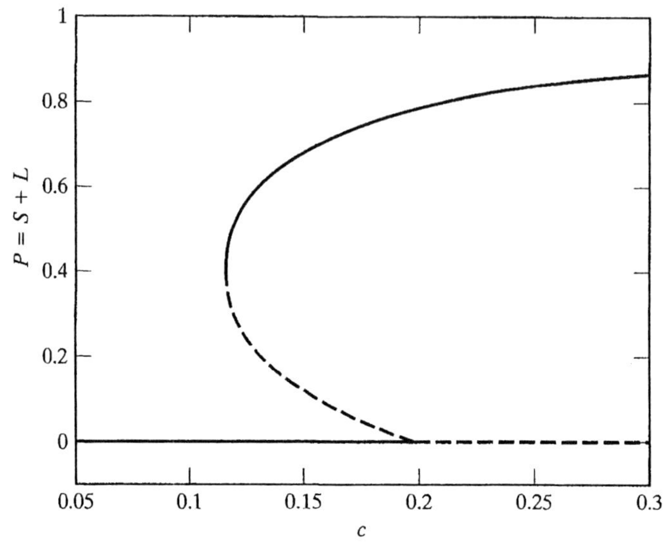


**Figure 6.14** Bifurcation diagram for model (6.14) showing stable (solid curves) and unstable equilibria (dashed curves) values for the proportion of occupied patches $P$ as a function of the colonization rate $c$. The remaining parameter values are $e_S = 1, e_L = 0.02, r = 0.1$, and $m = 0.5$. Bistability occurs for $c \in [0.116, 0.2]$.

the extinction equilibrium, $E = 1$ and $P = 0$, and the other two equilibria are positive equilibria, where $P > 0$. Only two of the equilibria are stable (bistability), the zero equilibrium and one of the positive equilibria. The bifurcation diagram in Figure 6.14 shows the stable and unstable equilibria as a function of the colonization rate $c$. Also see Exercise 13. Hanski and colleagues applied their metapopulation models to the Glanville fritillary butterfly.

One last metapopulation we consider is referred to as a spatially realistic Levins model. This model was formulated by Moilanen and Hanski (1995). The landscape is not dynamic but there are $n$ patches within the landscape. For each patch, there is a probability that it is occupied or not occupied. Let $p_i$ denote the probability patch $i$ is occupied. Then $1 - p_i$ is the probability that patch $i$ is not occupied. Let $p = (p_1, \dots, p_n)^T$ be the vector of patch occupancies. The $n$-patch model has the following form:

$$\frac{dp_i}{dt} = C_i(p)(1 - p_i) - E_i(p)p_i, \tag{6.15}$$

$i = 1, \dots, n$. The coefficient $C_i$ is the colonization rate and $E_i$ is the extinction rate for patch $i$. The colonization rate depends on patch areas $A_i$ and interpatch distances $d_{ij}$,

$$C_i(p) = c\sum_{j \neq i} e^{-ad_{ij}} A_j p_j. \tag{6.16}$$

The extinction rate is inversely proportional to patch area,

$$E_i(p) = \frac{e}{A_i}.$$

At the origin in model (6.15), $p_i = 0, i = 1, \dots, n$, all patches are not occupied by the species; the species is extinct. It can be shown that the extinction equilibrium is locally asymptotically stable if

$$\lambda_M < \frac{e}{c}, \tag{6.17}$$

where $\lambda_M$ is the largest eigenvalue of the following matrix:

$$M = \begin{pmatrix} 0 & e^{-ad_{12}}A_1A_2 & \cdots & e^{-ad_{1n}}A_1A_n \\ e^{-ad_{21}}A_2A_1 & 0 & \cdots & e^{-ad_{2n}}A_2A_n \\ \vdots & \vdots & \ddots & \vdots \\ e^{-ad_{n1}}A_nA_1 & e^{-ad_{n2}}A_nA_2 & \cdots & 0 \end{pmatrix}. \quad (6.18)$$

(See Exercise 14.) The eigenvalue $\lambda_M$ is a measure of the amount of suitable habitat and is referred to as the *metapopulation capacity* (Ovaskainen and Hanski, 2001). The fraction $e/c$ is the ratio of extinction and colonization rate parameters. When the inequality in (6.17) is reversed, then a unique positive equilibrium exists to system (6.15). The quantity $\lambda_M$ can be used to rank different landscapes in terms of their persistence or invasibility capability (DeWoody et al., 2005; Hanski and Ovaskainen, 2000).

The previous metapopulation models are referred to as *patch occupancy models* since they follow the dynamics of occupied patches, that is, whether the focal species is present or absent in a patch. The dynamics within a patch can be modeled also. For example, a two-species, two-patch model has the form

$$\frac{dN_1^i}{dt} = f^i(N_1^i, N_2^i) + d_{1i}(N_1^j - N_1^i),$$

$$\frac{dN_2^i}{dt} = g^i(N_1^i, N_2^i) + d_{2i}(N_2^j - N_2^i), \; i, j = 1, 2, i \neq j,$$

where $N_1^i$ and $N_2^i$ represent two species interacting and diffusing between two patches, $i = 1, 2$. The terms $d_{1i}$ and $d_{2i}$ represent rates of movement between the two patches. See, for example, Levin (1974) and Akçakaya et al. (2004) for models with multiple species and multiple patches. Spatially explicit metapopulation or patch models can become very complicated, especially when the number of patches is large.

## 6.7 Chemostat Model

A *chemostat* is a device used to grow nutrients, bacteria, and other substances in the laboratory. The dynamics of the substance are studied at the cellular level. The rate of growth of cells is modeled differently from populations. One of the most often used model that approximates the rate of cellular growth originates from what is known as Michaelis-Menten kinetics.

### 6.7.1 Michaelis-Menten Kinetics

An expression widely used to model the rate of nutrient uptake by a cell is derived. This rate follows what is known as *Michaelis-Menten kinetics.* The rate of change of the nutrient concentration $n(t)$ used by a cell for growth and development is modeled by the following differential equation:

$$\frac{dn}{dt} = -K(n) = -\frac{k_{\max}\, n}{k_n + n},$$

where $k_n, k_{\max} > 0$. The parameter $k_{\max}$ is the maximum rate of uptake by the cell of the nutrient and $k_n$ is the half-saturation constant, that is, the amount of nutrient so that $K(n) = k_{max}/2$. The rate of growth $K(n)$ arises from *Michaelis-Menten kinetics*, also referred to as *Michaelis-Menten-Monod kinetics* (Smith and Waltman, 1995). The names Michaelis and Menten refer to the German scientists whose work in enzyme kinetics during the early part of the of the twentieth century contributed much to this theory. Jacques L. Monod was a French scientist who made many contributions to the theory of microbial growth during the 1940s to 1960s (Monod, 1950).

We describe briefly how nutrients enter a cell and contribute to the growth of a microorganism such as bacteria (see, e.g., Edelstein-Keshet, 1988). Nutrient molecules $n$ (substrate) enter the bacterial cell membrane by attaching to membrane-bound receptors or enzymes $x$. If a nutrient molecule is captured by a cell, we denote it as $p$ for the product. The notation $x_0$ denotes a receptor (or enzyme) not occupied by a nutrient molecule, $x_1$ denotes a receptor occupied by a nutrient molecule (or a complex formed from the enzyme and the nutrient molecule), and the resulting product is denoted $p$. The following relationships summarize the direction and the rates of the reactions:

$$n + x_0 \underset{k_{-1}}{\overset{k_1}{\rightleftharpoons}} x_1, \quad x_1 \overset{k_2}{\longrightarrow} p + x_0,$$

where the constants $k_i, i = -1, 1, 2$ are the rate constants. The arrows indicate that the first reaction is reversible. An occupied receptor can lose the nutrient molecule before the nutrient molecule is captured by the cell. See Figure 6.15.

The reaction relationships between the molecules can be expressed as differential equations. We assume the law of mass action. That is, the rate of reaction between two quantities is proportional to the product of their concentrations. Let the variables $n$, $x_0$, $x_1$, and $p$ denote the concentrations (average number per unit volume) of nutrient, unbound receptors, bound receptors, and product, respectively. The differential equations for these four variables are given by

$$\frac{dn}{dt} = -k_1 n x_0 + k_{-1} x_1,$$

$$\frac{dx_0}{dt} = -k_1 n x_0 + k_{-1} x_1 + k_2 x_1,$$

$$\frac{dx_1}{dt} = k_1 n x_0 - k_{-1} x_1 - k_2 x_1,$$

$$\frac{dp}{dt} = k_2 x_1,$$

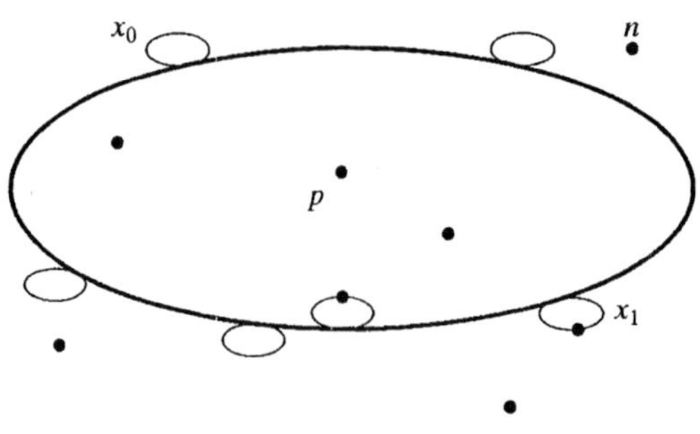


**Figure 6.15** A cell with unbound, $x_0$, and bound, $x_1$, receptor sites, nutrient, $n$, and product, $p$.

(see Edelstein-Keshet, 1988). It can be seen from the equations that the total concentration of receptor cells is constant: $dx_0/dt + dx_1/dt = 0$. Letting $r = x_0 + x_1$ and replacing $x_0$ in the differential equations for $n$ and $x_1$ by $x_0 = r - x_1$, it follows that only differential equations for $n$ and $x_1$ need to be considered. The differential equation for the product concentration $p$ can be solved after the solution for $x_1$ is found. Simplifying the differential equations for $n$ and $x_1$ yields

$$\frac{dn}{dt} = -k_1rn + (k_{-1} + k_1n)x_1,$$

$$\frac{dx_1}{dt} = k_1rn - (k_{-1} + k_2 + k_1n)x_1.$$

The nutrient concentration is usually much higher than the receptor concentration. The receptors are working at maximum capacity so that their occupancy rate is approximately constant. Therefore, it is reasonable to make the assumption that $dx_1/dt = 0$. This assumption is known as the *quasi-equilibrium hypothesis*. From this assumption the equation for Michaelis-Menten kinetics is generated. Let $dx_1/dt = 0$. Then

$$k_1rn - (k_{-1} + k_1n)x_1 = k_2x_1$$

and

$$x_1 = \frac{k_1rn}{k_{-1} + k_2 + k_1n}.$$

Making these substitutions into the differential equation for the nutrient concentration yields

$$\begin{aligned}\frac{dn}{dt} &= -k_2x_1 \\ &= -\frac{k_2k_1rn}{k_{-1} + k_2 + k_1n} \\ &= -\frac{k_2rn}{\dfrac{k_{-1} + k_2}{k_1} + n}.\end{aligned}$$

Therefore,

$$\frac{dn}{dt} = -\frac{k_{max}n}{k_n + n}, \tag{6.19}$$

where $k_{max} = k_2r$ and $k_n = (k_{-1} + k_2)/k_1$. It follows that the differential equation for the product satisfies

$$\frac{dp}{dt} = \frac{k_{max}n}{k_n + n}.$$

The rate of nutrient uptake is the same as the rate of the formation of the product. See Figure 6.16.

The form of the Michaelis-Menten uptake rate is used in many applications involving growth of microorganisms. In the next example, the Michaelis-Menten uptake rate is used to model growth of bacteria in a chemostat.

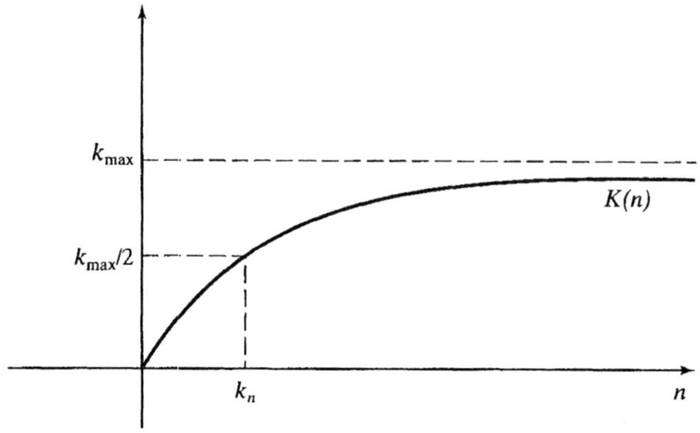


**Figure 6.16** The rate of nutrient uptake, $K(n) = k_{max}n/(k_n + n)$, where $\lim_{n\to\infty} K(n) = k_{max}$ and $K(k_n) = k_{max}/2$.

### 6.7.2 Bacterial Growth in a Chemostat

Growth in a continuous culture chemostat is studied. It is assumed that there is a continuous supply of nutrient and a continuous harvest of bacteria. This is in contrast to a batch culture chemostat, where a fixed quantity of nutrient is supplied and bacteria are harvested after a certain growth period. According to Smith and Waltman (1995), there are several reasons why the chemostat is important biologically.

(i) The chemostat can be used to study competition for a single limiting nutrient.

(ii) The chemostat plays an important role in certain fermentation processes, such as in the commercial production of insulin.

(iii) The chemostat represents a model of a simple lake or of a wastewater treatment process.

Please consult the book by Smith and Waltman (1995) for additional information about variations on the basic chemostat model.

Figure 6.17 shows a schematic of a chemostat. The variable N represents the nutrient container, C the culture container, and O the overflow container. The growth of bacteria in the culture container will be modeled. Let $n(t)$ be the concentration of nutrient in the culture container and let $b(t)$ be the concentration of bacteria. Several assumptions are made concerning the chemostat.

1. The culture container is well stirred, so that $n$ and $b$ are only functions of time $t$ and not functions of the spatial location within the container.
2. The nutrient $n$ is the single growth-limiting nutrient whose concentration determines the growth rate of the culture.

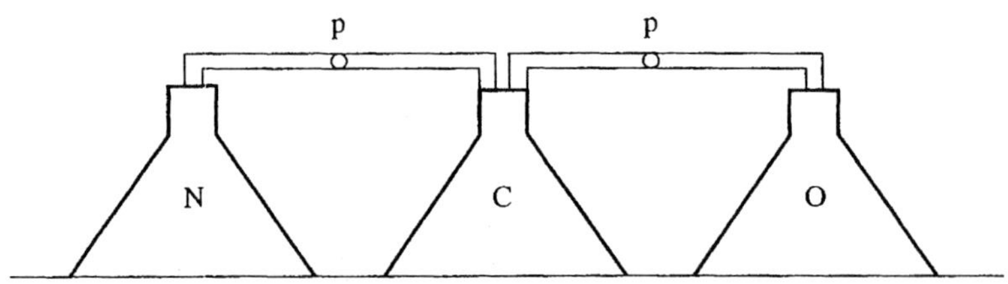


**Figure 6.17** Schematic diagram of a chemostat, N = nutrient container, C = culture container, O = overflow container, and p = pump (Smith and Waltman, 1995).

3. The growth rate of the bacterial population is a function of nutrient availability, $K(n)$.
4. Nutrient depletion occurs continuously as a result of reproduction.

We need to consider the dimensions of each of the variables to ensure that the model is dimensionally consistent. Let $V$ = volume (length$^3$), $F$ = flow rate (volume/time), and $n_0$ = input nutrient concentration (mass/volume). The dimensionally correct model has the following form:

$$\frac{db}{dt} = K(n)b - \frac{F}{V}b = b\left(\frac{k_{\max}n}{k_n + n} - D\right),$$

$$\frac{dn}{dt} = -\beta K(n)b - \frac{F}{V}n + \frac{F}{V}n_0 = D(n_0 - n) - \beta\frac{k_{\max}nb}{k_n + n},$$

where $D = F/V = (1/\text{time})$ is the dilution rate. It is assumed that the bacterial growth rate has a Michaelis-Menten form,

$$K(n) = \frac{k_{\max}n}{k_n + n},$$

where $k_{\max}$ is the maximum rate of growth, $k_n$ is the half-saturation constant, and $1/\beta$ is the yield. The nutrient-bacteria model has five parameters, $k_{\max}$, $k_n$, $\beta$, $D$, and $n_0$. Via a dimensional analysis the number of parameters can be reduced to 2.

Redefine the variables $n$, $b$, and $t$ in terms of dimensionless variables $x$, $S$, and $t^*$ as follows:

$$b = \frac{xn_0}{\beta}, \quad n = Sn_0, \quad t = \frac{t^*}{D}.$$

Then the differential equations expressed in terms of the dimensionless variables are

$$\frac{dx}{dt^*} = x\left(\frac{k_{\max}S/D}{[k_n/n_0] + S} - 1\right) = x\left(\frac{mS}{\alpha + S} - 1\right),$$

$$\frac{dS}{dt^*} = 1 - S - \frac{k_{\max}Sx/D}{[k_n/n_0] + S} = 1 - S - \frac{mSx}{\alpha + S},$$

where $m = k_{\max}/D$ and $\alpha = k_n/n_0$. Initial conditions satisfy $x(0) > 0$ and $S(0) > 0$. For notational convenience, in the following analysis, the star notation on $t$ is dropped. However, keep in mind that it is the dimensionless system that is being analyzed.

The chemostat model has interesting dynamics. Define $\Sigma(t) = 1 - x(t) - S(t)$. Then it follows that

$$\frac{d\Sigma}{dt} = -\Sigma.$$

Hence, $\Sigma(t) = \Sigma(0)\exp(-t) \to 0$ as $t \to \infty$. The total amount of nutrient and bacteria concentration, $x(t) + S(t)$, approaches 1, that is,

$$\lim_{t\to\infty}[x(t) + S(t)] = 1.$$

To study the dynamics of $x$ and $S$, we do a local stability analysis and a phase plane analysis. The dimensionless system has two equilibria, $(\bar{x}, \bar{S})$, given by

$$E_0 = (0, 1) \quad \text{and} \quad E_1 = \left(1 - \frac{\alpha}{m-1}, \frac{\alpha}{m-1}\right).$$

The Jacobian matrix of the dimensionless system satisfies

$$J = \begin{pmatrix} \dfrac{mS}{\alpha + S} - 1 & \dfrac{\alpha m x}{(\alpha + S)^2} \\ -\dfrac{mS}{\alpha + S} & -1 - \dfrac{\alpha m x}{(\alpha + S)^2} \end{pmatrix}.$$

The Jacobian matrix evaluated at $E_0$ satisfies

$$J(E_0) = \begin{pmatrix} \dfrac{m}{\alpha + 1} - 1 & 0 \\ -\dfrac{m}{\alpha + 1} & -1 \end{pmatrix}.$$

Notice that the eigenvalues of $J(E_0)$ are along the diagonal, $\lambda_1 = m/(\alpha + 1) - 1$ and $\lambda_2 = -1$. Whether the equilibrium $E_0$ is stable or unstable depends on the sign of $\lambda_1$. If $\lambda_1 < 0$, $E_0$ is a stable node, and if $\lambda_1 > 0$, $E_0$ is a saddle point. Therefore, the condition for local asymptotic stability of $E_0$ is

$$m < \alpha + 1.$$

The Jacobian matrix evaluated at $E_1 = (\bar{x}, \bar{S})$ satisfies

$$J(E_1) = \begin{pmatrix} 0 & \dfrac{\alpha m \bar{x}}{(\alpha + \bar{S})^2} \\ -1 & -1 - \dfrac{\alpha m \bar{x}}{(\alpha + \bar{S})^2} \end{pmatrix}.$$

If $\bar{x} > 0$ and $\bar{S} > 0$, then $\operatorname{Tr} J(E_1) < 0$ and $\det J(E_1) > 0$. In addition, we can apply two properties satisfied by eigenvalues $\lambda_1$ and $\lambda_2$, $\operatorname{Tr}(J) = \lambda_1 + \lambda_2$ and $\det(J) = \lambda_1 \lambda_2$. For our model,

$$\operatorname{Tr} J(E_1) = -1 - \frac{\alpha m \bar{x}}{(\alpha + \bar{S})^2} \quad \text{and} \quad \det J(E_1) = \frac{\alpha m \bar{x}}{(\alpha + \bar{S})^2}.$$

It follows that

$$\lambda_1 = -1 \quad \text{and} \quad \lambda_2 = -\frac{\alpha m \bar{x}}{(\alpha + \bar{S})^2}.$$

Both eigenvalues are negative provided the equilibrium $E_1$ is positive. Thus, $E_1$ is a stable node provided

$$m > \alpha + 1.$$

Next, we perform a phase plane analysis. The nullclines of the dimensionless chemostat system are

$$x\text{-nullclines:} \quad x = 0 \quad \text{and} \quad S = \frac{\alpha}{m-1}$$

$$S\text{-nullcline:} \quad x = \frac{(1-S)(\alpha+S)}{mS}.$$

There are three cases to consider:

(I) $m \le 1$,

(II) $1 < m \le \alpha + 1$,

(III) $m > \alpha + 1$.

In Cases (I) and (II), only the equilibrium $E_0$ is feasible. Because solutions are bounded, solutions must converge to $E_0$. However, in Case (III), $E_0$ and $E_1$ are feasible equilibria. Equilibrium $E_0$ is a saddle point with solutions approaching $E_0$ only along the $S$-axis. In addition, applying Dulac's criterion with the Dulac function $1/x$, it can be shown that there do not exist any periodic solutions. Because solutions are bounded, Poincaré-Bendixson theory implies that solutions in Case (III) converge to $E_1$. See Figure 6.18.

In terms of the original parameters, the inequality in Case (I) is equivalent to $k_{max} \le D$. The maximal growth rate of the bacteria is less than the washout rate. The bacteria are washed out faster than they can reproduce; solutions converge to $E_0$. The expression $\text{BE} = \alpha/(m-1)$ is referred to as the *break-even concentration* of the nutrient (Smith and Waltman, 1995). If $\text{BE} \ge 1$, then the amount of nutrient flowing into the culture container is insufficient for the bacteria to survive [Case (II)]; solutions converge to $E_0$. In terms of the original parameters and variables, BE is given by $\text{BE} = k_n D/(k_{max} - D)$, which can be found by solving $db/dt = 0$ when $n = n_0$. In Case (III), $1 > \text{BE}$, or in terms of the original parameters, $n_0 > \text{BE}$. In this last case, the bacteria survive; solutions converge to $E_1$.

**Example 6.5** Consider a chemostat model with two competing bacterial populations. The system in dimensionless form is

$$\frac{dx_1}{dt} = x_1\left(\frac{m_1 S}{\alpha_1 + S} - 1\right),$$

$$\frac{dx_2}{dt} = x_2\left(\frac{m_2 S}{\alpha_2 + S} - 1\right),$$

$$\frac{dS}{dt} = 1 - S - \frac{m_1 S x_1}{\alpha_1 + S} - \frac{m_2 S x_2}{\alpha_2 + S}.$$

Define $\Sigma(t) = 1 - S(t) - x_1(t) - x_2(t)$. Then, as in the simple chemostat model, $d\Sigma(t)/dt = -\Sigma(t)$, which implies $\lim_{t\to\infty} \Sigma(t) = 0$. Equivalently, $\lim_{t\to\infty}[S(t) + x_1(t) + x_2(t)] = 1$. It can be shown that this system is asymptotically autonomous so that the three-dimensional system can be reduced to two dimensions by replacing $S(t)$ with $1 - x_1(t) - x_2(t)$ (Smith and Waltman, 1995):

$$\frac{dx_1}{dt} = x_1\left(\frac{m_1(1 - x_1 - x_2)}{\alpha_1 + 1 - x_1 - x_2} - 1\right),$$

$$\frac{dx_2}{dt} = x_2\left(\frac{m_2(1 - x_1 - x_2)}{\alpha_2 + 1 - x_1 - x_2} - 1\right).$$

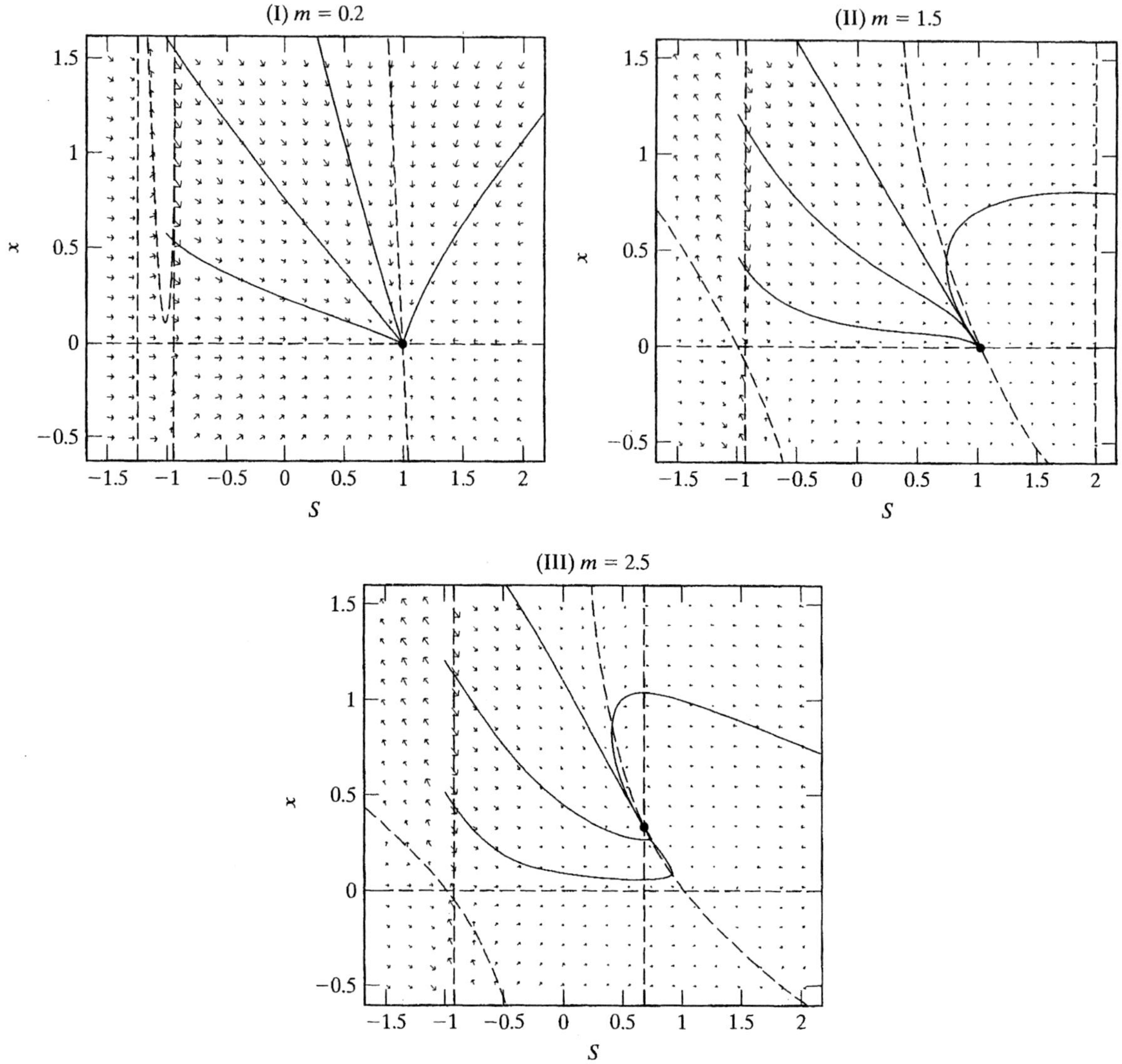


**Figure 6.18** Phase plane diagrams for the dimensionless chemostat model when $\alpha = 1$. (I) $m = 0.2$, (II) $m = 1.5$, and (III) $m = 2.5$. In each figure, the nullclines are identified by the dashed curves and several solution trajectories are identified by the solid curves. The stable equilibrium is identified by a solid dot.

In the case $\alpha_1/(m_1 - 1) \neq \alpha_2/(m_2 - 1)$ and $m_1, m_2 \neq 1$, the reduced system has three equilibria, $(\bar{x}_1, \bar{x}_2)$, where $\bar{S} = 1 - \bar{x}_1 - \bar{x}_2$:

$$E_0 = (0, 0), \quad E_1 = \left(1 - \frac{\alpha_1}{m_1 - 1}, 0\right), \quad \text{and} \quad E_2 = \left(0, 1 - \frac{\alpha_2}{m_2 - 1}\right).$$

A fourth equilibrium with both components positive is not possible under these assumptions. This example is considered further in the Exercises. This example exhibits the principle of competitive exclusion. If two species compete for one limiting nutrient, only the "dominant" one survives. In this case, the "dominant" species is the one that requires the least amount of nutrient, that is, the lowest level of the break-even concentration, $\alpha_i/(m_i - 1), i = 1, 2$. ■

The chemostat has many practical applications, and many variations of the basic model have been studied, including competition, predation, size structure, and delays (see, e.g., Li et al., 2000; Smith and Waltman, 1995; Smith and Zhao, 2001; Wolkowicz et al., 1997, 1999 and references therein).

# 6.8 Epidemic Models

## 6.8.1 SI, SIS, and SIR Epidemic Models

Infectious diseases such as measles, mumps, rubella, and chickenpox are modeled by classifying individuals in the population according to their status with respect to the disease: healthy, infected, and immune. Diseases caused by viruses or bacteria are not modeled directly at the population level, only indirectly through the number of infected individuals. The disease states, $S$, $I$, and $R$, are defined as follows:

$S$ = susceptible; individuals not infected but who are capable of contracting the disease and becoming infective.

$I$ = infected; individuals who are infected and infectious, capable of transmitting the disease to others.

$R$ = removed; individuals who have had the disease and have recovered, and who are permanently immune, or are isolated until recovery and permanent immunity occur.

Models with these three states are referred to as SIR epidemic models. There are variations of the SIR epidemic model, depending on whether individuals recover and develop immunity. They are denoted by various acronyms.

SI implies no recovery, $S \rightarrow I$.

SIS implies recovery, but no immunity $S \rightarrow I \rightarrow S$.

SIRS implies temporary immunity, $S \rightarrow I \rightarrow R \rightarrow S$.

A compartmental diagram of an SIR epidemic model is sketched in Figure 2.15 in Chapter 2.

Differential equations corresponding to an SIRS epidemic model are as follows:

$$\frac{dS}{dt} = -\frac{\beta}{N}SI - bS + bS + bI + bR + \nu R,$$

$$\frac{dI}{dt} = \frac{\beta}{N}SI - \gamma I - bI,$$

$$\frac{dR}{dt} = \gamma I - bR - \nu R,$$

where the initial conditions satisfy $S(0) > 0, I(0) > 0, R(0) \geq 0$, and $S(0) + I(0) + R(0) = N$. The parameters are defined as follows:

$\beta$ = average number of adequate contacts made by an infected individual per time.

$\frac{\beta}{N}S$ = average number of adequate contacts made by an infected individual resulting in an infection of a susceptible individual per time.

$\frac{\beta}{N}SI$ = number of infections caused by all infected individuals per time.

$\gamma$ = recovery rate, $1/\gamma$ = average infectious period.

$\nu$ = rate of loss of immunity, $1/\nu$ = average length of immunity.

$b$ = birth rate = death rate.

$N$ = total population size.

Recovery and loss of immunity are linear functions in the differential equations, $\gamma I$ and $\nu R$. The waiting time for recovery or loss of immunity can be assumed to be exponentially distributed with parameters $\gamma$ and $\nu$, respectively. Then it follows that $1/\gamma$ and $1/\nu$ are the mean waiting times for recovery and loss of immunity or the average infecctious period and the average length of immunity. Because we have assumed that the birth rate equals the death rate, the population size remains constant over time. That is, $dS/dt + dI/dt + dR/dt = 0$ implies $S(t) + I(t) + R(t) = N$.

The dynamics of the four basic epidemic models, SI, SIR, SIS, and SIRS, will be studied in the next four examples. For simplicity, in each example, it is assumed that there are no births and no deaths, $b = 0$. The epidemic dynamics occur on a faster time scale than the population dynamics.

**Example 6.6** **(SI Epidemic Model).** A simple SI epidemic model without births and deaths has the following form:

$$\frac{dS}{dt} = -\frac{\beta}{N}SI,$$

$$\frac{dI}{dt} = \frac{\beta}{N}SI,$$

where $S(0) > 0$, $I(0) > 0$, and $S(0) + I(0) = N$. Thus, $S(t) + I(t) = N$ and replacing $S$ by $N - I$ in the differential equation for $I$ results in

$$\frac{dI}{dt} = \beta I\left(1 - \frac{I}{N}\right).$$

This latter equation is logistic growth with carrying capacity given by $N$. A phase line diagram in Figure 6.19 shows that $I(t) \to N$. Eventually, everyone becomes infected. Such types of model are applicable for a highly infectious disease such as influenza, where a large proportion of the population becomes infected. The duration of the epidemic is only followed until infection but not recovery occurs. ■

**Example 6.7** **(SIS Epidemic Model).** A simple SIS model with no births and deaths has the following form:

$$\frac{dS}{dt} = -\frac{\beta}{N}SI + \gamma I,$$

$$\frac{dI}{dt} = \frac{\beta}{N}SI - \gamma I,$$

**Figure 6.19** Phase line diagram for the SIS epidemic model.

where $\gamma$ is the recovery rate in this model, $S(0) > 0$, $I(0) > 0$, and $S(0) + I(0) = N$. Again, $S(t) + I(t) = N$, and replacing $S$ by $N - I$ in the differential equation for $I$ results in

$$\frac{dI}{dt} = (\beta - \gamma)I\left(1 - \frac{\beta}{(\beta - \gamma)N}I\right)$$

for $\beta \neq \gamma$. If $\beta > \gamma$, the infected population approaches $(\beta - \gamma)N/\beta$ and the susceptible population approaches $\gamma N/\beta$. The disease remains endemic. However, if $\beta \leq \gamma$, then the infected population approaches zero; the epidemic does not persist. Models of SIS type have been applied to sexually transmitted diseases such as gonorrhea and syphilis, where individuals are treated but immediately become susceptible again. ■

The ratio $\beta/\gamma$ in the SIS epidemic model is referred to as the *basic reproduction number* and is often denoted as $\mathcal{R}_0$. We have seen this parameter in the epidemic models discussed in previous chapters. In models with $\beta$ redefined as $\beta = \lambda N$, the basic reproduction number is

$$\mathcal{R}_0 = \frac{\beta}{\gamma} = \frac{\lambda N}{\gamma}.$$

The expression $\beta SI/N$ is generally referred to as *standard incidence rate* whereas $\lambda SI$ is generally referred to as *mass action incidence rate*. In all of the epidemic models studied here, the population size is constant, so that there is no difference in the dynamics with mass action or standard incidence. As can be seen from the definitions, mass action implies that contact rates depend on the population size or density $N$, but for standard incidence, contact rates are independent of $N$. If $N$ is not constant but varies with time, then epidemic models with mass action and standard incidence can produce quite different dynamics.

The basic reproduction number $\mathcal{R}_0$ is the number of secondary infections ($\beta$) caused by one infective individual during his/her infectious period ($1/\gamma$) (Anderson and May, 1991; Hethcote, 2000). This is a very important parameter in epidemiology. Generally, if $\mathcal{R}_0 > 1$, introduction of infected individuals into the population results in an epidemic. For the SIS epidemic model, the value of $\mathcal{R}_0$ predicts a lot more about the epidemic dynamics. In particular, if $\mathcal{R}_0 > 1$, the disease becomes endemic, and if $\mathcal{R}_0 \leq 1$, the disease fades out.

**Example 6.8** **(SIR Epidemic Model).** In an SIR epidemic model without births and deaths, the model has the form

$$\frac{dS}{dt} = -\frac{\beta}{N}SI,$$

$$\frac{dI}{dt} = \frac{\beta}{N}SI - \gamma I = I\left(\frac{\beta}{N}S - \gamma\right),$$

$$\frac{dR}{dt} = \gamma I,$$

where $S(0) > 0$, $I(0) > 0$, $R(0) \geq 0$, and $S(0) + I(0) + R(0) = N$. Thus, $S(t) + I(t) + R(t) = N$. Since $R(t)$ can be found from $S(t)$ and $I(t)$, it is sufficient to consider only the variables $S$ and $I$. Note that

$$\frac{dI}{dS} = -1 + \frac{\gamma N}{\beta S},$$

which can be integrated to obtain

$$I(t) = N - R(0) - S(t) + \frac{\gamma N}{\beta} \ln \frac{S(t)}{S(0)}.$$

Also, note that $S(t)$ is decreasing and the maximum of $I(t)$ occurs when $S(t) = \gamma N/\beta$ (the $I$-nullcline). Since $S(t)$ is a decreasing function of $t$, if $S(0) > \gamma N/\beta$, then $I(t)$ will increase to a maximum (an epidemic occurs), then decrease to zero. On the other hand, if $S(0) \leq \gamma N/\beta$, then $I(t)$ will decrease to zero (no epidemic). The number of infected individuals must approach zero since if not $R(t) \to \infty$, a contradiction. The limiting value of $S(t)$ must satisfy the implicit equation

$$S(\infty) = N - R(0) + \frac{\gamma N}{\beta} \ln \frac{S(\infty)}{S(0)}.$$

The limiting value depends on initial conditions but it is always positive, $S(\infty) > 0$. The dynamics for an SIR model depend on the ratio

$$\mathcal{R} = \frac{\beta S(0)}{\gamma N} = \mathcal{R}_0\, x^*,$$

where $x^* = S(0)/N$ is the proportion of susceptible individuals and $\mathcal{R}_0 = \beta/\gamma$ is the basic reproduction number. The parameter $\mathcal{R}$ is sometimes referred to as the *effective rate* (Anderson and May, 1991). The phase plane diagram in Figure 6.20 illustrates the SIR dynamics. Note that every point on the $S$-axis is an equilibrium. ∎

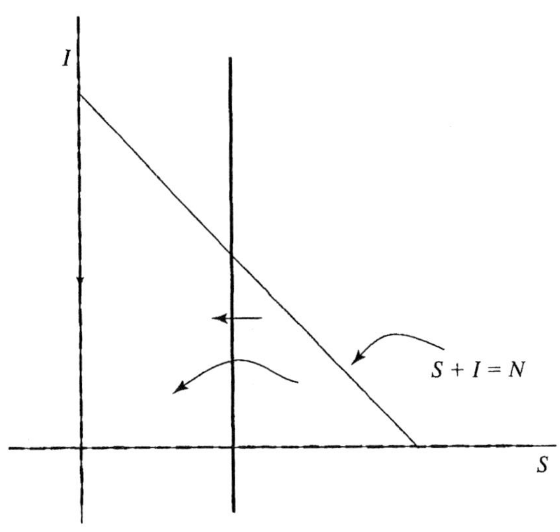


**Figure 6.20** Phase plane diagram for the SIR epidemic model. Every point on the positive $S$-axis is an equilibrium. Solutions tend to some point on the $S$-axis.

**Example 6.9** **(SIRS Epidemic Model).** The last of the four basic epidemic models is an SIRS epidemic model without births and deaths. The model has the following form:

$$\frac{dS}{dt} = -\frac{\beta}{N}SI + \nu R,$$

$$\frac{dI}{dt} = \frac{\beta}{N}SI - \gamma I = I\left(\frac{\beta}{N}S - \gamma\right),$$

$$\frac{dR}{dt} = \gamma I - \nu R,$$

where $S(0) > 0$, $I(0) > 0$, $R(0) \geq 0$, and $S(0) + I(0) + R(0) = N$. Thus, $S(t) + I(t) + R(t) = N$. Since $R(t)$ can be obtained from $S(t)$ and $I(t)$, it is sufficient to consider only the variables $S$ and $I$. The differential equations in $S$ and $I$ are given by

$$\frac{dS}{dt} = -\frac{\beta}{N}SI + \nu(N - S - I),$$

$$\frac{dI}{dt} = I\left(\frac{\beta}{N}S - \gamma\right).$$

The dynamics of this model differ from the SIR model. It is not always the case that the epidemic dies out. The loss of immunity by immune individuals allows the disease to become endemic. The basic reproduction number is a threshold which determines whether the disease becomes endemic. A more complete analysis of this model is left as an exercise. ■

Additional states can be included in the basic epidemic models to make the model more realistic. For example, a latent or exposed state is the period after exposure but prior to being infectious $I$. The latent state is often denoted as $E$, so that the model is referred to as a SEIR epidemic model. Another state often included in childhood diseases is a stage where newborns are initially immune to infection due to the maternal antibodies (from a mother who has had prior exposure to the disease). Eventually, newborns lose this immunity and become susceptible. This latter state is denoted as $M$. If all states are included in the model, it is referred to as an MSEIR epidemic model.

The effect of mass vaccination programs can be examined in the basic epidemic models. Vaccination reduces or eliminates the incidence of infection directly or indirectly. Vaccinated individuals are protected from direct infection, and fewer infectious individuals leads to a decreased likelihood that an unvaccinated susceptible will come in contact with the disease. This latter indirect effect is referred to as *herd immunity*. For a vaccination program to be effective, the fraction $p$ immunized must be such that the remaining population $(1 - p)N$ will be less than the threshold level necessary for the disease to continue. Recall that in an SIR epidemic model $\mathcal{R} = \mathcal{R}_0 x^*$, where $\mathcal{R}_0 = \beta/\gamma$ and $x^* = S(0)/N$. At the start of an epidemic, $S(0) \approx N$. Thus, if $pN$ of susceptible individuals are vaccinated, $S(0) \approx (1 - p)N$ and $\mathcal{R} = \mathcal{R}_0(1 - p)$. To prevent an epidemic,

$$\mathcal{R}_0(1 - p) < 1.$$

An estimate for the minimum value of $p$ is found by solving $\mathcal{R}_0(1 - p) = 1$ or

$$p = \frac{\mathcal{R}_0 - 1}{\mathcal{R}_0}.$$

In England and Wales (1956–1968), the value of $\mathcal{R}_0$ estimated for measles was $\mathcal{R}_0 \approx 13$ (May, 1983). For this value of $\mathcal{R}_0$, to prevent an epidemic, it would have been necessary to vaccinate at least

$$p = \frac{13 - 1}{13} = \frac{12}{13} \approx 92\%.$$

Mass vaccination can have negative consequences. After a mass vaccination program, the average age at which the disease is first acquired generally increases. Recall that

$$\frac{dS}{dt} = -\frac{\beta}{N}SI + \cdots \quad \text{or} \quad \frac{1}{S}\frac{dS}{dt} = -\frac{\beta}{N}I + \cdots.$$

The value $F = \beta I/N$ is the *per capita force of infection* or the rate of acquiring the disease for a population of $I$ infected individuals and transmission constant $\beta/N$ (1/time). Thus, the reciprocal of this value is the average time before acquiring the disease (an exponential distribution with parameter $F$) or the *average age of infection.* Of course, this assumes that the value of $I$ has stabilized at an equilibrium value. The average age of infection is $A = N/(\beta I)$. Vaccination decreases the number of infected individuals $I$ (at equilibrium) which increases the average age of infection, $A$ (see Anderson and May, 1991). If a disease is acquired as an adult rather than as a child, there can be more serious consequences. For example, rubella contracted by a pregnant female may result in congenital rubella syndrome in the newborn. Initiation of any mass vaccination program requires careful examination of all of the consequences.

Vaccination is a preventive strategy. However, during an outbreak, if there are no vaccines or treatments available for the disease, quarantine of suspected cases or isolation of those diagnosed with the disease are alternative strategies. Such control strategies were used in the outbreak of a new disease in 2003, SARS–Severe Acute Respiratory Syndrome. (See, e.g., Chowell et al., 2003, 2004; Gumel et al., 2004; Zhang et al., 2005; Zhou et al., 2004.) An excellent introduction to epidemic models can be found in the articles and books in the references (Anderson and May, 1991; Brauer and Castillo-Chávez, 2002; Hethcote, 1976, 2000; Thieme, 2003).

## 6.8.2 Cellular Dynamics of HIV

Human Immunodeficiency Virus (HIV) infection, which ultimately leads to Acquired Immune Deficiency Syndrome (AIDS) is one of the most serious and widespread of human diseases. At the end of the twentieth century, it was estimated that 50 million people have been infected by HIV, 15 million had died from AIDS, and 35 million are currently infected (Nowak and May, 2000).

Many mathematical models have been formulated based on basic laws that govern the spread of a virus within an individual. These models have been used to determine the impact of the virus on the immune system and to test the responsiveness of the immune system to treatment. HIV attacks certain white blood cells important to immune system function known as helper T cells (specifically, $CD4^+$ T cells). The helper T cells are responsible for enhancing the production of antibodies by B cells. T cells and B cells are produced in the bone marrow ($B$ = bone), but T cells migrate to the thymus ($T$ = thymus), where they mature. A simple model is presented for the effect of viral particles on healthy $CD4^+$ T cells. This model differs from the SIR-type epidemic models considered previously in that the T cell model follows the cellular dynamics within one individual and not the entire population. However, the T cell model is somewhat similar to an SI epidemic model because both uninfected and infected T cells are modeled.

The basic model for T cell and virus dynamics is a system of three differential equations representing the number of uninfected cells, $x$, infected cells, $y$, and free virus, $v$ (see Nowak and May, 2000; Perelson et al., 1993). The interactions among these three variables are described by the following system of differential equations:

$$\begin{aligned} \frac{dx}{dt} &= \gamma - d_x x - \beta x v, \\ \frac{dy}{dt} &= \beta x v - d_y y, \\ \frac{dv}{dt} &= k y - d_v v - \beta x v, \end{aligned} \tag{6.20}$$

where $x(0) > 0$, $v(0) > 0$, and $y(0) = 0$. In model (6.20) it is assumed that uninfected cells are produced by the immune system at a constant rate $\gamma$, and uninfected cells encounter free virus and become infected at a rate $\beta x v$. At the same rate free viral particles are lost (they enter the cell). Each infected cell is taken over by the virus and the virus produces, on the average, $N$ free virus particles, where $N \gg 1$. The per capita death rates of uninfected cells $x$, infected cells $y$, and virus particles $v$ are $d_x$, $d_y$, and $d_v$, respectively. The rate of production of free viral particles from one infected cell is $k = Nd_y$.

A healthy human adult has about $10^6$ $CD4^+$ T cells per milliliter of blood or $10^3$ per microliter ($mm^3$) (Nowak and May, 2000). The units of $x$, $y$, and $v$ are number of cells or particles per milliliter of blood. Time is measured in days. For example, the units of $\gamma$ are the number of cells per milliliter produced per day, cells $ml^{-1}day^{-1}$. The units of $\beta$ are $cells^{-1}$ $ml^{-1}$ $day^{-1}$, and the units of $d_x$ and $d_y$ are $ml^{-1}day^{-1}$. Only the primary phase of HIV is modeled, the first few weeks after infection. Progression to AIDS takes years. If infected with HIV, the viral load, $v(t)$, initially increases, then the circulating virus decreases up to 2 to 3 orders of magnitude over the next few weeks. These dynamics are represented in Figure 6.21.

In the absence of infection, the uninfected cells $x$ reach a level equal to $\gamma/d_x$ cells $ml^{-1}$. If infection is introduced and the viral reproduction $N$ is sufficiently large, then an individual will acquire HIV infection. For model (6.20), it can be shown that there exists a basic reproduction number $\mathcal{R}_0$, where

$$\mathcal{R}_0 = \frac{k\beta\gamma/d_x}{d_y(d_v + \beta\gamma/d_x)} = \frac{N\beta\gamma}{d_x d_v + \beta\gamma}$$

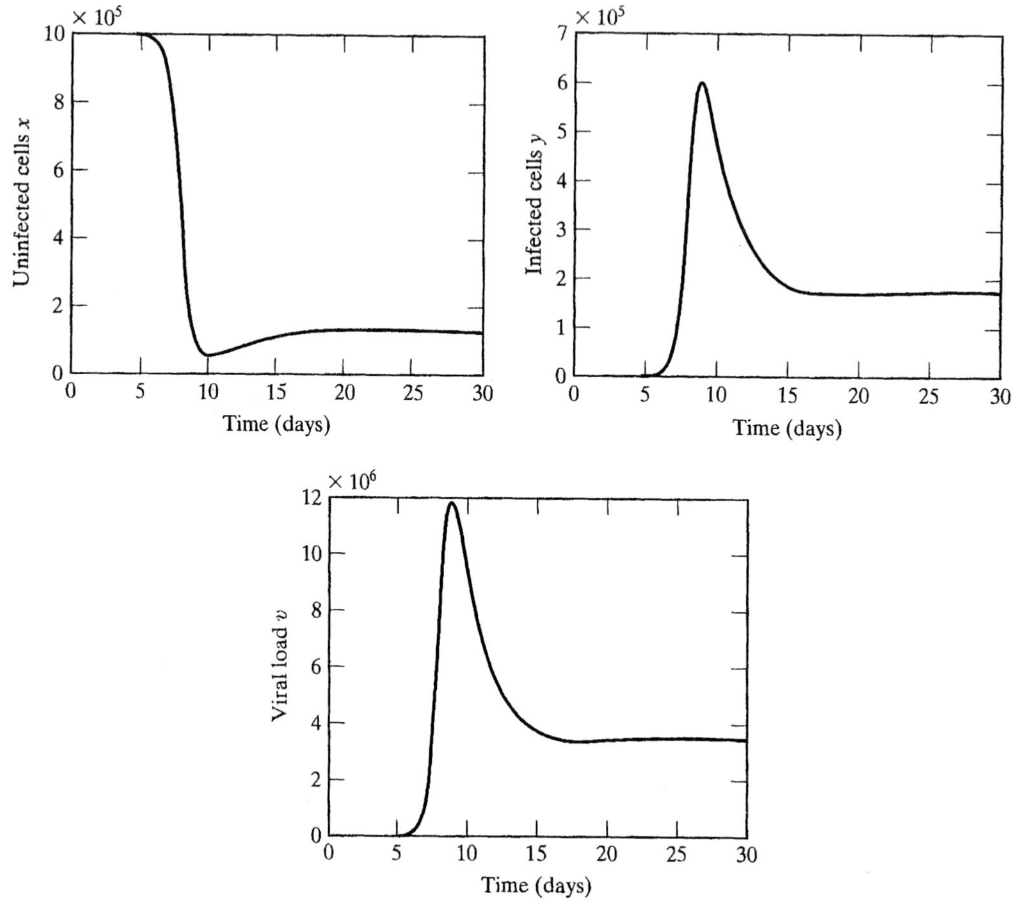


**Figure 6.21** Parameter values for model (6.20): $\gamma = 10^5$ cells day$^{-1}$, $d_x = 0.1$ day$^{-1}$, $d_y = 0.5$ day$^{-1}$, $d_v = 5$ day$^{-1}$, $\beta = 2 \times 10^{-7}$ cells$^{-1}$ ml$^{-1}$ day$^{-1}$, and $k = 100$ day$^{-1}$ (Nowak and May, 2000). Initial conditions: $x(0) = 10^6$, $y(0) = 0$, and $v(0) = 1$.

(see Exercise 20). If $\mathcal{R}_0 < 1$, then the virus will not become established. The parameter values used for Figure 6.21 yield a value of $\mathcal{R}_0 \approx 7.69$.

In Perelson et al., (1993), a similar type of T cell model is studied, one with logistic growth of the uninfected T cells and an additional group $u$ of latently infected T cells. The dynamics of their model is studied over a period of 10 years, until HIV progresses to AIDS. Other models include more detailed study of the immune system dynamics. Another type of white blood cell, known as cytotoxic T lymphocyte (CTL), is a major component of the immune system in fighting viral infections. In response to viral antigens, the CTL proliferate, attack, and kill the virus. Much more complex models than the model (6.20) have been studied that include the CTL response, the effect of drugs, development of drug resistance, delays, and many other modifications (see, e.g., Banks et al., 2003; Culshaw et al., 2003; Hyman et al., 1999; Nowak and May, 2000; Nelson and Perelson, 2002; Perelson and Nelson, 1999; Wodarz, 2001; and references therein). In addition, many epidemic models for the spread of HIV-AIDS in structured populations (structured by sexual activity) have been formulated and analyzed (see, e.g., Blower et al., 1998, 2000; Castillo-Chávez, 1989; Thieme and Castillo-Chávez, 1993).

## 6.9 Excitable Systems

In this last section, we discuss models that exhibit what has been called *excitability*. These types of models have been applied to physiological systems, where cells such as cardiac and skeletal muscle and nerve cells exhibit sudden bursts of activity. One of the most well-known models of an excitable physiological system is the Hodgkin-Huxley model, named after its developers, Alan Hodgkin and Andrew Huxley, two English scientists who received the Nobel Prize in medicine in 1963 for their work. Their model was developed to study the propagation of an electrical signal along an axon (branch) of a nerve cell (neuron). The Hodgkin-Huxley model and a simpler version of this model known as the FitzHugh-Nagumo model (named after FitzHugh and Nagumo) are some of the excitable systems that are discussed here. An excellent reference for mathematical models of excitable physiological systems is the book by Keener and Sneyd (1998).

### 6.9.1 Van der Pol Equation

Before discussing the biological models, we discuss an example from engineering that exhibits periodicity, the van der Pol equation. The van der Pol equation represents an electric circuit where resistance depends nonlinearly on current $u(t)$,

$$\frac{d^2 u}{dt^2} + k(u^2 - 1)\frac{du}{dt} + u = 0, \tag{6.21}$$

$k > 0$. Equation (6.21) has the general form of a Liénard equation,

$$\frac{d^2u}{dt^2} + g(u)\frac{du}{dt} + f(u) = 0. \tag{6.22}$$

In the special case of the van der Pol equation, $g(u) = k(u^2 - 1)$.

A change of variables is made in the van der Pol equation (6.21) to convert it to a first-order system. The new variables are $u$ and $v$, where $u$ is the same as the original variable and $v$ is defined as follows:

$$v = \frac{1}{k}\frac{du}{dt} + \left(\frac{u^3}{3} - u\right).$$

Then the following first-order system is obtained:

$$\frac{1}{k}\frac{du}{dt} = v - \left(\frac{u^3}{3} - u\right),$$

$$\frac{dv}{dt} = -\frac{1}{k}u. \tag{6.23}$$

Solutions to the van der Pol system (6.23) are bounded, the origin (0,0) is unstable, and a stable limit cycle exists. Due to the periodic behavior exhibited by the van der Pol equation, it is often referred to as a *van der Pol oscillator*.

Periodicity can be shown for the more general Liénard equation (6.22). Conditions exist on the functions $f$ and $g$ that guarantee the existence of periodic solutions.

**Theorem 6.3** *Suppose the functions $f$ and $g$ in the Liénard equation (6.22) are continuously differentiable for $-\infty < u < \infty$. Let $G(u) = \int_0^u g(s)\,ds$. In addition, suppose*

*(i)* $uf(u) > 0$ for $u \neq 0$,

*(ii)* $\lim_{|u|\to\infty} |G(u)| = +\infty$, *and*

*(iii) there exist constants $a, b > 0$ such that $G(u) < 0$ for $u < -a$ and $0 < u < b$, and $G(u) > 0$ for $-a < u < 0$ and $u > b$.*

*Then the Liénard equation (6.22) has a nontrivial periodic solution.* □

See Brauer and Nohel (1969) for a proof of Theorem 6.3. The proof depends on construction of a Liapunov function.

Theorem 6.3 can be applied to the van der Pol equation. Let $g(u) = k(u^2 - 1)$, $G(u) = k(u^3/3 - u)$, and $f(u) = u$. Then it is easy to show that conditions (i), (ii), and (iii) of Theorem 6.3 are satisfied (see Exercise 22). When $k$ is large, system (6.23) exhibits what is known as *relaxation oscillations*. In this case, when $v$ is close to $G(u)$ or when solutions are near the cubic nullcline ($\dot{u}/k \approx 0$), they change slowly through time. When solutions are not close to the cubic nullcline, then solutions change rapidly (*excitable*). See Figure 6.22.

### 6.9.2 Hodgkin-Huxley and FitzHugh-Nagumo Models

Hodgkin and Huxley performed a series of experiments with the giant squid axon and developed a model that governs the variation in sodium ions ($Na^+$) and potassium ions ($K^+$) across the cell membrane of a nerve cell (neuron). In the rest state, there is a transmembrane potential difference of about −70 millivolts (the ionic concentration inside the cell is slightly negative in relation to the outside of the cell). There are sodium and potassium pumps that maintain

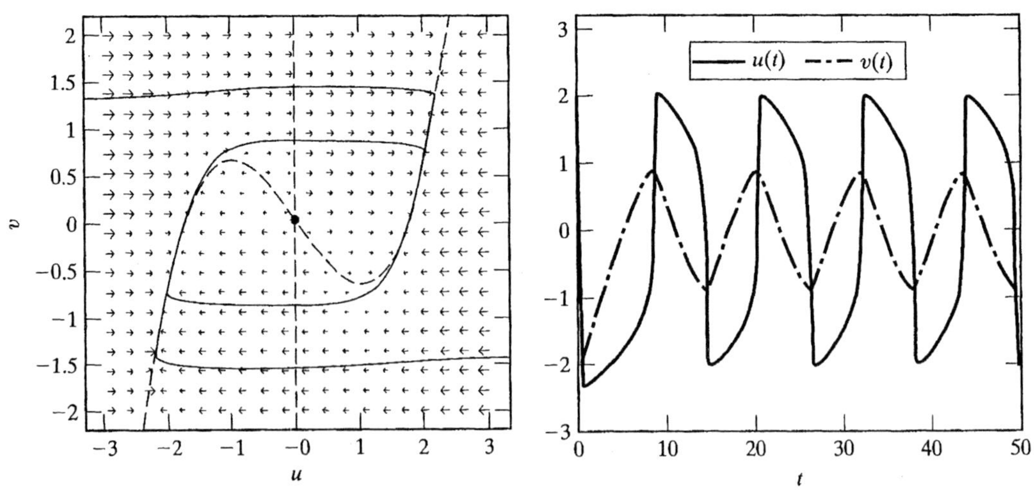


**Figure 6.22** Solutions to van der Pol system (6.23) in the phase plane and as a function of time, $k = 5$, $u(0) = 2$, and $v(0) = -2$. In the phase plane, the direction field and the nullclines are plotted (dashed curves).

this negative potential, which is known as the resting potential. An action potential or electrical pulse is triggered when the concentration gradient of sodium and potassium ions changes significantly from the resting potential. Hodgkin and Huxley formulated a model for the action potential in a neuron based on an electric-circuit analog in which physical properties such as ionic conductivities are represented as resistors. The voltage across the membrane corresponds to voltage across a collection of resistors, each one representing a set of ionic pores that selectively permit a limited current of ions (Edelstein-Keshet, 1988). Their model consists of four nonlinear equations that can be derived using physical laws concerning electrical circuits. We concentrate on the simpler, two-dimensional model independently derived by FitzHugh (1961) and Nagumo et al. (1962). A discussion about the biological background and the mathematical behavior of the Hodgkin-Huxley and FitzHugh-Nagumo models can be found in many books, including Edelstein-Keshet (1988), Keener and Sneyd (1998), Murray (1993, 2002), and Yeargers et al. (1996).

The FitzHugh-Nagumo equations are a system of two nonlinear equations with the following form:

$$\frac{dx}{dt} = f(x) - y + I,$$

$$\frac{dy}{dt} = x - c - bv,$$

where the variable $x$ represents the excitation variable (fast variable) and $y$ the recovery variable (slow variable). The function $f(x)$ is cubic (similar to the van der Pol equation) and $I$ is an input to the system (see, e.g., Britton, 1986; Keener and Sneyd, 1998; Murray, 1993, 2002). The function $f$ has a form similar to

$$f(x) = \beta x(x - \alpha)(1 - x), \quad \beta > 0, \quad 0 < \alpha < 1.$$

The class of models where the nullclines have the shape of a cubic represents excitable-oscillatory systems. Here, we consider the original equations formulated and studied by FitzHugh (1961). These equations are referred to as the FitzHugh-Nagumo model,

$$\begin{aligned} \frac{dx}{dt} &= c\left(y + x - \frac{x^3}{3} + z(t)\right) \\ \frac{dy}{dt} &= -\frac{x - a + by}{c}. \end{aligned} \tag{6.24}$$

System (6.24) is difficult to derive from physical principles, but the mathematical behavior is similar to that seen in an action potential. The parameters in the model are assumed positive with the following restrictions:

$$1 - \frac{2b}{3} < a < 1, \quad 0 < b < 1, \quad b < c^2.$$

The function $z(t)$ is a stimulus intensity which could be represented by a pulse, a step function, or a constant function. We consider the case when $z = 0$ and $z =$ constant.

System (6.24) has two nullclines, one of which is cubic when $z = 0$:

$$y = \frac{x^3}{3} - x \quad \text{and} \quad y = \frac{a - x}{b}.$$

The restrictions on the parameters guarantee that there is only one positive equilibrium, $(\bar{x}, \bar{y})$. The Jacobian matrix evaluated at the equilibrium is

$$J(\bar{x}, \bar{y}) = \begin{pmatrix} (1 - \bar{x}^2)c & c \\ -\dfrac{1}{c} & -\dfrac{b}{c} \end{pmatrix}.$$

For stability, the trace should be negative and the determinant positive,

$$(1 - \bar{x}^2)c - \frac{b}{c} < 0 \quad \text{and} \quad 1 - (1 - \bar{x}^2)b > 0. \tag{6.25}$$

Thus, the equilibrium $(\bar{x}, \bar{y})$ is stable provided the following conditions are *not* satisfied:

$$-\sqrt{1 - \frac{b}{c^2}} \leq \bar{x} \leq \sqrt{1 - \frac{b}{c^2}}.$$

Next, we consider what happens when the neuron is stimulated, when $z(t) \neq 0$. For $z < 0$, the cubic nullcline moves upward. For small impulses, the stability of the equilibrium does not change but for large impulses the equilibrium changes from a stable state to an unstable state (see Exercise 23). For small impulses, the model exhibits excitability because although the equilibrium is stable, solutions starting near the equilibrium may move far away from the equilibrium before returning (exhibiting the behavior of an action potential). For large impulses, the model exhibits oscillations. The equilibrium is unstable, solutions are bounded, and there exists a stable limit cycle (such as repetitive firing of the neuron). Graphs of solutions to the FitzHugh-Nagumo model in the phase plane and over time are given in Figure 6.23.

There are many biological models which exhibit behavior similar to the van der Pol system and to the FitzHugh-Nagumo model. For example, a model for red tides or blooms (explosive growth of phytoplankton) developed by Truscott and Brindley (1994) for phytoplankton and zooplankton interactions has an S-shaped nullcline, similar to the cubic nullcline. The model exhibits, in some cases, periodic behavior similar to the relaxation oscillations in van der Pol's equation (Truscott and Brindley, 1994). In dimensionless form, the model satisfies

$$\begin{aligned} \frac{dP}{dt} &= \beta P(1 - P) - Z\frac{P^2}{\nu^2 + P^2}, \\ \frac{dZ}{dt} &= \gamma Z\left(\frac{P^2}{\nu^2 + P^2} - \omega\right), \end{aligned} \tag{6.26}$$

where $P$ and $Z$ represent phytoplankton and zooplankton densities, respectively, and the parameters $\beta, \nu, \omega$, and $\gamma$ are positive. Although the nullcline is not cubic, $Z = (1 - P)(\nu^2 + P^2)/P$, its shape is similar to a cubic, which results in the excitability behavior. System (6.26) has a positive equilibrium. When this equilibrium is stable, phytoplankton density is at a low level. When zooplankton density is also low, phytoplankton rises rapidly (red tide or bloom) before returning to its low equilibrium value. The equilibrium is reached only after a large deviation in phase space. The analysis of model (6.26) is left as an exercise.

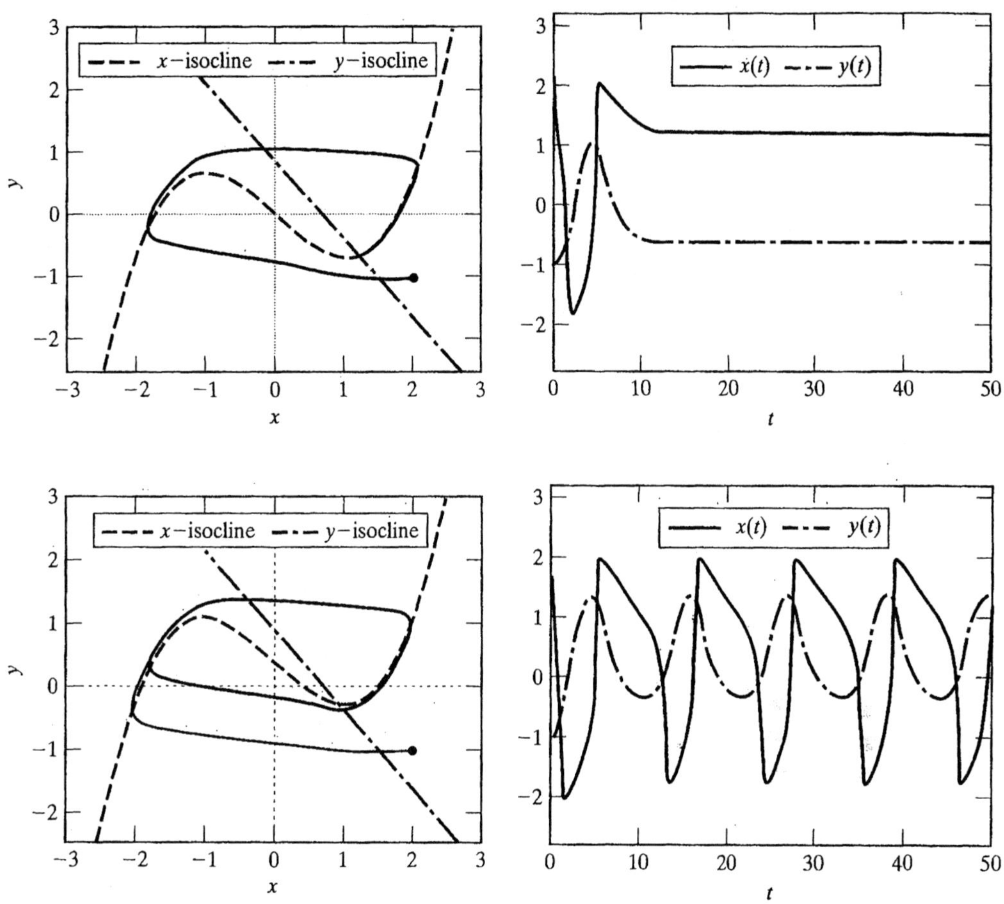


**Figure 6.23** The parameters $a = 0.7, b = 0.8$, and $c = 3$. In the top two figures there is no stimulation, $z(t) \equiv 0$, and solutions approach the stable equilibrium. There is one action potential. In the bottom two figures, the stimulus satisfies $z(t) = -0.4$; the equilibrium is unstable and there exists a stable limit cycle. There are multiple action potentials. The circle in the figures on the left represents the initial value, $(x(0), y(0)) = (2, -1)$.

## 6.10 Exercises for Chapter 6

**1.** For the harvesting model with constant yield, draw the phase line diagrams for three cases: $Y_0 < rK/4$, $Y_0 = rK/4$, and $Y_0 > rK/4$. Describe the model dynamics in these three cases.

**2.** Consider the Lotka-Volterra predator-prey model, where there is density dependence in the prey growth rate:

$$\frac{dx}{dt} = ax - fx^2 - bxy = x(a - fx - by)$$

$$\frac{dy}{dt} = -cy + exy = y(ex - c),$$

where $a, b, c, e, f > 0$.

(a) Put the model in a dimensionless form. Let $u = ex/c, v = by/a$, and $\tau = at$, and show

$$\frac{du}{d\tau} = u\left(1 - \frac{u}{\alpha_1} - v\right),$$

where $\alpha_1 = (ae)/(fc)$ and

$$\frac{dv}{d\tau} = \alpha_2\, v(u - 1),$$

where $\alpha_2 = c/a$.

(b) Find the three equilibria of the dimensionless model.

(c) Graph the nullclines and the direction of flow along the nullclines. There are two cases to consider: $\alpha_1 \le 1$ and $\alpha_1 > 1$.

(d) Calculate the Jacobian matrix of the dimensionless Lotka-Volterra equations and determine stability conditions for each of the equilibria in the two cases $\alpha_1 \le 1$ and $\alpha_1 > 1$.

**3.** Continue the analysis of the predator-prey model in Exercise 2 by showing the following.

(a) Use Dulac's criterion with the Dulac function $B(x, y) = 1/xy$ to show there do not exist any periodic solutions.

(b) Apply Poincaré-Bendixson trichotomy in the case $\alpha_1 < 1$ to show that solutions must approach an equilibrium. Which equilibrium do solutions approach?

(c) For $\alpha_1 > 1$, show that there are no cycle graphs; solutions approach an equilibrium. Which equilibrium do they approach?

(d) Express the stability conditions in terms of the original parameters: $\alpha_1 \le 1$ and $\alpha_1 > 1$. Explain the meaning of the inequalities. (*Hint:* The ratios $a/f$ and $c/e$ are the prey equilibria in the absence of the predator and in the presence of the predator, respectively.)

**4.** Predator data based on Canadian lynx fur pelts collected in Canadian provinces are given in the Appendix to this chapter.

(a) Assume the predator equilibrium $\bar{y} = a/b$ satisfies Theorem 6.2. Use the results of Theorem 6.2 and the lynx data in the Appendix to estimate $a/b$ assuming $T = 10$. [*Hint:* Estimate the integral $\frac{1}{30}\int_{1920}^{1950} y(t)\,dt$ using the trapezoidal rule.]

(b) Assume the prey equilibrium $\bar{x} = 2a/b$. Then use the estimate for the equilibrium $a/b$ to graph the solution of the following predator-prey equations:

$$\frac{dx}{dt} = cx(1 - y/[a/b])$$

$$\frac{dy}{dt} = cy(x/[2a/b] - 1),$$

where $c = 0.67$. The solution is graphed in Figure 6.24.

(c) For the linear approximation near the equilibrium solution $(2a/b, a/b)$, solutions $x$ and $y$ have the form of cosines and sines, $\cos(\beta t)$ and $\sin(\beta t)$, where $\pm\beta i$ are the eigenvalues of the linearization. Hence, an estimate for the period of the solution is $T \approx 2\pi/\beta$. For the system in part (b), estimate the period $T$.

**5.** Consider a mutualistic system, where each species benefits from the presence of the other species. One such model assumes logistic growth for each species in the absence of the other. Each species has a positive effect on the other species growth rate, that is,

$$\frac{dx}{dt} = \frac{r_1}{K_1}x(K_1 - x + y)$$

$$\frac{dy}{dt} = \frac{r_2}{K_2}y(K_2 - y + bx),$$

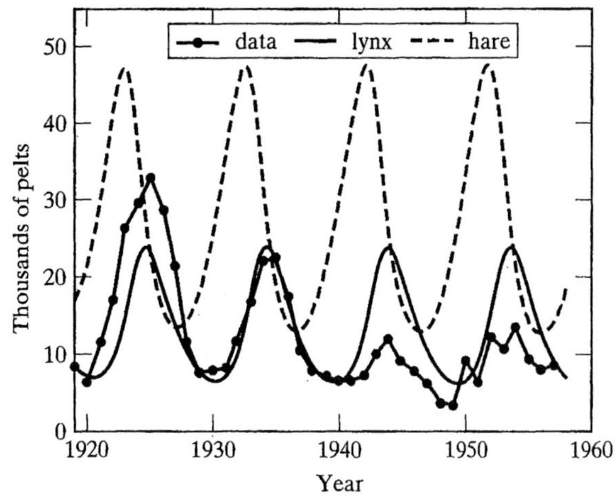


**Figure 6.24** Solution to the Lotka-Volterra predator-prey equations and data on lynx fur pelts.

$x(0)$ and $y(0)$ are positive. The type of mutualistic interaction is referred to as *facultative* mutualism (Kot, 2001). All of the parameters are positive. The parameters $K_i, i = 1, 2,$ represent the carrying capacities of each species. The parameters $r_i, i = 1, 2,$ are the intrinsic growth rates, and the positive terms $y$ and $bx$ measure the positive effect of species $y$ on $x$ and species $x$ on $y$, respectively.

(a) Find all equilibria for these equations.

(b) Draw the nullclines and the phase plane diagram for the case $0 < b < 1$. Use the Jacobian matrix and determine the stability of the nonnegative equilibria. What happens to $x(t)$ and $y(t)$ as $t \to \infty$?

(c) Draw the nullclines and the phase plane diagram for the case $b > 1$. Use the Jacobian matrix and determine the stability of the nonnegative equilibria. What happens to $x(t)$ and $y(t)$ as $t \to \infty$?

**6.** In another mutualistic system each species is unable to survive in the absence of the other one. This type of mutualistic interaction is referred to as *obligate* mutualism (Kot, 2001). Each species cannot survive without the presence of the other species (unlike the mutualistic system in the previous exercise). In this system each species is dependent on the other species for its survival. The model has the form

$$\frac{dx}{dt} = c_1 x(-K_1 - x + y)$$

$$\frac{dy}{dt} = c_2 y(-K_2 - y + bx),$$

$x(0), y(0) > 0$. All of the parameters are positive.

(a) Find all equilibria for these equations.

(b) Draw the nullclines and the phase plane diagram for the case $0 < b < 1$. Use the Jacobian matrix and determine the stability of the nonnegative equilibria. What happens to $x(t)$ and $y(t)$ as $t \to \infty$?

(c) Draw the nullclines and the phase plane diagram for the case $b > 1$. Use the Jacobian matrix and determine the stability of the nonnegative equilibria. What happens to $x(t)$ and $y(t)$ as $t \to \infty$?

**7.** Consult Harrison's 1979 paper as a reference and prove the global stability of the positive equilibrium $(\bar{x}, \bar{y})$ for the following general predator-prey system:

$$\frac{dx}{dt} = a(x) - f(x)b(y),$$

$$\frac{dy}{dt} = n(x)g(y) + c(y).$$

**8.** A fishery model with harvesting has the form

$$\frac{dN}{dt} = rN\left(1 - \frac{N}{K}\right) - H\frac{N}{A + N},$$

where $r, K, H$, and $A$ are positive (Strogatz, 2000). The variable $N$ is the population size and $HN/(A + N)$ is the harvest rate. The population follows logistic growth in the absence of harvesting. The harvest rate increases with $N$ because it is harder to catch fish when the population size is small.

(a) Show that the equation can be written in a dimensionless form as follows:

$$\frac{du}{d\tau} = u(1 - u) - h\frac{u}{a + u}$$

for suitably chosen variables $u$ and $\tau$ and parameters $a$ and $h$. What are these variables and parameters? This model has the same form as the model studied in Chapter 5, equation (5.25).

(b) Show that the equation in (a) can have one, two, or three equilibria, depending on the values of $a$ and $h$. Classify the stability of the equilibria in each case.

(c) Analyze the behavior near $u = 0$ and show that a bifurcation occurs when $h = a$. What is the type of bifurcation?

**9.** The following model represents two competing bacteria, $x_1$ and $x_2$, grown in a chemostat:

$$\frac{dx_1}{dt} = x_1\left[\frac{m_1(1 - x_1 - x_2)}{\alpha_1 + 1 - x_1 - x_2} - 1\right],$$

$$\frac{dx_2}{dt} = x_2\left[\frac{m_2(1 - x_1 - x_2)}{\alpha_2 + 1 - x_1 - x_2} - 1\right],$$

where $m_1, m_2 > 1$, $x_1(0) > 0$, $x_2(0) > 0$, and $0 < \alpha_1/(m_1 - 1) < \alpha_2/(m_2 - 1) < 1$. Assume that $x_1(t) + x_2(t) \le 1$.

(a) Show that there exist three nonnegative equilibria $(\bar{x}_1, \bar{x}_2)$:

$$E_0 = (0, 0), \quad E_1 = \left(1 - \frac{\alpha_1}{m_1 - 1}, 0\right), \quad \text{and} \quad E_2 = \left(0, 1 - \frac{\alpha_2}{m_2 - 1}\right).$$

(b) Perform a local stability analysis for all of the equilibria, then a phase plane analysis to show that all solutions with positive initial conditions converge to the equilibrium $E_1$ (asymptotically stable for all positive initial conditions). The ratio $\alpha_i/(m_i - 1)$ is referred to as the *break-even concentration*, the minimal amount of nutrient that is required for species $x_i$ to survive. In this case, the species that requires the smallest amount of nutrient survives.

**10.** A model for the growth of proliferating $P$ and quiescent $Q$ cancer cells assumes that the total cell population $N = P + Q$ satisfies the Gompertz growth equation (Kozusko and Bajzer, 2003). Here the total cell population size has been normalized to one, $N(0) = 1$. The differential equations for proliferating and quiescent cells satisfy

$$\frac{dP}{dt} = [\beta - \mu_p - r_o(N)]P + r_i(N)Q,$$

$$\frac{dQ}{dt} = r_o(N)P - [r_i(N) + \mu_q]Q.$$

All parameters are assumed to be positive. The parameters $\mu_p$ and $\mu_q$ are the death rates of proliferating and quiescent cells, respectively. The functions $r_o(N)$ and $r_i(N)$ are the transition rates between proliferating and quiescent cells. The parameter $\beta$ is the growth rate of the proliferating cells.

(a) Show that the differential equation for the total cell population can be expressed as

$$\frac{dN}{dt} = [\beta - \mu_p + \mu_q]P - \mu_q N.$$

(b) Let $dP/dt = [\beta - \mu_p]P - f(P,N)$, where $f(P,N) = r_o(N)P - r_i(N)[N - P]$. Use the fact that $N$ satisfies Gompertz growth, $dN/dt = ke^{-\alpha t}N = N(k - \alpha \ln N)$, to show that $f(P,N)$ can be expressed as a function of $N$:

$$f(P,N) = N\left[\frac{(\beta - \mu_p - \mu_q + \alpha + g(N))g(N) + \mu_q(\beta - \mu_p)}{\beta - \mu_p + \mu_q}\right],$$

where $g(N) = k - \alpha \ln(N) = ke^{-\alpha t}$. Because the solution for $N(t)$ is known and $g(N) = ke^{-\alpha t}$, information can be obtained about the transition rates $r_o(N)$ and $r_i(N)$.

(c) Substitute the derivative $dN/dt = ke^{-\alpha t}N$ into the differential equation for $N$ in part (a) to show that the solution for $P$ is

$$P(t) = \frac{\mu_q + ke^{-\alpha t}}{\beta - \mu_p + \mu_q} N(t).$$

Use the solution for $P$ to find the solution for $Q$. What is the limiting ratio of proliferating to quiescent cells?

(d) Use the solution for $N(t) = \exp(k/\alpha[1 - e^{-\alpha t}])$ to find the limiting solutions, $\lim_{t\to\infty} P(t)$ and $\lim_{t\to\infty} Q(t)$.

**11.** In a two-sex population model, denote the number of males as $M$ and the number of females as $F$. Let $B(M,F)$ be the birth rate. Then a simple model for the rate of change of males and females satisfies the following differential equations:

$$\frac{dM}{dt} = -\mu_M M + b_M B(M,F),$$

$$\frac{dF}{dt} = -\mu_F F + b_F B(M,F),$$

where $\mu_F, \mu_M, b_M$ and $b_F$ are positive. The parameters $\mu_M$ and $\mu_F$ are the per capita death rates of males and females, respectively, and $b_M B(M,F)$ and $b_F B(M,F)$ are the birth rates of males and females, respectively (Kot, 2001). The ratio of males to females at birth is $b_M/b_F$.

(a) Suppose $b_M = b_F = b =$ constant and $\mu_M = \mu_F = \mu =$ constant. Show that $M(t) - F(t) = [M(0) - F(0)]e^{-\mu t}$. Eventually, the number of males equals the number of females.

(b) Suppose the birth rate is female dominated, $B(M,F) = F$. Solve the differential equations for $F$ and $M$ and find the limit of the ratio $M(t)/F(t)$.

(c) Suppose there is a harmonic mean birth rate,

$$B(M,F) = \frac{2MF}{M+F}.$$

This is the form most often used by scientists because of its nice properties [e.g., $B(0,F) = 0 = B(M,0)$]. Suppose $M(t) = \bar{M}e^{\lambda t}$ and $F(t) = \bar{F}e^{\lambda t}$. There is exponential growth or decline of the population, depending on the value of $\lambda$. Show that there is exponential growth if

$$b_F(b_M - \mu_M) + b_M(b_F - \mu_F) > 0.$$

**12.** Consider the metapopulation model with a dynamic landscape formulated by Keymer et al. (2000),

$$\frac{dp_1}{dt} = \lambda(1 - p_1 - p_2) - \beta p_1 p_2 + \delta p_2 - e p_1,$$

$$\frac{dp_2}{dt} = p_2(\beta p_1 - \delta - e). \tag{6.27}$$

(a) Show that the system (6.27) has two equilibria, $(\bar{p}_1, \bar{p}_2)$, given by

$$\left(\frac{\lambda}{\lambda + e}, 0\right) \quad \text{and} \quad \left(\frac{\delta + e}{\beta}, 1 - \frac{e}{\lambda + e} - \frac{\delta + e}{\beta}\right).$$

At the first equilibrium the species is extinct and at the second equilibrium the species persists.

(b) Let

$$\mathcal{R}_0 = \frac{\beta\lambda}{(\lambda + e)(\delta + e)}.$$

Show that the first equilibrium is locally asymptotically stable if $\mathcal{R}_0 < 1$ and the second equilibrium is locally asymptotically stable if $\mathcal{R}_0 > 1$.

(c) Apply Dulac's criterion to show for $\mathcal{R}_0 > 1$ that system (6.27) does not have any periodic solutions. [*Hint:* Use the Dulac function $1/(p_1 p_2)$.]

(d) Show for $\mathcal{R}_0 > 1$ that no solutions with $p_i(0) > 0, i = 1, 2$ can approach the origin. Because solutions are bounded, $0 \leq p_i \leq 1$, and there are no periodic solutions, apply Poincaré-Bendixson theory to show that the positive equilibrium is globally asymptotically stable.

**13.** The metapopulation model (6.14) of Hanski (1985, 1999) exhibits what is known as a *backward bifurcation*. The bifurcation diagram in Figure 6.14 shows the stable and unstable equilibria as a function of the colonization rate parameter $c$. It can be seen that there are two bifurcations, a saddle node bifurcation at $c = 0.116$ and a transcritical bifurcation at $c = 0.2$. For values of $c$ within the region of bistability, either there is extinction or there is convergence to the positive equilibrium, depending on the initial values. If the value of $c$ is decreased past the saddle node bifurcation, then solutions at the positive equilibrium jump to the extinction equilibrium, and a population crash occurs. Let $e_S = 1, e_L = 0.02, r = 0.1$, and $m = 0.5$ in model (6.14).

(a) Reduce model (6.14) to a system of two equations in $S$ and $L$ by letting $E = 1 - (S + L)$. Show for $c \in [0.116, 0.2]$ that there exist two positive equilibria, for $c > 0.2$, only a single positive equilibrium, and for $c < 0.116$, only the extinction equilibrium.

(b) Show that the extinction equilibrium, $P = 0 = S + L$, is locally asymptotically stable if $c < 0.2$.

**14.** Consider the metapopulation model (6.15). In this exercise, it will be shown that local asymptotic stability of the extinction equilibrium, $p_i = 0, i = 1, \ldots, n$, depends on the dominant eigenvalue $\lambda_M$ of $M$, where $M$ is defined in (6.18). That is, the extinction equilibrium is locally asymptotically stable if inequality (6.17) holds:

$$\lambda_M < e/c.$$

(a) Matrix $M$ is a nonnegative matrix so that Perron-Frobenius theory (Chapter 1) can be applied. Show that $M$ is irreducible. Then conclude that $M$ has a positive dominant eigenvalue $\lambda_M$ that is greater in magnitude than all of its other eigenvalues.

(b) To show local stability of the extinction equilibrium, theory developed by Diekmann et al. (1990) and by van den Driessche and Watmough (2002) needs to be applied. Compute the Jacobian matrices $H$ and $D$ (evaluated at the extinction equilibrium) for the vectors

$$\mathcal{H} = (C_1(1 - p_1), \dots, C_n(1 - p_n))^T \quad \text{and} \quad \mathcal{D} = (ep_1/A_1, \dots, ep_n/A_n)^T,$$

where $C_i$, $i = 1, \dots, n$, is defined in (6.16). Matrix $H$ is computed from the colonization rate vector and matrix $D$ from the extinction rate vector. Then apply Theorem 6.4, given in the Appendix, to show that the extinction equilibrium is locally asymptotically stable if $\rho(D^{-1}H) < 1$, where $\rho$ is the spectral radius of the matrix $D^{-1}H$.

(c) Show that $\rho(D^{-1}H) < 1$ iff $\lambda_M < e/c$.

**15.** Perform a phase plane analysis of the following SIRS epidemic model:

$$\frac{dS}{dt} = -\frac{\beta}{N}SI + \nu(N - I - S),$$
$$\frac{dI}{dt} = I\left(\frac{\beta}{N}S - \gamma\right).$$

Find the equilibria and determine conditions for their local asymptotic stability. Consider two cases, $\mathcal{R}_0 > 1$ and $\mathcal{R}_0 \leq 1$.

**16.** The following SIS epidemic model includes disease-related deaths ($\alpha$):

$$\frac{dS}{dt} = -\frac{\beta}{N}SI + \gamma I,$$
$$\frac{dI}{dt} = \frac{\beta}{N}SI - (\gamma + \alpha)I,$$
$$\frac{dN}{dt} = -\alpha I.$$

(a) Show that the basic reproduction number is $\mathcal{R}_0 = \beta/(\gamma + \alpha)$.

(b) Show that if $\mathcal{R}_0 > 1$, then $\lim_{t\to\infty} N(t) = 0$. (*Hint*: Note that $dN/dt = -\alpha i N$ where $i$ = I/N.)

**17.** Suppose the total population size $N(t)$ satisfies a logistic-type growth, that is,

$$\frac{dN}{dt} = N(b - m(N)), \quad N(0) > 0, \quad b > 0,$$

where $bN$ is the birth rate and $Nm(N)$ is the mortality rate. The function $m(N)$ is a continuous, differentiable, and increasing function of $N$ satisfying $m(0) = 0$. There is a unique positive constant $K$ satisfying $m(K) = b$. ($K$ is the carrying capacity.) We will analyze the following SIS epidemic model:

$$\frac{dS}{dt} = bN - Sm(N) - \frac{\beta}{N}SI + \gamma I,$$
$$\frac{dI}{dt} = -Im(N) + \frac{\beta}{N}SI - \gamma I. \qquad (6.28)$$

(a) Let $s(t) = S(t)/N(t)$ and $i(t) = I(t)/N(t)$ denote the proportions susceptible and infected. Then show that the differential equations satisfied by $s(t)$ and $i(t)$ simplify to

$$\frac{ds}{dt} = b(1 - s) - \beta si + \gamma i,$$
$$\frac{di}{dt} = \beta si - (\gamma + b)i.$$

(b) Use the fact that $s(t) = 1 - i(t)$ to find conditions for the disease to become endemic. Can you define a basic reproduction number $\mathcal{R}_0$ (when $\mathcal{R}_0 > 1$, the disease becomes endemic)?

(c) Is the total population size $N(t)$ affected by disease?

**18.** The following SIS epidemic model includes *vertical transmission*, that is, children born to infected mothers are infected:

$$\frac{dS}{dt} = bS - Sm(N) - \frac{\beta}{N}SI + \gamma I,$$
$$\frac{dI}{dt} = bI - Im(N) + \frac{\beta}{N}SI - \gamma I,$$
$$\frac{dN}{dt} = N(b - m(N)). \tag{6.29}$$

Notice how the birth terms in model (6.29) differ from the model (6.28) in Exercise 17.

(a) Find the differential equations satisfied by the proportions $i(t) = I(t)/N(t)$ and $s(t) = S(t)/N(t)$.

(b) Then find a condition that ensures $\lim_{t\to\infty} i(t) = 0$. Show that this condition is equivalent to $\mathcal{R}_0 < 1$, where $\mathcal{R}_0$ is the basic reproduction number

$$\mathcal{R}_0 = \frac{\beta}{b + \gamma} + \frac{b}{b + \gamma}.$$

The basic reproduction number in this case is the sum of two terms resulting in cases from either horizontal transmission (direct contacts) or vertical transmission.

**19.** Suppose an SIS epidemic model with disease-related deaths and a growing population satisfies

$$\frac{dN}{dt} = N(b - cN) - \alpha I, \quad b, c, \alpha > 0.$$

(a) Find the differential equations satisfied by the proportions $i(t) = I(t)/N(t)$ and $s(t) = S(t)/N(t)$. (See Exercise 17.) Then find the basic reproduction number.

(b) Do the dynamics of $N(t)$ change with disease? Is it possible for $N(t) \to 0$? Note that $m(N) = cN$ and $dN/dt = N(b - cN - \alpha i)$.

**20.** Consider the model for the cellular dynamics associated with HIV infection,

$$\frac{dx}{dt} = \gamma - d_x x - \beta x v,$$
$$\frac{dy}{dt} = \beta x v - d_y y,$$
$$\frac{dv}{dt} = ky - d_v v - \beta x v,$$

where all of the parameters are positive and $k = Nd_y > 0$.

(a) In the absence of infection $y = 0 = v$, find conditions for the equilibrium $(\bar{x}, \bar{y}, \bar{v}) = (\gamma/d_x, 0, 0)$ to be locally asymptotically stable. Show that these conditions can be simplified to $\mathcal{R}_0 < 1$, where

$$\mathcal{R}_0 = \frac{N\beta\gamma}{d_x d_v + \beta\gamma}.$$

(b) Suppose $\gamma = 10^5$, $d_x = 0.1$, $d_y = 0.5$, $d_v = 5$, and $\beta = 2 \times 10^{-7}$. Find the minimal value, $N = N_{\min}$, such that the virus persists. Then perform several numerical simulations, $N < N_{\min}$, $N > N_{\min}$, and $N \gg N_{\min}$ and plot the solutions over time, $t \in [0, 30]$. (*Hint:* See Figure 6.21.)

**21.** A recent HIV model studies the effect of a genetically modified virus for controlling infection. The engineered virus enters HIV-infected cells and ultimately causes destruction of these cells. The model of Revilla and Garćia-Ramos (2003) adds two states to the susceptible-infected-HIV model $(x-y-v)$ in Exercise 20. Let $w$ be the density of the genetically modified virus and $z$ denote the density of cells infected with HIV and the genetically modified virus $w$. The model has the following form:

$$\frac{dx}{dt} = \gamma - d_x x - \beta x v,$$

$$\frac{dy}{dt} = \beta x v - d_y y - \alpha w y,$$

$$\frac{dv}{dt} = k y - d_v v,$$

$$\frac{dw}{dt} = c z - d_w w,$$

$$\frac{dz}{dt} = \alpha w y - d_z z.$$

The virus $w$ grows at a rate proportional to the number of infected cells $z$ and dies at a rate $d_w$. The doubly infected cells $z$ are formed at a rate $\alpha w y$ and die at a rate $d_z$. In this model, the terms $-\beta x v$ and $-\alpha w y$ are not included in the virus equations for $v$ and $w$, respectively. These terms represent a loss of the virus because it has entered a cell. Because the parameters $\beta$ and $\alpha$ are very small, they are assumed negligible in this model. All model parameters are assumed to be positive.

(a) There exist three equilibria for the new HIV model, a disease-free equilibrium, an equilibrium with virus $v$, and an equilibrium with viruses $v$ and $w$. Find the three equilibria and state conditions for each of them to be nonnegative.

(b) Determine the local asymptotic stability of the disease-free equilibrium.

**22.** (a) Show that the van der Pol equation,

$$\frac{d^2u}{dt^2} + k(u^2 - 1)\frac{du}{dt} + u = 0,$$

satisfies the conditions of Theorem 6.3. What are the values of $a$ and $b$ in part (iii) of Theorem 6.3?

(b) For the van der Pol equation expressed in terms of a first-order system,

$$\frac{1}{k}\frac{du}{dt} = v - \left(\frac{u^3}{3} - u\right),$$

$$\frac{dv}{dt} = -\frac{1}{k}u,$$

plot several solutions in the phase plane and over time.

**23.** For the FitzHugh-Nagumo model,

$$\frac{dx}{dt} = c\left(y + x - \frac{x^3}{3} + z(t)\right),$$

$$\frac{dy}{dt} = -\frac{x - a + by}{c},$$

assume $z(t)$ is a constant and that there exists a unique positive equilibrium $(\bar{x}, \bar{y})$.

(a) Find conditions on the parameters such that the positive equilibrium is locally asymptotically stable.

(b) Assume $a = 0.7, b = 0.8$, and $c = 3$. What restrictions must be placed on $z$ so that the positive equilibrium is stable?

(c) Perform some numerical simulations for various constant values of $z$ when $a = 0.7, b = 0.8$, and $c = 3$ (values when the positive equilibrium is stable and when it is unstable). Plot solutions in the phase plane and over time.

**24.** For the following phytoplankton-zooplankton model, find all of the equilibria and determine conditions for their local asymptotic stability:

$$\frac{dP}{dt} = \beta P(1 - P) - Z\frac{P^2}{\nu^2 + P^2},$$

$$\frac{dZ}{dt} = \gamma Z\left(\frac{P^2}{\nu^2 + P^2} - \omega\right).$$

There are at most three nonnegative equilibria (see Truscott and Brindley, 1994).

## 6.11 References for Chapter 6

Ackleh, A.S., D. F. Marshall, and H.E. Heatherly. 2000. Extinction in a generalized Lotka-Volterra predator-prey model. *J. Applied Mathematics and Stochastic Analysis* 13: 287–297.

Akçakaya, H. R., M. A. Burgman, O. Kindvall, C. Wood, P. Sjögren-Gulve, J. Hatfield, and M. A. McCarthy (Eds.) 2004. *Species Conservation and Management: Case Studies*. Oxford Univ. Press, New York.

Anderson, R. M. and R. M. May. 1991. *Infectious Diseases of Humans, Dynamics and Control*. Oxford University Press, Oxford.

Banks, H. T., D. M. Bortz, and S. E. Holte. 2003. Incorporation of variability in the mathematical modeling of viral delays in HIV infection dynamics. *Math. Biosci.* 183: 63–91.

Blower, S. M., H. B. Gershengorn, and R. M. Grant. 2000. A tale of two futures: HIV and antiretroviral therapy in San Francisco. *Science* 287: 650–654.

Blower, S. M., T. C. Porco, and G. Darby. 1998. Predicting and preventing the emergence of antiviral drug resistance in HSV-2. *Nature Med.* 4: 673–678.

Brauer, F. and C. Castillo-Chávez. 2001. *Mathematical Models in Population Biology and Epidemiology*. Springer-Verlag, New York.

Brauer, F. and J. A. Nohel. 1969. *Qualitative Theory of Ordinary Differential Equations*. W. A. Benjamin, Inc., New York. Reprinted: Dover, 1989.

Braun, M. 1975. *Differential Equations and Their Applications*. Springer-Verlag, New York.

Britton, N. F. 1986. *Reaction Diffusion Equations and Their Applications to Biology*, Academic Press, London, Orlando, New York.

Campbell M.J. and A. M. Walker. 1977. A survey of statistical work on the MacKenzie River series of annual Canadian lynx trappings for the years 1821–1934 with a new analysis. *J. Roy. Stat. Soc. A* 140: 432–436.

Castillo-Chávez C. (Ed.) 1989. *Mathematical and Statistical Approaches to AIDS Epidemiology*, Lecture Notes in Biomathematics 83. Springer-Verlag, Berlin, Heidelberg, New York.

Chi, C.-W., Hsu, S.-B., Wu, L.-I. 1998. On the asymmetric May-Leonard model of three competing species. *SIAM J. Appl. Math.* 58: 211–226.

Chowell, G., C. Castillo-Chávez, P. W. Fenimore, C. M. Kribs-Zaleta, L. Arriola, and J. M. Hyman. 2004. Model parameters and outbreak control for SARS. *Emerg. Infect. Dis.* 7: 1258–1263.

Chowell, G., P. W. Fenimore, M. A. Castillo-Garsow, and C. Castillo-Chávez. 2003. SARS outbreaks in Ontario, Hong Kong and Singapore: The role of diagnosis and isolation as a control mechanism. *J. Theor. Biol.* 224: 1–8.

Clark, C. W. 1976. *Mathematical Bioeconomics, the Optimal Control of Renewable Resources*. John Wiley & Sons, New York.

Culshaw, R. V., S. Ruan, and G. Webb. 2003. A mathematical model of cell-to-cell spread of HIV-1 that includes a time delay. *J. Math. Biol.* 46: 425–444.

DeWoody, Y. D., Z. Feng, and R. K. Swihart. 2005. Merging spatial and temporal structure within a metapopulation model. *Amer. Natur.* 166: 42–55.

Diekmann, O., J. A. P. Heesterbeek, and J. A. J. Metz. 1990. On the definition and the computation of the basic reproduction ratio $R_0$ in models for infectious diseases in heterogeneous populations. *J. Math. Biol.* 28: 365–382.

Edelstein-Keshet, L. 1988. *Mathematical Models in Biology*. The Random House/Birkhäuser Mathematics Series, New York.

Elton, C. and M. Nicholson. 1942. The ten-year cycle in numbers of the lynx in Canada. *J. Anim. Ecol.* 11: 215–244.

FitzHugh, R. 1961. Impulses and physiological states in theoretical models of nerve membrane. *Biophysical Journal* 1: 445–466.

Gause, G. F. 1932. Experimental studies on the struggle for existence. I. Mixed population of two species of yeast. *J. Exp. Biol.* 9: 389–402.

Gilpin, M. E. 1975. Limit cycles in competition communities. *Amer. Nat.* 109: 51–60.

Gumel, A. B., S. Ruan, T. Day, J. Watmough, F. Brauer, P. van den Driessche, D. Gabrielson, C. Bowman, M. E. Alexander, S. Ardal, J. Wu, and B. M. Sahai. 2004. Modelling strategies for controlling SARS outbreaks. *Proc. Roy. Soc. B* 271: 2223–2232.

Hallam, T. G. and S. A. Levin (Eds.) 1986. *Mathematical Ecology: An Introduction*. Biomathematics Vol. 17. Springer-Verlag, Berlin, Heidelberg.

Hanski, I. 1985. Single-species spatial dynamics may contribute to long-term rarity and commonness. *Ecology* 66: 335–343.

Hanski, I. 1999. *Metapopulation Ecology*. Oxford Univ. Press, Oxford, U.K.

Hanski, I. and O. Ovaskainen. 2000. The metapopulation capacity of a fragmented landscape. *Nature* 404: 755–758.

Hanski, I. A. and M. E. Gilpin. 1997. *Metapopulation Biology: Ecology, Genetics and Evolution*. Academic Press. San Diego, Calif.

Harrison, G. W. 1979. Global stability of predator-prey interactions. *J. Math. Biol.* 8: 159–171.

Hassell, M. P. 1978. *The Dynamics of Arthropod Predator-Prey Systems*. Princeton University Press, Princeton, N. J.

Hengeveld, R. 1989. *Dynamics of Biological Invasions*. Chapman and Hall Ltd., London.

Hethcote, H. W. 1976. Qualitative analyses of communicable disease models. *Math. Biosci.* 28: 335–356.

Hethcote, H. W. 2000. The mathematics of infectious diseases. *SIAM Review*. 42: 599–653.

Hofbauer J. and J. W.-H. So. 1994. Multiple limit cycles for three dimensional competitive Lotka-Volterra equations. *Appl. Math. Lett.* 7: 65–70.

Huisman, J and F. J. Weissing. 2001. Biological conditions for oscillations and chaos generated by multispecies competition. *Ecology* 82: 2682–2695.

Hyman, J. M., J. Li, and E. A. Stanley. 1999. The differential infectivity and staged progression models for the transmission of HIV. *Math. Biosci.* 155: 77–109.

Keener, J. and J. Sneyd. 1998. *Mathematical Physiology*. Springer-Verlag, New York, Berlin, and Heidelberg.

Keith, L. B. 1963. *Wildlife's Ten-Year Cycles*. Univ. Wisc. Press, Madison, Wisc.

Keymer, J. E., P. A. Marquet, J. X. Velasco-Hernández, and S. A. Levin. 2000. Extinction thresholds and metapopulation persistence in dynamic landscapes. *Amer. Natur.* 156: 478–494.

Kot, M. 2001. *Elements of Mathematical Ecology*. Cambridge Univ. Press, Cambridge, U.K.

Kozusko, F. and Ž. Bajzer. 2003. Combining Gompertzian growth and cell population dynamics. *Math. Biosci.* 185: 153–167.

Krebs, C. J., R. Boonstra, S. Boutin, and A. R. E. Sinclair. 2001. What drives the 10-year cycle of the snowshoe hares? *BioScience* 51: 25–35.

Kuang, Y. and E. Beretta. 1998. Global qualitative analysis of a ratio-dependent predator-prey system. *J. Math. Biol.* 36: 389–406.

Levin, S. A. 1974. Dispersion and population interactions. *Amer. Natur.* 108: 207–225.

Levins, R. 1969. Some demographic and genetic consequences of environmental heterogeneity for biological control. *Bull. Entomol. Soc. Am.* 15: 227–240.

Levins, R. 1970. Extinction. *Lecture Notes Math.* 2: 75–107.

Li, B., Wolkowicz, G. S. K., and Y, Kuang. 2000. Global asymptotic behavior of a chemostat model with two perfectly complementary resources and distributed delay. *SIAM J. Appl. Math.* 60: 2058–2086.

Logan, J. A. and J. A. Powell. 2001. Ghosts forests, global warming and the mountain pine beetle (*Coleoptera: Scolytidae*). *Amer. Entomol.* 47: 160–173.

Logan, J. A., P. White, B. J. Bentz, and J. A. Powell. 1998. Model analysis of spatial patterns in mountain pine beetle outbreaks. *Theor. Pop. Biol.* 53: 236–255.

Lotka, A. J. 1956. *Elements of Mathematical Biology*. Dover Publications, Inc., New York. Republication of the first edition, *Elements of Physical Biology*, The Williams and Wilkins Co., Inc., 1924.

Ludwig, D, D. D. Jones, and C. S. Holling. 1978. Qualitative analysis of insect outbreak systems: the spruce badworm and forest. *J. Anim. Ecol.* 47: 315–332.

May, R. M. 1976. *Theoretical Ecology Principles and Applications*. W. B. Saunders Co., Philadelphia.

May, R. M. 1983. Parasitic infections as regulators of animal populations. *Amer. Scientist* 71: 36–45.

May, R. M. and W. J. Leonard. 1975. Nonlinear aspects of competition between three species. *SIAM J. of Appl. Math.* 29: 243–253.

Moilanen, A. and I. Hanski. 1995. Habitat destruction and coexistence of competitors in a spatially realistic metapopulation model. *J. Anim. Ecol.* 64: 141–144.

Monod, J. 1950. La technique de culture continue; theorie et applications. *Annals de l'Institut Pasteur*. 79: 390–410.

Morris, R. F. (Ed.) 1963. The dynamics of epidemic spruce budworm populations. *Memoirs Entomol. Soc. Canada*. no. 31, 332 pages.

Murray, J. D. 1993. *Mathematical Biology*. 2nd ed. Springer-Verlag, Berlin, Heidelberg, New York.

Murray, J. D. 2002. *Mathematical Biology I: An Introduction*. 3rd ed. Springer-Verlag, New York.

Nagumo, J., S. Arimoto, and S. Yoshizawa. 1962. An active pulse transmission line simulating nerve axon. *Proc. Inst. Radio Eng.* 50: 2061–2070.

Nelson, P. W. and A. S. Perelson. 2002. Mathematical analysis of delay differential equation models of HIV-1 infection. *Math. Biosci.* 179: 73–94.

Nowak, M. A. and R. M. May. 2000. *Virus Dynamics*. Oxford Univ. Press, New York.

Ortega, J. M. 1987. *Matrix Theory: A Second Course*. Plenum Press, New York and London.

Ovaskainen, O. and I. Hanski. 2001. Spatially structured metapopulation models: global and local assessment of metapopulation capacity. *Theor. Pop. Biol.* 60: 281–302.

Perelson, A. S. and P. W. Nelson. 1999. Mathematical analysis of HIV-1 dynamics in vivo. *SIAM Review* 41: 3–44.

Perelson, A. S., D. E. Kirschner, and R. de Boer. 1993. Dynamics of HIV infection of $CD4^+$ T cells. *Math. Biosci.* 114: 81–125.

Powell, J. A., J. Jenkins, J. A. Logan, and B. J. Bentz. 2000. Seasonal temperature alone can synchronize life cycles. *Bull. Math. Biol.* 62: 977–998.

Revilla, T. and G. Garćia-Ramos. 2003. Fighting a virus with a virus: a dynamic model for HIV-1 therapy. *Math. Biosci.* 185: 191–203.

Schuster, P., K. Sigmund, and R. Wolff. 1979. On $\omega$-limit sets for competition between three species. *SIAM J. Appl. Math.* 37: 49–54.

Smith, H. L. and P. Waltman. 1995. *The Theory of the Chemostat*. Cambridge Univ. Press, Cambridge, U.K.

Smith, H. L. and X.-Q. Zhao. 2001. Competitive exclusion in a discrete-time size-structured chemostat model. *Discrete and Contin. Dyn. Sys.* 1: 183–191.

Stenseth, N. C., W. Falck, O. N. Bjornstad, and C. J. Krebs. 1997. Population regulation in showshoe hare and Canadian lynx: Asymmetric food web configurations between hare and lynx. *Proc. Nat. Acad. Sci.* 94: 5147–5152.

Strogatz, S. H. 2000. *Nonlinear Dynamics and Chaos With Applications to Physics, Biology, Chemistry and Engineering*. Perseus Pub., Cambridge, Mass.

Thieme, H. and C. Castillo-Chávez. 1993. How may infection-age-dependent infectivity affect the dynamics of HIV/AIDS? *SIAM J. Appl. Math.* 53: 1447–1479.

Thieme, H. R. 2003. *Mathematics in Population Biology*. Princeton Univ. Press, Princeton and Oxford.

Truscott, J. E. and J. Brindley. 1994. Ocean plankton populations as excitable media. *Bull. Math. Biol.* 56: 981–998.

Tuchinsky, P. M. 1981. *Man in Competition with the Spruce Budworm*. UMAP Expository Monograph Series. Birkhauser, Boston.

van den Driessche, P. and J. Watmough. 2002. Reproduction numbers and sub-threshold endemic equilibria for compartmental models of disease transmission. *Math. Biosci.* 180: 29–48.

van den Driessche, P. and M. L. Zeeman. 1998. Three-dimensional competitive Lotka-Volterra systems with no periodic orbits. *SIAM J. Appl. Math.* 58: 227–234.

Wodarz, D. 2001. Helper-dependent vs. helper-independent CTL responses in HIV infection; implications for drug therapy and resistance. *J. Theor. Biol.* 213: 447–459.

Wolkowicz, G. S. K., H. Xia, and S. Ruan. 1997. Competition in the chemostat: A distributed delay model and its global asymptotic behavior. *SIAM J. Appl. Math.* 57: 1281–1310.

Wolkowicz, G. S. K., H. Xia, and J. Wu. 1999. Global dynamics of a chemostat competition model with distributed delay. *J. Math. Biol.* 38: 285–316.

Yeargers, E. K., R. W. Shonkwiler, and J. V. Herod. 1996. *An Introduction to the Mathematics of Biology*. Birkhäuser, Boston.

Zeeman, M. L. 1993. Hopf bifurcations in competitive three-dimensional Lotka-Volterra systems. *Dynamics and Stability of Systems*. 8: 189–217.

Zhang, J., J. Lou, Z. Ma, and J. Wu. 2005. A compartmental model for the analysis of SARS transmission patterns and outbreak control in China. *App. Math. Computation* 162: 909–924.

Zhou, Y., Z. Ma, and F. Brauer. 2004. A discrete epidemic model for SARS transmission and control in China. *Math. Computer Modeling* 40: 1491–1506.

## 6.12 Appendix for Chapter 6

### 6.12.1 Lynx and Fox Data

Data on lynx (*L. canadensis*) and red fox (*V. fulva*) fur pelts in Canadian provinces from 1919 to 1957 are shown in Table 6.2 and 6.3 (Keith, 1963).

### 6.12.2 Extinction in Metapopulation Models

The extinction equilibrium of a metapopulation model is locally asymptotically stable if all of the eigenvalues of the corresponding Jacobian matrix are negative or have negative real part. An important result simplifies this computation. Without determining the eigenvalues directly, it can be shown that a bound on the spectral radius of a nonnegative matrix determines local asymptotic stability. This result can be applied to metapopulation and to epidemic models. In epidemic models, the extinction equilibrium refers to the disease-free state. This result (stated in the following theorem for a simple case) was verified by Diekmann et al. (1990) and van den Driessche and

**Table 6.2** Lynx fur pelts collected in eight Canadian provinces from 1919 through 1957.

| Year | Number | Year | Number | Year | Number | Year | Number |
|---|---|---|---|---|---|---|---|
| 1919 | 8378 | 1929 | 7572 | 1939 | 7411 | 1949 | 3714 |
| 1920 | 6456 | 1930 | 7957 | 1940 | 6583 | 1950 | 9592 |
| 1921 | 11617 | 1931 | 8410 | 1941 | 6979 | 1951 | 6653 |
| 1922 | 17202 | 1932 | 11916 | 1942 | 7544 | 1952 | 12636 |
| 1923 | 26381 | 1933 | 16781 | 1943 | 10164 | 1953 | 10876 |
| 1924 | 29529 | 1934 | 22012 | 1944 | 12259 | 1954 | 13876 |
| 1925 | 33027 | 1935 | 22448 | 1945 | 9306 | 1955 | 9660 |
| 1926 | 28619 | 1936 | 17534 | 1946 | 8129 | 1956 | 8397 |
| 1927 | 21363 | 1937 | 10523 | 1947 | 6548 | 1957 | 8958 |
| 1928 | 11582 | 1938 | 8079 | 1948 | 4083 | | |

Watmough (2002) for general epidemic models. The stability result depends on forming a matrix, denoted as $HD^{-1}$, known as the *next generation matrix*.

First, the variables in the model must be separated into occupied and not occupied states (or new infections versus all other states). In the metapopulation model (6.15), all of the states $p_i$ represent occupied states; the unoccupied states $1 - p_i$ are

**Table 6.3** Fox fur pelts collected in seven Canadian provinces from 1919 through 1957.

| Year | Number | Year | Number | Year | Number | Year | Number |
|---|---|---|---|---|---|---|---|
| 1919 | 6181 | 1929 | 8179 | 1939 | 16346 | 1949 | 5144 |
| 1920 | 4144 | 1930 | 11441 | 1940 | 31413 | 1950 | 10751 |
| 1921 | 6853 | 1931 | 18175 | 1941 | 65727 | 1951 | 5639 |
| 1922 | 16085 | 1932 | 33246 | 1942 | 83791 | 1952 | 4167 |
| 1923 | 29412 | 1933 | 57104 | 1943 | 88183 | 1953 | 3549 |
| 1924 | 40203 | 1934 | 57985 | 1944 | 39149 | 1954 | 2131 |
| 1925 | 34481 | 1935 | 28404 | 1945 | 23079 | 1955 | 1639 |
| 1926 | 17023 | 1936 | 19448 | 1946 | 12731 | 1956 | 857 |
| 1927 | 9049 | 1937 | 17052 | 1947 | 6889 | 1957 | 2429 |
| 1928 | 7993 | 1938 | 13088 | 1948 | 6172 | | |

not needed in this model. Next, the rates of colonization are denoted by the vector $\mathcal{H}$ and the rates of extinction denoted by the vector $\mathcal{D}$, that is,

$$\mathcal{H} = (C_1(1 - p_1), \dots, C_n(1 - p_n))^T \quad \text{and} \quad \mathcal{D} = (ep_1/A_1, \dots, ep_n/A_n)^T.$$

In general, vector $\mathcal{H}$ represents the rates of new occupancies (new infections) and vector $\mathcal{D}$ the rates of extinction (removals from infection). Denote the Jacobian matrices of $\mathcal{H}$ and $\mathcal{D}$, evaluated at the extinction equilibrium, as $H$ and $D$, respectively. For example, $D = \text{diag}(e/A_1, \dots, e/A_n)$. The dominant eigenvalue of the nonnegative matrix $HD^{-1}$ determines whether the extinction equilibrium is locally asymptotically stable. The following theorem is stated only for model (6.15) but it applies to more general models (Diekmann et al., 1990; van den Driessche and Watmough, 2002).

**Theorem 6.4.** *The extinction equilibrium, $p_i = 0$, $i = 1, \dots, n$, of the metapopulation model (6.15) is locally asymptotically stable if $\rho(HD^{-1}) < 1$ and unstable if $\rho(HD^{-1}) > 1$.*

The eigenvalues of the product of two $n \times n$ matrices $A$ and $B$, $AB$, are the same as the eigenvalues of $BA$. Therefore, it follows that $\rho(D^{-1}H) = \rho(HD^{-1})$.

Chapter

7

# PARTIAL DIFFERENTIAL EQUATIONS: THEORY, EXAMPLES, AND APPLICATIONS

## 7.1 Introduction

Partial differential equations arise in biological systems because the quantity being modeled not only changes continuously with respect to time but changes continuously with respect to another variable such as age or spatial location. The state variables in a partial differential equation model are functions of two or more variables. For example, a population $n$ in which age and spatial location are important can be a function of time $t$, age $a$, and spatial location $x$, $n(t, a, x)$.

We begin by studying a continuous age-structured model in Section 7.2. The age-structured model is a first-order partial differential equation. Thus, prior to the study of this model, the solution method for a first-order partial differential equation is reviewed. A quantity known as the inherent net reproductive number $R_0$ is defined for the age-structured model. This quantity is the continuous analogue of the inherent net reproductive number defined for the Leslie matrix model in Chapter 1.

Movement of cells, animals, or other living organisms is particularly important in the study of biological systems. The assumption of random diffusion with population growth leads to second-order partial differential equations known as reaction-diffusion equations. Reaction-diffusion equations are studied in Section 7.3. Equilibrium and traveling wave solutions of reaction-diffusion equations are studied in Section 7.4. Traveling wave solutions illustrate spread of a particular quantity throughout a spatial domain. A classical problem introduced by Fisher in 1937 was to find traveling wave solutions in a population genetics model. In this application, a traveling wave solution represents a dominant gene that is spreading throughout a population. This problem is studied in Section 7.6.

Another classical problem, important in spatial ecology, is the minimum spatial region needed for population survival. This minimum region is sometimes referred to as the *critical patch size*. Critical patch size was first introduced in a biological setting by Kierstead and Slobodkin (1953) and Skellam (1951). This problem is studied in Section 7.5.

Another biological application for which partial differential equations with time and spatial variation are important is pattern formation. An extensive variety of problems have been studied in relation to pattern formation. In Section 7.7, we discuss some problems that relate to patterns of growth in an embryo (morphogenesis) and to patterns in animal coats (spots on leopards, strips on tigers, etc.).

In this chapter, we concentrate on partial differential equation models, where both the dependent and the independent variables are continuous. Models where at least one of the independent variables is discrete can be a mixture of discrete and continuous models. When all of the independent variables are discrete, the models take the form of difference equations (Chapters 1–3), such as Leslie's matrix model. Examples of discrete-time and -space models are cellular automata models, where the dynamics governed by each cell depend on the state of neighboring cells (see, e.g., Kaplan and Glass, 1995). If the time variable is continuous but the spatial variable is discrete, then the model can be formulated as a system of ordinary differential equations in the temporal domain but a system of difference equations in the spatial domain. The metapopulation and patch models discussed in Chapter 6 are examples of these latter types of models. On the other hand, if the spatial variable is continuous but the time variable is discrete, then the model can be formulated as a system of integrodifference equations. Integrodifference equation models are discussed in the last section of this chapter. There are a variety of modeling formats to study the dynamics of biological processes and systems with respect to two or more independent variables. The appropriate choice of the modeling format must coincide with the one most accurately depicting the underlying biology of the problem being studied.

## 7.2 Continuous Age-Structured Model

One of the first continuous age-structured models expressed in terms of partial differential equations was studied by Sharpe and Lotka in 1911. This model was more thoroughly studied in connection with biological systems in the work of McKendrick and Von Foerster in 1926 and 1959, respectively. Therefore, the age-structured model studied in this section is often referred to as the McKendrick–Von Foerster equation. The model is the continuous analogue of the Leslie matrix model.

Let $n(t, a)$ denote the population density at time $t$ and age $a$, $b(a)$ the birth rate of individuals of age $a$, and $\mu(a)$ the death rate. The following linear, first-order, hyperbolic partial differential equation is known as the *McKendrick–Von Foerster equation*:

$$\frac{\partial n}{\partial t} + \frac{\partial n}{\partial a} + \mu(a)n(t, a) = 0. \tag{7.1}$$

The preceding equation is known as a conservation equation. As the population ages, changes in the population density occur due to deaths of individuals. Individuals are added to the population at birth, age 0. Births are modeled by the following boundary condition:

$$n(t, 0) = \int_0^\infty b(a)n(t, a)\, da. \tag{7.2}$$

Specification of the initial age distribution completes the formulation of the age-structured model,

$$n(0, a) = f(a).$$

The age-structured population model consists of the partial differential equation (7.1), the birth function $n(t, 0)$, and the initial age distribution, $n(0, a)$.

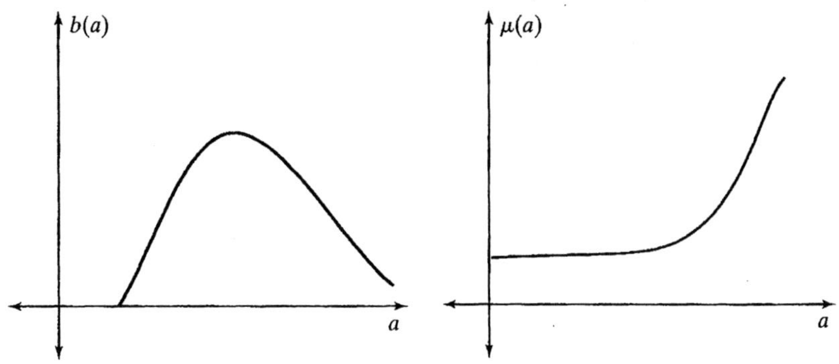


**Figure 7.1** The birth rate $b(a)$ and the death rate $\mu(a)$ as a function of age $a$.

Reasonable assumptions for the birth and the death rate functions, $b(a)$ and $\mu(a)$, and the initial distribution $f(a)$ are that they are nonnegative and continuous (or piecewise continuous) on the interval $[0, \infty)$. A typical shape for $b(a)$ and $\mu(a)$ as functions of age $a$ are graphed in Figure 7.1.

Recall that the Leslie matrix model has the following form (Chapter 1):

$$n(t_i, a_i) = s(a_{i-1})n(t_{i-1}, a_{i-1}), \quad n(t_{i+1}, 0) = \sum_{i=0}^{\infty} \beta(a_i)n(t_i, a_i). \tag{7.3}$$

The functions $s$ and $\beta$ and the time and the age intervals satisfy the following assumptions:

(1) $\Delta t = t_{i+1} - t_i = a_{i+1} - a_i = \Delta a$,

(2) $s(a_i) =$ fraction of the age group $a_i$ that survives to age $a_{i+1}$,

(3) $2\beta(a_i) =$ the average number of newborns produced by one female in the age group $a_i$ that survive through the time interval in which they were born.

Although the summation in the number of births extends to $\infty$, the sum only contains a finite number of terms since $\beta(a_i) = 0$ for $a_i > M$, where $M$ is the maximum age attained by the population. The male/female sex ratio is assumed to be one-to-one. Recall that the Leslie matrix model only follows the females in the population.

By taking a limiting form of the Leslie matrix model (7.3), the continuous age-structured model can be obtained. Let

$$n(t_{i+1}, a_i) = s(a_{i-1})n(t_i, a_{i-1}) \approx [1 - \mu(a_{i-1})\Delta a]n(t_i, a_{i-1})$$

so that

$$n(t_{i+1}, a_i) - n(t_i, a_{i-1}) \approx -\mu(a_{i-1})n(t_i, a_{i-1})\Delta a.$$

For small $\Delta a$, the proportion of the population surviving in an exponentially declining population in time $\Delta a$ is $s(a_{i-1}) = e^{-\mu(a_{i-1})\Delta a} \approx [1-\mu(a_{i-1})\Delta a]$. Then

$$\begin{aligned} n(t_{i+1}, a_i) - n(t_i, a_i) + n(t_i, a_i) - n(t_i, a_{i-1}) &= \Delta_t n + \Delta_a n \\ &\approx -\mu(a_{i-1})n(t_i, a_{i-1})\Delta a. \end{aligned}$$

Dividing both sides of the preceding equation by $\Delta t$ $(= \Delta a)$ and letting $\Delta t \to 0$, the continuous age-structured model ($a_i = a$ and $t_i = t$) is obtained:

$$\frac{\partial n}{\partial t} + \frac{\partial n}{\partial a} = -\mu(a)n(t, a).$$

Next, we show how the birth functions of the discrete and continuous age-structured models are related. We begin with the continuous birth function (7.2).

Integrating the birth function from $t_j$ to $t_{j+1}$ gives the total numbers of births during that interval,

$$\begin{aligned} B &= \text{Births } (t_j \text{ to } t_{j+1}) \\ &= \int_0^{\infty} \int_{t_j}^{t_{j+1}} b(a) n(x, a)\, dx\, da \\ &\approx \int_0^{\infty} b(a) \left[ \frac{n(t_{j+1}, a) + n(t_j, a)}{2} \right] \Delta t\, da \\ &\approx \frac{1}{2} \sum_{i=0}^{\infty} b(a_i) \left[ n(t_{j+1}, a_i) + n(t_j, a_i) \right] \Delta t \Delta a \\ &= \frac{1}{2} \sum_{i=0}^{\infty} b(a_i) n(t_j, a_i) \Delta t \Delta a + \frac{1}{2} \sum_{i=1}^{\infty} b(a_i) s(a_{i-1}) n(t_j, a_{i-1}) \Delta t \Delta a \\ &= \frac{1}{2} \sum_{i=0}^{\infty} [b(a_i) \Delta t \Delta a + b(a_{i+1}) s(a_i) \Delta t \Delta a] n(t_j, a_i). \end{aligned}$$

Individuals born during the interval $t_j$ to $t_{j+1}$ could be born close to time $t_j$ and must survive the interval $\Delta t$ to be counted as births or be born at the middle of the time interval and must survive the interval $\Delta t/2$ to be counted as births, and so on. Therefore, it seems reasonable to assume that individuals born during the interval $t_j$ to $t_{j+1}$ must survive on the average an interval of length $\Delta t/2$. Let $l$ be the probability that newborns survive for a period of $\Delta t/2\ (= \Delta a/2)$. Then

$$n(t_{j+1}, 0) \approx lB = \frac{l}{2} \sum_{i=0}^{\infty} [b(a_i) \Delta t \Delta a + b(a_{i+1}) s(a_i) \Delta t \Delta a] n(t_j, a_i).$$

Thus, the birth function $\beta(a_i)$, defined in (7.3), satisfies

$$\beta(a_i) \approx \frac{l}{2} [b(a_i) \Delta t \Delta a + b(a_{i+1}) s(a_i) \Delta t \Delta a].$$

Because the population contains males and females but only females give birth, we need to use $\beta(a_i)$ rather than $2\beta(a_i)$ as defined in (7.3).

The continuous age-structured model can be solved by the *method of characteristics*. The first-order partial differential equation is reduced to an ordinary differential equation on the characteristic curves. Before continuing the discussion of the the age-structured model, we study the method of characteristics.

### 7.2.1 Method of Characteristics

Suppose $u(t, x)$ satisfies the following first-order linear partial differential equation:

$$a(t, x) \frac{\partial u}{\partial t} + b(t, x) \frac{\partial u}{\partial x} + c(t, x) u = 0, \qquad (7.4)$$

where $-\infty < x < \infty, 0 < t < \infty$. The equation is linear in $u$ since the coefficients $a$ and $b$ only depend on $t$ and $x$ and do not depend on $u$ or its derivatives. The initial condition is given by

$$u(0, x) = \phi(x).$$

For this problem, boundary conditions in the variable $x$ are not required because $-\infty < x < \infty$. For example, in the particular case $a(t, x) = 1$, $b(t, x) = v$, and $c(t, x) = 0$, the variable $u(t, x)$ can be thought of as a concentration of some medium moving along a stream with velocity $v$: $\partial u/\partial t = -v\,\partial u/\partial x$. The solution in this simple case is called a *traveling wave*, $u(t, x) = \phi(x - vt)$. This latter form of the solution can be verified by differentiation, $\partial u/\partial x = \phi'(x - vt)$ and $\partial u/\partial t = -v\phi'(x - vt) = -v\,\partial u/\partial x$.

In the method of characteristics, the partial differential equation can be expressed as an ordinary differential equation along characteristic curves; curves expressed in terms of auxiliary variables $s$ and $\tau$. Along these curves $\tau$ is constant so that $\tau$ can be thought of as a parameter rather than a variable. Assume that

$$u(t, x) \equiv u(t(s, \tau), x(s, \tau)) \equiv u(s, \tau).$$

The characteristic curves are found by solving

$$\frac{dt}{ds} = a(t, x) \quad \text{and} \quad \frac{dx}{ds} = b(t, x)$$

with corresponding initial conditions

$$t(0, \tau) = 0, \quad x(0, \tau) = \tau, \quad \text{and} \quad u(0, \tau) = \phi(\tau).$$

Then

$$\frac{du}{ds} = \frac{\partial u}{\partial t}\frac{dt}{ds} + \frac{\partial u}{\partial x}\frac{dx}{ds} = a(t, x)\frac{\partial u}{\partial t} + b(t, x)\frac{\partial u}{\partial x}.$$

The partial differential equation (7.4) and its initial condition are replaced by the following ordinary differential equation and its associated initial condition:

$$\frac{du}{ds} + c(s, \tau)u = 0, \quad 0 < s < \infty,$$

$$u(0, \tau) = \phi(\tau).$$

The solution to the ordinary differential equation is found along the characteristic curves, that is, where $\tau$ is constant (Figure 7.2).

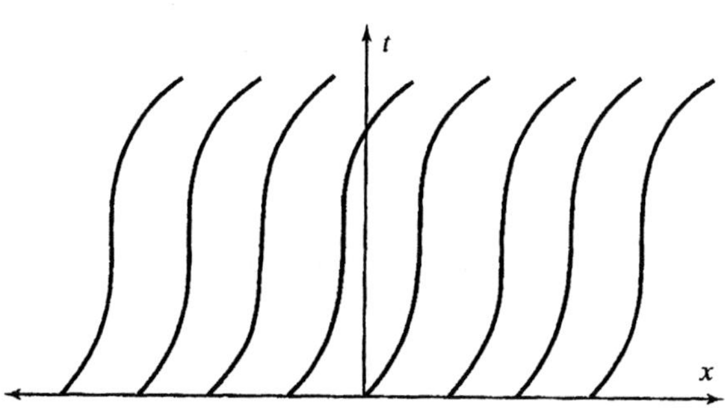


**Figure 7.2** Characteristic curves in the $x$-$t$ plane, where $\tau$ is constant.

**Example 7.1** Solve the following initial value problem by the method of characteristics:

$$\frac{\partial u}{\partial t} + v\frac{\partial u}{\partial x} = 0, \quad -\infty < x < \infty, \quad 0 < t < \infty,$$

$$u(0, x) = \phi(x), \quad -\infty < x < \infty.$$

This equation is known as the *advection equation*, where the flux is proportional to $u$, the rate of change of flux with respect to $x$ equals $v\, \partial u/\partial x$, and $v$ is the velocity.

To solve the advection equation, first, the characteristic curves are found by solving the following differential equations:

$$\frac{dt}{ds} = 1 \quad \text{and} \quad \frac{dx}{ds} = v$$

with initial conditions $t(0, \tau) = 0$ and $x(0, \tau) = \tau$. Integrating the differential equation for $t$ and applying the initial condition, it follows that $t = s$. Integrating the differential equation for $x$, $x = vs + x_0$. But at $s = 0$, $x = \tau$, so that $x = vs + \tau$ or

$$\tau = x - vt,$$

the *characteristic curves*. The differential equation and initial condition for $u(s, \tau)$ satisfy

$$\frac{du}{ds} = 0, \quad u(0, \tau) = \phi(\tau)$$

$(c(t, x) = 0)$. Therefore, $u(s, \tau) = \phi(\tau)$. The variable $s$ and parameter $\tau$ are then replaced by expressions in $x$ and $t$, $s = t$ and $x - vt = \tau$. Therefore, the solution $u(t, x)$ to the partial differential equation with the specified initial condition is given by

$$u(t, x) = \phi(x - vt).$$

This solution is known as a *traveling wave solution*. For a fixed time $t$, solutions have the shape of the initial curve, $\phi(x)$. As time increases these curves move to the right with speed $v$. ∎

**Example 7.2** Solve the following initial value problem by the method of characteristics:

$$\frac{\partial u}{\partial t} + \frac{\partial u}{\partial x} = -2u, \quad -\infty < x < \infty, \quad 0 < t < \infty,$$

$$u(0, x) = \sin(x), \quad -\infty < x < \infty.$$

This is the advection equation with decay. In the solution, it will be seen that there is an exponential decay rate of $e^{-2t}$ (corresponding to the solution of $du/dt = -2u$ when there is no flux).

The characteristic curves are found by solving

$$\frac{dx}{ds} = 1 \quad \text{and} \quad \frac{dt}{ds} = 1$$

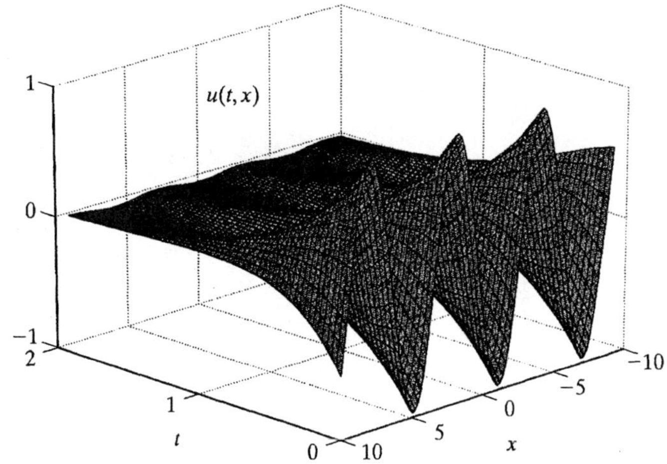


**Figure 7.3** The solution $u(t, x) = \sin(x-t)e^{(-2t)}$ to the first-order partial differential equation in Example 7.2 is graphed in the region $(t, x) \in [0, 2] \times [-10, 10]$.

with initial conditions $t(0, \tau) = 0$ and $x(0, \tau) = \tau$. Integrating and applying the initial condition yields $x = s + \tau$ and $t = s$ with characteristic curves given by $\tau = x - t$. The differential equation and initial condition satisfied by $u$ are

$$\frac{du}{ds} = -2u \quad u(0, \tau) = \sin(\tau).$$

Solving for $u$ by separation of variables, we obtain

$$u(s, \tau) = \sin(\tau)e^{(-2s)}.$$

Expressing $u$ in terms of $x$ and $t$, the solution to the initial value problem is

$$u(t, x) = \sin(x - t)e^{(-2t)}.$$

Note the exponential decay $e^{-2t}$ of the traveling wave solution $\sin(x - t)$. The answer can always be checked by differentiation. The solution is graphed in Figure 7.3. ■

**Example 7.3** Solve the following initial value problem by the method of characteristics:

$$x\frac{\partial u}{\partial x} + \frac{\partial u}{\partial t} + tu = 0, \quad -\infty < x < \infty, \quad 0 < t < \infty,$$

$$u(0, x) = \phi(x), \quad -\infty < x < \infty.$$

The characteristic curves are found by solving

$$\frac{dx}{ds} = x \quad \text{and} \quad \frac{dt}{ds} = 1$$

with initial conditions $x(0, \tau) = \tau$ and $t(0, \tau) = 0$. Integrating yields $x = \tau e^s$ and $t = s$ with characteristic curves $\tau = xe^{-t}$. The differential equation and initial condition satisfied by $u$ are

$$\frac{du}{ds} = -tu = -su, \quad u(0, \tau) = \phi(\tau).$$

The solution is

$$u(s, \tau) = \phi(\tau)\exp(-s^2/2).$$

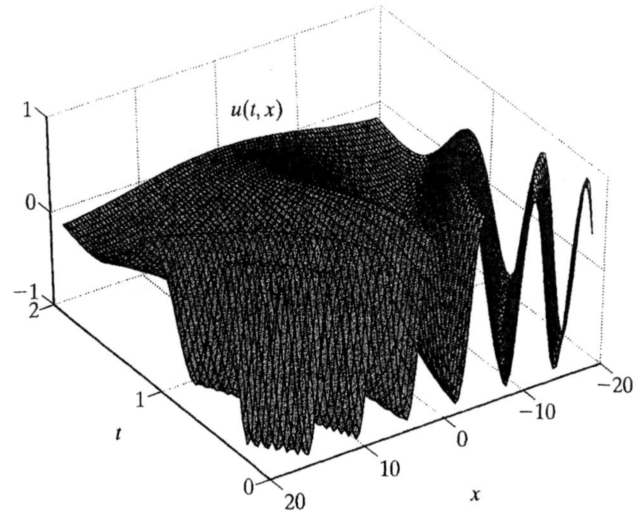


**Figure 7.4** The solution $u(t,x) = \cos(xe^{-t})\exp(-t^2/2)$ to the first-order partial differential equation in Example 7.3 is graphed in the region $(t,x) \in [0,2] \times [-20,20]$.

Expressing $u$ in terms of the original variables $x$ and $t$ gives the solution to the initial value problem,

$$u(t,x) = \phi(xe^{-t})\exp(-t^2/2).$$

For the particular case $\phi(x) = \cos(x)$, the solution is $u(x,t) = \cos(xe^{-t})$ $\exp(-t^2/2)$. This solution is graphed in Figure 7.4. ■

## 7.2.2 Analysis of the Continuous Age-Structured Model

The method of characteristics is used to analyze the continuous age-structured model,

$$\frac{\partial n}{\partial t} + \frac{\partial n}{\partial a} = -\mu(a)n(t,a), \quad 0 < a < \infty, \quad 0 < t < \infty,$$

$$n(t,0) = \int_0^\infty b(a)n(t,a)\,da, \quad 0 < t < \infty,$$

$$n(0,a) = f(a), \quad 0 < a < \infty.$$

The solution to this problem is more difficult than the examples in the previous section because the domain in the variable $a$ does not extend from $-\infty$ to $\infty$ as was true for the variable $x$ in the previous examples. The solution to the age-structured model is the solution to an initial boundary value problem rather than an initial value problem.

The method of characteristics is used to reduce the partial differential equation to an ordinary differential equation. The characteristic curves are found by integrating

$$\frac{dt}{ds} = 1 \quad \text{and} \quad \frac{da}{ds} = 1,$$

with initial conditions

$$a > t: \quad t(0,\tau) = 0 \quad \text{and} \quad a(0,\tau) = \tau$$

and

$$a < t: \quad t(0,\tau) = \tau \quad \text{and} \quad a(0,\tau) = 0.$$

The solutions are

$$a > t: \quad t = s \quad \text{and} \quad a = s + \tau, \quad \text{so that } \tau = a - t.$$

$$a < t: \quad t = s + \tau \quad \text{and} \quad a = s, \quad \text{so that } \tau = t - a.$$

The partial differential equation simplifies to the following ordinary differential equation:

$$\frac{dn}{ds} = -\mu(a)n$$

with initial conditions

$$a > t: \quad n(0, \tau) = f(\tau) \quad \text{and} \quad a < t: \quad n(\tau, 0) = \int_0^\infty b(a)n(\tau, a)\, da.$$

We need to solve the differential equation for two separate cases: $a > t$ and $a < t$.

**Case 1:** $a > t$

$$n(s, \tau) = n(0, \tau)\exp\left[-\int_0^s \mu(s' + \tau)\, ds'\right]$$

The solution in this case is

$$n(s, \tau) = f(\tau)\exp\left[-\int_0^s \mu(s' + \tau)\, ds'\right]$$

$$= f(a - t)\exp\left[-\int_0^t \mu(s' + a - t)\, ds'\right], \quad s'' = s' + a - t$$

$$n(t, a) = f(a - t)\exp\left[-\int_{a-t}^a \mu(s'')\, ds''\right]$$

**Case 2:** $a < t$

$$n(s, \tau) = n(\tau, 0)\exp\left[-\int_0^s \mu(s')\, ds'\right], \quad \tau = t - a,$$

$$n(t, a) = n(t - a, 0)\exp\left[-\int_0^a \mu(s')\, ds'\right]$$

However, this last equation does not give an explicit solution for $n(t, a)$ since the right-hand side involves the births $n(t - a, 0)$, which is unknown because $n(t, a)$ is unknown. We can write an integral equation for $n(t - a, 0)$ (which can be solved iteratively) by using the age boundary condition: $n(t, 0) = \int_0^\infty b(a)n(t, a)\, da$ and the preceding solutions in Cases 1 and 2:

$$n(t, 0) = \int_0^t b(a)n(t - a, 0)\exp\left[-\int_0^a \mu(s)\, ds\right] da$$
$$+ \int_t^\infty b(a)f(a - t)\exp\left[-\int_{a-t}^a \mu(s)\, ds\right] da.$$

The solution for $n(t, 0)$ can be solved using an iterative method. The preceding equation can be expressed as $n(t, 0) = \int_0^t g(t - z)n(z, 0)\, dz + h(t, a)$, where $g$ and $h$ are known functions. Given an initial guess $n_0(t, 0) = n_0$, a sequence of functions $\{n_k(t, 0)\}$ can be formed as follows: $n_{k+1}(t, 0) = \int_0^t g(t - z)n_k(z, 0)\, dz + h(t, a)$. If $g$ is continuous on $[0, t]$, then Picard's method of iteration can be applied (a technique from ordinary differential equations) to show that the sequence $\{n_k(t, 0)\}$ converges uniformly to $n(t, 0)$ on the interval $[0, t]$. This iterative procedure is not applied here.

Of interest is the asymptotic behavior of $n$, the behavior as $t \to \infty$. If there is no age dependence, then the structured model simplifies to an exponential growth model. Therefore, it is reasonable to seek solutions $n(t, a)$ of the form $\exp(\lambda t)r(a)$, where $0 \le r(a) < \infty$. These solutions are referred to as *similarity solutions* or *separable solutions* (superposition of such types of solutions forms the solution to $n(t, a)$). If $\lambda < 0$, then $\lim_{t\to\infty} n(t, a) = 0$, and if $\lambda > 0$, then $\lim_{t\to\infty} n(t, a) = \infty$, provided $r(a) > 0$. If $\lambda = 0$, then $n(t, a) = r(a)$ is an equlibrium or steady-state solution.

A simple expression, the characteristic equation, determines whether $\lambda$ is positive or negative. First, we solve for $r(a)$. Substituting the separable solution into the partial differential equation yields

$$\lambda \exp(\lambda t)r(a) + \exp(\lambda t)r'(a) = -\mu(a)\exp(\lambda t)r(a)$$

or, equivalently,

$$r'(a) = -[\mu(a) + \lambda]r(a).$$

Separating variables and integrating yields

$$r(a) = r(0)\exp\left[-\lambda a - \int_0^a \mu(s)\,ds\right] > 0$$

for $r(0) > 0$. Substituting $n(t, a) = \exp(\lambda t)r(a)$ into the integral birth equation leads to

$$n(t, 0) = \exp(\lambda t)r(0) = \int_0^\infty b(a)n(t, a)\,da$$

$$= \int_0^\infty b(a)\exp(\lambda t)r(0)\exp\left[-\lambda a - \int_0^a \mu(s)\,ds\right]da.$$

Finally, eliminating $\exp(\lambda t)r(0)$, the remaining expression leads to the following *characteristic equation*:

$$1 = \int_0^\infty b(a)\exp\left[-\lambda a - \int_0^a \mu(s)\,ds\right]da.$$

Let $\phi(\lambda) = \int_0^\infty b(a)\exp[-\lambda a - \int_0^a \mu(s)\,ds]\,da$ denote the function on the right side of the preceding equation. Then $R_0 = \phi(0)$ is known as the *inherent net reproductive number*,

$$R_0 = \int_0^\infty b(a)\exp\left[-\int_0^a \mu(s)\,ds\right]da.$$

It can be seen easily that $\phi(\lambda)$ is a decreasing function of $\lambda$ on $(-\infty, \infty)$. In addition, $\lim_{\lambda\to-\infty}\phi(\lambda) = +\infty$ and $\lim_{\lambda\to+\infty}\phi(\lambda) = 0$. (The function $\phi(\lambda)$ can be bounded above and below by functions of the form $c\int_{a_1}^{a_2} e^{-\lambda a}da$ for some positive constants $c$, $a_1$, and $a_2$.) Consequently, $R_0 < 1$ iff the solution $\lambda_0$ to $\phi(\lambda) = 1$ satisfies $\lambda_0 < 0$. In addition, $R_0 > 1$ iff $\lambda_0 > 0$. Recall that a similar definition for the inherent net reproductive number determined exponential growth or decline in the Leslie matrix model in Chapter 1. Thus, we have shown that if $R_0 < 1$, then for any age $a$, the separable solution $n(t, a) = \exp(\lambda t)r(a) \to 0$ as $t \to \infty$. In addition, if $R_0 > 1$, then for $r(a) > 0$, $n(t, a) = \exp(\lambda t)r(a) \to \infty$. These results are summarized in the next theorem.

**Theorem 7.1** *Assume solutions to the continuous age-structured model are of the form $n(t,a) = e^{\lambda t}r(a)$. If $R_0 < 1$, then $\lim_{t\to\infty} n(t,a) = 0$, and if $R_0 > 1$, then $\lim_{t\to\infty} n(t,a) = \infty$.* □

**Example 7.4** Let $b(a) = 10$ for $1 < a < 10$ and $b(a) = 0$ otherwise. Let $\mu(a) = \sqrt{a}$ for $a > 0$. Then calculating the inherent net reproductive number,

$$R_0 = \int_1^{10} 10\left(\exp\left[-\int_0^a \sqrt{s}\,ds\right]\right) da = \int_1^{10} 10\exp(-2a^{3/2}/3)\,da \approx 4.02.$$

Since $R_0 > 1$, solutions approach infinity as $t \to \infty$. If the birth rate $b(a)$ is reduced to 2, then $R_0 = 0.804$ and solutions approach zero. ■

The continuous age-structured model can be made more realistic by making certain assumptions about the birth or death rates. For example, the birth or death rates may depend on the total population size, $N(t) = \int_0^\infty n(t,a)\,da$ (Cushing, 1994, 1998). Age-structured models have been applied to models of infectious diseases in humans. Age structure is especially important in immunization programs. See, for example, Rouderfer et al. (1994), Anderson and May (1991), Allen and Thrasher (1998), Hethcote (2000), Schuette and Hethcote (1999), and Thieme (2003).

## 7.3 Reaction-Diffusion Equations

Partial differential equations that model population growth with random diffusion are often referred to as *reaction-diffusion equations*. Suppose $N(t,x)$ represents the density of the population at time $t \in [0,\infty)$ and spatial position $x \in \Omega$. The spatial domain $\Omega$ may be a bounded or an infinite subset of $\mathbf{R}$, $\mathbf{R}^2$, or $\mathbf{R}^3$. In the cases we consider here, the domain is restricted to subsets of $\mathbf{R}$. A simple reaction-diffusion equation has the following form:

$$\frac{\partial N}{\partial t} = f(N) + D\frac{\partial^2 N}{\partial x^2}. \tag{7.5}$$

The term $f(N)$ is the population growth rate (reaction rate) and the term $D\,\partial^2 N/\partial x^2$ is the diffusion rate (random motion). An excellent introduction to the mathematical theory of reaction-diffusion equations including applications to biology is the book by Britton (1986). A more comprehensive discussion about reaction-diffusion equations in biology is the book by Cantrell and Cosner (2003).

We will derive the *diffusion equation* (without population growth) from first principles. The diffusion equation has the following form:

$$\frac{\partial N}{\partial t} = D\frac{\partial^2 N}{\partial x^2}, \tag{7.6}$$

where $t \in [0,\infty)$ and $x \in \mathbf{R}$. Equation (7.6) is also referred to as the *heat equation* in physical applications. Random diffusion is one of the simplest types of spatial movement.

To derive equation (7.6), the spatial domain $\mathbf{R}$ is divided into intervals of length $\Delta x$ and the time domain $[0,\infty)$ into intervals of length $\Delta t$. Let $N(t,x)$ be the population size at time $t$ and position $x$. Let $\lambda_r$ be the probability of moving to the right a distance of $\Delta x$ and $\lambda_l$ be the probability of moving to the left a distance of $\Delta x$ during a time period $\Delta t$, $0 \le \lambda_r + \lambda_l \le 1$. Then, in a time period $\Delta t$, the population size at time $t$ and position $x$ is given by

$$N(t, x) = (1 - \lambda_r - \lambda_l)N(t - \Delta t, x) + \lambda_l N(t - \Delta t, x + \Delta x) + \lambda_r N(t - \Delta t, x - \Delta x),$$

where the first term on the right-hand side is the proportion of the population that stays at position $x$, the second term is the proportion of the population that moves left, from $x + \Delta x$ to $x$, and the last term is the proportion of the population that moves right, from $x - \Delta x$ to $x$. If the motion is unbiased, so that $\lambda_r = 1/2 = \lambda_l$, then

$$N(t, x) = \frac{1}{2}N(t - \Delta t, x + \Delta x) + \frac{1}{2}N(t - \Delta t, x - \Delta x).$$

Subtracting $N(t - \Delta t, x)$ from both sides and dividing by $\Delta t$ leads to

$$\frac{N(t, x) - N(t - \Delta t, x)}{\Delta t} = \frac{(\Delta x)^2}{2\Delta t}\left[\frac{N(t - \Delta t, x + \Delta x) - 2N(t - \Delta t, x) + N(t - \Delta t, x - \Delta x)}{(\Delta x)^2}\right].$$

The expression on the left-hand side is a difference equation approximation to $\partial N/\partial t$, and the expression in square brackets on the right-hand side is a difference equation approximation to $\partial^2 N/\partial x^2$. Thus, if $\Delta t \to 0$, $\Delta x \to 0$, and $(\Delta x)^2/(2\Delta t) \to D =$ constant, then the diffusion equation (7.6) is obtained. The coefficient $D$ is known as the *diffusion coefficient*. The units of $D$ are (distance)$^2$/time.

The diffusion equation in two spatial dimensions, $N(t, x, y)$, has the following form:

$$\frac{\partial N}{\partial t} = D\left(\frac{\partial^2 N}{\partial x^2} + \frac{\partial^2 N}{\partial y^2}\right) = D\,\Delta N, \tag{7.7}$$

where $\Delta = \nabla^2$ is the Laplace operator. The diffusion equations (7.6) and (7.7) assume only spatial movement; there is no population growth. If population growth is included in the diffusion equation (7.6), then a term for the growth rate, $f(N)$, is added to the right-hand side. The differential equation $dN/dt = f(N)$ represents a population growth model, such as the models studied in Chapters 4–6. Examples of reaction-diffusion equations with particular population growth rates include the following:

Exponential growth: $$\frac{\partial N}{\partial t} = rN + D\frac{\partial^2 N}{\partial x^2}.$$

Logistic growth: $$\frac{\partial N}{\partial t} = rN\left(1 - \frac{N}{K}\right) + D\frac{\partial^2 N}{\partial x^2}.$$

Two-species population growth:

$$\frac{\partial N_1}{\partial t} = f(N_1, N_2) + D_1\frac{\partial^2 N_1}{\partial x^2},$$

$$\frac{\partial N_2}{\partial t} = g(N_1, N_2) + D_2\frac{\partial^2 N_2}{\partial x^2}.$$

The reaction-diffusion equation (7.5) is first order in time $t$ and second order in space $x$. It is classified as a *parabolic* partial differential equation.

Reaction-diffusion equations can have more complex forms. For example, there can be cross-diffusion terms such as

$$\frac{\partial N_1}{\partial x}\frac{\partial N_2}{\partial x}$$

or diffusion rates that depend on the population size,

$$\frac{\partial N}{\partial t} = D\frac{\partial}{\partial x}\left[\left(\frac{N}{N_0}\right)^m \frac{\partial N}{\partial x}\right],$$

$m > 0$ and $N_0 > 0$. This latter equation was used to model insect dispersal (Murray, 1993). Here, we restrict the discussion on reaction-diffusion equations to models with simple random diffusion with population growth such as exponential, logistic, and predator-prey interactions.

If the domain $\Omega$ is finite in extent, knowledge about the behavior of solutions on the boundary of $\Omega, \partial\Omega$, is required. To solve a partial differential equation such as (7.5) on the spatial domain $[a, b], -\infty < a < b < \infty$, it is necessary to specify two spatial constraints (boundary conditions at $x = a$ and $x = b$) and one time constraint (initial condition at $t = 0$). Even for infinite domains, there is generally a constraint put on $\pm\infty$. Several types of boundary conditions that have been applied in modeling population and cellular dynamics are described.

## Types of Boundary Conditions (BCs):

***Infinite domain:*** $N(t, x) \to 0$ as $x \to \pm\infty$. The population size approaches zero at $\pm\infty$. This assumption provides a bound on the solution at the ends of the domain.

***BCs of the first kind (Dirichlet):*** $N(t, 0) = N_0(t)$ and $N(t, L) = N_L(t)$. The domain is finite in length, $x \in [0, L]$ and the population size is prescribed at the ends of the domain. If $N_0(t) = 0 = N_L(t)$, the boundary conditions are said to be *homogeneous Dirichlet* BCs.

***BCs of the second kind (Neumann):*** $\partial N(t, 0)/\partial x = N_0(t)$ and $\partial N(t, L)/\partial x = N_L(t)$. The domain is finite in length, $x \in [0, L]$. If the population is enclosed or fenced (e.g., pond or rangeland) and no individuals can leave or enter the enclosed area, $[0, L]$, $N_0(t) = 0 = N_L(t)$, then the BCs are said to be *zero flux* BCs:

***BCs of the third kind (Robin):***

$$\frac{\partial}{\partial x}N(t, 0) = h[N(t, 0) - N_0(t)]$$

at the boundary $x = 0$, where $h$ is constant. BCs of the third kind are a combination of Dirichlet and Neumann BCs. The flux across the boundary is proportional ($h$ is a constant of proportionality) to the difference between the population size and $N_0(t)$.

***Periodic BCs:*** $N(t, x) = N(t, x + L)$ for $x$ on the boundary. The domain is circular with a circumference $L$. The solutions must be equal every distance $L$ along the boundary.

Robin BCs are not applied as frequently as Dirichlet or Neumann BCs in biological problems. However, the selection of BCs depends on the shape of the domain and the assumptions concerning boundary behavior.

There are various methods that can be applied to solve reaction-diffusion equations such as (7.6) or (7.7) when the spatial domain is relatively simple. If the spatial domain is infinite in extent, then transform techniques can be employed (Laplace transform if the domain is semi-infinite, $[0, \infty)$, or Fourier transform if the domain is infinite, $(-\infty, \infty)$). If the spatial domain is finite, such as a line segment, $[0, L]$, a rectangular region, $[0, L_1] \times [0, L_2]$, or a circular region, then separation of variables can be applied with Fourier series or other types of series and special functions. Some explicit solutions to various initial boundary value problems (IBVPs) and initial value problems (IVPs) are presented in the following examples. We sketch the methods of solution. Detailed methods of solution can be found in many textbooks on partial differential equations (see, e.g., Andrews, 1986; Farlow, 1982; John, 1975; Logan, 2004; Trim, 1990).

**Example 7.5** Consider the following IVP:

$$\frac{\partial N}{\partial t} = D\frac{\partial^2 N}{\partial x^2}, \quad x \in \mathbf{R}, \quad t \in (0, \infty),$$

$$N(0, x) = N_0(x), \quad x \in \mathbf{R}.$$

The IVP is called a *Cauchy problem*. The spatial domain is infinite, and to ensure that unique solutions exist to this problem, we assume that the initial condition is bounded and $N(t, x) \to 0$ as $x \to \pm\infty$ (Britton, 1986). We apply a Fourier transform in the variable $x$. The Fourier transform of $N(t, x)$ is defined as

$$\mathcal{F}[N] = \mathcal{N}(t, s) = \frac{1}{\sqrt{2\pi}}\int_{-\infty}^{\infty} N(t, x)e^{isx}\, dx.$$

Some important properties of the Fourier transform include

$$\mathcal{F}[\partial N/\partial x] = -is\mathcal{N}(t, s)$$

and

$$\mathcal{F}[\partial^2 N/\partial x^2] = -s^2\mathcal{N}(t, s),$$

where we assume that $N(t, x) \to 0$ and $\partial N(t, x)/\partial x \to 0$ as $x \to \pm\infty$. Thus, applying Fourier transforms to the partial differential equation and interchanging differentiation and integration yields

$$\frac{\partial \mathcal{N}(t, s)}{\partial t} = -Ds^2\, \mathcal{N}(t, s).$$

The transformed initial condition is

$$\mathcal{N}(0, s) = \mathcal{N}_0(s).$$

The partial differential equation is converted to an ordinary differential equation in $t$ and the IBVP is converted to an IVP. The solution to the IVP is

$$\mathcal{N}(t, s) = \mathcal{N}_0(s)e^{-Ds^2 t}, \tag{7.8}$$

$s \in \mathbf{R}$. The inverse Fourier transform and the convolution theorem of Fourier (Andrews, 1986) are applied,

$$\mathcal{F}^{-1}[\mathcal{A}(s)\, \mathcal{B}(s)] = \frac{1}{\sqrt{2\pi}}\int_{-\infty}^{\infty} A(v)B(x - v)\, dv. \tag{7.9}$$

In our case, the Fourier transforms are $\mathcal{A}(s) = \mathcal{N}_0(s)$ and $\mathcal{B}(s) = e^{-Ds^2 t}$. The inverse Fourier transform of $\mathcal{A}(s)$ is $A(v) = N_0(v)$. The inverse Fourier transform of $e^{-Ds^2 t}$ is

$$\mathcal{F}^{-1}[e^{-Ds^2 t}] = \frac{1}{\sqrt{2\pi}} \int_{-\infty}^{\infty} e^{-Ds^2 t} e^{-isx}\, ds = \frac{1}{\sqrt{2Dt}} \exp\left(-\frac{x^2}{4Dt}\right),$$

which can be found in tables of Fourier transforms or via complex integration. Thus, applying the inverse Fourier transform to equation (7.8) and the identity (7.9), the solution $N(t, x) = \mathcal{F}^{-1}[\mathcal{N}(t, s)]$ is given by

$$N(t, x) = \frac{1}{2\sqrt{D\pi\, t}} \int_{-\infty}^{\infty} N_0(v) \exp\left(-\frac{(x-v)^2}{4Dt}\right) dv. \tag{7.10}$$

The solution (7.10) simplifies if the initial condition is a single point release. Let

$$N(0, x) = N_0 \delta(x - x_0),$$

where $\delta$ is a Dirac delta function satisfying $\int_{-\infty}^{\infty} \delta(x - x_0)\, dx = 1$ and $\delta(x - x_0) = 0$ for $x \neq x_0$. Then the solution to the IVP is

$$N(t, x) = \frac{N_0}{2\sqrt{D\pi\, t}} \exp\left(-\frac{(x - x_0)^2}{4Dt}\right). \tag{7.11}$$

■

In the Cauchy problem of Example 7.5, the total population size is constant for all time. That is, $\int_{-\infty}^{\infty} N(t, x)\, dx = \int_{-\infty}^{\infty} N_0(x)\, dx =$ constant. Here, we have used the fact that $\int_{-\infty}^{\infty} e^{-x^2}\, dx = \sqrt{\pi}$. This behavior is due to the fact that there is no growth and the diffusion represents simply a redistribution of the population over the spatial region $(-\infty, \infty)$.

**Example 7.6** Consider the following IBVP:

$$\begin{aligned} \frac{\partial N}{\partial t} &= D\frac{\partial^2 N}{\partial x^2}, \quad x \in (0, L), \quad t \in (0, \infty), \\ N(0, x) &= N_0(x), \quad x \in [0, L], \\ N(t, 0) &= 0 = N(t, L), \quad t \in (0, \infty). \end{aligned} \tag{7.12}$$

The IBVP has a finite domain $[0, L]$ with homogeneous Dirichlet boundary conditions. The population is absent from the boundaries. This latter assumption is sometimes interpreted as the environment exterior to the domain is hostile to the population. Individuals in the population cannot survive outside of the domain $[0, L]$.

The solution to this IBVP can be found via separation of variables and Fourier series. For example, assume $N(t, x) = T(t)X(x)$. Then substituting this expression into the differential equation yields

$$\frac{T'(t)}{DT(t)} = \frac{X''(x)}{X(x)} = -k = \text{constant}.$$

This leads to two differential equations,

$$T'(t) + kDT(t) = 0,$$

$$X''(x) + kX(x) = 0,$$

where $X(0) = 0 = X(L)$. The constant $k$ is chosen to be positive because otherwise the solution for $X$ is the trivial one. There is an infinite number of solutions to $X(x)$ known as *eigenfunctions*,

$$X_n(x) = \sin(\sqrt{k_n}\, x), \quad k_n = \frac{(n\pi)^2}{L^2}, \quad n = 1, 2, \ldots.$$

The solutions $k_n = (n\pi)^2/L^2$ are the *eigenvalues*. Summing all of these solutions leads to the final solution,

$$N(t, x) = \sum_{n=1}^{\infty} B_n \sin\left(\frac{n\pi x}{L}\right) \exp\left(-D\left(\frac{n\pi}{L}\right)^2 t\right), \tag{7.13}$$

where

$$B_n = \frac{2}{L}\int_0^L N_0(x)\sin(n\pi x/L)\,dx,$$

$n = 1, 2, \ldots$, are the coefficients in a Fourier sine series. ■

**Example 7.7** Consider the IBVP with zero flux boundary conditions:

$$\frac{\partial N}{\partial t} = D\frac{\partial^2 N}{\partial x^2}, \; x \in (0, L), \; t \in (0, \infty),$$

$$N(0, x) = N_0(x), \; x \in [0, L],$$

$$\frac{\partial}{\partial x} N(t, 0) = 0 = \frac{\partial}{\partial x} N(t, L), \; t \in (0, \infty).$$

The solution to this IBVP can be found via separation of variables and Fourier series similar to the previous example,

$$N(t, x) = \frac{A_0}{2} + \sum_{n=1}^{\infty} A_n \cos\left(\frac{n\pi x}{L}\right) \exp\left(-D\left(\frac{n\pi}{L}\right)^2 t\right),$$

where

$$A_n = \frac{2}{L}\int_0^L N_0(x)\cos(n\pi x/L)\,dx,$$

$n = 0, 1, 2, \ldots$, are the coefficients in a Fourier cosine series. ■

A reaction-diffusion equation with exponential growth, $f(N) = rN$, can be converted to a random diffusion equation.

**Example 7.8** Suppose exponential growth is included in the random diffusion equation,

$$\frac{\partial N}{\partial t} = D\frac{\partial^2 N}{\partial x^2} + rN, \quad x \in (0, L), \quad t \in (0, \infty),$$

$$N(0, x) = N_0(x), \quad x \in [0, L],$$

$$N(t, 0) = 0 = N(t, L), \quad t \in (0, \infty).$$

The change of variable,

$$P(t, x) = N(t, x)e^{-rt} \quad \text{or} \quad N(t, x) = P(t, x)e^{rt},$$

leads to the following partial differential equation for the variable $P$:

$$\frac{\partial P}{\partial t} = D\frac{\partial^2 P}{\partial x^2}.$$

The boundary and initial conditions, expressed in terms of $P$, are

$$P(0, x) = N_0(x), \quad x \in [0, L],$$
$$P(t, 0) = 0 = P(t, L), \quad t \in (0, \infty).$$

Hence, the solution to $P(t, x)$ is given by (7.13). The solution to $N(t, x)$ is just the solution to $P(t, x)$ multiplied by $e^{rt}$,

$$N(t, x) = \sum_{n=1}^{\infty} B_n \sin\left(\frac{n\pi x}{L}\right)\exp\left(rt - D\left(\frac{n\pi}{L}\right)^2 t\right). \qquad \blacksquare$$

Examples 7.5–7.8 are linear partial differential equations defined on a one-dimensional spatial domain. We have shown that explicit solutions can be found to these linear equations. However, for nonlinear partial differential equations on more complicated domains, explicit solutions generally cannot be found. Other methods are used to study solution behavior. First, we state some sufficient conditions on the initial condition $N_0(x)$ and the growth rate $f(N)$ that guarantee a unique bounded solution exists to reaction-diffusion equations of the form (7.5) for $x \in \Omega$ an open connected subinterval of $\mathbf{R}$ and for $t \in [0, \infty)$.

Suppose that $N_0(x)$ is continuous for $x \in \bar{\Omega}$ or $x \in \mathbf{R}$. In addition, suppose there exist constants $a$ and $b$ such that $a \le N_0(x) \le b$, for $x \in \bar{\Omega}$, $f(a) \ge 0$, $f(b) \le 0$, and that $f$ is uniformly Lipschitz continuous, that is, there exists a constant $c$ such that

$$|f(u) - f(v)| \le c|u - v|$$

for all values $u, v \in [a, b]$. Then for the Cauchy problem or the IBVP (7.5) with homogeneous Dirichlet or zero flux BCs, there exists a unique bounded solution $N(t, x)$ for $x \in \Omega$ or $x \in \mathbf{R}$ and $t \in (0, \infty)$. In addition, the solution $N(t, x) \in [a, b]$. See, for example, Britton (1986) and Cantrell and Cosner (2003) for other existence and uniqueness conditions for more general reaction-diffusion equations and for more complex domains. The uniqueness results can be applied to systems of reaction-diffusion equations as well, such as

$$\frac{\partial N_1}{\partial t} = f(N_1, N_2) + D_1\frac{\partial^2 N_1}{\partial x^2},$$
$$\frac{\partial N_2}{\partial t} = g(N_1, N_2) + D_2\frac{\partial^2 N_2}{\partial x^2},$$

where $f$ and $g$ satisfy a uniform Lipschitz condition in $N = (N_1, N_2)$.

The following example shows that the Cauchy problem with logistic growth satisfies the uniqueness conditions.

**Example 7.9** Suppose the Cauchy problem satisfies

$$\frac{\partial N}{\partial t} = N(1 - N) + D\frac{\partial^2 N}{\partial x^2}, \quad x \in \mathbf{R}, \quad t \in (0, \infty),$$
$$N(0, x) = N_0(x), \quad x \in \mathbf{R}.$$

The initial condition $N_0(x)$ is continuous and $0 \le N_0(x) \le 1$ for $x \in \mathbf{R}$. Then $f(N) = N(1 - N)$ satisfies $f(0) = 0, f(1) = 0$, and

$$\begin{aligned} |f(u) - f(v)| &= |u(1 - u) - v(1 - v)| \\ &= |(u - v)[1 - (u + v)]| \\ &\le |u - v| \end{aligned}$$

for $u, v \in [0, 1]$. The Lipschitz constant is $c = 1$. Hence, according to the uniqueness conditions, this IVP has a unique bounded solution $N(t, x) \in [0, 1]$. ■

For nonlinear reaction-diffusion equations, insight into the qualitative behavior of solutions can be obtained by methods other than solving the equation. Some methods are similar to the methods used in the analysis of ordinary differential equations (e.g., stability of equilibrium solutions). These latter methods are employed in detecting diffusive instabilities discussed in a later section. Some mathematical techniques applicable to nonlinear reaction-diffusion equations are introduced in the following applications. First, we review the types of solutions that are useful in nonlinear equations of the form (7.5).

## 7.4 Equilibrium and Traveling Wave Solutions

Consider the following nonlinear partial differential equation:

$$\frac{\partial N}{\partial t} = f(N) + D\frac{\partial^2 N}{\partial x^2}, \quad x \in \Omega \subset \mathbf{R}, \quad t \in (0, \infty). \tag{7.14}$$

We assume equation (7.14) with prescribed initial and boundary conditions (IVP or IBVP) has a unique solution for $x \in \Omega$ and $t \in [0, \infty)$.

**Definition 7.1.** An *equilibrium solution* of equation (7.14) is a time-independent solution $\bar{N}(x)$ satisfying

$$D\frac{d^2\bar{N}}{dx^2} + f(\bar{N}) = 0, \quad x \in \Omega,$$

and all applicable BCs. If $\bar{N}(x)$ depends explicitly on $x$, then the equilibrium solution is said to be *spatially nonuniform* or spatially *heterogeneous*. If $\bar{N} =$ constant, then the equilibrium is said to be *spatially uniform* or *spatially homogeneous*.

A spatially homogeneous equilibrium solution $\bar{N}$ is a constant solution. A constant solution $\bar{N}$ of (7.14) is a solution satisfying $f(\bar{N}) = 0$.

Equilibrium solutions can be defined for more than one spatial dimension and for multispecies models. For example, for the two-species system,

$$\begin{aligned} \frac{\partial N_1}{\partial t} &= f(N_1, N_2) + D_1\frac{\partial^2 N_1}{\partial x^2}, \\ \frac{\partial N_2}{\partial t} &= g(N_1, N_2) + D_2\frac{\partial^2 N_2}{\partial x^2}, \end{aligned}$$

an *equilibrium solution* is a solution $(\bar{N}_1(x), \bar{N}_2(x))$ satisfying

$$D_1\frac{d^2\bar{N}_1}{dx^2} + f(\bar{N}_1, \bar{N}_2) = 0,$$

$$D_2\frac{d^2\bar{N}_2}{dx^2} + g(\bar{N}_1, \bar{N}_2) = 0,$$

and all applicable BCs.

**Example 7.10** We will determine the equilibrium solutions to the following IBVP:

$$\frac{\partial N}{\partial t} = D\frac{\partial^2 N}{\partial x^2}, \quad x \in (0, L), \quad t \in (0, \infty),$$

$$N(t, 0) = k_1,$$

$$N(t, L) = k_2,$$

$$N(0, x) = N_0(x), \quad x \in [0, L].$$

Equilibrium solutions $\bar{N}(x)$ must satisfy

$$D\frac{\partial^2 \bar{N}}{\partial x^2} = 0, \quad \bar{N}(0) = k_1, \quad \bar{N}(L) = k_2.$$

Integrating the preceding differential equation twice and applying the boundary condition yields the following equilibrium solution:

$$\bar{N}(x) = \frac{k_2 - k_1}{L}x + k_1.$$

■

If, in Example 7.10, the BCs are zero flux BCs, $\partial N/\partial x = 0$ at $x = 0, L$, then there exists an infinite number of equilibrium solutions. Any constant solution is an equilibrium solution.

Stability analyses for equilibrium solutions of partial differential equations are more complicated than for ordinary differential equations because instabilities may arise in the spatial domain. We study local stability of solutions via linearization techniques as in previous chapters. Stability of solutions for partial differential equations can be defined in a manner similar to ordinary differential equations.

**Definition 7.2.** Let $N(t, x)$ be a solution of an IVP or IBVP satisfying

$$\frac{\partial N}{\partial t} = f(N) + D\frac{\partial^2 N}{\partial x^2}, \quad x \in \Omega, \quad t \in (0, \infty),$$

$$N(0, x) = N_0(x), \quad x \in \Omega.$$

For a finite domain, BCs are specified on $\partial\Omega$. Then $N(t, x)$ is said to be a *stable solution* of the IVP or IBVP if, given any $\epsilon > 0$, there exists $\delta > 0$ such that whenever $\Phi(0, x) = \Phi_0(x)$ satisfies

$$\|N_0(\cdot) - \Phi_0(\cdot)\| < \delta,$$

the solution $\Phi(t, x)$ to the same IVP or IBVP with initial condition $\Phi_0(x)$ satisfies

$$\|N_0(t, \cdot) - \Phi(t, \cdot)\| < \epsilon,$$

for all $t > 0$. If not, the solution $N(t, x)$ is said to be *unstable*. The solution $N(t, x)$ is said to be *locally asymptotically stable* if it is *stable* and, in addition,

$$\|N(t, \cdot) - \Phi(t, \cdot)\| \to 0,$$

as $t \to \infty$.

In Definition 7.2, the norm $\|\cdot\|$ is over the $x$-variable. For example, for continuous functions $f(x)$ we could use the following norm:

$$\|f(\cdot)\| = \sup_{x \in \bar{\Omega}} |f(x)|.$$

Local asymptotic stability of an equilibrium solution $\bar{N}$ implies $\|\bar{N} - \Phi(t, \cdot)\| \to 0$ as $t \to 0$.

Diffusion may give rise to what has been called *diffusive instabilities*, where constant solutions become unstable. These types of instabilities were first investigated by Turing (1952) in connection with the formation of patterns.

Another type of solution that is interesting from a biological and a mathematical point of view is a traveling wave solution. Figure 7.5 illustrates the concept of a traveling wave. Snapshots in time show that the solution curve at a particular time $t$, $N(t, x)$, shifts to the right at a later time, traveling at a finite speed.

The graphs in Figure 7.5 show that the solution curve at position $x_1$ and time $t_1$, $N(t_1, x_1)$, is the same as the solution curve at position $x_1 - vt_1$ at time 0, $N(0, x_1 - vt_1)$. The solution is traveling at a finite rate of speed $v$. Traveling wave solutions are of interest in epidemics and population genetics, where infection or gene frequency change in a wavelike manner throughout the spatial region. A well-known application of traveling wave solutions to population genetics will be discussed in Section 7.6.

**Definition 7.3.** A *traveling wave solution* of (7.14) is a solution that can be expressed in terms of a single variable $z = x - vt$,

$$N(t, x) \equiv N(z) = N(x - vt),$$

where the constant $v$ is known as the *wave speed*.

**Figure 7.5** A traveling wave solution.

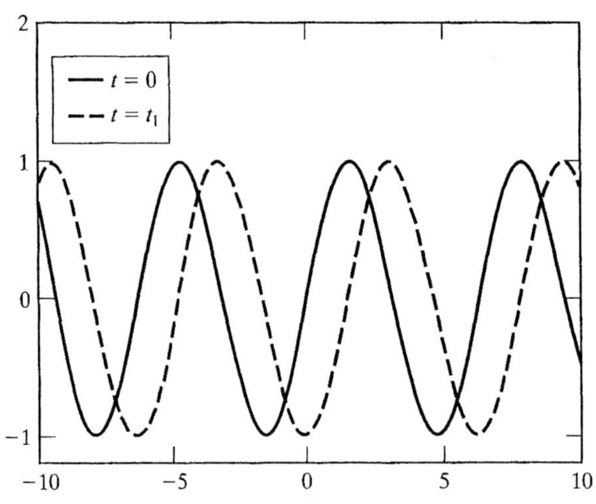

To find a traveling wave solution, we make the assumption that the solution is a function of one variable, $N(z)$, where $z = x - vt$. Substituting $N(z)$ into the partial differential leads to an ordinary differential equation because $N$ is a function of a single variable, $z$. Differentiating $N(z)$ yields the following:

$$\frac{\partial N}{\partial t} = \frac{dN}{dz}\frac{\partial z}{\partial t} = -v\frac{dN}{dz},$$

$$\frac{\partial N}{\partial x} = \frac{dN}{dz}\frac{\partial z}{\partial x} = \frac{dN}{dz},$$

$$\frac{\partial^2 N}{\partial x^2} = \frac{d^2 N}{dz^2}.$$

The partial differential equation (7.14) expressed in terms of $N(z)$ is

$$D\frac{d^2N}{dz^2} + v\frac{dN}{dz} + f(N) = 0, \tag{7.15}$$

a second-order ordinary differential equation.

The differential equation (7.15) in $z$ can be expressed as a first-order system. Define a new variable $P$, where $dN/dz = -P$. Then

$$\frac{d^2N}{dz^2} = -\frac{dP}{dz} \quad \text{but} \quad \frac{d^2N}{dz^2} = -\frac{v}{D}\frac{dN}{dz} - \frac{1}{D}f(N).$$

Thus, the following first-order system is obtained:

$$\frac{dN}{dz} = -P,$$

$$\frac{dP}{dz} = -\frac{v}{D}P + \frac{1}{D}f(N). \tag{7.16}$$

Traveling wave solutions can be studied by analyzing the first-order system. In an application related to the spread of advantageous genes, traveling wave solutions to system (7.16) are analyzed (Section 7.6).

In the next sections, several classical examples of reaction- diffusion equations are discussed. Three applications that model the behavior of biological systems in a spatial setting and some of the generalizations resulting from them are discussed. The three applications of reaction-diffusion models relate to critical patch size (Kierstead and Slobodkin, 1953), spread of advantageous genes (Fisher, 1937), and pattern formation (Turing, 1952).

## 7.5 Critical Patch Size

The first application of reaction-diffusion equations is to population biology. A model for growth and spread of a population is used to estimate the minimal size of the spatial domain needed for population survival. This minimum size is often referred to as the *critical patch size*. In a classical paper by Kierstead and Slobodkin (1953), the critical patch size was determined for a simple reaction-diffusion equation with exponential growth. Their model was applied to growth of *phytoplankton*. Phytoplankton are microscopic plants living in the ocean that represent the bottom of the marine food chain.

In the model of Kierstead and Slobodkin, the spatial domain is $[0, L]$ and the BCs are homogeneous Dirichlet BCs. The population does not live outside of the interval $[0, L]$, meaning there is only a small water mass suitable for growth. The IBVP takes the form

$$\frac{\partial N}{\partial t} = D\frac{\partial^2 N}{\partial x^2} + rN, \quad x \in (0, L), \quad t \in (0, \infty),$$

$$N(0, x) = N_0(x), \quad x \in [0, L],$$

$$N(t, 0) = 0 = N(t, L), \quad t \in (0, \infty),$$

where $r > 0, D > 0$, and $N_0$ is continuous on $[0, L]$. The solution to this IBVP is given in Example 7.6,

$$N(t, x) = \sum_{n=1}^{\infty} B_n \sin\left(\frac{n\pi x}{L}\right)\exp\left(rt - D\left(\frac{n\pi}{L}\right)^2 t\right),$$

where

$$B_n = \frac{2}{L}\int_0^L N_0(x)\sin(n\pi x/L)\,dx.$$

The question posed by Kierstead and Slobodkin was, "What is the minimal size of the spatial domain so that the population persists?" We turn this question around and determine conditions on the spatial domain so that solutions of the IBVP approach zero (extinction). Examination of the solution shows that

$$\|N(t, \cdot)\| = \sup_{x\in[0,L]} |N(t, x)| \le \sum_{n=1}^{\infty} \hat{B}\exp\left(\left[r - D\left(\frac{n\pi}{L}\right)^2\right]t\right),$$

where $\hat{B} = 2\hat{N}_0$ and $\hat{N}_0 = \sup_{x\in[0,L]}|N_0(x)|$. If

$$r < D\left(\frac{n\pi}{L}\right)^2$$

holds for all $n$, then $\lim_{t\to\infty}\|N(t, \cdot)\| = 0$ (the zero equilibrium is stable). However, the preceding inequality holds for all $n$ iff it holds for $n = 1$. Thus, population extinction occurs if

$$r < D\left(\frac{\pi}{L}\right)^2. \tag{7.17}$$

It is interesting to note that $(\pi/L)^2$ is the smallest eigenvalue of the following BVP (also referred to as an eigenvalue problem):

$$X'' + \lambda X = 0 \text{ in } (0, L)$$

$$X(0) = 0 = X(L),$$

that is, $\min \lambda = (\pi/L)^2$.

Returning to Kierstead and Slobodkin's question, in order for the population to survive, the inequality (7.17) should be reversed, $r \ge D\pi^2/L^2$. This latter inequality defines the minimum patch size necessary for population survival. That is, when the inequality is replaced by an equality and solved for $L$, the *critical patch size*, denoted as $L_c$, is defined as

$$L_c = \pi\sqrt{\frac{D}{r}}. \tag{7.18}$$

The population size increases if the domain size $L > L_c$, but decreases to zero if $L < L_c$. In terms of phytoplankton, nutrient-rich waters must be at least of size $L_c$ for growth to counteract the loss due to diffusion across the boundary.

Skellam (1951) applied the same reaction-diffusion equation but on an infinite domain, $x \in \mathbf{R}^2$. Skellam was interested in the expansion of the population over time from a small initial population size. He modeled two different biological populations with an IVP having an initial point release. The two problems were the spread of muskrats over central Europe and the spread of oak forests over Britain.

Cantrell and Cosner (1989) studied a diffusive logistic equation in a more general setting than Kierstead and Slobodkin or Skellam and found conditions for population persistence and extinction. These conditions depend on the domain size and shape. In their model,

$$\frac{\partial N}{\partial t} = D\,\Delta N + rN\left(1 - \frac{N}{K}\right), \quad r, K > 0, \quad x \in \Omega, \quad t \in (0, \infty),$$

$$N(0, x) = N_0(x), \quad x \in \Omega,$$

$$0 = \beta N + (1 - \beta)\frac{\partial N}{\partial \eta}, \quad x \in \partial\Omega,$$

where $\partial\Omega$ is the boundary of $\Omega$ and $\partial N/\partial\eta$ is the outward normal derivative of $N$ in the $\eta$-direction. The parameter $\beta \in [0, 1]$. If $\beta = 0$, the BCs are zero flux and if $\beta = 1$, the BCs are homogeneous Dirichlet. Cantrell and Cosner (1989) showed that population persistence and extinction depend on the smallest eigenvalue $\lambda_1$ of the following eigenvalue problem:

$$\Delta u + \lambda u = 0 \quad \text{in} \quad \Omega,$$

$$\beta u + (1 - \beta)\frac{\partial u}{\partial \eta} = 0 \quad \text{on} \quad \partial\Omega.$$

The eigenvalue $\lambda_1$ depends on the parameter $\beta$ and the domain $\Omega$. Their condition for population extinction is $r < D\lambda_1$, which simplifies to the Kierstead and Slobodkin result (7.17) if $\lambda_1 = (\pi/L)^2$.

The concept of a critical patch size has been extended to other types of domains and to discrete patch models as well (Allen, 1987; Allen et al., 1990). The books by Okubo (1980) and Okubo and Levin (2001) describe a variety of applications where the critical patch size plays an important role in ecological problems.

## 7.6 Spread of Genes and Traveling Waves

R. A. Fisher (1890–1962) made many contributions to the theory of genetics, natural selection, and statistics. In 1937, Fisher developed a model for a population in which individuals within the population have one of three particular genotypes, $AA$, $Aa$, or $aa$. Those individuals carrying the $A$ allele have an advantage (e.g., greater fitness). For example, they have a greater reproduction capability. Fisher's model is described next.

Let $p(t, x)$ denote the proportional density of the population carrying allele $A$ located at position $x$ at time $t$. Let $q(t, x) = 1 - p(t, x)$ be the proportional density of the population carrying allele $a$. Assuming random mating, no migration from outside the domain, and no selection, we can apply the Hardy-Weinberg laws of genetics, so that

$$(p + q)^2 = p^2 + 2pq + q^2 = 1.$$

The terms $p^2, 2pq$, and $q^2$ represent the proportion of gene pairs $AA, aA$, and $aa$ in the next generation. If each allele is equally likely to be selected, the proportion of alleles stays the same in the next generation. However, suppose we assume that allele $A$ has an advantage. The proportion of $A$ alleles increases at a rate proportional to $pq = p(1 - p)$, that is, $rp(1 - p), r > 0$. Assuming random movement and growth rate $rp(1 - p)$, the differential equation for the rate of change of the proportional density $p$ is given by

$$\frac{\partial p}{\partial t} = D\frac{\partial^2 p}{\partial x^2} + rp(1 - p), \quad x \in \mathbf{R}, \quad t \in (0, \infty), \tag{7.19}$$

where $r$ is the intensity of selection. Equation (7.19) is often referred to as *Fisher's equation*. When $r = 0$, the proportion of alleles stays the same from generation to generation. The initial proportion of the population with the $A$ allele satisfies

$$p(0, x) = p_0(x), \quad x \in \mathbf{R},$$

where $p_0(x)$ is continuous on $\mathbf{R}$, $p_0(x) \equiv 0$ for $x \notin [a, b] \subset \mathbf{R}$ and $0 \leq p_0(x) \leq 1$ ($p_0(x) \not\equiv 0$). Fisher's problem is a Cauchy problem.

Notice that in the absence of diffusion, the differential equation for $p$ is a logistic equation with a stable equilibrium at $p = 1$. Thus, it is reasonable to suppose that the spatially dependent solution $p(t, x)$ should eventually approach one. This is in fact true. It can be verified that $\lim_{t\to\infty} \|p(t, x)\| = 1$, for each $x \in \mathbf{R}$. Here convergence to the equilibrium is pointwise for each $x \in \mathbf{R}$ (Britton, 1986). Eventually the advantageous gene $A$ with proportion $p$ dominates the entire population.

To illustrate the spread of the advantageous allele in the population, we are interested in finding traveling wave solutions for Fisher's IVP. Let $p(t, x) = N(z)$, where $z = x - vt$. Apply the method in Section 7.4 to convert the partial differential equation to a system of first-order ordinary differential equations in $N$ and $P$, where $P = -dN/dz$. See system (7.16). The system of equations for $N$ and $P$ satisfies

$$\begin{aligned} \frac{dN}{dz} &= -P = f(N, P), \\ \frac{dP}{dz} &= -\frac{v}{D}P + \frac{r}{D}N(1 - N) = g(N, P). \end{aligned} \tag{7.20}$$

For traveling wave solutions to Fisher's problem, we impose the following additional restrictions on the solution $N(z)$: $0 \leq N(z) \leq 1$, $N(z) \to 1$ as $z \to -\infty$, and $N(z) \to 0$ as $z \to \infty$.

A phase plane and local stability analysis are performed for system (7.20). First, note that there are two constant equilibrium solutions given by $P = 0$ and $N = 0$ or $N = 1$. The $N$-nullcline is $P = 0$ and the $P$-nullcline is $P = \dfrac{r}{v}N(1 - N)$. The Jacobian matrix for system (7.20) is

$$J(N, P) = \begin{pmatrix} 0 & -1 \\ \dfrac{r}{D}(1 - 2N) & -\dfrac{v}{D} \end{pmatrix}.$$

Evaluating the Jacobian matrix at the equilibrium (0, 0) yields

$$J(0,0) = \begin{pmatrix} 0 & -1 \\ \dfrac{r}{D} & -\dfrac{v}{D} \end{pmatrix}.$$

The trace is negative and the determinant, $r/D$, is positive. Thus, equilibrium (0, 0) is locally asymptotically stable. The discriminant satisfies $v^2/D^2 - 4r/D$. If the discriminant is positive, then the equilibrium (0, 0) is a stable node. Therefore, the origin is a stable node provided the wave speed $v$ satisfies

$$v > 2\sqrt{r\,D}.$$

Next, evaluating the Jacobian matrix at the equilibrium (1, 0) yields

$$J(1,0) = \begin{pmatrix} 0 & -1 \\ -\dfrac{r}{D} & -\dfrac{v}{D} \end{pmatrix}.$$

The trace is negative and the determinant, $-r/D$, is also negative. The equilibrium (1, 0) is unstable. Because the determinant is negative, the equilibrium (1, 0) is a saddle point. One eigenvalue is positive and one is negative, $\lambda_1 < 0$ and $\lambda_2 > 0$. The corresponding eigenvectors have the form $(1, -\lambda_i)^T$, $i = 1, 2$. The phase plane diagram and the nullclines are graphed in Figure 7.6. Note that in the case $v < 2\sqrt{rD}$, solutions cycle around (0, 0) and the value of $N$ can become negative.

All of the solutions $N(z)$ for the case $v > 2\sqrt{rD}$, graphed in the phase plane in Figure 7.6, represent traveling wave solutions. Thus, the minimal wave speed for existence of traveling wave solutions is

$$v_{\min} = 2\sqrt{rD}. \tag{7.21}$$

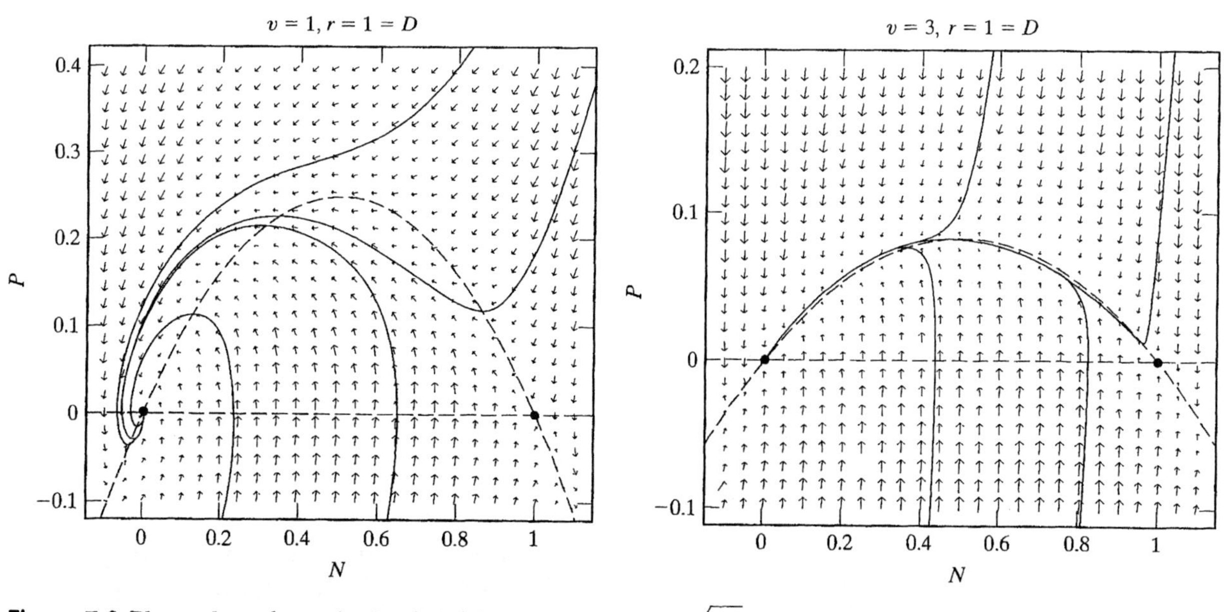


**Figure 7.6** Phase plane dynamics for (7.20) in two cases: $v < 2\sqrt{rD}$ ($v = 1, r = 1 = D$) and $v > 2\sqrt{rD}$ ($v = 3, r = 1 = D$).

**Figure 7.7** A traveling wave solution to Fisher's equation.

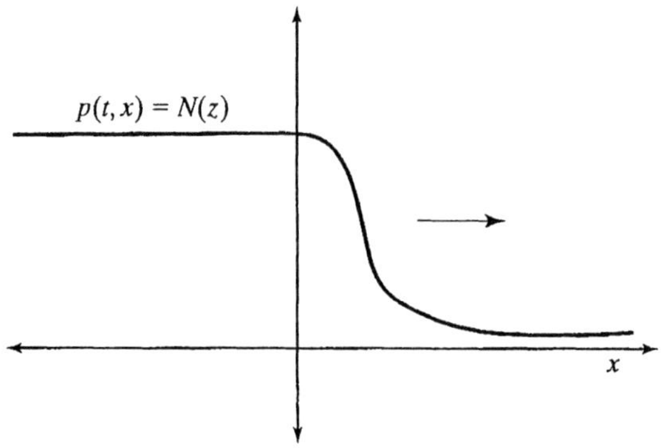


Note that there is one solution, a heteroclinic trajectory, that satisfies the following conditions:

(1) $0 \le N(z) \le 1$,
(2) $N(z) \to 1$ as $z \to -\infty$ (or $x - vt \to -\infty$),
(3) $N(z) \to 0$ as $z \to \infty$ (or $x - vt \to \infty$).

All solutions approach the heteroclinic trajectory as $t \to \infty$. This trajectory represents the traveling wave. Therefore, along the traveling wave solution, the proportional density $p(t, x) \to 1$ (pointwise in $x$) as $t \to \infty$. This result follows from the preceding condition (2). The advantageous allele $A$ becomes dominant as the wave sweeps through the population. A sketch of a traveling wave solution $p(t, x)$ that moves to the right can be seen in Figure 7.7. Another example with two traveling wave solutions is discussed in Exercise 11.

Fisher's equation has been studied by many scientists. Some early investigations include those of Kolmogoroff, Petrovsky, and Piscounoff in 1937. They proved that if $p(0, x)$ satisfies

$$p(0, x) = \begin{cases} 1, & \text{if } x \le x_1 \\ 0, & \text{if } x \ge x_2 \end{cases},$$

where $x_1 < x_2$ and $p(0, x)$ is continuous for $(x_1, x_2)$, then the solution to Fisher's equation evolves to a traveling wavefront solution with minimum wave speed $v_{\min} = 2\sqrt{rD}$, equation (7.21). Because of their work on this equation, Fisher's equation is often referred to as the *KPP equation*.

Numerous biological examples of traveling waves and other types of waves (wave trains and spiral waves) are studied by Fife (1979), Keener and Sneyd (1998), and Murray (1993, 2002, 2003). Traveling wave solutions are important in the study of invasions of nonnative species (weeds, moths, birds) and spread of epidemics (rabies, plague, hantavirus) (see, e.g., Abramson et al., 2003; Anderson et al., 1981; Andow et al., 1990; Hengeveld, 1989; Murray, 1993, 2003; Shigesada and Kawasaki, 1997).

The last application that we discuss relates to pattern formation. One of the first papers in this area was by Alan Turing (1952) on the chemical basis for morphogenesis. The term *morphogenesis* refers to the study of shape, pattern, and form, especially in the developing embryo. Diffusion causes a stable spatially homogeneous solution to become unstable and from this instability arise patterns in space.

## 7.7 Pattern Formation

Turing's work has had a tremendous impact on theoretical biology. Edelstein-Keshet (1988) stated that Turing's contribution is "one of the most important contributions mathematics has made to the realm of developmental biology" (p. 497). In reference to Turing's work, Murray (1993) stated that Turing's paper is "one of the most important papers in theoretical biology this century" (pp. 237, 238). Alan Turing (1912–1954) is known as the founder of computer science.

Random mixing tends to homogenize any heterogeneities that arise in space. Therefore, random diffusion was generally thought to give rise to spatially homogeneous solutions. However, Turing discovered that diffusion may have the opposite effect. Diffusion may cause inhomogeneous patterns or spatial nonuniformities referred to as *diffusive instabilities* or *Turing instabilities*. There are numerous biological applications demonstrating the formation of pattern in space due to Turing instabilities. For example, Murray (1988, 1993, 2002, 2003) discusses many applications of pattern formation, including animal coat patterns (spots on leopards, stripes on tigers, patterns on snake skins, patterns on butterfly wings), neural models (stripe formation in the visual cortex, brain mechanism underlying visual hallucination patterns, shell patterns), feather patterns, and cartilage formation in the vertebrate limb.

To show how diffusive instabilities arise, we consider two quantities that can be considered as two species, two chemicals, or two morphogens that interact and diffuse. *Morphogens* are substances present in embryonic development that are believed to control growth patterns. Let $N_1(t, x)$ and $N_2(t, x)$ denote two quantities satisfying the following reaction-diffusion equations:

$$\begin{aligned} \frac{\partial N_1}{\partial t} &= f(N_1, N_2) + D_1 \frac{\partial^2 N_1}{\partial x^2}, \\ \frac{\partial N_2}{\partial t} &= g(N_1, N_2) + D_2 \frac{\partial^2 N_2}{\partial x^2}, \end{aligned} \tag{7.22}$$

where the diffusion coefficients $D_1$ and $D_2$ are positive. At this point we make no assumptions about the BCs. However, the domain $\Omega$ is finite, and generally, the BCs are zero flux. We shall see that various patterns are generated because of the shape of the domain. We assume there exists a constant equilibrium solution $\bar{N}_1$ and $\bar{N}_2$ satisfying

$$f(\bar{N}_1, \bar{N}_2) = 0 \quad \text{and} \quad g(\bar{N}_1, \bar{N}_2) = 0.$$

Note that constant equilibrium solutions are consistent with zero flux BCs.

Turing's idea simply stated is as follows: If the spatially homogeneous equilibrium $(\bar{N}_1, \bar{N}_2)$ is stable in the absence of diffusion but becomes unstable in the presence of diffusion, then spatially nonhomogeneous patterns may evolve. Therefore, the conditions for diffusive instability require that the equilibrium $(\bar{N}_1, \bar{N}_2)$ be locally asymptotically stable in the reaction part of the system:

$$\begin{aligned} \frac{dN_1}{dt} &= f(N_1, N_2), \\ \frac{dN_2}{dt} &= g(N_1, N_2). \end{aligned} \tag{7.23}$$

But with diffusion, the stability changes. The equilibrium is unstable in the reaction-diffusion system.

First, we derive conditions for local asymptotic stability of $(\bar{N}_1, \bar{N}_2)$ in the reaction system. Linearizing the system about the equilibrium $(\bar{N}_1, \bar{N}_2)$ leads to

$$\frac{du_1}{dt} = a_{11}u_1 + a_{12}u_2,$$

$$\frac{du_2}{dt} = a_{21}u_1 + a_{22}u_2,$$

where $u_i = N_i - \bar{N}_i, i = 1, 2$. The matrix $J = (a_{ij})$ is the Jacobian matrix for system (7.23) evaluated at the equilibrium $(\bar{N}_1, \bar{N}_2)$. The eigenvalues of matrix $J$ have negative real part and the linear solutions approach zero iff $\text{Tr}(J) < 0$ and $\det(J) > 0$ or, equivalently,

$$a_{11} + a_{22} < 0 \quad \text{and} \quad a_{11}a_{22} - a_{12}a_{21} > 0.$$

These latter conditions are necessary for diffusive instability.

Next, in the reaction-diffusion system (7.22), we linearize about the equilibrium. Let $v_i = N_i - \bar{N}_i$ for $i = 1, 2$. Linearizing the system leads to

$$\frac{\partial v_1}{\partial t} = a_{11}v_1 + a_{12}v_2 + D_1\frac{\partial^2 v_1}{\partial x^2},$$

$$\frac{\partial v_2}{\partial t} = a_{21}v_1 + a_{22}v_2 + D_2\frac{\partial^2 v_2}{\partial x^2}, \tag{7.24}$$

where $J = (a_{ij})$ denotes the Jacobian matrix for the reaction system (7.23). In matrix notation, system (7.24) can be expressed as

$$\frac{\partial V}{\partial t} = JV + D\frac{\partial^2 V}{\partial x^2}, \tag{7.25}$$

where $V = (v_1, v_2)^T$ and $D = \text{diag}(D_1, D_2)$.

To determine conditions for local asymptotic stability in the linearized system (7.24), we first find the time-independent solution, $W(x)$ (Murray, 1993, 2002). Recall in Example 7.6 that solutions were sums of terms of the form $X(x)T(t)$. However, in this case $X(x)$ is a vector $W(x) = (w_1(x), w_2(x))^T$. Assume the time-independent solution $W(x)$ satisfies the following eigenvalue problem:

$$W'' + k^2 W = \mathbf{0}, \quad x \in \Omega, \tag{7.26}$$

with associated BCs defined on $\partial\Omega$. There is a set of eigenvalues $k_n^2$ and associated eigenfunctions $W_n$ (see Examples 7.6 and 7.7). Here we have used a constant $k^2$ rather than $k$. (Generally, $k^2 > 0$ for nontrivial solutions to the eigenvalue problem.) To find the eigenfunctions, assume solutions are sums of the form $W_n(x)T_n(t)$, where $T_n(t) = e^{\sigma t}$ and $\sigma$ depends on $k_n$. The exponential assumption is reasonable because the linearized reaction system $dV/dt = JV$ has exponential solutions. Therefore, we assume solutions satisfy

$$\sum A_n e^{\sigma t} W_n(x).$$

The equilibrium $(\bar{N}_1, \bar{N}_2)$ of the linearized reaction-diffusion system (7.24) is locally asymptotically stable if the eigenvalues $\sigma$ are negative or have negative real part. However, for diffusive instabilities, we want the equilibrium to be unstable in the presence of diffusion. Therefore, there should be at least one eigenvalue that is positive or has positive real part. The value of $\sigma$ depends on $k$, $\sigma \equiv \sigma(k^2)$.

Substituting the solutions $v_i = c_i e^{\sigma t} w_i(x)$, $i = 1, 2$ into the linearized system leads to

$$\frac{\partial v_i}{\partial t} = c_i \sigma e^{\sigma t} w_i(x) = \sigma v_i$$

and

$$\frac{\partial^2 v_i}{\partial x^2} = -c_i\, k^2 e^{\sigma t} w_i(x) = -k^2 v_i.$$

The latter identity follows from the eigenvalue problem (7.26). Expressing the preceding derivatives for the linearized reaction-diffusion system (7.25) in matrix form, $\sigma V = JV - k^2\, DV$ or, equivalently,

$$[J - k^2 D - \sigma I]\, V = \mathbf{0},$$

where $V = (v_1, v_2)^T$, $J = (a_{ij})$, and $D = \operatorname{diag}(D_1, D_2)$.

The values of $\sigma$ are the eigenvalues of the matrix $J - k^2 D$. They are negative or have negative real part iff $\operatorname{Tr}(J - k^2 D) < 0$ and $\det(J - k^2 D) > 0$. However, for diffusive instabilities, we require that at least one of the inequalities be reversed. Thus, $\operatorname{Tr}(J - k^2 D) > 0$ or $\det(J - k^2 D) < 0$, expressed in terms of the elements of matrices $J$ and $D$, are

(1) $(a_{11} + a_{22} - D_1 k^2 - D_2 k^2) > 0.$
(2) $(a_{11} - D_1 k^2)(a_{22} - D_2 k^2) - a_{12}a_{21} < 0.$

Condition (1) is never satisfied because $\operatorname{Tr}(J) < 0$ and the parameters $D_i, k^2 \geq 0$. (Here, we used the assumption $k^2 \geq 0$.) Thus, condition (2) must be satisfied. Denote the quantity in condition (2) as $B(k^2)$. That is,

$$B(k^2) = D_1 D_2 k^4 - (D_1 a_{22} + D_2 a_{11})k^2 + (a_{11}a_{22} - a_{12}a_{21}),$$

a quadratic function of $k^2$. Therefore, in order for $B$ to be negative it is necessary that its vertex lie below the $k^2$-axis. Denote the vertex of $B$ as $(k^2_{\min}, B(k^2_{\min}))$. Then diffusive instability requires that $B(k^2_{\min}) < 0$. The vertex is given by

$$\left(\frac{1}{2}\left(\frac{a_{22}}{D_2} + \frac{a_{11}}{D_1}\right), a_{11}a_{22} - a_{12}a_{21} - \frac{(D_1 a_{22} + D_2 a_{11})^2}{4 D_1 D_2}\right).$$

Thus, $B(k^2_{\min}) < 0$ if

$$a_{11}a_{22} - a_{12}a_{21} - \frac{(D_1 a_{22} + D_2 a_{11})^2}{4 D_1 D_2} < 0$$

or, equivalently,

$$a_{22}D_1 + a_{11}D_2 > 2\sqrt{D_1 D_2 (a_{11}a_{22} - a_{12}a_{21})}.$$

We summarize these results in the following theorem.

**Theorem 7.2** *Suppose the reaction system (7.23) linearized about the equilibrium $(\bar{N}_1, \bar{N}_2)$ has a Jacobian matrix $J = (a_{ij})$. Then the minimal requirements for the reaction-diffusion system (7.22) to exhibit diffusive instabilities are the following three conditions:*

*(i)* $a_{11} + a_{22} < 0$,
*(ii)* $a_{11}a_{22} - a_{12}a_{21} > 0$, *and*
*(iii)* $a_{22}D_1 + a_{11}D_2 > 2\sqrt{D_1 D_2 (a_{11}a_{22} - a_{12}a_{21})}$. □

It is straightforward to derive some additional necessary conditions for diffusive instability from Theorem 7.2.

**Corollary 7.1** *Suppose the reaction system (7.23) linearized about the equilibrium $(\bar{N}_1, \bar{N}_2)$ has a Jacobian matrix $J = (a_{ij})$. If the reaction-diffusion system (7.22) exhibits diffusive instabilities, then*

(a) $a_{11}a_{22} < 0$,

(b) $a_{12}a_{21} < 0$, *and*

(c) $D_1 \neq D_2$.

**Proof** Condition $(a)$ follows from the conditions (i) and (iii) in Theorem 7.2: $a_{11} + a_{22} < 0$ and $a_{11}D_2 + a_{22}D_1 > 0$. Condition $(b)$ follows from condition (ii) in Theorem 7.2 and condition $(a)$: $a_{11}a_{22} - a_{12}a_{21} > 0$. Suppose $D_1 = D = D_2$. From condition (iii) in Theorem 7.2, it follows that $a_{11}D_2 + a_{22}D_1 = (a_{11} + a_{22})D > 0$. But this contradicts condition (i). ◀

It follows from Corollary 7.1 that the signed matrix corresponding to the Jacobian matrix, $Q = \text{sign}(J)$, must have a particular form. For diffusive instabilities to occur, the signs of matrix $Q$ must have one of the following forms:

$$\begin{pmatrix} + & - \\ + & - \end{pmatrix}, \quad \begin{pmatrix} - & + \\ - & + \end{pmatrix}, \tag{7.27}$$

or

$$\begin{pmatrix} + & + \\ - & - \end{pmatrix}, \quad \begin{pmatrix} - & - \\ + & + \end{pmatrix}. \tag{7.28}$$

Reaction-diffusion systems whose matrices $J$ have signs as in (7.27) are referred to *activator-inhibitor systems* whereas matrices as in (7.28) are referred to as *positive feedback systems* (see, e.g., Edelstein-Keshet, 1988; Murray, 1993, 2003; Segel and Jackson, 1972).

The next example shows that a Lotka-Volterra competitive system cannot exhibit diffusive instabilities.

**Example 7.11** Consider a competitive reaction-diffusion system (7.22) with the reaction functions $f$ and $g$ given by

$$f(N_1, N_2) = N_1(a_{10} - a_{11}N_1 - a_{12}N_2),$$
$$g(N_1, N_2) = N_2(a_{20} - a_{21}N_1 - a_{22}N_2),$$

where the parameters $a_{ij} > 0$, $i = 0, 1, 2$ and $j = 1, 2$. Suppose there exists a positive equilibrium at $(\bar{N}_1, \bar{N}_2)$. Then the Jacobian matrix evaluated at the equilibrium is

$$J = \begin{pmatrix} -a_{11}\bar{N}_1 & -a_{12}\bar{N}_1 \\ -a_{21}\bar{N}_2 & -a_{22}\bar{N}_2 \end{pmatrix}.$$

All of the elements of the Jacobian matrix are negative because the equilibrium is positive. The system is neither an activator-inhibitor system nor a positive feedback system. ■

If the Lotka-Volterra system in Example 7.11 is a predator-prey system, then it may exhibit diffusive instabilities (see Exercise 16). Segel and Jackson (1972) and Mimura and Murray (1978) give examples of predator-prey models that exhibit diffusive instabilities.

The conditions in Theorem 7.2 give restrictions on the parameters that must be satisfied for the onset of diffusive instability, but they do not give information on the types of patterns that might occur. The types of patterns depend on the domain $\Omega$ and the BCs. It is the eigenvalues $k_n$ and the eigenfunctions $W_n$ that determine the spatial patterns.

**Example 7.12** Suppose in the eigenvalue problem (7.26), the BCs are zero flux BCs: $\partial N_i(L,t)/\partial x = 0 = \partial N_i(0,t)/\partial x$. Then the solution to the eigenvalue problem (7.26) has eigenvalues $k_n^2 = (n\pi/L)^2$, $n = 0, 1, 2, \dots$ and eigenfunctions of the form $w_i(x) = \cos(k_n x)$, $n = 0, 1, 2, \dots$. Hence, the solution to the linear system (7.24) has the form

$$v_i(x,t) = \sum_{n=0}^{\infty} A_n e^{\sigma t} \cos(k_n x) \approx N_i(x,t) - \bar{N}_i, \quad i = 1, 2,$$

where $\sigma \equiv \sigma(k_n^2)$ and $k_n = n\pi/L$. If for some mode $n$, $B(k_n^2) < 0$, then $\sigma(k_n^2)$ will be positive or have positive real part. (Recall the definition of $B(k_n^2)$ given prior to Theorem 7.2.) Then solutions will be amplified at this particular wave number. Patterns will appear with a period approximately $2\pi/k_n$. The solution $N_i(x,t)$ is above or below the equilibrium value $\bar{N}_i$ if the components of the eigenfunction $W$ are positive or negative, respectively. Patterns where $\cos(n\pi x/L) > 0$ are different from those where $\cos(n\pi x/L) < 0$. Patterns generated from solutions where diffusive instability occurs at modes $n = 2$ and $n = 4$ are graphed in Figure 7.8. ■

**Example 7.13** Consider a system with two species or chemicals in a rectangular spatial domain, $[0, L_x] \times [0, L_y]$. Suppose the reaction-diffusion system satisfies

$$\frac{\partial N_1}{\partial t} = f(N_1, N_2) + D_1 \Delta N_1$$

$$\frac{\partial N_2}{\partial t} = g(N_1, N_2) + D_2 \Delta N_2,$$

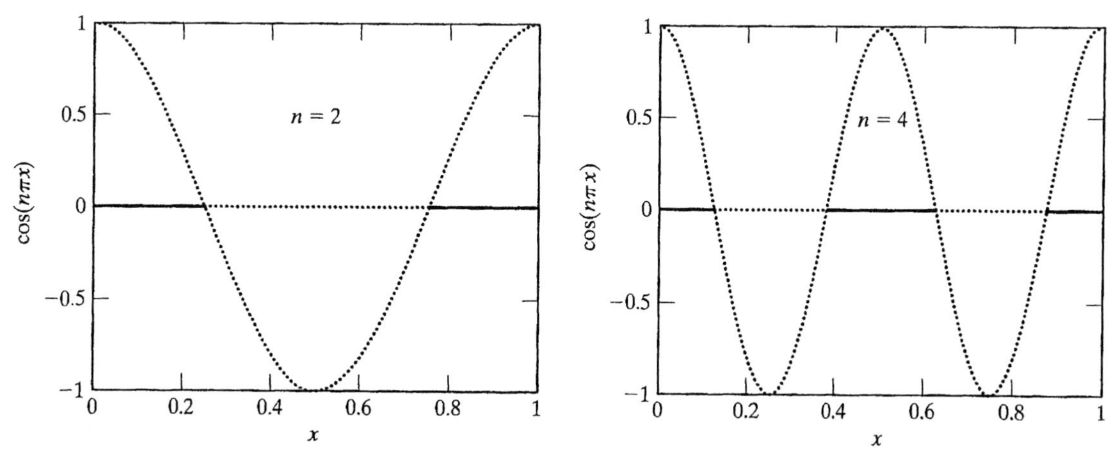


**Figure 7.8** Patterns generated along the line segment $[0, 1]$, $L = 1$, when the instability occurs at modes $n = 2$ and $n = 4$. For example, when $\cos(n\pi x) > 0$, the pattern is dark, and when $\cos(n\pi x) < 0$, the pattern is light.

where $N_i \equiv N_i(x, y, t), i = 1, 2, \Delta N_i = \partial^2 N_i/\partial x^2 + \partial^2 N_i/\partial y^2$ with zero flux BCs:

$$\frac{\partial N_i}{\partial x} = 0, \quad x = 0, L_x \quad \text{and} \quad \frac{\partial N_i}{\partial y} = 0, \quad y = 0, L_y.$$

The eigenvalue problem has the following form:

$$\Delta W + k^2 W = \mathbf{0}$$

with zero flux BCs:

$$\frac{\partial W_i}{\partial x} = 0, \quad x = 0, L_x,$$

$$\frac{\partial W_i}{\partial y} = 0, \quad y = 0, L_y.$$

For this rectangular domain the eigenvalues and eigenfunctions can be found explicitly. The eigenfunctions are $W = (c_1, c_2)^T \cos(q_{1n}x)\cos(q_{2m}y)$, where $q_{1n} = n\pi/L_x$ and $q_{2m} = m\pi/L_y$, $m, n = 0, 1, 2, \ldots$. The eigenvalues are $k_{nm}^2 = k_{1n}^2 + k_{2m}^2$, $m, n = 0, 1, 2, \ldots$. When there are diffusive instabilities, the double cosine series generates a checkerboard pattern, $\cos(q_{1n}x)\cos(q_{2m}y) > 0$, versus $\cos(q_{1n}x)\cos(q_{2m}y) < 0$. See Exercise 14. ■

Consult the references for a wealth of interesting examples on pattern formation in biological systems.

## 7.8 Integrodifference Equations

Another type of model that incorporates spatial variation is an integrodifference equation, where the time variable is discrete and the spatial variable is continuous. Let $N_t(x)$ denote the population size at time $t$ and spatial position $x$. Then a simple example of an integrodifference equation for the population size at time $t + 1$ is given by

$$N_{t+1}(x) = \int_{\mathbf{R}} f(N_t(y))k(x, y)\, dy.$$

The function $k$ is known as the *dispersal kernel*. A dispersal kernel of the form $k(|x - y|)$ means that the probability of dispersing from $y$ to $x$ depends only on the distance $|x - y|$ between the two positions. The function $f$ determines the rate of population growth. For example, a Beverton-Holt form for $f$ is

$$f(N_t) = \frac{rKN_t}{K + (r - 1)N_t}, \quad r > 1$$

and a Ricker form is

$$f(N_t) = N_t \exp\left[r\left(1 - \frac{N_t}{K}\right)\right], \quad 0 < r < 2.$$

In the integrodifference equation, population growth and dispersal occur in two stages, growth followed by dispersal. Such types of models have been studied by Kot (1992), Kot et al. (1996), and Kot and Schaffer (1986). Integrodifference equation models have been shown to exhibit Turing instabilities and traveling wave solutions (e.g., Kot, 1989, 1992; Kot et al., 1996; Kot and Schaffer, 1986;

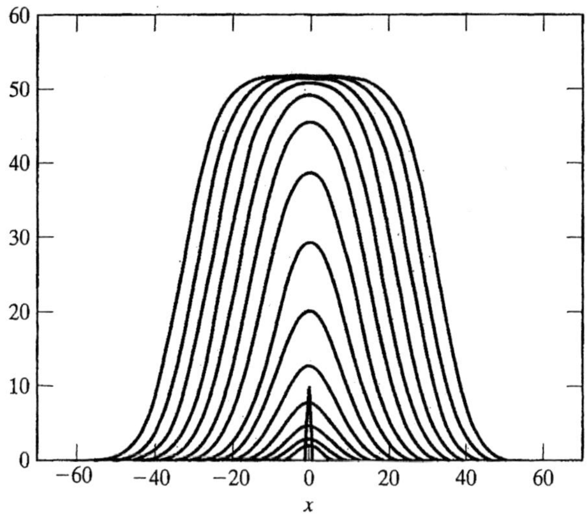


**Figure 7.9** Traveling wave solution of an integrodifference equation modeling the spread of infection (Allen and Ernest, 2001). Each curve represents the solution at different times. The traveling wave solution begins at the center $x = 0$ and spreads outward in both directions over time. The rate of spread in this example is constant.

Neubert et al., 1995, 2002). An example of a traveling wave solution exhibited by an integrodifference equation is graphed in Figure 7.9. Some questions of interest relate to the wave speed. For example, if the dispersal kernel is leptokurtic (fat tails), then the wave speed may be an increasing function of time (Kot et al., 1996).

## 7.9 Exercises for Chapter 7

**1.** (a) Find the solution to the following first-order partial differential equation by the method of characteristics.

$$\frac{\partial u}{\partial t} + \frac{\partial u}{\partial x} + u = 0, \quad -\infty < x < \infty, \ t > 0$$
$$u(0, x) = x^2, \quad -\infty < x < \infty$$

(b) Graph the solution for times $t = 0, 1,$ and 2.

**2.** Consider the partial differential equation,

$$t\frac{\partial u}{\partial t} + x\frac{\partial u}{\partial x} + 2u = 0, \quad -\infty < x < \infty, \ 1 < t < \infty$$
$$u(1, x) = \sin(x), \quad -\infty < x < \infty.$$

Note that time starts at $t = 1$.

(a) Find the characteristic curves; then solve the differential equation.

(b) Graph the solution for times $t = 1, 2,$ and 3.

(c) Describe the solution as $t \to \infty$.

**3.** Suppose

$$\frac{\partial n}{\partial t} + \frac{\partial n}{\partial a} = -\mu(a)\, n,$$
$$n(t, 0) = \int_0^\infty b(a) n(t, a)\, da,$$
$$n(0, a) = n_0(a),$$

where solutions are of the form $n(t, a) = e^{\lambda t} r(a)$. Use the inherent net reproductive number $R_0$ and Theorem 7.1 to verify the following results.

(a) If $0 < b(a) < \mu(a)$ for all ages $a > 0$ (birth rate is less than death rate), show that $\lim_{t\to\infty} n(t, a) = 0$.

(b) If $0 < \mu(a) < b(a)$ for all ages $a > 0$ (death rate is less than birth rate) and $\int_0^\infty \mu(a)\, da = \infty$, show that $\lim_{t\to\infty} n(t, a) = \infty$ for those ages such that $r(a) > 0$.

**4.** Suppose

$$\frac{\partial n}{\partial t} + \frac{\partial n}{\partial a} = -\mu(a)n, \ \mu(a) \ge 0,$$

$$n(t, 0) = \int_0^\infty b(a)n(t, a)\, da, \ b(a) \ge 0,$$

$$n(0, a) = n_0(a),$$

where solutions are of the form $n(t, a) = e^{\lambda t} r(a)$. Use the inherent net reproductive number $R_0$ and Theorem 7.1 to verify the following results.

(a) If $\int_0^\infty b(a)\, da < 1$, show that $\lim_{t\to\infty} n(t, a) = 0$.

(b) Suppose

$$b(a) = \begin{cases} 4, & a \in [1, 3] \\ 0, & a \notin [1, 3] \end{cases} \quad \text{and} \quad \mu(a) = \begin{cases} 1, & a \in [0, 3] \\ 0, & a \notin [0, 3]. \end{cases}$$

Show $n(t, a)$ does not approach zero as $t \to \infty$ (i.e., $R_0 > 1$).

**5.** Find the solution to the following IBVP via separation of variables and Fourier series:

$$\frac{\partial N}{\partial t} = D\frac{\partial^2 N}{\partial x^2}, \ x \in (0, 1), \ t \in (0, \infty),$$

$$N(0, x) = \cos(\pi x) + 3\cos(2\pi x), \ x \in [0, 1],$$

$$\frac{\partial N(t, 0)}{\partial x} = 0 = \frac{\partial N(t, 1)}{\partial x}, \ t \in (0, \infty).$$

**6.** Find the solution to the following Cauchy problem via Fourier transforms.

$$\frac{\partial N}{\partial t} = D\frac{\partial^2 N}{\partial x^2}, \ x \in \mathbf{R}, \ t \in (0, \infty)$$

$$N(0, x) = e^{-x^2}, \ x \in \mathbf{R}.$$

**7.** Find the solution to the following IBVP via separation of variables and Fourier series.

$$\frac{\partial N}{\partial t} = D\frac{\partial^2 N}{\partial x^2}, \ x \in (0, 1), \ t \in (0, \infty),$$

$$N(0, x) = 1, \ x \in [0, 1],$$

$$N(t, 0) = 0 = N(t, 1), \ t \in (0, \infty).$$

**8.** For the following IBVP, show that the only equilibrium solution is the zero solution, $\bar{N} = 0$.

$$\frac{\partial N}{\partial t} = D\frac{\partial^2 N}{\partial x^2} + rN, \ x \in (0, L), \ t \in (0, \infty),$$

$$\frac{\partial N\,(t, 0)}{\partial x} = 0 = \frac{\partial N\,(t, L)}{\partial x}, \ t \in (0, \infty),$$

$$N(0, x) = N_0(x), \ x \in [0, L],$$

where $r \ne 0$. For $r > 0$, assume $L\sqrt{r} \ne n\pi, n \ne 1, 2, \ldots.$

**9.** Consider the diffusive exponential growth model with an initial point source,

$$\frac{\partial N}{\partial t} = D\frac{\partial^2 N}{\partial x^2} + rN, \ x \in \mathbf{R}, \ t \in (0, \infty),$$

$$N(0, x) = N_0\delta(x), \ x \in \mathbf{R}.$$

(a) Show that the solution to this IVP is given by

$$N(t, x) = \frac{N_0}{2\sqrt{D\pi t}} \exp\left(rt - \frac{x^2}{4Dt}\right).$$

(b) Let $N(t, x) = \bar{N}$ = constant in part (a); then solve for $x$. Show that

$$\frac{x}{t} = \pm\sqrt{4rD - \frac{2D}{t}\ln t - \frac{4D}{t}\ln\left(\bar{N}\frac{\sqrt{2\pi D}}{N_0}\right)}.$$

Then show that $x \approx \pm 2\sqrt{rD}\,t$ as $t \to \infty$. The asymptotic wave speed is $v = 2\sqrt{rD}$. [*Hint:* See equation (7.11) and Example 7.8.]

**10.** Suppose fishing is regulated within a zone of $L$ kilometers from a country's shore (taken to be a straight line), but outside of this zone overfishing is so excessive that the population is zero. Assume that fish reproduction follows a logistic curve, dispersal is by diffusion within the regulated zone, and fish are harvested with a constant effort $E$ (Murray, 1993). Then we have the following model for the fish population $N(t, x)$ at time $t$ and position $x$ from the shore:

$$\frac{\partial N}{\partial t} = rN\left(1 - \frac{N}{K}\right) - EN + D\frac{\partial^2 N}{\partial x^2}, \quad x \in (0, L), \ t \in (0, \infty),$$

$$N(t, L) = 0, \frac{\partial N\,(t, 0)}{\partial x} = 0, \ N(0, x) = N_0(x),$$

where $K, E, D > 0$ and $r > E$.

(a) Show that there are two constant solutions $\bar{N} = 0$ and $\bar{N} > 0$. Then linearize the differential equation about $\bar{N} = 0$ $(v = N - \bar{N})$ and show that the linearization satisfies

$$\frac{\partial v}{\partial t} = (r - E)v + D\frac{\partial^2 v}{\partial x^2}.$$

(b) Solutions to the linearized equation in $v(t, x)$ are sums of terms of the form

$$v_n(t, x) = e^{\sigma t}\cos(k_n x), \ k_n = \frac{(2n + 1)\pi}{2L}, \quad n = 0, 1, 2, \ldots,$$

$v = \sum_{n=0}^{\infty} v_n$, where $\sigma$ depends on $n$. Show that $\partial v/\partial x = 0$ for $x = 0$ and $v(t, L) = 0$. If $\sigma < 0$, then $\lim_{t\to\infty} v(t, x) = 0$, that is, the zero equilibrium is locally stable and the fish stock collapses. Show that to prevent the fish stock from collapsing, the fishing zone must satisfy $L \geq \pi\sqrt{D/(4r - 4E)}$. [*Hint:* Substitute the solutions $v_n$ into the linearized equation in $v$ and for each $n$ obtain the values of $\sigma$. To prevent collapse, $\sigma \geq 0$ for every $n$ (i.e., for every possible initial condition).]

**11.** Suppose, in the population genetics problem, the proportional density for the advantageous allele satisfies

$$\frac{\partial p}{\partial t} = D\frac{\partial^2 p}{\partial x^2} + rp(\alpha - p)(p - \beta),$$

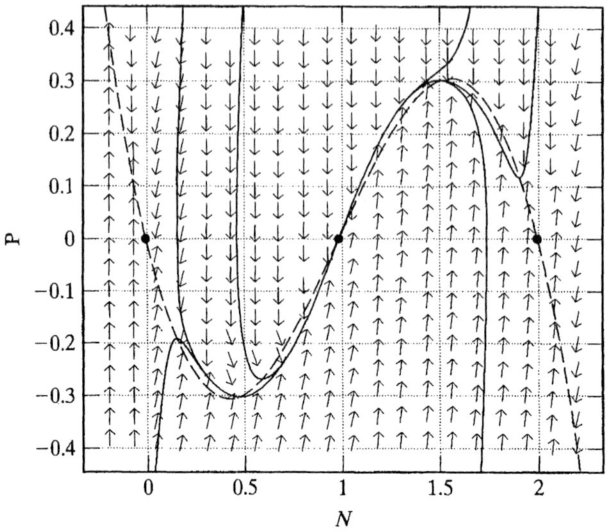


**Figure 7.10** Solutions to the $N$-$P$ system graphed in the phase plane, $v = 5, r = 4$, $D = 1 = \alpha, \beta = 2$, and $v > 2\sqrt{r\, D\alpha(\beta - \alpha)}$.

where $0 < \alpha < \beta$, $r > 0$ and $D > 0$. In the absence of diffusion, there are three equilibria, $p = 0, \alpha, \beta$. We will investigate the existence of traveling wave solutions in this problem.

(a) Show that $p = 0$ and $p = \beta$ are locally asymptotically stable and $p = \alpha$ is unstable in the model without diffusion:

$$\frac{dp}{dt} = rp(\alpha - p)(p - \beta).$$

This model has an Allee effect in the temporal dynamics.

(b) Make the change of variable $N(z) = p(t, x)$ and $P = -dN/dz$, where $z = x - vt$, $v > 0$, to express the reaction-diffusion equation in $p$ as a first-order system in $N$ and $P$ as in (7.16). Then show that there are three equilibria for this new system, $(\bar{N}, \bar{P}) = (0, 0), (\alpha, 0)$, and $(\beta, 0)$.

(c) Show that equilibria (0,0) and $(\beta, 0)$ are saddle points in the linearized $N$-$P$ system. In addition, show that $(\alpha, 0)$ is a stable node provided

$$v > 2\sqrt{r\, D\alpha(\beta - \alpha)}. \tag{7.29}$$

(d) The condition (7.29) guarantees that there are two heteroclinic trajectories in the phase plane connecting the following equilibria: $(0, 0) \to (\alpha, 0)$ and $(\beta, 0) \to (\alpha, 0)$. Solutions converge to one of these trajectories, depending on the initial conditions. See Figure 7.10. Explain the dynamics as $t \to \infty$ in terms of the two traveling wave solutions.

**12.** The phytoplankton-zooplankton model discussed in Chapter 6 includes random movement of both species. Assume a positive equilibrium $(\bar{P}, \bar{Z})$ exists. Show that it is not possible for diffusive instability to occur in this model.

$$\frac{\partial P}{\partial t} = \beta P(1 - P) - Z\frac{P^2}{\nu^2 + P^2} + \frac{\partial^2 P}{\partial x^2}$$

$$\frac{\partial Z}{\partial t} = \gamma Z\left(\frac{P^2}{\nu^2 + P^2} - \omega\right) + \delta\frac{\partial^2 Z}{\partial x^2}$$

**13.** At least two chemicals or species are needed for diffusion-driven instability. Consider the following reaction-diffusion equation:

$$\frac{\partial N}{\partial t} = f(N) + D\frac{\partial^2 N}{\partial x^2},$$

with zero flux boundary conditions and a positive equilibrium at $\bar{N}$. The system linearized about $\bar{N}$ is

$$\frac{\partial v}{\partial t} = a_{11}v + D\frac{\partial^2 v}{\partial x^2}.$$

Assume $a_{11} = f'(\bar{N}) < 0$. Let $v = e^{\sigma t}\cos(kx)$ and show that it is impossible for the linearized system to become unstable. The eigenvalues of the linearized system are always negative.

**14.** Consider Example 7.13. Draw the checkerboard patterns that are generated in the unit square $[0, 1] \times [0, 1]$, $L_x = L_y = 1$, for the following unstable modes.

(a) $m = 2 = n$.

(b) $m = 3, n = 5$.

**15.** Turing instability is applied to the markings on animal tails (Murray, 1993). Assume the tail is modeled by a long cone, where the variables for length $z$, radius $r$, and angle $\theta$ are significant. The cone has a fixed length $s$, $0 \le z \le s$. We only model the surface of the cone. The radius $r$ is a parameter which reflects the thickness of the cone at a given length $z$. In this case, the solution to the linearized reaction-diffusion problem is $v_i \equiv v_i(\theta, z, t)$, where $v_i$ is a superposition of solutions of the form

$$e^{\sigma t}\cos(q_{1n}\theta)\cos(q_{2m}\, z), \quad n, m = 0, 1, 2, \ldots,$$

where the eigenvalues

$$k_{nm}^2 = q_{1n}^2 + q_{2m}^2 = \frac{n^2}{r^2} + \frac{m^2\pi^2}{s^2}.$$

Describe the patterns that are possible for large values of $r$ and small values of $r$.

**16.** A predator-prey reaction diffusion model was studied by Mimura and Murray (1978) with the reaction terms satisfying

$$f(N_1, N_2) = N_1\left(\frac{35 + 16N_1 - N_1^2}{9} - N_2\right),$$

$$g(N_1, N_2) = N_2(N_1 - 0.4N_2 - 1),$$

and zero flux boundary conditions on the domain $[0, L]$.

(a) Show that the reaction system,

$$\frac{dN_1}{dt} = f(N_1, N_2),$$

$$\frac{dN_2}{dt} = g(N_1, N_2),$$

has a unique stable positive equilibrium at (5, 10).

(b) Show that reaction-diffusion system with $D_1 \neq D_2$ may exhibit diffusive instabilities by showing that the signed matrix has the correct configuration.

(c) Let $D_1 = 0.0125$ and $D_2 = 1$. Show that conditions (i)–(iii) of Theorem 7.2 are satisfied.

(d) For $D_1 = 0.0125$ and $D_2 = 1$, perform some numerical simulations of the reaction-diffusion system.

## 7.10 References for Chapter 7


Abramson, G., V. M. Kenkre, T. L. Yates, and R. R. Parmenter. 2003. Traveling waves of infection in the hantavirus epidemics. *Bull. Math. Biol.* 65: Global: 519–534.

Allen, L. J. S. 1987. Persistence, extinction, and critical patch number for island populations. *J. Math. Biol.*, 24: 617–625.

Allen, L. J. S., M. P. Moulton, and F. L. Rose. 1990. Persistence in an age-structured population for a patch-type environment. *Nat. Res. Modeling* 4: 197–214.

Allen, L. J. S. and R. K. Ernest. 2001. The impact of long-range dispersal in population and epidemic models. In: *Mathematical Approaches for Emerging and Reemerging Infectious Diseases: Models, Methods and Theory*. C. Castillo-Chavez with Sally Blower, P. van den Driessche, D. Kirschner, and A.-A. Yakubu (Eds.). IMA Vol. 125: 183–197.

Anderson, R. M. and R. M. May. 1991. *Infectious Diseases of Humans, Dynamics and Control*. Oxford University Press, Oxford, U.K.

Anderson, R. M., H. C. Jackson, R. M. May, and A. M. Smith. 1981. Population dynamics of fox rabies in Europe. *Nature* 289: 765–771.

Andow, D., P. Kareiva, S. Levin, and A. Okubo. 1990. Spread of invading organisms. *Landscape Ecology*. 4: 177–188.

Andrews, L. C. 1986. *Elementary Partial Differential Equations with Boundary Value Problems*. Academic Press, Inc., London.

Britton, N. F. 1986. *Reaction-Diffusion Equations and Their Applications to Biology*. Academic Press, London, Orlando, New York.

Cantrell, R. S. and C. Cosner. 1989. Diffusive logistic equations with indefinite weights: population models in disrupted environments. *Proc. Roy. Soc. Edinburgh* 112A: 293–318.

Cantrell, R. S. and C. Cosner. 2003. *Spatial Ecology via Reaction-Diffusion Equations*. John Wiley & Sons, New York.

Cushing, J. M. 1994. The dynamics of hierarchical age-structured populations. *J. Math. Biol.* 32: 705–729.

Cushing, J. M. 1998. *An Introduction to Structured Population Dynamics*, CBMS-NSF Regional Conference Series in Applied Mathematics # 71. SIAM, Philadelphia, Pa.

Edelstein-Keshet, L. 1988. *Mathematical Models in Biology*. The Random House/Birkhäuser Mathematics Series, New York.

Farlow, S. J. 1982. *Partial Differential Equations for Scientists & Engineers*. John Wiley & Sons, New York.

Fife, P. C. 1979. *Mathematical Aspects of Reacting and Diffusing Systems*. Lecture Notes in Biomathematics. Vol. 28. Springer-Verlag, New York.

Fisher, R. A. 1937. The wave of advance of advantageous genes. *Ann Eugen.* (London) 7: 355–369.

Hengeveld, R. 1989. *Dynamics of Biological Invasions*. Chapman and Hall, London.

Hethcote, H. W. 2000. The mathematics of infectious diseases. *SIAM Review* 42: 599–653.

John, F. 1975. *Partial Differential Equations*. 2nd ed. Applied Mathematical Sciences, Vol. 1. Springer-Verlag, New York.

Kaplan, D. and L. Glass. 1995. *Understanding Nonlinear Dynamics*. Spring-Verlag, New York, Berlin.

Keener, J. and J. Sneyd. 1998. *Mathematical Physiology*. Springer-Verlag, New York, Berlin, and Heidelberg.

Kierstead, H. and L. B. Slobodkin. 1953. The size of water masses containing plankton blooms. *J. Mar. Res.* 12: 141–147.

Kolmogoroff, A., I. Petrovsky, and N. Piscounoff. 1937. Étude de l'équation de la diffusion avec croisssance de la quantité de matière et son application à un problème biologique. *Moscow Univ. Bull. Math.* 1: 1–25.

Kot, M. 1989. Diffusion-driven period-doubling bifurcations. *BioSystems* 22: 279–287.

Kot, M. 1992. Discrete-time travelling waves: ecological examples. *J. Math. Biol.* 30: 413–436.

Kot, M., M. A. Lewis, and P. van den Driessche. 1996. Dispersal data and the spread of invading organisms. *Ecology*. 77: 2027–2042.

Kot, M. and W. M. Schaffer. 1986. Discrete-time growth-dispersal models. *Math. Biosci.* 80: 109–136.

Logan, J. D. 2004. *Applied Partial Differential Equations*. 2nd ed. Springer-Verlag, New York, Berlin.

McKendrick, A. G. 1926. Applications of mathematics to medical problems. *Proc. Edin. Math. Soc.* 44: 98–130.

Mimura, M. and J. D. Murray. 1978. On a diffusive prey-predator model which exhibits patchiness. *J. Theor. Biol.* 75: 249–262.

Murray, J. D. 1988. How the leopard gets its spots. *Scientific American* 258: 80–87.

Murray, J. D. 1993. *Mathematical Biology*. 2nd ed. Springer-Verlag, Berlin, Heidelberg, New York.

Murray, J. D. 2002. *Mathematical Biology I: An Introduction*. 3rd ed. Springer-Verlag, New York.

Murray, J. D. 2003. *Mathematical Biology II: Spatial Models and Biomedical Applications*. 3rd ed. Springer-Verlag, New York.

Neubert, M. G., H. Caswell, and J. D. Murray. 2002. Transient dynamics and pattern formation: Reactivity is necessary for Turing instabilities. *Math. Biosci.* 175: 1–11.

Neubert, M. G., M. Kot, and M. A. Lewis. 1995. Dispersal and pattern formation in a discrete-time predator-prey model. *Theor. Pop. Biol.* 48: 7–43.

Okubo, A. 1980. *Diffusion and Ecological Problems: Mathematical Models*. Springer-Verlag, New York.

Okubo, A. and S. A. Levin. 2001. *Diffusion and Ecological Problems: Modern Perspectives*. 2nd ed. Springer-Verlag, New York.

Rouderfer, V., N. G. Becker, and H. W. Hethcote. 1994. Waning immunity and its effects on vaccination schedules. *Math. Biosci.* 124: 59–82.

Schuette, M. C. and H. W. Hethcote. 1999. Modeling the effects of varicella vaccination programs on the incidence of chickenpox and shingles. *Bull. Math. Biol.* 61: 1031–1064.

Segel, L. and J. Jackson. 1972. Dissipative structure: an explanation and an ecological example. *J. Theor. Biol.* 37: 545–559.

Sharpe, F. R. and A. J. Lotka, 1911. A problem in age distribution. *Philos. Mag.* 21: 435–438.

Shigesada, N. and K. Kawasaki. 1997. *Biological Invasions: Theory and Practice.* Oxford Univ. Press, Oxford, U.K.

Skellam, J. G. 1951. Random dispersal in theoretical populations. *Biometrika* 38: 196–218.

Thieme, H. R. 2003, *Mathematics in Population Biology.* Princeton Univ. Press, Princeton and Oxford.

Trim, D. W. 1990. *Applied Partial Differential Equations.* Prindle, Weber, and Schmidt-Kent Pub. Co., Boston.

Turing, A. M. 1952. The chemical basis for morphogenesis. *Phil. Trans. Roy. Soc.* (London) 237: 37–72.

Von Foerster, H. 1959. Some remarks on changing populations. In: *The Kinetics of Cellular Proliferation.* F. Stohlman, (Ed.). Grune and Stratton, New York pp. 382–407.

# INDEX

## A



## B



## C

**F**

## N